Freshwater Ecology and Conservation

Techniques in Ecology and Conservation Series

Series Editor: William J. Sutherland

Freshwater Ecology and Conservation

Approaches and Techniques

Edited by
Jocelyne M.R. Hughes

OXFORD
UNIVERSITY PRESS

Great Clarendon Street, Oxford, OX2 6DP,
United Kingdom

Oxford University Press is a department of the University of Oxford.
It furthers the University's objective of excellence in research, scholarship,
and education by publishing worldwide. Oxford is a registered trade mark of
Oxford University Press in the UK and in certain other countries

First Edition published in 2019

Impression: 1

Published in the United States of America by Oxford University Press
198 Madison Avenue, New York, NY 10016, United States of America

British Library Cataloguing in Publication Data

Data available

Library of Congress Control Number: 2018948522

ISBN 978–0–19–876638–4 (hbk.)
ISBN 978–0–19–876642–1 (pbk.)

DOI: 10.1093/oso/9780198766384.001.0001

Printed in Great Britain by
Bell & Bain Ltd., Glasgow

Preface

Water underpins all life on Earth, and in particular, hydrology and water chemistry play a signature role in maintaining the structure and function of freshwater ecosystems. Despite the manifest importance of freshwater, and the human dependence on healthy freshwater ecosystems, wetlands have continued to be degraded, polluted, and drained, and freshwater species have declined at an alarming rate—for example, the well-documented decline of global amphibians. This book aims to provide a comprehensive and concise synthesis of the vast literature on the techniques used in freshwater ecology and conservation that is currently dispersed globally in manuals, toolkits, journals, handbooks, 'grey' literature, and websites. It should be regarded as a 'book of books', and it is on this basis that the late Brian Moss encouraged me to carry out the project. The book is intended to provide a thorough understanding of different approaches and techniques needed for successful research, management, and conservation of freshwater ecosystems. Freshwater conservationists and practitioners often need to understand hydrochemical storages and fluxes in wetlands, the physical processes influencing freshwaters at the catchment and landscape scale, and the hydrochemical processes that maintain species assemblages and their dynamics. Conservation builds on a sound ecological framework, in which each species must be examined at the individual, community, and catchment level of interaction, and in which human interactions with freshwaters are a critical part. The survey approaches and techniques used, and the subsequent analytical methods, directly affect the interpretation of results, and it is essential to understand their limitations or biases in order to make informed decisions on conservation and management. This book cannot list all techniques used to survey or sample freshwaters—that is an impossible task; rather, it presents some of the practical approaches that are used in freshwater ecological research and surveys, and, using examples from international studies, synthesises why certain methods are used in preference to others for particular types of freshwater habitat or particular types of organism, or for measuring, documenting, and evaluating the dynamics and management of the ecosystem. Central to all freshwater surveys is the identification of organisms to the species level. This book does not attempt to cover identification and taxonomy although passing reference is made to new methods in species identification in some chapters.

The book is divided into three parts. The first part is an overview, presenting an outline of the diversity of freshwater ecosystems; the importance of hydrology and water chemistry in driving ecosystem processes; the different approaches required to answer questions in freshwater ecology and provide effective solutions to conservation problems, including the essential nexus between scientists, social scientists, and stakeholder participation; and an overview of sampling strategies and protocols used in freshwater investigations. The second part describes the techniques needed to quantify the individual physical and biological components of the ecosystem, including, for example, water quantity, water chemistry, aquatic plants, aquatic animals, and algae. This section summarises the huge array of approaches and techniques used for surveying and measuring the multitude of organisms to be found in freshwaters and their physical environment, and that are needed to answer questions on—for example—species richness and abundance, species hydrochemical preferences and tolerances, or the impacts of human populations on freshwater ecosystems. The final part deals with the bigger picture in conservation and management and the approaches needed to answer complex questions involving different stakeholders. It brings together the different approaches and techniques needed to understand interactions between the physical and biological components of freshwater ecosystems, changes over time and ecosystem dynamics, human dependency on and interactions with freshwaters, and the conservation, management, and evaluation of these ecosystems. Individual chapters synthesise approaches and methods drawn from science, economics, landscape management, and engineering; for example, evaluating the restoration of eutrophic waters, monitoring the spread and control of non-native invasive species, and quantifying ecosystem services.

I would like to thank the people who assisted in the production of this book—in reviewing, editing, discussing the contents, making suggestions, and providing moral support: Flora Botsford, David Bradley, Ana Castro Castellon, C. Ken Dodd Jr., the late Mike Edmunds, François Edwards, Max Finlayson, Dustin Garrick, Thomas Hesselberg, Francine Hughes, Lizzy Jeffers, Tim Johns, Curt Lamberth, the late Brian Moss, Liz Sanders, Carl Sayer, Rebecca Tharme, Jennie Whinam; seven anonymous referees; and the fantastic team of lead authors and co-authors who contributed to this book and helped with the reviewing. I would like to thank my previous department at Oxford—the Department for Continuing Education—that supported a six-month sabbatical that enabled me to focus on developing this book. I am so grateful to the inspirational, international students I teach on the MSc in Water Science, Policy and Management, the Postgraduate Certificate in Ecological Survey Techniques, and the Postgraduate Diploma in International Wildlife Conservation Practice at the University of Oxford, who are justification alone for this project and have taught me the vital importance of training early career researchers and practitioners in field skills, practical surveys, and analytical methods. I would like to thank Lydia Shinoj and Paul Nash at SPi Global for their infinite patience. At Oxford University Press I would like to thank Bethany Kershaw, Lucy Nash, and Ian Sherman; and the overall editor of the Techniques in Ecology and Conservation Series, Bill Sutherland, for giving me the chance of steering this project. Finally, I could not have completed this volume without the infinite support of my beloved family: my life-long partner Charles, and our children George, Madeleine, and Fifi; and my parents, Arlette and Gren, and parent-in-law, Alison, who all three passed away during the production of the book and to whom I dedicate it.

Jocelyne Hughes
Department of Geography and the Environment,
University of Oxford

Contents

3. Sampling Strategies and Protocols for Freshwater Ecology and Conservation   48

Leon A. Barmuta

Part II Measuring the Component Parts

4. Water Quantity and Hydrology   67

Matthew McCartney

5. Chemical Determinands of Freshwater Ecosystem Functioning   89

Nic Pacini, Libor Pechar, and David M. Harper

9. Wetland Plants and Aquatic Macrophytes   173

Jocelyne M.R. Hughes, Beverley R. Clarkson, Ana T. Castro-Castellon,
and Laura L. Hess

Part III Ecosystem Dynamics, Conservation, and Management

12. Freshwater Populations, Interactions, and Networks   257
David M. Harper and Nic Pacini

13. Changes Over Time   283
Peter A. Gell, Marie-Elodie Perga, and C. Max Finlayson

List of Contributors

Mike C. Acreman Centre for Ecology and Hydrology, Maclean Building, Crowmarsh Gifford, Wallingford, OX10 8BB, UK.

Leon A. Barmuta School of Zoology, University of Tasmania, Hobart, 7001 Tasmania, Australia.

Helen Bennion Pond Restoration Research Group, Environmental Change Research Centre, Department of Geography, University College London, Gower Street, WC1E, UK.

Vanessa Bremerich Leibniz-Institute of Freshwater Ecology and Inland Fisheries, Müggelseedamm 310, 12587 Berlin, Germany.

Andrew S. Buxton Durrell Institute of Conservation and Ecology, School of Anthropology and Conservation, Marlowe Building, University of Kent, Canterbury, Kent, CT2 7NR, UK.

Ana T. Castro-Castellon Thames Water, Farmoor WTW, Cumnor Road, Farmoor, Oxford, OX2 9NS, UK.

Beverley R. Clarkson Landcare Research, Private Bag 3127, Hamilton 3240, New Zealand.

Julie A. Coetzee Centre for Biological Control, Department of Botany, Rhodes University, Grahamstown, 6140, South Africa.

John Conallin IHE Delft Institute for Water Education, Westvest 7, 2611 AX Delft, PO Box 3015, 2601 DA Delft, The Netherlands & Institute for Land Water and Society, Charles Sturt University, Elizabeth Mitchell Drive, Albury NSW, 2640, Australia.

Aaike De Wever Royal Belgian Institute of Natural Sciences, Vautierstreet 29, 1000 Brussels, Belgium.

C. Max Finlayson Institute for Land, Water and Society, Charles Sturt University, Elizabeth Mitchell Drive, Albury NSW, 2640, Australia & IHE Delft, Institute for Water Education, Westvest 7, 2611 AX Delft, PO Box 3015, 2601 DA Delft, The Netherlands.

Joerg Freyhof Leibniz-Institute of Freshwater Ecology and Inland Fisheries, Müggelseedamm 310, 12587 Berlin, Germany.

Peter A. Gell Water Research Network, Federation University Australia, PO Box 663, Ballarat, 3350 Victoria, Australia.

Gillian Gilbert Royal Society for the Protection of Birds, South and West Scotland Regional Office, 10 Park Quadrant, Glasgow, G3 6BS, UK.

Danielle L. Gilroy School of Earth and Environmental Sciences, University of Manchester, Williamson Building, Oxford Road, Manchester, M13 9PL, UK.

Emma Goodyer IUCN UK Peatland Programme, c/o Harbour House, 110 Commercial Street, Edinburgh, EH6 6NF, UK.

Stephen E.W. Green Centre for Applied Zoology, Cornwall College Newquay, Wildflower Lane, Trenance Gardens, Newquay, Cornwall, TR7 2LZ, UK.

Rudolph S. de Groot Environmental Systems Analysis Group, Department of Environmental Sciences, Wageningen University and Research, 6700AA Wageningen, The Netherlands.

Angela Gurnell School of Geography, Queen Mary University of London, Mile End Road, E1 4NS, UK.

David M. Harper Emeritus Professor, University of Leicester, Adrian Building, University Road, LE1 7RH, UK.

Lauren A. Harrington Wildlife Conservation Research Unit, Department of Zoology, University of Oxford, Recanati-Kaplan Centre, Tubney House, Abingdon Road, Tubney, Abingdon, OX13 5QL, UK.

Elizabeth Heagney School of Environment, Science and Engineering, Southern Cross University, Lismore, New South Wales, 2480, Australia & Economic and Strategic Analysis Branch, NSW Office of Environment and Heritage, Sydney, New South Wales, 2000, Australia.

Laura L. Hess Earth Research Institute, 6832 Ellison Hall, University of California, Santa Barbara, CA 93106-3060, USA.

Martin P. Hill Centre for Biological Control, Department of Zoology and Entomology, Rhodes University, Grahamstown, 6140, South Africa.

Francine M.R. Hughes Animal and Environment Research Group, Department of Life Sciences, Anglia Ruskin University, Cambridge, CB1 1PT, UK.

Jocelyne M.R. Hughes School of Geography and the Environment, University of Oxford, South Parks Road, Oxford, OX1 3QY, UK.

Andreas Hussner Förderverein Feldberg-Uckermärkische Seen, 17268 Templin, Germany.

Donovan Kotze Centre for Water Resources Research, University of KwaZulu-Natal, Scottsville 3209, South Africa.

Curt Lamberth University of Oxford, Wytham Woods, Woods Sawmill Yard, Wytham, Oxford, OX2 8QQ, UK.

Richard Lindsay Sustainability Research Institute, University of East London, Docklands Campus, University Way, London, E16 2RD, UK.

Simon Linke Australian Rivers Institute, Nathan Campus, Griffith University, 170 Kessels Road, Queensland 4111, Australia.

Richard Marchant Department of Entomology, Museum Victoria, GPO Box 666, Melbourne, Victoria 3001, Australia.

Matthew McCartney International Water Management Institute, c/o National Agriculture and Forestry Research Institute (NAFRI), Ban Nongviengkham, Xaythany District, Vientiane, Lao PDR.

G. Randy Milton Department of Natural Resources, Kentville, Nova Scotia, B4N 4E5, Canada & Institute for Land, Water and Society, Charles Sturt University, Elizabeth Mitchell Drive, Albury, 2640 NSW, Australia.

Ana L. Nunes Centre for Invasion Biology, Department of Botany and Zoology, University of Stellenbosch, Stellenbosch, South Africa; Centre for Invasion Biology, South African Institute for Aquatic Biodiversity, Grahamstown, 6139, South Africa & INNS Programme, South African National Biodiversity Institute, Kirstenbosch Research Centre, Cape Town, South Africa.

Thierry Oberdorff Centre National de la Recherche Scientifique (CNRS), Institut de Recherche pour le Developpement, Laboratoire Évolution & Diversité Biologique, Université Toulouse III Paul Sabatier, 118 route de Narbonne, F-31062 Toulouse, France.

Nic Pacini Department of Environmental and Chemical Engineering, University of Calabria, Arcavacata di Rende, Italy & School of Geography, Geology and the Environment, University of Leicester, Leicester, UK.

Libor Pechar University of South Bohemia, Faculty of Agriculture, Applied Ecology Laboratory, Studentská 13, 370 05 Cöeské Budeöjovice, Czech Republic.

Marie-Chantale Pelletier School of Environment, Science and Engineering, Southern Cross University, Lismore, NSW 2480, Australia & Economic and Strategic Analysis Branch, NSW Office of Environment and Heritage, Sydney, 2000 NSW, Australia.

Marie-Elodie Perga Faculté des Géosciences et de l'Environnement, Institut des Dynamiques de la Surface Terrestre, University of Lausanne, CH-1015 Lausanne, Switzerland.

Jamie Pittock Fenner School of Environment and Society, The Australian National University, 48 Linnaeus Way, Acton, ACT 2600, Australia.

Julia Reiss Department of Life Sciences, Whitelands College, University of Roehampton, London, SW15 4JD, UK.

Rosie D. Salazar Science Area, Research Services, University of Oxford, Robert Hooke Building, Parks Road, Oxford, OX1 3PR, UK.

Carl Sayer Pond Restoration Research Group, Environmental Change Research Centre, Department of Geography, University College London, Gower Street, WC1E, UK.

Astrid Schmidt-Kloiber Institute of Hydrobiology and Aquatic Ecosystem Management, University of Natural Resources and Life Sciences, 1180 Vienna, Austria.

David C. Sigee School of Earth and Environmental Sciences, University of Manchester, Manchester, M13 9PL, UK.

Caroline A. Sullivan School of Environment, Science and Engineering, Southern Cross University, Lismore, NSW 2480, Australia & Marine Ecology Research Centre, Southern Cross University, Lismore, NSW 2480, Australia.

Rebecca E. Tharme Riverfutures Ltd, 48 Middle Row, Cressbrook, Derbyshire, SK17 8SX, UK & Australian Rivers Institute, Griffith University Nathan Campus, Queensland 4111, Australia.

David Tickner WWF-UK, Rufford House, Brewery Road, Woking, Surrey, GU21 4LL, UK.

Olaf L.F. Weyl Centre for Invasion Biology, South African Institute for Aquatic Biodiversity, Grahamstown, 6139, South Africa.

Catherine M. Yule Faculty of Science, Health, Education and Engineering, University of the Sunshine Coast, Locked Bag 4, Maroochydore DC, Queensland 4558, Australia.

Lauren Zielinski Zielinski Environmental Monitoring and Evaluation LLC, 6 Riverwalk, Hampton, NH 03842, USA & IHE Delft Institute for Water Education, Westvest 7, 2611 AX Delft, PO Box 3015, 2601 DA Delft, The Netherlands.

Plate 1 Examples of wetlands, left to right, top to bottom: a – riverine floodlplain, Wisconsin River (USA); b – Pantanal riverine marsh and forest, Rio Claro (Brazil); c – lowland tropical dipterocarp peat swamp forest with encroaching *Pandanus helicopus* clumps, Tasek Bera (Malaysia);); d – *Phragmites* marsh in lower reaches of the Danube River delta (Romania); e – Cypress forest showing pneumatophores, Florida (USA); f – Yellow Water lagoon, Kakadu National Park (Australia); g – Waikoropupu Springs, a karst wetland, Takaka, South Island (New Zealand); h – prairie pothole marshes in the Missouri Coteau, southern Saskatchewan (Canada); i – dense cattail *Typha sp* at Delta marsh, Manitoba (Canada); j – Floodplain marsh (Kenya); k –Domed *Sphagnum* blanket bogs and fens in highlands Cape Breton Island (Canada); l – Rougerai plateau peatland (China). Photo credits: a, b, d, e, g, i - G.R. Milton ©; c - C.R. Prentice ©; h - © Ducks Unlimited Canada; k - S. Basquill © Nova Scotia Department of Natural Resources; f, j, l - C.M. Finlayson © Rights remain with the authors.

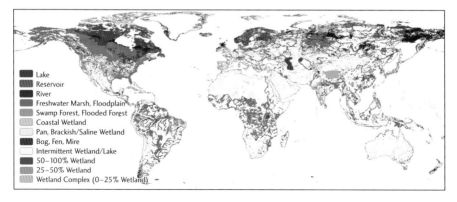

Plate 2 Generalized distribution of wetland types (Reprinted from Journal of Hydrology, Vol 296, B. Lehner and P. Döll, Development and validation of a global database of lakes, reservoirs and wetlands, page 16, 2004, with permission from Elsevier).

Part I

Overall Considerations

1

Diversity of Freshwater Ecosystems and Global Distributions

G. Randy Milton and C. Max Finlayson

Corresponding author: gordon.milton@novascotia.ca

1.1 Introduction

Freshwater ecosystems occur on all continents and have been described and classified by a diversity of terms and definitions (Finlayson and van der Valk 1995; Mitsch and Gosselink 2015; Gerbeaux et al. 2018). These are dynamic systems in which the hydrology and geomorphic setting can be affected by natural and anthropogenic activities interacting at multiple spatial and temporal scales to influence the components, biogeochemical processes, functions, and ecosystem services they provide (Millennium Ecosystem Assessment 2005). Defining and classifying freshwater ecosystems into categories (e.g., rivers and streams, lakes and reservoirs, and wetlands) has generally been based on a combination of features, including the landform and geomorphic setting, physical dimensions, vegetation, water quality and water regime, soils, and as habitat for specific fauna (Semeniuk and Semeniuk 1997). The term wetland is often used for inland, coastal, and marine habitats defined by the Ramsar Convention on Wetlands as '…areas of marsh, fen, peatland or water, whether natural of artificial, permanent or temporary, with water that is static or flowing, fresh, brackish or salt, including areas of marine water the depth of which at low tide does not exceed six metres' (Ramsar Convention Secretariat 2010) and at other times to more narrowly describe shallow vegetated systems, such as peatlands, marshes, and swamps. This narrower description typically distinguishes wetlands as occurring in the transition between the upland and deep water; and are characterised by the presence of water at the surface or within the root zone for at least part of the year, soils that are saturated, flooded, or ponded during the growing season long enough to develop anaerobic conditions in the upper horizons (hydric soils), and vegetation adapted to survive in wet conditions (Mitsch and Gosselink 2015). Although adaptations are not addressed in this chapter, it is important to note that aquatic animals also exhibit physiological and behavioural adaptations to deal with low oxygen concentrations, anoxic substrates, changes in salinity, and fluctuating hydroperiods (Mendelssohn and Batzer 2006). The terms 'freshwater' or 'freshwater ecosystem' are used here unless specific habitat types are unambiguously referred to in the source material.

Milton, G. R. and Finlayson, C. M., *Diversity of freshwater ecosystems and global distributions*. In: *Freshwater Ecology and Conservation: Approaches and Techniques*. Edited by Jocelyne M. R. Hughes: Oxford University Press (2019).
© Oxford University Press 2019. DOI: 10.1093/oso/9780198766384.003.0001

1.2 Distribution and extent

Representation of the global distribution and estimates of surficial freshwater area are inferred from two primary sources: digital maps and attribute data sources (Finlayson et al. 1999; Lehner and Döll 2004; Messager et al. 2016) and satellite remote sensing (Prigent et al. 2001; Verpoorter et al. 2014; Chen et al. 2015). Fluet-Chiounard et al. (2015) report remote sensing inventories and cartographic surveys reasonably constrain estimates of the abundance and surface area of large lakes and reservoirs. Developed by compiling multiple water-related data sets from the two primary sources, the Global Wetland Database (GLWD—Lehner and Döll 2004) underestimates but displays greater agreement with ground-truthed wetlands data than satellite-only-based products for open water and other wetlands (Frey and Smith 2007); and was assessed by Nakaegawa (2012) as the best among six 1-km global water-related (snow and ice, wetlands, and open water) land cover mapping initiatives. These initiatives have documented the latitudinal and longitudinal distribution of freshwater ecosystems as well as a general distribution of wetland types (Plate 1 and Plate 2). Despite the effort to describe and map freshwater ecosystems using combinations of primary source data, a comprehensive and accurate global inventory and map does not exist due in part to differences in definitions and classifications (Finlayson et al. 2018), complexity and variability within and among wetlands even when similarly classified (Milton et al. 2018), and limitations in observing smaller size classes (Lewis 2011; MacDonald et al. 2012; Seekell et al. 2013).

The general pattern that emerges is that the latitudinal distribution of surficial wetland area largely parallels the distribution of lakes with a preponderance in boreal and arctic latitudes (50–70° N) (Lehner and Döll 2004). This distribution corresponds to the extent of the last glacial maximum in northern North America and Scandinavia including parts of northern Russia (Messager et al. 2016); the abundance is much lower in southern latitudes where the continental area is lower and 50 per cent of lake area is located at elevations below 500 m above sea level (Verpoorter et al. 2014). Fluet-Chouinard et al. (2015) compare the latitudinal and longitudinal distribution of inundated area from their downscaling of the Global Inundation Extent from Multi-Satellites (GIEMS) with the GLWD (Lehner and Döll 2004) and Global Land Cover 2000 (Bartholomé and Belward 2005) and find substantial departure occurs from the generally expected pattern over parts of Asia due to GIEMS seasonal documentation of inundated rice cultivation.

Milton and Finlayson (2018) reviewed the estimates of surficial freshwater extent by various studies to obtain a composite global estimate of 12,543.5–14,443.5 × 10³ km² for permanent, seasonal, and intermittent freshwater ecosystems ≥1 ha, that was higher than many previous estimates. This estimate excludes the unknown area of hundreds of freshwater lakes of Antarctic's coastal oases (Laybourn-Parry and Wadham 2014) and its 400+ subglacial lakes (Siegert et al. 2016), and that of subterranean karst groundwater circulation (see Beltram 2018). This higher estimate is not unexpected as Junk et al. (2013) acknowledge that their synthesis of global wetland coverage of 11,521 × 10³ km² underestimated the true extent for several major regions, including South America, Africa, and Russia. Applying a downscaling approach to the Global Inundation Extent from Multi-Satellites (GIEMS) data set (excludes Antarctica), Fluet-Chouinard et al. (2015)

provide an approximation to temporal and spatial variability in coverage due to climate and hydrology. Seasonal lows and highs (6,538–12,089 × 10^3 km^2) and long-term maximum (17,255 × 10^3 km^2) suggest the higher estimate above is realistic and probably still underestimates the true extent given the lower resolution used in the analysis.

1.3 Freshwater hydrology

The hydrologic regime influences the many physical and chemical properties of freshwater systems, such as soil and water salinity, soil anaerobiosis, nutrient availability, pH and the influx, deposition, and removal of sediment (Mitsch and Gosselink 2015). It has both a temporal, varying from ephemeral to perennial, and a dynamic dimension for flowing (rivers—lotic systems) and standing waters (lakes and ponds—lentic systems) and others (often described as wetlands) that may experience permanent flooding or, at times, large fluctuations in water depths seasonally, periodically, intermittently, or episodically (Finlayson and D'Cruz 2005). River basins are the primary hydrological unit and the basis for understanding the interactions with hydrological processes as they are characterised by unique combinations of geomorphology, climate, basin connectivity, and timing, volume, and duration of flows (Revenga and Tyrell 2018). River basins encompass the entire area drained by a major river system or one of its tributaries to eventually flow into another river or the ocean unless it is a closed (endorheic) basin. Rain that falls onto endorheic basins can be sufficient to create downstream flow that forms seasonal wetlands where water eventually evaporates or seeps into the ground (e.g., Okavango Delta, Kati Thanda—Lake Eyre) or form a land-locked water body (e.g., Caspian Sea) (Revenga and Tyrell 2018).

Through feedback mechanisms, changes in the physical and chemical properties and interactions with its biotic components can influence and alter the hydrological regime (Thompson and Finlayson 1999; Mitsch and Gosselink 2015). Sediment deposition can, for example, affect basin inflows and outflows, or change the basin geometry providing new areas for plant growth which in turn slows water movement, promotes sediment deposition, gradually decreases water depth, and favours animals adapted to shallow water. This process can be reversed by extreme hydrological events that either erode or deposit additional sediment preventing aquatic plant development and habitat. Animals can also affect hydrological conditions by damming flows and flooding low lying areas (e.g., beaver—*Castor canadensis*) or by grazing, wallowing, and trampling of aquatic vegetation to create depressions and open water areas.

The water balance is commonly used to understand a system's hydrology, expressed as the relationship between hydrological inputs and outputs, and provides insight into the key processes determining the functioning of the system as a whole (Baker et al. 2009). The water budget can be expressed by the basic components in the following equation and expanded as required to consider the possible contributors to each (Table 1.1).

Input = Output + Δ Storage

Each of these components varies in importance between the different types of freshwater ecosystem and not all the terms of the water budget need apply due to variation in geomorphology, geology, and seasonality.

Table 1.1 *The hydrological units associated with the basic input, output, and storage components used in constructing a water budget. (Adapted from Baker et al. 2009)*

	Water Budget Components		
	Inputs	Outputs	Δ Storage
Hydrological Units	Precipitation	Evapotranspiration	Open water
	Overland flow	Channel outflow	Saturated soils
	Inundation	Overland flow	Unsaturated soils
	Subsurface stormflow	Vegetation interception	
	Groundwater discharge	Groundwater recharge	

Water level regimes demonstrate tremendous variability; and specific hydroperiods can typify certain wetland types (Mitsch and Gosselink 2015), although prevailing hydrometeorological conditions introduces inter-annual variation in the patterns (Baker et al. 2009). Cowardin et al. (1979) classified water level regimes on non-tidal wetlands using eight categories based upon the duration of standing water and frequency of inundation (Table 1.2). Tidal freshwater wetlands in contrast occur where energy from the tidal wave in an estuary acting against the constant input from the river creates flooding regimes in freshwater systems which are tied to daily tidal amplitude changes of the lunar cycle (Barendregt 2018). The water level regime is a major factor controlling ecosystem structure and its ecological processes, including plant establishment and growth, diversity of species, oxidation and reduction conditions, nutrient transformations and cycling, organic accumulation, and system productivity (Thompson and Finlayson 1999).

Topography is the primary factor that controls how the water balance influences a wetland's water regime with fluctuations in the vertical depth, horizontal extent (area),

Table 1.2 *Water level regimes on non-tidal wetlands (after Cowardin et al. 1979)*

Water Level Regime	Definition
Permanently flooded	Flooded throughout the year in all years
Intermittently exposed	Flooded throughout the year except in years of extreme drought
Semi-permanently flooded	Flooded throughout the growing period in most seasons, when surface water is absent the water table is at or very near the surface
Seasonally flooded	Flooded for extended periods in the growing season, surface water is usually absent by end of growing but water table at or near the surface
Temporarily flooded	Flooded for brief periods during the growing season, but water table is otherwise well below the surface
Intermittently flooded	Surface is usually exposed with surface water present for variable periods without detectable periodicity; intervals between periods of inundation are indeterminate and may last years
Artificially flooded	Amount and duration of flooding is controlled by artificial means in combination with dikes and dams
Saturated	Substrate is saturated to the surface for extended periods during the growing season but surface water is seldom present

and volume of stored surface water (Thompson and Finlayson 1999; Baker et al. 2009). At varying scales, topographic features can act to restrain water, promote surface water storage, and affect downstream discharge and flow velocity. For example, channel dimensions, bank elevation, and the presence, dimensions, and elevations of features (e.g., ditches) linking rivers to floodplains controls when a river meets its threshold discharge and begins to inundate a floodplain; and flood extent and depth of surface water is controlled by the topography of the floodplain and its surroundings (Baker et al. 2009). Water level–area–volume relationships are evident in the feedback between the water balance and water level regime due to changes in open water precipitation inputs and evaporation and seepage losses (Baker et al. 2009).

1.4 Freshwater ecosystem diversity

The increasing usage of the term 'wetland' in recent decades has to some extent masked the array of freshwater ecosystems that exist. There is an increasing amount of information about individual wetlands and wetland types, although efforts to collate this to compile a global inventory and descriptions have been constrained by differences in definitions and classifications (Finlayson et al. 2018). With this in mind a description of broad wetland classes is provided below; namely, for rivers, lakes and reservoirs, marshes and swamps, peatlands—mires, and freshwater karst systems (Plate 1).

1.4.1 Rivers

Rivers and streams (hereafter used interchangeably) are self-organized and complex ecosystems that are heterogeneous and dynamic, responding to changes in hydrologic flow regimes across multiple spatial and temporal scales within a broader terrestrial landscape (Everard and Powell 2002; Wiens 2002; Harris and Heathwaite 2012; Tadaki et al. 2014). Interactions between hydrological and geomorphological conditions and ecological processes in river corridors connect aquatic, riparian, and floodplain ecosystems and influence river morphology (e.g., width, depth, planform pattern); chemistry and physical processes (e.g., bank erosion/aggradation, sediment transport/retention); species distribution, assemblages, and interactions; nutrient cycling; and energy flow (Ward and Tockner 2001; Wiens 2002; Grabowski and Gurnell 2016; Gurnell et al. 2016).

The global surface area of rivers exclusive of ephemeral and intermittent streams is an estimated 624,000 km^2 (Raymond et al. 2013) although flooding in tropical areas increases the areal extent of tropical rivers 100–1,000 fold (Melack and Hess 2009). Numerous river classification systems have been developed to meet a specific purpose or need: ecological inventory and comparative assessment of river condition; modelling ecological functions to prioritise conservation actions; prescription of river restoration practices; and place-based typologies of river forms and processes for catchment-scale planning. Criteria used to distinguish different river classes include hydrological, physioclimatic, geomorphological, chemical, biological, and anthropogenic parameters (Tadaki et al. 2014).

Thorp et al. (2006) review the numerous theories and models proposed to explain the structural and functional complexity that can vary spatially and temporally within river networks. This includes the River Continuum Concept (RCC; Vannote et al. 1980) which emphasises the longitudinal continua of ecosystem process through the downstream flows of water and material in unperturbed rivers (Wiens 2002). As noted by Vannote et al. (1980), additional studies were needed to test and refine this concept and question its general applicability. Rather than the RCC's expected changes in biotic response along a continuous, longitudinal gradient in physical conditions, Stazner and Higler (1986), for example, identified species assemblages responding to discontinuities in water flow and substrate size; and Thorp et al. (2006) portrayed rivers as arrays of large hydrogeomorphic patches formed by catchment geomorphology and flow characteristics that differ in physical and chemical conditions and biocomplexity with boundaries that may be sharp or indistinct. Additional to the longitudinal flows, essential components of river ecosystems include *inter alia* vertical transitions and interactions between surface and subsurface water and riparian systems (Ward et al. 1998), lateral and temporal variability described in the 'flood pulse' concept (Junk et al. 1989), interactions and feedbacks between hydrogeomorphological processes and vegetation (Gurnell et al. 2016), and the temporal importance of natural stream flow variability in regulating ecological processes in river ecosystems (Poff et al. 1997). Human activities can interrupt or alter the structural and functional interactions within a riverscape resulting in ecological simplification with a loss of landscape complexity and ecological integrity (Peippoch et al. 2015).

1.4.2 Lakes and reservoirs

Naturally occurring lakes and ponds are characterized by standing water in depressions formed in the landscape principally from tectonic, glacial, or fluvial processes, although they can also occur through volcanic activity, by solution in karst terrain, damming of rivers by landslides or biological activity (e.g., beaver dams), or wind action that erodes or transports sediment forming irregularities in the terrain (Hutchinson 1957). The oldest lakes were formed by tectonic processes (e.g., African rift lakes, Caspian Sea) but most lakes are relatively young in the order of tens of thousands of years originating with the glacial recession following the last ice age. Human-made lakes and reservoirs have existed for millennia and vary greatly in size and depth depending upon the size of the constructed dam, river basin morphology, and upstream catchment area. Large impoundments and reservoirs (dam height >15 m) significantly increased from 5,000 to 50,000 between the middle and end of the twentieth century (Berga et al. 2006). In addition to their origin, lakes can be characterised by their hydrology, trophic status, and seasonal thermal stratification and mixing regime.

The hydrology of lakes and reservoirs is the sum of water inflows and outflows associated with their basin morphology, climate, and form of surrounding landscape. Lakes with no surface outflows are described as terminal or *endorheic* lakes with surface area and volume varying in response to increases in either water inputs or outputs through evaporation and groundwater flow. Ephemeral and intermittent systems are filled irregularly with freshwater when recently flooded and the salinity gradually increases as the water evaporates; these lakes may remain dry for long periods. Most lakes, however, have water levels that vary in response to surface water inflows and outflows and can be further characterized by the thermal conditions in temperate, warm-temperate, and tropical zones that can vary seasonally or daily to affect the frequency of stratification and mixing processes (Hutchinson 1957; Wetzel and Likens 1991). Thermal stratification is identified by the establishment of a thermocline, a steep temperature gradient, that occurs in the *metalimnion* between surface (*epilimnion*) and deeper (*hypolimnion*) layers of water. Depending upon the frequency of mixing, lakes are described as *monomictic*, *dimictic*, or *polymictic* and affect the entire water column (*holomictic* lakes). Lakes that do not experience complete circulation due to salinity or turbidity gradients establishing density concentration profiles that stabilize and isolate their lower layers, referred to as *meromictic* lakes, are uncommon (Hall and Northcote 2012).

The level of nutrients within lakes is the basis for determining their trophic status (Hutchinson 1957; Wetzel and Likens 1991). Low concentrations of nutrients, generally clear water, and low plant or phytoplankton production characterise oligotrophic lakes. In contrast, eutrophic lakes experience high nutrient concentrations, turbid water, and high plant production and possibly blooms of phytoplankton. Mesotrophic lakes are intermediate between these two states although there is variation in scales used to determine these states. Lakes with low pH values, coloured water from organic acids, and typically low in calcium, phytoplankton, and fish production are dystrophic. Nitrogen and phosphorus are important nutrients for plant growth and the lake's production; and most primary production occurs in the upper photic zone where there is sufficient light for photosynthesis by phytoplankton and higher plants. Irradiance and light attenuation, temperature, and the mixing of the water also influence primary production.

1.4.3 Marshes and swamps

Freshwater swamps and marshes comprise an array of non-tidal wetlands with swamps generally considered as being dominated by trees and marshes by emergent herbaceous plants (Thompson and Finlayson 1999; Mitsch et al. 2009). They have hydric soils but do not contain large amounts of peat, and are found from the temperate zone to the tropics. While the terms 'swamps' and 'marshes' are widely used, many compilations of ecological information, such as those provided by the global freshwater biodiversity assessment (Balian et al. 2008) and freshwater ecoregions (Abell et al. 2008), do not specifically differentiate between them. In some respects this may reflect the mix of vegetation types that occur in many freshwater wetlands, especially the larger floodplain systems or wetland complexes. Lehner and Döll (2004) using multiple data sources provided estimates of the areal extent of freshwater marshes as $2,529 \times 10^3$ km^2 and swamp forest as $1,165 \times 10^3$ km^2. The generalized distribution of these wetlands is shown in Plate 2.

As described by Mitsch et al. (2009), marshes are typically found in areas that are frequently or continuously inundated with water and with mineral soils. Dominant plants include a mix of reeds, rushes, grasses, and sedges. Swamps, on the other hand, may be permanently or intermittently covered with water and are generally dominated by trees or shrubs, with those associated with rivers typically having inorganic substrates. Swamps with little or no connection to flowing streams may accumulate small amounts of peat, although generally not in sufficient quantities to be classed as peat-lands. Tropical freshwater marshes and swamps are typically found on floodplains and may have a complex mix of woodland, forest, and herbaceous vegetation. Junk et al. (2006), in a comparative analysis of large wetlands, illustrate the vegetation mixes that occur and also describe the variety in structure and ecological interactions.

Given their widespread distribution and mix of vegetation types it is not surprising that the water quality of freshwater swamps and marshes also varies across these as well as seasonally and diurnally. The water quality of the wetlands assessed by Junk et al. (2006) provide a guide to the range of conditions, but cannot by themselves be seen as representative of such wetlands. The latter is also influenced by the occurrence of brackish or even highly saline conditions that can occur under drying conditions.

The water regimes in these wetlands also vary considerably, and include those that are permanently inundated as well as those that are ephemeral and intermittently inundated, including experiencing long periods of drought and sudden flooding. The importance of water regimes in wetlands has been widely discussed with reference to ecological processes, such as nutrient cycling, and how the frequency and duration of flooding, including the importance of flood pulsing, influences the occurrence and life cycles of many species, whether resident or migratory (Thompson and Finlayson 1999). As a consequence of these variable and different conditions the species diversity across the range of swamps and marshes is high, as illustrated in a number of detailed descriptions (Thieme et al. 2005; Abell et al. 2008; Groombridge and Jenkins 2000; Balian et al. 2008). These, however, covered freshwater ecosystems and did not specif-ically differentiate swamps and marshes. Similarly, in their comparative examination of six well-investigated large freshwater wetlands, Junk et al. (2006) note overall species diversity is high; and although the trend does not hold true for all groups and sites, diversity generally increases from high latitudes to the equator. Data was, however, insufficient to assess many lower groups of plants and animals.

1.4.4 Peatlands—mires

Wetlands characterised by the accumulation of surficial peat, defined as *in situ* formed material consisting of at least 30 per cent (dry mass) organic material derived from decaying plant material under almost permanent water saturation and absence of oxy-gen (Joosten and Clark 2002; Schurmann and Joosten 2008), at least 30 cm thick are peatlands (with or without vegetation). Mire is a term typically used in Eurasia for any wetland currently forming peat, regardless of depth, to differentiate them from peat-lands where peat accumulation has stopped (e.g., because of drainage (Joosten and Clarke 2002)). The typology associated with mires and peatlands is large due to numerous classification approaches, but there is no definitive classification system (Lindsay 2018). A hierarchical approach based upon hydro-morphological features has been described by Charman (2002); and the hierarchical Tope System includes traditional approaches of dominant hydrological processes, morphology, and associ-ated landform with additional features such as interconnectedness of mires, surface patterns, and micro-topography (Lindsay 2018). The source of water that leads to saturation is the key factor in distinguishing systems whereby peat-forming vegetation is entirely dependent upon precipitation to supply its water and solutes (ombrotrophic), and those receiving solute-enriched ground or surface water (minerotrophic) (Joosten and Clark 2002; Lindsay 2018). The terms 'bog' and 'fen' have increasingly been used respectively to describe peats that are higher than their surroundings and ombrotrophic and those that occur in landscape depressions and minerotrophic (Joosten and Clark 2002; Schurmann and Joosten 2008). In tropical regions, bogs and fens are typically covered by rainforest and are often called peat swamp forest (Page et al. 1999).

As described by Rydin and Jeglum (2006) a peatland's overall water balance consists of inflows, outflows, and storage. Although variations do occur in a peatland's water balance, a positive water storage is required to produce anaerobic conditions which prevent the complete decomposition of plant biomass. The rate of water movement within a peatland is determined using Darcy's law that relates the hydraulic conductivity of the peat to the hydraulic gradient. Related to the degree of humification and bulk density, hydraulic conductivity often decreases with depth and *Sphagnum* peats are generally dense with small pore spaces and less permeable than peats derived from sedges and wood. The aerobic acrotelm is the 'active' layer of loose, living vegetation lying above the lowest level of a peatland's oscillating water table. The 'inactive' and constantly anoxic lower layer, or catotelm, is usually more humified with denser peat and comprises most of the volume of the peatland. The distinction between these layers in peat swamp forests is generally less clear than it is in temperate and boreal peatlands.

The water chemistry of a peatland's water source(s) has a large influence on the vegetation, chemical processes, and character of peat accumulation. Precipitation has low concentrations of all elements while the chemistry of minerotrophic water depends upon its interaction with the geology, soils, and vegetation which it flows through or over. Nutrient deficiencies are typical in peatlands and, in general, nutrient levels tend to increase along the trophic gradient of bog-fen-swamp and increasing depth. In boreal regions, growing *Sphagnum* species continuously create cation exchange sites and can thereby shape the habitat and shift peatland dynamics by acidifying its substrate and being resistant to decay. Organic matter has a high cation exchange capacity for hydrogen ions and as the influence of cations derived from mineral soils declines, peat accumulation leads to increasing acid and nutrient-poor conditions. Increasing acidity inversely affects the reduction potential (or redox) and can result in the production of toxic substances (e.g., Mn^{2+}, Fe^{2+}). Under anaerobic conditions there is an increase in the peatland's redox and bacterial activity that can result in the reduction of nitrate and carbon dioxide to nitrous oxide and methane respectively and release to the atmosphere (Rydin and Jeglum 2006).

Low oxygen availability, mobilisation of toxic elements, low nutrient availability, and high acidity greatly affect the species they support, with many needing specific adaptations to extreme conditions (Rydin and Jeglum 2006). Minayeva et al. (2008) in their peatland biodiversity review noted peatland diversity is generally low but with highly characteristic and obligate species that can show disjunct, azonal distribution patterns compared with nearby upland ecosystems in the same biogeographic zone.

The extreme conditions and low species diversity have led to the co-evolution of strong inter-specific dependencies and interactions in peatland communities. Moreover, the expression of intraspecific genetic and phenotypic diversity is high in peatlands and reflects local conditions, as well as geographical isolation and the island character of many peatlands. In addition, trans-specific diversity, the expression of similar mor-phological or functional traits by different species, is a typical adaptive mechanism in peatlands. Many species occur seasonally, at particular stages of their life cycle, or as the last refuge in altered landscapes.

Joosten (2010) summarised information available in the International Mire Conservation Global Peatland Database ('Global Peatland Database' website) to pro-vide a global peatland (peat >30 cm) area estimate of approximately 4 million km^2. Although found in most countries, a preponderance (nearly 75 per cent) are distrib-uted over mid-high latitudes of North America and Eurasia. Page et al. (2011) note South East Asia has the largest share (56 per cent) of tropical peatlands estimated at 441,025 km^2, followed by South America (24 per cent), and Africa (13 per cent). The 145,500 km^2 estimate by Dargie et al. (2017) for the Cuvette Central in the Congo Basin is 2.6 times the estimate of Page et al. (2011) Africa estimate and the most exten-sive peatland complex in the tropics. The uncertainty in peatland estimates is due to gaps in data coverage, reliance on remote sensing with limited ground truthing, and the use of different definitions and classifications in global and national compilations (Joosten 2010; Montanarella 2014).

In their review, Sirin and Laine (2008) describe peatlands playing an important role in the global climate cycle through the storage of atmospheric carbon and as a source of greenhouse gases, including carbon dioxide, methane, and nitrous oxide. Although peatlands can emit methane, the amount of carbon stored in peatlands, despite uncer-tainties in the data, exceeds that of global vegetation (Joosten and Couwenberg 2008) and may be of similar magnitude to the atmospheric carbon pool (Turetsky et al. 2015). The drainage and use of peatlands for agriculture and forestry has, however, resulted in the oxidation of peat and the release of carbon dioxide to the atmosphere (Joosten and Couwenberg 2008). For example, fires in drained and degraded peat for-ests in Indonesia are estimated to have released in 1997 the equivalent of 13–40 per cent of the 6.4 $GtCyr^{-1}$ global annual (2002) emissions from fossil fuels (Page et al. 2002); and CO_2 emissions from peatland drainage in South East Asia is contributing the equivalent of 1.3 per cent to 3.1 per cent of current global CO_2 emissions from the combustion of fossil fuels (Hooijer et al. 2010). The importance of peatlands as global carbon stores and sinks is increasingly recognised, and the conservation and restoration of peatlands is identified as an adaptive measure to mitigate climate change (Biancalani and Avagyan 2014; Hooijer et al. 2010; Joosten 2010; Joosten et al. 2012).

1.4.5 Freshwater karst systems

Outcropping and subterranean water soluble and porous carbonate rock is estimated to occur over 14 per cent of the world's land area and develops characteristic landscapes and features with surficial and subterranean standing and flowing water. Precipitation dissolving rock (referred to as dissolution by precipitation absorbing CO_2 to form a weak solution of carbonic acid) is the key factor in karst landscape development but other natural processes, such as river erosion and glaciation, contribute. Karst development is most intensive in the humid tropics and temperate zones but exhibits global variability due to differences in lithography, geological structure, and climate. Williams (2008a) briefly describes the origin of the word with karst landscapes and features linked to the hydrological cycle as water passes into, flows through, and emerges from karst terrains.

The three-dimensional structure of a karst landscape and the complex hydrology are inter-related. Meteoric water (originating from rain or snow) enters through the surface (exokarst), percolating through dissolutionally enlarged micro and mesoscale fractures of permeable carbonate bedrock in the epikarst down 10–30 m to the underlying endokarst and its deeper components (e.g., caves, voids, conduits, and passages) (White et al. 1995; Williams 2008 a,b; Stokes et al. 2010). Subterranean water flow paths are difficult to follow and can vary from tiny conduits to underground rivers that may have originated as concentrated inflows sinking underground at lateral or vertical swallow holes (Beltram 2018).

The saturated (phreatic) zone lies below the partially water-filled fractures in the upper unsaturated (vadose) zone of the karst aquifer. Porosity in the epikarst decreases with depth and water may be temporarily stored in the rock matrix, producing an epikarst aquifer in the vadose zone, before infiltrating into larger conduits below or emerging as small perennial springs on hillsides (Williams 2008b). Most water passing though karst conduits originates from outside rather than within the karst boundary to re-emerge as springs. Underground residence time of meteoric water is typically short (a few days to a year) with longer-term water storage in the rock matrix and fracture porosity compared to conduits (Williams 2008a; Stokes et al. 2010). Deep circulating waters may remain underground for a decade or more (Williams 2008a).

Surficial shallow depressions in karst landscapes are linked to subterranean karst geomorphology landscapes and function as sinks or springs depending on water level; and may dry out during low precipitation periods. Fed mainly by rising groundwater, they are interdependent with the subterranean system using two general conceptual models: the flow-through system and the surcharged tank. In the former, inflow and outflow occur simultaneously and largely independently providing a constant flow of groundwater. Acting as overflow storage for the underlying karst flow network, sites exhibiting the surcharged tank model accumulate excess groundwater that cannot be accommodated in the subterranean conduit network. During the flood period, the surficial landscape resembles and functions like shallow lakes (Irvine et al. 2018).

It is not unusual for karst systems to have several discharge points, and subsurface flow routes vary greatly in length (e.g., 10–100 km^2 in the Yucatan peninsula—Beltram 2018). Ground water level fluctuations are characteristic as most karst conduits have limited capacity to expand their surficial area and respond to flood pulses by filling and over-flowing into higher level passages, raising the aquifer's water table by as much as 100+ m in several hours during a severe flood (White et al. 1995; Beltram 2018). A consequence of the limitation on conduit flow volume on discharge areas is lower, broader flood crests than streams of similar discharge draining insoluble rocks (White et al. 1995).

Surficial depressions exhibit characteristic wetland vegetation zonal communities and seasonal succession of phytoplankton, invertebrate, and vertebrate communities responding to variations in water depth, extent, and duration of flooding (Irvine et al. 2018). Subterranean systems are dependent upon nutrients imported by percolating water or sinking steams to support a much-reduced biotic diversity with low populations and high endemism, although communities of autotrophic, mainly sulphur-oxidizing micro-biota, may develop with mineral water influx (Sket 2018).

1.5 Summary statement

Freshwater ecosystems, comprising many different types and sizes, are found on all continents and global mapping initiatives have documented their latitudinal and longitudinal distribution as well as displayed the general distribution of wetland types. They are especially widespread over the mid-high latitudes of North America and Eurasia and areas of excess moisture. Estimates indicate that they globally cover an area of $12,543.5$–$14,443.5 \times 10^3$ km^2. The temporal and dynamic dimension of the hydrologic regime is a key influence of the many physical and chemical features of freshwater ecosystems, and organisms exhibit physiological and behavioural adaptations to deal with low oxygen concentrations, anoxic substrates, changes in salinity, and fluctuating hydroperiods.

The diversity of freshwater habitats and the challenges facing their conservation and management requires an array of approaches and methods in order to describe, quantify, and monitor their structure, productivity, functions, and services. The approaches and methods employed must be appropriate to the scope (e.g., policy development, ecosystem dynamics, impact assessment, restoration) and scale (e.g., site, watershed, global) of the conservation and management issue(s) being addressed (Part 3 of this volume). Approaches may entail community engagement and citizen science, accessing existing databases, field surveys, remote sensing, or hypothesis-based research *in situ* or in a laboratory (Chapters 2 and 3). Sampling techniques to describe the physical, chemical, and biological components of freshwater ecosystems and their temporal and spatial interactions at site, catchment, and landscape scales may be well established or novel (Parts 2 and 3 of this volume).

The application of Earth Observation technologies provides many options for determining the area and condition of wetlands, and when accompanied by appropriate hydrological modelling can also enable increasingly more reliable estimates of the water volume and flows to be provided. The importance of ground-truthing data obtained using these technologies should not be underestimated. Davidson et al. (2018) have shown that a trend of increasing wetland extent reported in the literature is a consequence of mapping technologies rather than a real increase in wetland area. Further, it is likely that estimates of global wetland area persist in underestimating the extent of wetlands, noting that the 'grand challenge' of obtaining a global wetland inventory covering all types of freshwater wetlands at high spatial resolution has yet to be achieved.

References

Abell, R., Thieme, M., Revenga, C., Bryer, M., Kottelat, M., Bogutskaya, N., Coad, B., et al. (2008). Freshwater ecoregions of the world: a new map of biogeographic units for freshwater biodiversity conservation. *Bioscience*, 58, 403–14.

Baker, C., Thompson, J.R., and Simpson, M. (2009). Hydrological dynamics I: surface waters, flood and sediment dynamics. In: E. Maltby and T. Barker (eds) The Wetlands Handbook. Wiley-Blackwell, Chichester, U.K.

Balian, E.V., Leveque, C., Segers, H., and Martens, K. (eds) (2008). Freshwater Animal Diversity Assessment. Springer, Dordrecht.

Barendregt, A. (2018). Tidal freshwater wetlands, the fresh dimension of the estuary. In C.M. Finlayson, G.R. Milton, R.C. Prentice, and N.C. Davidson (eds) The Wetland Book II: Distribution, Description and Conservation. Springer, Dordrecht.

Bartholomé, E. and Belward, A.S. (2005). GLC2000: A new approach to global land cover mapping from earth observation data. *International Journal of Remote Sensing,* 26, 1959–77.

Beltram, G. (2018). Karst wetlands. In: C.M. Finlayson, G.R. Milton, R.C. Prentice, and N.C. Davidson (eds), The Wetland Book II: Distribution, Description and Conservation, Springer, Dordrecht.

Berga, L., Buil, J.M., Bofill, E., De Cea, C., Manueco, G., Polimon, J., Soriano, A., and Yague, J. (eds) (2006). Dams and reservoirs, societies and environment in the 21st century. In: Proceedings of the International Symposium on Dams in Societies of the 21st Century, Barcelona, Spain. Taylor and Francis Group, London, U.K.

Biancalani, R. and Avagyan, A. (eds) (2014). Towards Climate-Responsible Peatlands Management. Food and Agriculture Organization of the United Nations (FAO), Rome.

Charman, D.J. (2002). Peatlands and Environmental Change. John Wiley, West Sussex, U.K.

Chen, J., Chen, J., Liao, A., Cao, X., Chen, L., Chen, X., He, C., Han, G., Peng, S., Lu, M., Zhang, W., Tong, X., and Mills, J. (2015). Global land cover mapping at 30 m resolution: A POK-based operational approach. *Journal of Photogrammetry and Remote Sensing,* 103, 7–27.

Cowardin, L.M., Carter, V., Golet, F.C., and LaRoe, E.T. (1979). Classification of Wetlands and Deepwater Habitats of the United States. FWS/OBS-79/31. U.S. Fish and Wildlife Service, Washington (DC).

Dargie, G.C., Lewis, S.L., Lawson, I.T., Mitchard, E.T.A., Page, S.E., Bocks, Y.E., and Ifo, S.A. (2017). Age, extent and carbon storage of the central Congo Basin peatland complex. *Nature*, 54, 86–90.

Davidson, N.C., Fluet-Chouinard, E., and Finlayson, C.M. (2018). Global extent and distribution of wetlands: trends and issues. *Marine and Freshwater Research*, 69, 620–27.

Everard, M. and Powell, A. (2002). Rivers as living systems. *Aquatic Conservation: Marine and Freshwater Ecosystems*, 12, 329–37.

Finlayson, C.M. and D'Cruz, R. (2005). Inland water systems. In: R. Hassan, R. Scholes, and N. Ash (eds) Ecosystems and Human Well-Being: Current State and Trends: Findings of the Condition and Trends Working Group. Island Press, Washington, DC.

Finlayson, C.M., Davidson, N.C., Spiers, A.G., and Stevenson, N.J. (1999). Global wetland inventory—status and priorities, *Marine and Freshwater Research*, 50, 717–27.

Finlayson, C.M., Milton, G.R., and Prentice, R.C. (2018). Wetland Types and Distribution. In: C.M. Finlayson, G.R. Milton, R.C. Prentice, and N.C. Davidson (eds) The Wetland Book II: Distribution, Description and Conservation, Springer, Dordrecht.

Finlayson, C.M. and van der Valk, A.G. (eds) (1995). Classification and Inventory of the World's Wetlands. Advances in Vegetation Science, Volume 16. Kluwer Academic Publishers, Dordrecht.

Fluet-Chouinard, E., Lehner, B., Rebelo, L.-M., Papa, F., and Hamilton, S.K. (2015). Development of a global inundation map at high spatial resolution from topographic downscaling of coarse-scale remote sensing data. *Remote Sensing of the Environment,* 158, 348–61.

Frey, K.E. and Smith, L.C. (2007). How well do we know northern land cover? Comparison of four global vegetation and wetland products with a new ground-truth database for West Siberia. *Global Biogeochemical Cycles*, 21, GB1016.

Gerbeaux, P., Finlayson, C.M., and van Dam, A.A. (2018). Wetland classification: Overview. In: C.M. Finlayson, M. Everard, K. Irvine, R.J. McInnes, B.A. Middleton, A.A. van Dam, and N.C. Davidson (eds) The Wetland Book I: Structure and Function, Management and Methods. Springer, Dordrecht.

Grabowski, R.C. and Gurnell, A.M. (2016). Hydrogeomorphology—ecology interactions in river systems. *River Research and Applications*, 32, 139–41.

Groombridge, B. and Jenkins, M.D. (2000). Global Biodiversity: Earth's Living Resources in the 21st century. World Conservation Press, Cambridge.

Gurnell, A.M., Corenblit, D., García de Jalón, D., González del Tánago, M., Grabowski, R.C., O'Hare, M.T., and Szewczyk, M. (2016). A conceptual model of vegetation-hydrogeomorphology interactions within river corridors. *River Research and Applications*, 32, 142–63.

Hall, K.J. and Northcote, T.G. (2012). Meromictic lakes. In: L. Bengtsson, R.W. Herschy, and R.W. Fairbridge (eds) Encyclopedia of Lakes and Reservoirs. Springer, Dordrecht.

Harris, G.P. and Heathwaite, L. (2012). Why is achieving good ecological outcomes in rivers so difficult? *Freshwater Biology*, 57, 91–107.

Hooijer, A., Page, S., Canadell, J.G., Silvius, M., Kwadijk, J., Wösten, H., and Jauhiainen, J. (2010). Current and future CO_2 emissions from drained peatlands in Southeast Asia. *Biogeosciences*, 7, 1505–14.

Hutchinson, G. (1957). A Treatise on Limnology: Vol I. Geography, physics and chemistry. John Wiley and Sons, New York.

Irvine, K., Coxon, C., Gill, L., Kimberley, S., and Waldren, S. (2018). Turloughs (Ireland). In: C.M. Finlayson, G.R. Milton, R.C. Prentice, and N.C. Davidson (eds) The Wetland Book II: Distribution, Description and Conservation. Springer, Dordrecht.

Joosten, H. (2010). The Global Peatland CO_2 Picture: Peatland Status and Drainage Related Emissions in all Countries of the World. Wetlands International, Ede, The Netherlands.

Joosten, H. and Clarke, D. (2002). Wise Use of Mires and Peatlands. International Mire Conservation Group and International Peat Society. NHBS, Totness, Devon.

Joosten, H. and Couwenberg, J. (2008). Peatlands and Carbon. In: F. Parish, A. Sirin, D. Charman, H. Joosten, T. Minayeva, M. Silvius, and L. Stringer (eds) Assessment on Peatlands, Biodiversity and Climate Change: Main Report. Global Environment Centre, Kuala Lumpur (Malaysia) and Wetlands International, Wageningen, The Netherlands.

Joosten, H., Tapio-Biström, M.L., and Tol, S. (eds) (2012). Peatlands-Guidance for Climate Change Mitigation through Conservation, Rehabilitation and Sustainable Use. Food and Agriculture Organization of the United Nations and Wetlands International, Wageningen, The Netherlands.

Junk, W.J., Bayley, P.B., and Sparks, R.E. (1989). The flood-pulse concept in river-floodplain systems. In: D.P. Dodge (ed), Proceedings of the International Large River Symposium (LARS). Canadian Special Publication in Fisheries and Aquatic Sciences, 106. Department of Fisheries and Oceans, Ottawa.

Junk, W.J., Brown, M., Campbell, I.C., Finlayson, C.M., Gopal, B., Ramberg, L., and Warner, B.G. (2006). Comparative biodiversity of large wetlands: a synthesis. *Aquatic Sciences*, 68, 400–14.

Junk, W.J., Shuqing, A., Finlayson, C.M., Gopal, B., Květ, J., Mitchell, S.A., Mitsch, W.J., and Robarts, R.D. (2013). Current state of knowledge regarding the world's wetlands and their future under global climate change: a synthesis. *Aquatic Sciences*, 75, 151–67.

Laybourn-Parry, J. and Wadham, J.L. (2014). Antarctic Lakes, Oxford University Press, New York.

Lehner, B. and Döll, P. (2004). Development and validation of a global database of lakes, reservoirs and wetlands, *Journal of Hydrology*, 296, 1–22.

Lewis Jr, W.M. (2011). Global primary production of lakes: 19th Baldi Memorial Lecture. *Inland Waters*, 1, 1–28.

Lindsay, R. (2018). Peatland/Mire category descriptions based on origin and behaviour of water, peat genesis, landscape position and climate. In: C.M. Finlayson, G.R. Milton, R.C. Prentice, and N.C. Davidson (eds) The Wetland Book, Volume II: Distribution, Description and Conservation. Springer, Dordrecht.

McDonald, C.P., Rover, J.A., Stets, E.G., and Striegl, R.G. (2012). The regional abundance and size distribution of lakes and reservoirs in the United States and implications for estimates of global lake extent. *Limnology and Oceanography*, 57, 597–606.

Melack, J.M. and Hess, L.L. (2009). Remote sensing of the distribution and extent of wetlands in the Amazon basin. In: W.J. Junk, M.T.F. Piedade, J. Schöngart, and P. Parolin (eds) Amazonian Floodplain Forests: Ecophysiology, Ecology, Biodiversity and Sustainable Management. Springer, Dordrecht.

Mendelssohn, I.A. and Batzer, D.P. (2006). Abiotic constraints for wetland plants and animals. In: D.P. Batzer and R.R. Sharitz (eds) Ecology of Freshwater and Estuarine Wetlands. University of California Press, Berkley.

Messager, M.L., Lehner, B., Grill, G., Nedava, I., and Schemitt, O. (2016). How much water resides in lakes? Estimating the abundance and age of global lake volume. *Nature Communications*, 7, 13603.

Millennium Ecosystem Assessment (2005). Ecosystems and Human Well-Being: Wetlands and Water Synthesis. World Resources Institute, Washington DC.

Milton, G.R. and Finlayson, C.M. (2018). Freshwater ecosystem types and extents. In: C.M. Finlayson, A. Arthington, and J. Pittock (eds) Freshwater Ecosystems in Protected Areas: Conservation and Management. Taylor and Francis, Oxford, U.K.

Milton, G.R, Prentice, R.C., and Finlayson, C.M. (2018). Wetlands of the world. In: C.M. Finlayson, G.R. Milton, R.C. Prentice, and N.C. Davidson (eds) The Wetland Book II: Distribution, Description and Conservation. Springer, Dordrecht.

Minayeva, T., Bragg, O., Cherednichenko, O., Couwenberg, J., van Duinen, G-J., Giesen, W., Grootjans, A.B., Grundling, P-L., Nikolaev, V., and van der Schaaf, S. (2008). Peatlands and biodiversity. In F. Parish, A. Sirin, D. Charman. H. Joosten, T. Minayeva, M. Silvius, and L. Stringer (eds) Assessment on Peatlands, Biodiversity and Climate Change: Main Report. Global Environment Centre, Kuala Lumpur (Malaysia) and Wetlands International, Wageningen, The Netherlands.

Mitsch, W.J. and Gosselink, J.G. (2015). Wetlands. 5th Edition, John Wiley and Sons, Hoboken, New Jersey.

Mitsch, W.J., Gosselink, J.G., Anderson C.J., and Zhang, L. (2009). Wetland Ecosystems. John Wiley and Sons, Hoboken, New Jersey.

Montanarella, L. (2014). Mapping of Peatlands. In: R. Biancalani and A. Avagyan (eds) Towards climate-responsible peatlands management. Food and Agriculture Organization of the United Nations (FAO), Rome.

Nakaegawa, T. (2012). Comparison of water-related land cover types in six 1-km global land cover datasets. *Journal of Hydrometeorology*, 13, 649–64.

Page, S.E., Rieley, J.O., and Banks, C.J. (2011). Global and regional importance of the tropical peatland carbon pool. *Global Change Biology*, 17, 798–818.

Page, S.E., Rieley, J.O., Shotyk, O.W., and Weiss, D. (1999). Interdependence of peat and vegetation in a tropical peat swamp forest. *Philosophical Transactions of the Royal Society B*, 354, 1885–97.

Page, S.E., Siegert, F., Rieley, J.O., Boehm, H-D.V., Jaya, A., and Limin, S. (2002). The amount of carbon released from peat and forest fires in Indonesia in 1997. *Nature*, 420, 61–5.

Peipoch, M., Brauns, M., Hauer, F.R., Weitere, M., and Valett, H.M. (2015). Ecological simplification: human influences on riverscape complexity. *Bioscience*, 65, 1057–65

Poff, N.L., Allan, J.D., Bain, M.B., Karr, J.R., Prestegaard, K.L., Richter, B.D., Sparks, R.E., and Stromberg, J.C. (1997). The natural flow regime: a paradigm for river conservation and restoration. *BioScience*, 47, 769–84.

Prigent, C., Matthews, E., Aires, F., and Rossow, W.B. (2001). Remote sensing of global wetland dynamics with multiple satellite data sets. *Geophysical Research Letters*, 28, 4631–4.

Ramsar Convention Secretariat (2010). Designating Ramsar Sites: Strategic Framework and Guidelines for the Future Development of the List of Wetlands of International Importance. Ramsar Handbooks for the Wise Use of Wetlands, 4th edition, Volume 17. Ramsar Convention Secretariat, Gland, Switzerland.

Raymond, P.A., Hartmann, J., Lauerwald, R., Sobek, S., McDonald, C., Hoover, M., Butman, D., Striegl, R., Mayorga, E., Humborg, C., Kortelainen, P., Dürr, H., Meybeck, M., Cias, P., and Guth, P. (2013). Global carbon dioxide emissions from inland waters. *Nature*, 503, 355–9.

Revenga, C. and Tyrell, T. (2018). Major river basins of the world. In: C.M. Finlayson, G.R. Milton, R.C. Prentice, and N.C. Davidson (eds) The Wetland Book II: Distribution, Description and Conservation, Springer, Dordrecht.

Rydin, H. and Jeglum, J. (2006). The Biology of Peatlands. Oxford University Press, Oxford, U.K.

Schumann, M., and Joosten, H. (2008). Global Peatland Restoration Manual. Institute of Botany and Landscape Ecology, Greifswald University, Germany. Available online.

Seekel, D.A., Pace, M.L., Tranvik, L.J., and Verpoorter, C. (2013). A fractal-based approach to lake size-distributions. *Geophysical Research Letters*, 40, 517–21.

Semeniuk, V. and Semeniuk, C.A. (1997). A geomorphic approach to global classification for natural inland wetlands and rationalization of the system used by the Ramsar Convention—a discussion. *Wetlands Ecology and Management*, 5, 145–58.

Siegert, M.J., Ross, R., and Le Brocq, A.M. (2016). Recent advances in understanding Antarctic subglacial lakes and hydrology. *Philosophical Transactions Royal Society A*, 374, 20140306.

Sirin, A. and Laine, J. (2008). Peatlands and Greenhouse Gases. In: F. Parish, A. Sirin, D. Charman, H. Joosten, T. Minayeva, M. Silvius, and L. Stringer (eds) Assessment on Peatlands, Biodiversity and Climate Change: Main Report. Global Environment Centre, Kuala Lumpur (Malaysia) and Wetlands International, Wageningen, The Netherlands.

Sket, B. (2018). Subterranean (hypogean) habitats in karst and their fauna. In: C.M. Finlayson, G.R. Milton, R.C. Prentice, and N.C. Davidson (eds) The Wetland Book II: Distribution, Description and Conservation. Springer, Dordrecht.

Statzner, B. and Higler, R. (1986). Stream hydraulics as a major determinant of benthic invertebrate zonation patterns. *Freshwater Biology*, 16, 127–39.

Stokes, T., Griffiths, P., and Ramsey, C. (2010). Karst geomorphology, hydrology and management. In: R.G. Pike, T.E. Redding, R.D. Moore, R.D. Winkler, and K.D. Bladon (eds) Compendium of Forest Hydrology and Geomorphology in British Columbia. Land Management Handbook 66, BC Government and FORREX, Government Publications Services, Victoria, BC.

Tadaki, M., Brierley, G., and Cullum, C. (2014). River classification: theory, practice, politics. *Wiley Interdisciplinary Reviews: Water*, 1, 349–67.

Thieme, M.L., Abell, R., Stiassny, M.L.J., Skelton, P., Lehner, B., Teugels, G.G., Dinerstein, E., Kamden Tohan, A., Burgess, N., and Olson, D. (2005). Freshwater Ecoregions of Africa and Madagascar. Island Press, Washington, DC.

Thompson, J. and Finlayson, C.M. (1999). Freshwater wetlands. In: A. Warren and J.R. French (eds) Conservation and the Physical Environment. John Wiley and Sons, London, U.K.

Thorp, J.H., Thoms, M.C., and Delong, M.D. (2006). The riverine ecosystem synthesis: biocomplexity in river networks across space and time. *River Research and Applications*, 22, 123–47.

Turetsky, M.R., Benscoter, B., Page, S., Rein, G., van der Werf, G.R., and Watts, A. (2015). Global vulnerability of peatlands to fire and carbon loss. *Nature Geoscience*, 8, 11–14.

Vannote, R.L., Minshall, G.W., Cummins, K.W., Sedell, J.R., and Cushing C.E. (1980). The river continuum concept. *Canadian Journal of Fisheries and Aquatic Sciences*, 37, 130–7.

Verpoorter, C., Kutser, T., and Tranvik, L. (2014). Automated mapping of water bodies using Landsat multispectral data. *Limnology and Oceanography Methods*, 10, 1037–50.

Ward, J.V., Bretschko, G., Brunke, M., Danielopol, D., Gibert, J., Gonser, T., and Hildrew, A.G. (1998). The boundaries of river systems: the metazoan perspective. *Freshwater Biology*, 40, 531–69.

Ward, J.V. and Tockner, K. (2001). Biodiversity: towards a unifying theme for river ecology. *Freshwater Biology*, 46, 807–19.

Wetzel, R.G. and Likens, G.E. (1991). Limnological Analyses. Springer, New York.

White, W.B., Culver, D.C., Herman, J.S., Kane, T.C., and MyIroie, J.E. (1995). Karst lands. *American Scientist*, 83, 450–9.

Wiens, J.A. (2002). Riverine landscapes: taking landscape ecology into the water. *Freshwater Biology*, 47, 501–15.

Williams, P. (2008a). World Heritage Caves and Karst. IUCN, Gland.

Williams, P.W. (2008b). The role of the epikarst in karst and cave hydrogeology: a review. *International Journal of Speleology*, 37, 1–10.

2

Approaches to Freshwater Ecology and Conservation

Rebecca E. Tharme, David Tickner, Jocelyne M.R. Hughes, John Conallin, and Lauren Zielinski

Corresponding author: rebeccatharme@riverfutures.com

2.1 Introduction

2.1.1 The situation—freshwater ecosystems, conservation status, and trends

Freshwater ecology and conservation emerged as a distinct discipline in the first half of the twentieth century (Tickner and Acreman 2013). Since then, substantial advances have been made in freshwater science, policy, and practice, with applied ecology (Cadotte et al. 2017) and the role of social sciences (Stern 2018; Horne et al. 2017a) particularly growing in prominence and relevance. Increasing public interest, new technologies, and enhanced capacity have helped to generate large volumes of data and knowledge on the health of freshwater ecosystems globally. Policy developments have included laws, policies, and regulations that, at least on paper, safeguard rivers and wetlands in many countries (e.g., Le Quesne et al. 2010; Horne et al. 2017b). River, lake, and wetland restoration approaches have advanced, especially since the 1980s, and are now multi-billion dollar industries (Speed et al. 2016; Chapter 18 in this volume). And, in one form or another, freshwater conservation has been championed by institutions as diverse as the World Bank (see Projects and Reports on the World Bank website; World Bank 2016), humanitarian agencies (e.g., Wetlands International 2017), the intelligence community (International Community Assessment 2012), and the Organisation for Economic Co-operation and Development (Sadoff et al. 2015).

Despite such progress, freshwater ecosystems around the world remain degraded or threatened (see Chapter 1). Studies have repeatedly demonstrated worrying downward trends in freshwater species populations (WWF 2016), biodiversity (Vörösmarty et al. 2010; Dudgeon et al. 2006), and ecosystem services (Naiman and Dudgeon 2011; Finlayson and D'Cruz 2005; Chapter 15 in this volume). Global patterns of freshwater ecosystem health show rapid declines due to the combined effects of pollution and nutrient enrichment, over-abstraction of water, construction of water infrastructure, invasive species, and habitat loss. While there are some good news stories about species

Tharme, R. E., Tickner, D., Hughes, J. M. R., Conallin, J., and Zielinski, L., *Approaches to freshwater ecology and conservation*. In: *Freshwater Ecology and Conservation: Approaches and Techniques*. Edited by Jocelyne M. R. Hughes: Oxford University Press (2019). © Oxford University Press 2019. DOI: 10.1093/oso/9780198766384.003.0002 recoveries and reintroductions and broader ecosystem restoration, rivers, lakes, and other types of wetlands all face continued conservation challenges in every region of the world. Future pressures from an increasing global human population include rising water use (United Nations World Water Assessment Programme (WWAP) 2015; World Bank 2016; World Economic Forum (WEF) 2016); new waves of dam building (e.g., Winemiller et al. 2016; Zarfl et al. 2015; Opperman et al. 2015); and the combination of familiar and novel pollution sources (Moss 2018; OECD 2017). Such issues are likely to exacerbate the situation, especially in rapidly developing regions of the world. Climate change is disrupting patterns of rainfall and river runoff in all climatic zones, intensifying extreme hydrological events, degrading ecosystems (World Economic Forum (WEF) 2017), and adding new complexities to freshwater conservation challenges everywhere (Poff et al. 2017, 2016; Field et al. 2014).

If freshwater ecosystems are to survive and thrive through the twenty-first century and beyond, approaches to their conservation will be required that are strategic and policy relevant, innovative, and more rapidly effective (e.g., Berghöfer et al. 2016; Strayer and Dudgeon 2010). Quite simply, all human populations rely on freshwater ecosystems directly (or indirectly) for their security, from high-quality water supplies, food, transport, energy, to recreation and a multitude of other services. Without these healthy functioning systems and the natural capital they represent, many of the water-associated benefits people rely on would be lost or compromised (Gilvear et al. 2017; Lloyd et al. 2013; Arthington et al. 2010; Scholes et al. 2010). This chapter provides a critical overview of some of the numerous approaches to freshwater ecology in use, with a particular emphasis on techniques that link research with the applied, interdisciplinary practice of freshwater conservation and with current and likely future policy needs.

2.1.2 The human factor in freshwater conservation and management

There is one consistent factor pertinent to most, if not all, freshwater ecosystems—the human one. Human societies have been coupled with and have evolved with a dependence on freshwater ecosystems over thousands of years (Moss 2018). People, therefore, are, or arguably should be, at the centre of freshwater conservation and management. Freshwaters are now recognised as complex, dynamic, and heterogenous socio-ecological systems at multiple spatial and temporal scales characterised by interlinked social and ecological processes and feedbacks (Gavin et al. 2015; McGinnis and Ostrom 2014). The pressures on freshwater ecosystems—indeed, the *raison d'être* for freshwater conservation—stem from human activity. Equally, it is well documented that rivers, lakes, and other wetlands provide a wide range of important, though often under-valued, goods and services to communities and economies (Finlayson and D'Cruz 2005; Finlayson 2012; Parker and Oates 2016; Chapter 15 this volume). Approaches to the sustainable management of freshwater systems, including ecosystem and biodiversity conservation, should always be designed according to the specific environmental and socio-cultural contexts; there is no single blueprint that will work across contexts or geographies (Gavin et al. 2015; Irvine et al. 2016). Moreover, now that the world has entered the era of the Anthropocene (Cadotte et al. 2017; Kopf et al. 2015; Lewis and Maslin 2015; Poff 2014), it is essential that social, political, and economic considerations become common threads running through freshwater conservation research, policy, and practice (Tickner et al. 2017).

Conventionally, freshwater conservation research and practice have focused primarily on biophysical aspects. This is reflected in the contents of this book where many of the chapters provide advice on approaches and techniques for understanding ecosystem structure and processes and for understanding the behaviour, spread, and interaction of species, and human impacts on freshwaters. Conservationists need to understand how ecosystems work in order to understand the potential range of interventions to maintain or restore species and habitats. But given the 'perfect storm' of anthropogenic pressures acting on many rivers, lakes, and wetlands, and the profoundly political nature of the real-life decisions which affect freshwater ecosystems and water resources (Berbés-Blázquez et al. 2016), it is insufficient to focus solely on the biophysical. Today, therefore, practices need to encompass a broad spectrum of approaches which vary in the extent to which they balance the objectives of biodiversity conservation with human well-being (Gavin et al. 2015). Innovation will need to become an essential element of research, practice, and policy, particularly with regards to engaging people, businesses, and political decision-makers.

A full exploration of the social science approaches that might aid freshwater conservation and management is beyond the scope of this book, although the chapters in Part III touch on relevant topics. Freshwater conservationists of the twenty-first century need to be able to span disciplinary boundaries, including those which separate the biophysical and the social sciences, and to be prepared to work within interdisciplinary teams of practitioners and stakeholders (Section 2.5). It will be necessary to read further resources, besides this volume, to gain the full breadth of approaches and techniques needed (e.g., Stern 2018).

2.2 Framing freshwater conservation challenges

2.2.1 Scoping, objective setting, and guiding principles

(a) What challenges or questions need answering and why? Clarity in this step is key to success, and that comes from avoidance of framing the question, or similarly, the proposed solution, as a purely scientific or biophysical one. Nearly all freshwater conservation challenges are set within the broader context of a social-ecological system (McGinnis and Ostrom 2014). Delineating and understanding the system from this perspective typically allows for a better understanding of the problem, the information that needs to be collected to inform decision-making, and the potential impediments or solutions that might arise (Biggs et al. 2015).

There are three main kinds of challenges or questions to be addressed. Firstly, there are the questions that ecologists pose about freshwater organisms and habitats to better understand the underlying processes that drive these ecosystems. The fundamental and vitally important scientific research this generates provides the data and evidence to inform decision-making, management, and policy. Secondly, there are the questions and problems that result from human impacts on freshwaters and the consequent need to protect, restore, or adaptively manage them. Broadly, the issues can be divided into: biodiversity and habitat loss; pollution and eutrophication; invasive species; water-food-energy and environmental security and climate change; managing extreme hydrological events; and the exploitation of fisheries and other resources. These problems and questions arise from people's use and consumption of water and other natural resources (e.g., impacts of agriculture, industry, water infrastructure); the engineering and technology adaptations to an increase in extreme events (e.g., flood defences, overuse of groundwater); the use of wetlands and water bodies as sinks for nutrients and pollutants (e.g., agricultural runoff, industrial wastewater, eutrophication); and therefore the ensuing conflict of interests of different stakeholders. Finally, there is the problem of how to inform policy and politics (locally, nationally, and internationally) of the value of wetlands and freshwater functions when many national governments have populations that live in extreme poverty, lack basic nutrition and clean water, and need to tackle social and economic inequalities. This problem can only be resolved by embedding the intrinsic links between societal well-being, poverty reduction, and healthy wetland ecosystems into local decision-making involving local communities and educators, in an attempt to influence national policy.

(b) Vision and objectives. The development and agreement with the identified key stakeholders of an agreed vision and objectives fosters a people-centred approach, and allows the establishment of the fundamentals of understanding, trust and ownership (see also Section 2.2.2). The approach and the potential solutions it is expected to deliver need to be explicitly oriented to the question or challenge at hand. They should also line up as far as possible with any relevant management visions or objectives that are in place or proposed. For example, some of the environmental flow assessments being undertaken for the Mekong region of South East Asia, Tanzania, and Kenya are being aligned with basin visions and national water management objectives, and a stakeholder-focussed approach used. Historically, the tendency has been to focus purely on the biophysical aspects of the freshwater ecosystem. However, investing the resources required to clearly understand both the biophysical and the social dimensions is critical to success. Taking a more social-ecological-based approach allows for both social and ecological objectives to be understood, and trade-offs to be recognised and negotiated (Berkes et al. 2003); Ziv et al. (2012) provide an example for the Mekong.

(c) Guiding considerations. There are various general considerations to keep in mind when selecting an approach, beyond those principles that underpin the science. An obvious yet important one is the use of pragmatism. In many situations nowadays, especially in the developing regions of the world, it is necessary to work within present resource constraints, whether they be limitations in data availability, financial support, or technical or institutional capacity. In this way, the knowledge, evidence base and expertise can be progressively built and adaptively refined over time. Data and research gaps are inevitable, and an acuity of focus is required to ensure targeted gap-filling. The advantages of engaging early in any attempt to address a freshwater challenge, arguably far outweigh the disadvantages inherent in delaying efforts until there are 'sufficient data' to make the case with the highest possible degree of certainty.

Notwithstanding this, scientific rigour, repeatability, and robustness are important to increase confidence levels (see Chapter 3), and well-established methods exist to aid the process (e.g., snowballing as a technique for triangulation). There is often a tension between rapid, typically coarse, and low-resolution approaches that carry a risk of being overly simplistic; and higher-resolution, resource-intensive approaches that tend to be more comprehensive in scope, engendering greater confidence. Poff et al. (2017) describe this tension in terms of existing methods for environmental flow assessment (see also Section 2.3.7 for an example of ecological monitoring).

Particularly when framing complex issues, such as the 'wicked problems' the competing interests and demands of water management present, a clear argument can be made for considering the tenets of requisite simplicity (as discussed in Stirzaker et al. 2010); river basin management in the Ebro Basin, Spain, is a case in point (see 'articles' on Global Water Forum website). As uncertainty and risk routinely accompany such problems, and particularly where extensive scientific knowledge is lacking, it is prudent to apply a precautionary approach (or precautionary principle; e.g., Myers 1993). Confounding and unknown factors, unintended consequences, and complete surprises are features that should be transparently recognised and approaches designed to accommodate them. Alongside the scientific defensibility of the results, is the critical importance of their social acceptance (Section 2.2.2).

2.2.2 Working with stakeholders from the inception

In a collective design approach, science is not aimed at 'making' decisions, but 'building' decisions (Berthet et al. 2016). Co-design through to implementation and evaluation is key to a successful freshwater conservation programme, as ultimately people choose what decisions are made. Due to the increasing pressure from humans on freshwater ecosystems, and the fact that there are few freshwater ecosystems on earth not affected by them (Section 2.1.1), people need to be as much at the centre of the solution as they are part of the challenge. Providing inclusive co-design processes for relevant stakeholders is needed to further legitimise freshwater conservation programmes, foster understanding between different stakeholders, gain ownership and trust, and help to ensure sustainability within these systems (de Vente et al. 2016). As Reed et al. (2009) observed 'only by understanding who has a stake in an initiative, and through understanding the nature of their claims and inter-relationships with each other, can appropriate stakeholders be effectively involved in environmental decision-making'. A stakeholder is anyone who has a 'stake' or interest in the situation being considered, from a local landholder to an international non-governmental organisation (NGO). Without adequate stakeholder engagement directed at building understanding, trust, and ownership (Rogers 2006), freshwater conservation and management initiatives are likely to miss the actual challenges and solutions, fail altogether or, at a minimum, suffer from a disrupted process.

There are several practical methods available for stakeholder engagement (Reed et al. 2009) but all are dependent on the level of engagement needed. Arnstein's (1969) ladder provides a good example of the possible levels of engagement, ranging from the bottom of the ladder where non-participation is the aim, to central rungs where communication and stakeholder input are valued, but where stakeholders are still not decision-makers, to the top rungs on the ladder where stakeholders drive the process. It is essential that the levels of engagement are decided on in collaboration with the stakeholders themselves, and that it is recognised that these levels can change throughout the process. Stakeholder engagement must be a central theme of any freshwater conservation effort. Firstly, gaining an understanding of who are the stakeholders involved or affected and what is their stake, and how that relates to other stakeholders is important. Leading on from this, a stakeholder engagement strategy or plan is essential, and should be kept pragmatic and implementation focused. Execution of the plan and continually engaging at the agreed level during the project is needed, and lastly, as new information becomes available or attitudes change, engagement strategies should also adapt (Conallin et al. 2017). Figure 2.1 provides an understanding of the different steps needed to include a stakeholder engagement process within freshwater conservation programmes.

Adaptive management (AM) frameworks provide an effective way to include stakeholder engagement as a central component to a freshwater conservation effort (Allen and Garmestani 2015). Adaptive management provides a structure that works towards integrating social, management, and scientific aspirations into decision-making and doing so in a learning-by-doing manner. If stakeholder engagement is central within AM, then it provides a basis for co-design of programmes and co-knowledge

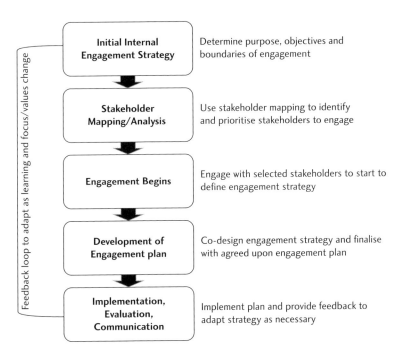

Figure 2.1 Stakeholder engagement steps within freshwater conservation programmes. Adapted from Reed (2009) and Conallin et al. (2017).

generation, combining both informal knowledge (e.g., indigenous or traditional knowledge) gathered from a diverse array of stakeholders, and formal knowledge gathered from scientific studies (e.g., experiments, empirical field surveys). Combining different forms of knowledge allows for multiple lines of evidence to be used and helps reduce uncertainty (Section 2.2.1 (c)). Most importantly though, it actively involves an array of different stakeholders in the process of knowledge generation and analysis, thereby increasing acceptability of the results (Conallin et al. 2018a).

Strategic AM (SAM; Roux and Foxcroft 2011) is a version of adaptive resource management that includes stakeholder engagement as a key aspect of the planning, implementation, and evaluation stages of programmes. A central component is setting a stakeholder-derived vision of the system being managed and then deconstructing it into a hierarchy of objectives of increasing focus and rigour, culminating in measurable end-points. Within stakeholder engagement it is important to be able to show how high-level objectives are linked to more local level ones, and to provide inclusive, understandable endpoints that relate to those objectives. For example, in Australia, in the Edward-Wakool System of the Murray Darling Basin, a SAM process was set up for delivering environmental water to reach both social and ecological objectives for native fish. Stakeholders established a vision for the system, and objectives for key fish groups, and environmental water was managed to meet those objectives. Measurable endpoints were derived so that monitoring could inform the AM process (Conallin et al. 2018b).

2.2.3 Sources of data, information, and knowledge

All freshwater conservation projects will involve an initial period of research, or desktop study, to scope the project and assess the availability of existing data and information. This will involve finding and collating information sources that are published (e.g., peer-reviewed journals, books, published and open-source data sets; see Chapter 14) or unpublished (e.g., grey literature, websites, databases; maps, photographs, and digital spatial data in hard copy or electronic format), remotely sensed data (Chapter 9), and local knowledge and information sources from stakeholders and managers. Other secondary data sources that can be used by wetland conservationists include, for instance, human health data from district clinics for risk mapping of waterborne diseases; fisheries catch data for freshwater condition assessments, invasive species mapping (see also Chapter 16), and environmental flow studies; and land use–land cover changes in assessing likely trends in water quality in rivers and lakes (UNEP and UNU-EHS 2016). These existing data or sources (secondary data) need to be quality assured for accuracy and reliability, and any permissions, copyrights, or intellectual property rights need to be addressed. In addition, it is vital to check that ethical clearance has been sought for information or data that were obtained from humans (e.g., interviews and focus groups), or by using direct techniques for surveying animals (e.g., handling or trapping). For a review of the use of secondary data, databases and depositories in freshwater ecology and conservation, see Chapter 14 of this volume. Gaps in secondary data will inform the project of any primary data that need collecting or generating, using methods ranging from experiments and field surveys, through to modelling (e.g., scenario simulation, forecasting, or hindcasting). The IUCN's Integrated Wetland Assessment Toolkit (on the IUCN website) is an example of how to integrate scientific data, information on local livelihoods, and wetland valuation (Brooks et al. 2014). A further application for the use of acquired data, knowledge, and information concerns knowledge transfer, sharing, and adaptation (e.g., the Environment Live (science and data for people) webpages on the UN Environment Programme website); or sharing information across transboundary rivers or lakes for freshwater management (e.g., the World Bank's Lake Victoria Environmental Management Plan).

Particular advantages can be gained by combining sources of traditional ecological or indigenous knowledge with scientific information, from confirming field survey results to generating knowledge new to both science and the communities. This was demonstrated in the Daly River, northern Australia, where indigenous knowledge influenced the conceptual models developed by scientists to understand fish assemblage-flow ecological interactions, as well as the risk-assessment tools designed to understand the vulnerability of particular fish species to low-flow scenarios (Jackson et al. 2014).

2.2.4 The importance of scale

The scale of the question or problem may strongly influence the approach selected and its scale of application, as the following examples demonstrate, from local to global level, and from site or project scale to that of the entire system or landscape (see chapters 3 and 13; 'scale' is a theme considered throughout the book). Clear delineation of the biophysical and social boundaries of the system for study (the study area) is important and in the case of freshwaters, the catchment is commonly adopted as the functional unit of study (see Chapter 19). Temporal scale is also highly relevant (e.g., Ward (1989) characterises the spatiotemporal dimensions of rivers) and may range from hourly to interannual, decadal, or longer, depending on the nature of the assessment (Chapter 13). For instance, a study of the ecological impacts of peaking hydropower may require analyses of diurnal discharge fluctuations, while an assessment of climate change implications for ecological species shifts will necessarily require simulations over far longer time frames.

(a) From local to system and basin scales. The majority of freshwater approaches are aimed at the local scale of an individual site or project, generally over fairly narrow temporal windows of a single sampling effort to at most a few years of study. Examples include environmental impact assessment studies for individual dams and river reaches, urban wetland restoration projects, and Ramsar wetland situation assessments. Increasingly, however, efforts are being directed at whole system, landscape, or catchment scales. Integrated conservation and development planning is one area where this is evident; for instance, in basin-scale analyses aimed at optimising the spatial configuration of proposed hydropower projects within river networks, to maintain energy generation while maximising system functional connectivity and health—see examples in Opperman et al. (2015). Decision-support tools as an aid to basin-scale management are receiving renewed attention, including for permitting procedures, stakeholder interactions, and the generation of scenarios of the cumulative effects of water

abstraction and infrastructure development. One example is the Decision Support System for the Colombian Magdalena Macro-Basin (or SIMA for its acronym in Spanish; see sima-magdalena website) for sharing perspectives and promoting dialogue on cumulative impacts on freshwater ecosystems stemming from, among others, river flow alteration due to navigation and hydropower. A basin management simulation game provides an additional tool designed to bring stakeholders and scientists closer together (see simbasin.hilab.nl website). Freshwater protected areas are also being examined more closely as spatially and temporally interconnected networks in the landscape, including in terms of their optimal design features for biodiversity conservation and management actions (e.g., Roux et al. (2008) for the rivers of the Kruger National Park, South Africa; Pittock et al. 2015).

(b) Regional, national, and global scales. Issues and associated approaches focused at larger scales dictate the use of quite different sets of data and methods. The use of existing national data sets, even where such data are limited in their completeness or distribution, can provide valuable first insights. In Mexico, for instance, a first national ecohydrological diagnosis of the state of the country's rivers and, thus, of their relative priority for future conservation attention, generated using a model based on multiple instream, riparian, and catchment indicators, highlighted concerning trends in river condition and its potential implications for human water security (Garrido et al. 2010). In the USA, catchment integrity approaches have been developed at a national scale (e.g., Flotemersch et al. 2016) and several state-wide or basin-wide environmental flow assessments have been conducted by scaling up local level data using a regional framework for setting flow standards. See the Ecological Limits of Hydrological Alteration, ELOHA, framework (Poff et al. 2017, 2010; the ELOHA toolbox for flow-ecology relationships on The Nature Conservancy website) as an example of how existing empirical data and information can be used to determine environmental flows at a regional scale.

New applications of big data sets to address global scale questions and drivers of change, identify emerging policy issues, and fill major data gaps are notable growth areas for freshwaters. The BioFresh freshwater biodiversity data portal (data.freshwaterbiodiversity.eu website) is one good example of the consolidation and leveraged use of existing freshwater biodiversity data sets (metadata) and tools for a wide range of purposes across different regions. Vörösmarty et al. (2010) showcase the use of global-scale data to generate new measures for addressing the interrelated challenges of safeguarding freshwater biodiversity conservation and human water security. Analyses of the international distribution of existing and planned hydropower dams, and their anticipated ecological implications, further illustrate the potential usefulness of existing data sources (Zarfl et al. 2015; Opperman et al. 2015).

2.3 Finding an approach appropriate for the context

2.3.1 Selecting from among the different types of approach

While there are many sub-disciplines within freshwater ecology, a consistent aim throughout any project is selecting or tailoring an appropriate approach(es) to investigate the specific issue or question at hand; generic or 'blueprint'-type approaches may prove of limited application value. There is no one correct way to do this, and the approach chosen is often dependent on the desired final products, the scale and complexity of the issue, data availability, and the resources available in terms of finances and technical and institutional capacity.

If the goal is to investigate a specific ecological research question or hypothesis, or to understand fundamental processes, the approach might include specialised field surveys and laboratory experiments that are targeted and isolated. If the investigation is to provide information for future policy or management decisions, a single lab or field experiment will be inappropriate. A broader field investigation may be required, which incorporates all ecosystem interactions, whether known or unknown. There are many factors to consider when creating such a survey or experimental design (e.g., see Chapter 6 for physical variables; and Chapter 9 for freshwater vegetation). In South Africa, for instance, toxicological studies of the effects of different single and combined mixtures of pollutants on native aquatic organisms and their tolerance ranges were conducted both in laboratory experimental channels and in the field, and used to establish a comprehensive database of toxicity responses (Palmer et al. 2004); this evidence base continues to be augmented and applied in methods and guidelines to protect local water resources.

According to Stewart et al. (2013), mesocosms have become increasingly popular experimentally, representing a valuable bridge between smaller, more tightly controlled, microcosm experiments, which are of limited realism and the greater biological complexity inherent in natural systems, where mechanistic relationships may be more difficult to identify. Stewart et al. (2013) demonstrate the contributions made by mesocosm applications in lakes, estuaries, and other freshwater environments in relation to climate change, describing their merits and limitations (in terms of realism, reproducibility, and control) and future potential. Real-world, adaptive experiments are also growing in significance among the mix of approaches in use, such as the synthesis and discussion of the potential of experimental flow releases in Olden et al. (2014). Indeed, large-scale quantitative synthesis and meta-analysis have an important function in supporting freshwater research efforts. Ghermandi et al. (2010) provide a meta-analysis of the values of wetlands, both natural and human-made. As integrators of local catchment and regional processes, freshwater ecosystems are highly sensitive to the net effects of multiple stressors (i.e., novel and extreme environmental changes; Jackson et al. 2016). Meta-analyses of stressors can assist with knowledge distillation, understanding cumulative synergistic and antagonistic effects of different stressors, or even in being prepared for ecological surprises (e.g., Matthaei and Lange (2016) for freshwater fish; and Jackson et al. (2016) who analysed data from 88 papers, including 286 responses of freshwater ecosystems to stressors).

A multitude of advances are also being made in statistical approaches, including statistical modelling using packages in R (e.g., AquaEnv, *mizer*, streamMetabolizer) and predictive modelling and optimisation tools. In a recent example, Bond and Kennard (2017) tested the ability to spatially extrapolate hydrologic metrics calculated from gauged streamflow data to ungauged sites in the Murray–Darling, Australia's largest river basin. Stewart-Koster et al. (2010) used Bayesian networks to help elicit stakeholder knowledge and guide investments in river flow and catchment restoration. In addition, Olden et al. (2008) observed that machine learning methods (a family of statistical techniques with origins in the field of artificial intelligence) hold promise for the advancement of understanding and prediction about ecological phenomena, and provide a primer on their use.

Biocultural approaches to conservation, which 'draw lessons from previous work on biocultural diversity and heritage, social-ecological systems theory, and different models of people-centered conservation' (Gavin et al. 2015) continue to open up new opportunities in the area of freshwater. It is worth considering the following principles of successful initiatives based on such approaches (Gavin et al. 2015): (1) acknowledge that conservation can have multiple objectives and stakeholders; (2) recognise the importance of intergenerational planning and institutions for long-term adaptive governance; (3) recognise that culture is dynamic, and this dynamism shapes resource use and conservation; (4) tailor interventions to the social-ecological context; (5) devise and draw upon novel, diverse, and nested institutional frameworks; (6) prioritise the importance of partnership and relation building for conservation outcomes; (7) incorporate the distinct rights and responsibilities of all parties; and (8) respect and incorporate different world views and knowledge systems into conservation planning.

2.3.2 Illustrations of the diversity of approaches

The following four short examples illustrate some of the incredible breadth and diversity of approaches that may be taken in freshwater studies—in terms of research and conservation frameworks, methods and tools—to frame up and address the equally varied types of problems, issues, and questions that confront the practitioner. Many other examples are well covered in the other chapters of this volume, or in the published literature.

(a) Roles of conceptual frameworks. Conceptual frameworks play a number of indispensable roles, as elegantly outlined by McGinnis and Ostrom (2014), and help to organise descriptive, prescriptive, and diagnostic enquiry. They assist in integrating multiple sets of viewpoints, knowledge or concepts; conceptualising complex and/or highly interconnected freshwater challenges; and in organising ecosystem components and processes to reflect potential causal linkages, interrelationships, and feedback loops. Tomich et al. (2005) discuss conceptual frameworks for ecosystem assessment.

Some recent examples that take an interesting and interdisciplinary perspective on, or can be applied in the context of different aspects of water resources management, include: a common socio-ecological system framework as the basis for local customisation (McGinnis and Ostrom 2014); a generic framework conceptualising the integration of environmental flows and ecosystem services assessments (Gopal 2016); a socio-ecohydrological conceptual framework to guide site selection and data collection efforts for an environmental flow study in the Rufiji Basin, Tanzania, where rural livelihoods are strongly dependent on healthy river assets (USAID 2016); and an innovative sociohydrological model for more realistically considering human-flood interactions, for flood risk management (Di Baldassarre et al. 2013).

(b) Approaches for freshwater biodiversity conservation. Numerous encouraging stories of biodiversity conservation success exist in the literature from which to draw lessons on what constitutes a suitable approach(es) and the enabling conditions for attaining the vision and objectives laid out. One of many ecologically oriented examples is the decades-long effort to actively restore and monitor European otter populations in multiple UK catchments (Mason and Macdonald 2004).

In a case example from the region of South East Asia, many of the fish species of conservation, commercial, and subsistence value are migratory and need to be able to move up and down rivers and laterally into wetlands, or even to the sea, to complete their life cycles. This migration is being disrupted through river and wetland regulation, and key species are now reducing in abundance or being lost, decreasing biodiversity and simultaneously increasing poverty and malnutrition in rural communities (Lynch et al. 2017). In this region, the majority of the human population is reliant on fish to supply their everyday protein and micronutrients. This is even more pronounced in the poor, rural areas where inland fisheries and paddy rice production are daily food sources. A novel technique (Section 2.3.5 gives other examples), the use of otolith microchemistry, is increasingly being used in fish conservation work, not only to age fish, but also to assess natal spawning grounds, migration patterns, diet, and relationships to parameters such as flow regime (Baumgartner 2016). The method is being applied across parts of South East Asia to determine which fish are migrating, to where and when, and how this information can be used in planning for infrastructure (e.g., dams, weirs, floodgates, and road culverts; Walther et al. 2017). For example, in Myanmar, research is addressing the migration patterns of hilsa shad, an economically important species, to the entire Bay of Bengal. Otolith microchemistry is just one element of the approach to determine if and how far these fish species are migrating inland (Lunn et al. 2017).

(c) Use of indirect detection methods. Environmental DNA (eDNA), which uses genetic material obtained directly from environmental samples taken from soil, sediment, water, etc. (Thomsen and Willerslev 2015), is increasingly being used to detect the presence of different organisms within freshwater ecosystems for the purpose of freshwater research and conservation: see 'eDNA' on the Freshwater Habitats Trust website and Taberlet et al. (2018) for details of eDNA techniques used in freshwater ecosystems. Sources of eDNA include gametes, shed skin, hair and carcasses, and secreted faeces and mucous. The DNA of a range of aquatic organisms can be detected in water samples at very low concentrations (e.g., using quantitative Polymerase Chain Reaction methods). In aquatic environments, eDNA can persist for 7–21 days in the water, and once trapped in sediments it can be preserved for thousands of years.

Researchers are especially interested in using eDNA for detecting rare and endangered species that traditional sampling methods are likely to miss (Stoeckle et al. 2016). For example, in a UK amphibian conservation project (Biggs et al 2014), a primer was first developed and tested (i.e., length of artificial DNA which specifically binds to and amplifies the DNA of the target organism) to detect Great Crested Newt eDNA successfully in water samples. An intensive study at 35 ponds was then undertaken to compare the ability of eDNA and traditional survey methods (torch counting, bottle trapping and egg searches) to detect newts during the breeding season. Volunteer surveyors also collected single eDNA samples from 239 ponds known to be used by Great Crested Newts across England, Wales, and Scotland. Volunteer data (an example of citizen science in action) were used to test whether eDNA detection was affected by variations in pond physical and chemical environmental factors, and also to assess the practicality of the technique for use by volunteers. Further examples of the use of eDNA can be found in chapters 9, 10, and 12.

In an example from South Asia, farming practices were used as a tool for the indirect detection and quantification of potential diffuse agrochemical pollution. Field surveys documenting rice farming systems, including detailed agricultural practice calendars and agrochemical sales at local centres, were used as an indirect means of measuring the impacts of diffuse agricultural pollution on small streams of southern Sri Lanka (Howarth et al. 2006). In such small systems in this tropical area, the diffuse nature of the run-off from rice fields, dilution factors due to high water volumes in the wet season, and complexity of agrochemical types and application cycles meant that instream chemical sampling seldom detected significant pollutant levels. However, clear evidence of high, potentially biologically harmful agrochemical inputs was obtained using the combined indirect methods, supplementing evidence gained from rapid bioassessment of benthic invertebrate communities. Rapid bioassessment represents another category of diverse and valuable tools for freshwater study, as exemplified by the many successful Rapid Assessment Programs (RAPs) undertaken by Conservation International (see Chapter 12).

(d) Applied toolboxes for environmental water management. In the field of environmental flows, a suite of interdisciplinary methods and tools, appropriate at different levels of application and resolution, are well documented and readily available for use in scientific determination of environmental flows (as described in Poff et al. 2017, 2010). Additionally, for the purposes of implementation, various mechanisms have been applied, singly or in concert, to secure and allocate water for environmental purposes (Horne et al. 2017b): caps on consumptive water use; licence conditions for water abstractors; conditions on storage operators or water resource managers (e.g., Warner et al. 2014; Olden et al. 2014); ecological or environmental reserves; environmental water rights; and various economic instruments. Practitioners can draw on this toolbox when addressing issues of environmental and social sustainability in water resources planning and adaptive management. Also see the ELOHA regional environmental flow online toolbox and IUCN Integrated Wetland Assessment Toolkit mentioned in sections 2.2.4 and 2.2.3, toolboxes for ecosystem services and their valuation (e.g., the TESSA toolkit for rapid ecosystem services assessment at sites of biodiversity conservation importance; Peh et al. 2013; Chapter 15 in this volume), and the online Global Water Partnership toolbox on integrated water resources management (IWRM) (see gwp.org website) as further examples.

2.3.3 Monitoring and evaluation, including a role for citizen science

Monitoring and evaluation (M&E), generally defined as the periodic gathering and analysis of information in relation to a desired objective or set of objectives, can provide a better understanding of current hydrological and ecological conditions, as well as of short-term and long-term status and trends in a freshwater system. Once this information is known, it can then be evaluated against the overall objectives to assess compliance or progress, for improved decision-making and management (Stem et al. 2005). A monitoring and evaluation programme is often developed, which details how monitoring activities will be carried out and the resolution.

For freshwater ecosystems, the level of M&E resolution applied can be thought of in a variety of ways. One way is in terms of physical resolution or scale. River basins can be large and are often broken down into smaller units to properly study or manage them. At the largest unit, the basin encompasses the entire river catchment and all its ecosystems. This unit is very useful for determining water balances and making connections between upstream and downstream areas/users. When broken down into reaches, or sections of a river with similar characteristics, different ecosystem types can be studied or managed according to their unique features. Within a reach, there are multiple sites that can be evaluated. A site can be influenced by localised pressures or contain unique conditions which may not be present in other parts of the reach or basin. Although not large in area, site condition can play a major role in the scientific understanding or management of a river basin.

Another form of resolution is data resolution, and there are often many options to choose from. Data on freshwater systems can vary in terms of ecology (biotic components vs. abiotic components), detail (broad or specific), temporal scale (continuous, frequently, rarely), and physical scale (basin to site). Combining all of these factors, there is a vast set of potential data combinations. In addition, there are practical concerns about effort and cost associated with collecting different types of data. To help decide what data should be collected as part of a focused and feasible monitoring and evaluation program, the following should be considered:

1) Monitoring effort should be linked back to the scientific or management objectives. The intent for collecting monitoring data is that it can be evaluated to see if there is progress towards or achievement of the objectives. Without a link back to a scientific or management objective, the usefulness of the data collection can become diminished.

2) The objectives should drive the selection of the scale and data resolution required for the monitoring and evaluation plan. An objective of protecting salmon migration in a river will require different monitoring data than an objective of conserving a waterfall-dependent ecosystem.

3) Objectives can be broken down to identify the important ecological topics, or key components. These key components may be biotic (e.g., fish, amphibians, vegetation) or abiotic (e.g., water quality, geomorphology, bathymetry). The key components for one objective may be different to another. There may also be overlap of key components for different objectives, allowing for multiple objectives to be monitored using only a few key components.

4) Each key component can be broken down further into different levels of data resolution. This can be done using a multi-level system, which categorises

monitoring activities into different levels of detail, temporal resolution, scale, and resources required for data collection (as illustrated using the Rufiji Basin example below). From this multi-level system, data collection methods can be identified to inform progress against the overall objectives while also working within resource limitations.

In the Rufiji River Basin in Tanzania, for instance, a tiered, three-level system is being designed to monitor social, ecological, and hydrological responses to environmental flow implementation (Figure 2.2). Level 1 (simplest) will use citizen science methods to collect simple hydrological, ecological, and social data at different points in the river basin; Level 2 (intermediate) has the regional water resources management authority collecting data using simple but standardised protocols a few times a year at key locations in the river basin; and Level 3 (most complex) requires the hiring of hydrological, ecological, and social experts to conduct a full, in-depth review of the basin at key sites. The information collected at each level contributes to the overall understanding of the condition of the river and how different key ecosystem components are responding to the implementation of environmental flows. In addition, the monitoring data help to guide management decisions by providing a 'trigger value' or a warning sign for managers when conditions in the river basin may be of concern, initiating action through adaptive management. Monitoring methods for each level were determined for seven different key components (fish, vegetation, macroinvertebrates, water quantity, water quality, geomorphology, and social), which relate back to the objective of effectively implementing environmental flows in the basin. Reporting on system overall health, and status and trends, is valuable for documenting and sharing the outcomes

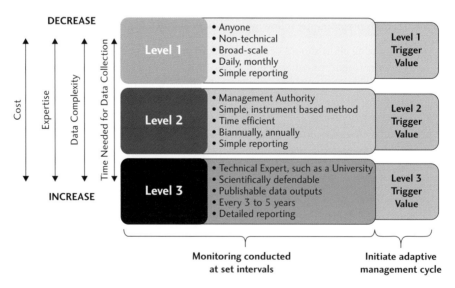

Figure 2.2 A multi-level system used in the Rufiji River Basin in Tanzania to help build a monitoring and evaluation programme that can be linked back to scientific or management objectives (adapted from Zielinski 2016).

of such M&E programmes, among other uses. Report cards are tools that enhance research, monitoring, and management in several ways, by assessing and communicating products that compare ecological, social, and/or economic information against predefined goals or objectives.

River basin report cards provide performance-driven metrics that reflect basin status on a regular basis. They integrate, synthesise, and distil large, often complex, information sets into simple measures that make communication to decision-makers and the general public more effective and transparent. Costanzo et al. (2017) present consolidated guidance for practitioners on how to develop river basin report cards for such purposes, drawing on recent illustrative case studies. These studies cover a range of socio-political settings, ecosystem types, and regions; namely: Mississippi River Watershed and Chesapeake Bay, USA; Orinoco River Basin, South America; Moreton Bay and catchments, Australia; Laguna de Bay Lake, Philippines; and Lake Chilika, India.

2.4 Emerging approaches, sectors, and research arenas in freshwater ecology and conservation

Human impacts during the Anthropocene, including those associated with climate change, have led to extensive modification of freshwater systems, to the extent that new research and practices have emerged dealing with topics ranging from sustainable water management under future uncertainty; the potential of optimised designer flows to meet ecological and human needs in novel and hybrid river ecosystems (e.g., Poff et al. 2016; Acreman et al. 2014); and biotic homogenisation and the implications of non-native species in such systems (e.g., Moore and Olden 2017; Early et al. 2016).

The inclusion of social sciences and socio-cultural approaches in freshwater ecology practice continues to increase, with calls for more attention to be given to this topic (e.g., Horne et al. 2017a; Wantzen et al. 2016; Lokgariwar et al. 2014; Pahl-Wostl et al. 2013; Finn and Jackson 2011). Engagement and collaboration with sectors, such as humanitarian and development agencies, is increasing in the face of growing water insecurity, environmental degradation, natural hazards and disaster risk reduction, as well as with greater attention focused on water, sanitation, hygiene, and other aspects of human health. For instance, there has been recent exploration of the inter-linkages between the condition of Sahelian wetlands, the state of water resources, and patterns of human migration (Wetlands International 2017). Such sectors are also identified as a future avenue of partnership in environmental flow practice (Poff et al. 2017). Studies of the economics of ecosystems and biodiversity (TEEB 2013) continue to emerge in the area of freshwater, even extending to assessments of the value of whole river systems to economies. Recent examples include the Kafue River-floodplain system of Zambia (WWF-Zambia 2016) and the Mekong River (WWF-Greater Mekong 2016).

The arts and humanities present still under-explored approaches for achieving effective freshwater ecology, conservation, and management, although examples exist from oral history research projects through to creative projects by ecological artists that engage communities on key issues (see 'ecological artists and river restoration' on the National Geographic website). Citizen science techniques are now able to contribute significant volumes of useful high-quality data and science outputs on freshwater systems (see 2016 special issue of *Science of the Total Environment*; Kobori et al. 2016). They also actively engage in civil society: examples include the more than a decade of removal by numerous volunteers of invasive crayfish from Pine Lake, Washington State, USA; engagement with Earthwatch's FreshWater Watch project; and participation in the Freshwater Habitat Trust's Clean Water for Life project. Freshwater Links is an initiative empowering volunteers from local communities, NGOs, government organisations, and researchers to work together and share data sets to improve the condition of the Thames Catchment, UK.

Two examples of emerging areas explored in greater depth below are around technological advances in data collection (Section 2.4.2) and one of several fruitful areas of growing collaboration with the private sector (Section 2.4.3).

2.4.1 Emerging technologies to collect ecological data at different scales

As new technologies emerge and old ones are refined, there are simple and rapid ways to gather ecological data. One of the most quickly adopted and fast-moving technologies is the use of smart devices (e.g., mobile phones, tablets) (Pocock et al. 2017). Thanks to the wide availability of the Internet, mobile data, and affordable devices, many practitioners and volunteers globally have access to free or affordable tools to assist with their research or conservation efforts. Having such tools available to ecologists has some advantages: data can be collected, consolidated, and analysed from almost anywhere on earth using multiple devices; it allows for the inclusion of non-scientists and local residents into a research or conservation effort; and the data collected can be shared with other researchers, environmental groups, and government agencies to promote collaboration and improve outcomes (Kobori et al. 2016). Through the use of smartphone applications or texting capability, data collection can occur on a spatial and temporal scale that was never possible before. Through a global innovative monitoring and modelling programme (iMoMo, on the imomohub website), smartphone applications have been developed to measure discharge in irrigation canals across large river basins in Tanzania and Central Asia, helping with water resources management. Mobile phone applications are being used to conduct quick and affordable water quality sampling to track drinking water and sanitation conditions in remote areas around the globe (mWater, on the mwater website). And with cameras included on many smartphones, high-quality and geo-referenced photographs can be collected for photo-point monitoring. These data can then be sent back to a main data management centre through mobile phone or data networks, which can simplify the tasks of data collection and management. Even without a human present, expanded cell phone networks have allowed automatic data collectors to send back real-time data via Bluetooth or telemetry systems (Aitkenhead et al. 2014).

Remote sensing is another well-established, still rapidly advancing field for understanding threats and pressures on freshwater ecosystems, providing rich tools and opportunities for collecting ecologically (and socially) relevant information (Chapter 9). Remotely sensed data comes from satellites orbiting Earth, which collect data on a wide variety of parameters, including vegetation, soil moisture, rainfall, and land cover. Depending on the information source, these parameters can be tracked across large areas and over time to analyse long-term changes in hydrological, ecological, and geomorphological conditions (Kerr and Ostrovsky 2003). Many of these satellites are launched by national governments and are freely available to the public for download and use, including the Sentinel and PROBA-V satellites launched by the European Space Agency, and the Landsat series of satellites launched by the United States Geological Survey. While specialised knowledge and the use of geographical information systems (GIS) are required to use the spatial data collected via remote sensing, the amount of information available is vast and can help analyse ecological trends on a customisable spatial and temporal scale (e.g., Ogilvie et al. 2015). An interesting new initiative under the US National Aeronautics and Space Administration (NASA), the Surface Water and Ocean Topography project (SWOT) when launched in 2021, will generate the first ever global survey of earth's surface water, providing comprehensive coverage and invaluable new data on the planet's freshwater ecosystems.

There are limitations to using remote sensing data, including cloud cover and poor resolution at certain scales. One way around these limitations is though the use of unmanned aerial vehicles (UAVs) or drones (Chapter 9). UAVs are small, portable vehicles that can be directed or programmed to cover specific areas and can carry equipment to collect photos, video, lidar, and other types of data. Since the data collection systems can be customised and the UAV is flown below any cloud cover, the resolution at which they can gather data is much greater than those obtained by remote sensing. Unmanned aerial vehicles have been used to classify physical river habitats in great detail along kilometres of river length, providing critical information to both scientists and managers (Woodget et al. 2017). They are also becoming more popular and affordable for use in ecological conservation activities, such as wildlife counts and vegetation surveys (Koh and Wich 2012), and in restoration efforts (e.g., the UK Moors of the Future project is using a UAV to monitor and analyse the impact of conservation management). Guidelines for UAV use exist that vary by state and by country in terms of local operational requirements or restrictions. Challenges remain in developing suitable methods for rapidly analysing such readily acquired, but large data sets.

2.4.2 Private sector water stewardship as a growing area for ecosystem conservation

In recent years, an increasing number of voices from the private sector have begun to express concern about the potential impacts of water-related risks to business and the economy (e.g., JP Morgan, Coca-Cola, SABMiller, WEF, International Council on Mining and Metals). Such risks can take the form of physical scarcity of water (which means that factories or farms are unable to operate); shifting regulation of water use and pollution control (leading to uncertainty over costs and hampering business planning); and perceptions among some communities that the private sector is partly or largely to blame for dwindling water supplies (leading to reputational damage to corporate brands) (Pegram 2010). Environmental NGOs (e.g., WWF 2013) have sought to encourage private sector actors to mitigate these risks by engaging with civil society and other stakeholders to encourage better management of water resources and freshwater ecosystems. The theory is that the causes of water-related risk to businesses are the same as those which pose risks to habitats and biodiversity (i.e., over-abstraction, pollution, poorly planned infrastructure and inadequate governance and management regimes). As such, businesses and ecosystems (and other water users) are exposed to shared water risks.

Water stewardship is a loosely defined term which describes the mechanisms that businesses, primarily, can potentially use to promote management of these shared risks, including through conservation of water resources and protection or restoration of freshwater ecosystem health. WWF (2013) summarised water stewardship as 'a progression of increased improvement of water use and a reduction in the water-related impacts [on ecosystems] of internal and value chain operations. More importantly, it is a commitment to the sustainable management of shared water resources in the public interest through collective action with other businesses, governments, NGOs and communities.' While businesses can take important measures internally, such as improving water efficiency within factories or reducing effluent discharge to rivers and lakes, steps are needed 'beyond the factory fenceline' to address shared risks and to promote enhanced water policy and governance, as follows (WWF 2013): (1) increase awareness of water issues and risks; (2) enhance knowledge of company impacts on freshwater systems; (3) take action within the company to reduce direct impacts; (4) engage river basin stakeholders to promote collective conservation; and (5) advocate for improved water policy and governance.

There are a number of examples of water stewardship theory being translated, at least partly, into practice, including some at the company scale (especially with food and beverage sector companies) and in specific locations. For instance, at Lake Naivasha, a Ramsar site in Kenya and the focus of a government-led alliance of companies (including European retailers who source cut flowers and vegetables from the Naivasha Basin), public agencies, and civil society actors have joined forces to address freshwater conservation challenges (see Imarisha Naivasha website). However, water stewardship is still a relatively novel approach to freshwater conservation and there is, as yet, relatively little evidence of substantial improvements in the condition of specific freshwater ecosystems as a direct or indirect result of private sector engagement. Critical assessment of water stewardship efforts will be necessary if the broad approach is to evolve into a widely applicable tool for achieving conservation outcomes and if perverse outcomes are to be avoided (Newborne and Dalton 2016; Hepworth 2012).

2.5 Concluding remarks

There are major challenges to overcome in any effort to develop and implement an approach to managing and conserving freshwater ecosystems, from within or across the public, private, or civil society sectors, if we are to ensure that the world will still contain an array of healthy freshwater ecosystems into the middle of the twenty-first century and beyond. Conservationists and ecologists, working alongside other disciplines and professions, need to make a compelling political and socioeconomic case for sustainable water management that incorporates freshwater ecosystem conservation at its core. Simply providing evidence and technical assessments of the state of freshwater ecosystems is unlikely to be sufficient; powerful stories of conservation successes, and why they are important, will be critical to motivating stakeholders and decision-makers to act. Implementing sustainable water management will require strengthened, cross-cutting links with water resource management communities of practice.

Knowledge co-production with stakeholders, including with indigenous and local communities, will need to become more prominent, complementing standard and technocentric views and approaches. Research frontiers and sophisticated, novel technologies should continue to be advanced, but not at the expense of fundamental knowledge. This is especially important in the context of the developing world, where the immediacy of development pressures diverts most attention and funding support to applied work; basic research still needs to occur, otherwise, a lack of understanding of basic processes and ecosystem functions will hamper future application. For instance, knowledge of species taxonomy is important in areas as diverse as the control of freshwater invasive species, aquaculture-based food production, and the prediction of ecological responses to climate change and flow regime alteration. A retrospective on the history of freshwater research in South Africa underscored several shortcomings associated with inadequate investment in fundamental research alongside problem-oriented and consultancy projects (Ashton et al. 2012).

In these days of social media, the blogosphere, and YouTube, publishing scientific papers predominantly read within the same scientific community is no longer sufficient. The readability of science is also declining, rendering the knowledge less accessible for non-specialists, such as policy-makers, journalists, and the wider public, as well as for other specialists (Plavén-Sigray et al. 2017). Communication and education tools and approaches that reach out to a range of stakeholders of broader demographic characteristics and backgrounds need to increase (e.g., Jacobson et al. 2015). There are encouraging efforts in this direction (e.g., Facebook projects, journal lay summaries, interactive online primers and story maps, e-learning short online courses, and report cards) promoting clearer, more succinct two-way knowledge exchange and learning. Science communication is a critical element of scientific practice (Nielsen 2013) and, as an emerging field, is helping address the ways in which science can effectively be used to communicate ideas to a wider audience (Fischhoff 2013).

The importance of multidisciplinary and interdisciplinary approaches is continuing to grow, particularly with the recognition that ecology and conservation approaches to social-ecological systems are ineffective in isolation of their socio-cultural context (e.g., Gavin et al. 2015). Transdisciplinarity and team-based problem-solving approaches are becoming the norm. More time needs to be apportioned for the systematic application of these kinds of cross-learning approaches and for sharing the knowledge gained from these forms of exchange. Boundary practitioners have a vital and now explicitly recognised role to play as bridges between science, practice, and policy, spurring the emergence of institutions and networks with this focus (Leith et al. 2016). Strategic influencing and other leadership skills are becoming increasingly important to bring about change based on good science.

The roles of freshwater scientists are undoubtedly changing. Scientists have increasingly greater responsibility to provide solutions that will help solve particular problems experienced by society. Moreover, they are no longer expected to only collect and analyse data, and report to government or private industry, but also be available to the broader public for questioning. This approach opens up a new and dynamic space for researchers, for which natural scientists in particular have traditionally been unprepared. The requisite skills and approaches are not universally developed within current university curriculums, but will need to be included in future to better equip the next generation of freshwater conservation scientists.

References

Acreman, M., Arthington, A.H., Colloff, M.J., Couch, C., Crossman, N.D., Dyer, F., Overton, I., Pollino, C.A., Stewardson, M.J., and Young, W. (2014). Environmental flows for natural, hybrid, and novel riverine ecosystems in a changing world. *Frontiers in Ecology and the Environment*, 12, 466–73.

Aitkenhead, M., Donnelly, D., Coull, M., and Hastings, E. (2014). Innovations in environmental monitoring using mobile phone technology—A review. *International Journal of Interactive Mobile Technologies*, 8, 42–50.

Allen, C.R. and Garmestani, A.S. (eds) (2015). Adaptive Management of Social-Ecological Systems. Springer, Netherlands.

Arnstein, S.R. (1969). A ladder of citizen participation. *J. Am. Inst. Plan.*, 35, 216–24.

Arthington, A.H., Naiman, R.J., McClain, M.E., and Nilsson, C. (2010). Preserving the biodiversity and ecological services of rivers: new challenges and research opportunities. *Freshwater Biology*, 55, 1–16.

Ashton, P.J., Roux, D.J, Breen, C.M., Day, J.A., Mitchell, S.A., Seaman, M.T., and Silberbauer, M.J. (2012). The freshwater science landscape in South Africa, 1900–2010: Overview of Research Topics, Key Individuals, Institutional Change and Operating Culture. WRC Report No. TT 530/12. Water Research Commission, Pretoria, SA.

Baumgartner, L. (2016). Adaptive management in action: using chemical marking to advance fish recovery programs in the Murray–Darling Basin. *Marine and Freshwater Research,* 67, i-iii http://dx.doi.org/10.1071/MF15353.

Berbés-Blázquez, M., González, J.A., and Pascual, U. (2016). Towards an ecosystem services approach that addresses social power relations. *Current Opinion in Environmental Sustainability*, 19, 134–43.

Berghöfer, A., Brown, C., Bruner, A., Emerton, L., Esen, E., Geneletti, D., Kosmus, M., Kumar, R., Lehmann, M., Leon Morales, F., Nkonja, E., Pistorius, T. Rode, J., Slootweg, R., Tröger, U., Wittmer, H., Wunder, S., and van Zyl, H. (2016). Increasing the policy impact of ecosystem service assessments and valuations—insights from practice. Helmholtz-Zentrum für Umweltforschung (UFZ) GmbH, Leipzig, and Deutsche Gesellschaft für Internationale Zusammenarbeit (GIZ) GmbH, Eschborn, Germany. Available online.

Berkes, F., Colding, J., and Folke, C. (2003). Navigating Social-Ecological Systems: Building Resilience for Complexity and Change. Cambridge University Press, Cambridge, UK.

Berthet, E.T., Segrestin, B., and Hickey, G.M. (2016). Considering agro-ecosystems as ecological funds for collective design: New perspectives for environmental policy. *Environmental Science and Policy*, 61, 108–15.

Biggs, J., Ewald, N., Valentini, A., Gaboriaud, C., Griffiths, R.A., Foster, J., Wilkinson, J., Arnett, A., Williams, P., and Dunn, F. (2014). Analytical and Methodological Development for Improved Surveillance of the Great Crested Newt. Defra Project WC1067, Freshwater Habitats Trust, Oxford.

Biggs, R., Rhode, C., Archibald, S., Kunene, L.M., Mutanga, S.S., Nkuna, N., Ocholla, P.O., and Phadima, L.J. (2015). Strategies for managing complex social-ecological systems in the face of uncertainty: examples from South Africa and beyond. *Ecology and Society*, 20, 52.

Bond, N.R. and Kennard, M.J. (2017). Prediction of hydrologic characteristics for ungauged catchments to support hydroecological modeling. *Water Resources Research*, 53, 8781–94.

Brooks, E.G.E., Smith, K.E., Holland, R.A., Poppy, G.M., and Eigenbrod, F. (2014). Effects of methodology and stakeholder disaggregation on ecosystem service valuation. *Ecology and Society*, 19, 18.

Cadotte, M.W., Barlow, J., Nuñez, M.A., Pettorelli, N., and Stephens, P.A. (2017). Solving environmental problems in the Anthropocene: the need to bring novel theoretical advances into the applied ecology fold. *Journal of Applied Ecology*, 54, 1–6.

Conallin, J., Dickens, C., Hearne, D., and Allan, C. (2017). Stakeholder engagement in environmental water management. In: Horne, A.C., Webb, J.A., Stewardson, M.J, Richter, B., and Acreman, M. (eds) Water for the Environment: From Policy and Science to Implementation and Management. Academic Press, London.

Conallin, J., McLoughlin, C.A., Campbell, J., Knight, R., Bright, T. and Fisher, I. (2018b). Stakeholder participation in freshwater monitoring and evaluation programs: Applying thresholds of potential concern within environmental flows. *Environmental Management*, 61, 408–20.

Conallin, J., Wilson, E., and Campbell, J. (2018a). Implementation of environmental flows for intermittent river systems: Adaptive management and stakeholder participation facilitate implementation. *Environmental Management*, 61, 497–505.

Costanzo, S.D., Blancard, C., Davidson, S., Dennison, W.C., Escurra, J., Freeman, S., Fries, A., Kelsey, R.H., Krchnak, K., Sherman, J., Thieme, M., and Vargas-Nguyen, V. (2017). Practitioner's Guide to Developing River Basin Report Cards. IAN Press, Cambridge, MD, USA.

De Vente, J., Reed, M.S., Stringer, L.C., Valente, S., and Newig, J. (2016). How does the context and design of participatory decision making processes affect their outcomes? Evidence from sustainable land management in global drylands. *Ecology and Society*, 21, 24.

Di Baldassarre, G., Viglione, A., Carr, G., Kuil, L., Salinas, J.L., and G. Blöschl, G. (2013). Socio-hydrology: conceptualising human-flood interactions. *Hydrol. Earth Syst. Sci.*, 17, 3295–303.

Dudgeon, D., Arthington, A.H., Gessner, M.O., Kawabata, Z., Knowler, D.J., Leveque, C., Naiman, R.J., Prieur-Richard, A.H., Soto, D., Stiassny, M.L., and Sullivan, C.A. (2006). Freshwater biodiversity: importance, threats, status and conservation challenges. *Biological Reviews of the Cambridge Philosophical Society*, 81, 163–82.

Early, R.I., Bradley, B.A., Dukes, J.S., Lawler, J.J., Olden, J.D., Blumenthal, D.M., D'Antonio, C.M., Gonzalez, P., Grosholz, E.D., Ibanez, I., Miller, L.P., Sorte, C.J.B., and Tatem, A.J. (2016). Global threats from invasive alien species in the 21st Century and national response capacities. *Nature Communications* 7, 12485. doi:10.1038/ncomms12485.

Field, C.B., Barros, V.R., Mach, K., and Mastrandrea, M. (2014). Climate Change 2014: Impacts, Adaptation, and Vulnerability. Cambridge University Press, Cambridge and New York.

Finlayson, M.C. (2012). Forty years of wetland conservation and wise use. *Aquatic Conservation: Marine and Freshwater Ecosystems*, 22, 139–43.

Finlayson, C. and D'Cruz, R. (2005). Inland water systems. Ecosystems and Human Well-being: Current State and Trends. Island Press, Washington, DC.

Finn, M. and Jackson, S. (2011). Protecting indigenous values in water management: a challenge to conventional environmental flow assessments. *Ecosystems*, 14, 1232–48.

Fischhoff, B. (2013). The sciences of science communication. *PNAS*, 110, 14033–9.

Flotemersch, J.E., Leibowitz, S.G., Hill, R.A., Stoddard, J.L., Thoms, M.C., and Tharme, R.E. (2016). A watershed integrity definition and assessment approach to support strategic management of watersheds. *River Research and Applic*ations, 32, 1654–71.

Garrido, A., Cuevas, M.L., Cotler, H., González, D.I., and Tharme, R. (2010). El estado de alteración ecohidrológica de los ríos de México, 108–11. In: Cotler Ávanos, H. (Coord.). Las cuencas hidrográficas de México: diagnóstico y priorización. Instituto Nacional de Ecología-Fundación Gonzalo Río Arronte I.A.P., México City, México.

Gavin, M.C., McCarter, J., Mead, A., Berkes, F., Stepp, J.R., Peterson, D., and Tang, R. (2015). Defining biocultural approaches to conservation. *Trends in Ecology and Evolution,* 30, 140–5.

Ghermandi, A., van den Bergh, J.C.J.M., Brander, L.M., de Groot, H.L.F., and Nunes, P.A.L.D. (2010). Values of natural and human-made wetlands: a meta-analysis. *Water Resources Research*, 46, W12516.

Gilvear, D., Beevers, L., O'Keeffe, J., and Acreman, M. (2017). Environmental water regimes and natural capital—Free-flowing ecosystem services. In: Horne, A.C., Webb, J.A., Stewardson, M.J, Richter, B., and Acreman, M. (eds) Water for the Environment: From Policy and Science to Implementation and Management. Academic Press, London.

Gopal, B. (2016). A conceptual framework for environmental flows assessment based on ecosystem services and their economic valuation. *Ecosystem Services,* 21, 53–8.

Hepworth, N.D. (2012). Open for business or opening Pandora's box? A constructive critique of corporate engagement in water policy: An introduction. *Water Alternatives*, 5, 543–62.

Horne, A., O'Donnell, E., Acreman, M., McClain, M.E., Poff, L., Webb, J.A., Stewardson, M.J, Bond, N., Richter, B., Arthington, A., Tharme, R., Garrick, D., Daniell, K., Conallin, J., Thomas, G.A., and Hart, B.T. (2017a). Moving forward—the implementation challenges for environmental water management. In: Horne, A.C., Webb, J.A., Stewardson, M.J, Richter, B., and Acreman, M. (eds) Water for the Environment: From Policy and Science to Implementation and Management. Academic Press, London.

Horne, A.C., O'Donnell, E., and Tharme, R.E. (2017b). Mechanisms to allocate environmental water. In: Horne, A.C., Webb, J.A., Stewardson, M.J, Richter, B., and Acreman, M. (eds) Water for the Environment: From Policy and Science to Implementation and Management. Academic Press, London.

Howarth, S., Tharme, R.E., Jinapala, K., Clemett, A., Meilhac, C., Turner, S., Somaratne, P.G., and Abeysekera, T. (2006). Identifying sustainable options for the mitigation of diffuse agricultural pollution, Final Report. Mott MacDonald, Cambridge, UK.

International Community Assessment (2012). Global Water Security. ICA 2012–08, 2 February 2012, Office of the Director of National Intelligence, USA.

Irvine, K., Castello, L., Junqueira, A., and Moulton, T. (2016). Linking ecology with social development for tropical aquatic conservation. *Aquatic Conservation: Marine and Freshwater Ecosystems*, 26, 917–41.

Jackson, M.C., Loewen, C.J., Vinebrooke, R.D., and Chimimba, C.T. (2016). Net effects of multiple stressors in freshwater ecosystems: a meta-analysis. *Global Change Biology*, 22, 180–9.

Jackson, S.E., Douglas, M.M., Kennard, M.J. Pusey, B.J., Huddleston, J., Harney, B., Liddy, L., Liddy, M. Liddy, R., Sullivan, L., Huddleston, B., Banderson, M., McMah, A., and Allsop, Q. (2014). 'We like to listen to stories about fish': integrating indigenous ecological and scientific knowledge to inform environmental flow assessments. *Ecology and Society*, 19, 43.

Jacobson, S.K., McDuff, M., and Monroe, M. (2015). Conservation Education and Outreach Techniques, Second Edition, Techniques in Ecology and Conservation Series. Oxford University Press, Oxford.

Kerr, J.T. and Ostrovsky, M. (2003). From space to species: ecological applications for remote sensing. *Trends in Ecology and Evolution*, 18, 299–305.

Kobori, H., Dickinson, J.L., Washitani, I., Sakurai, R., Amano, T., Komatsu, N., Kitamura, W., Takagawa, S., Koyama, K., Ogawara, T., and Miler-Rushing, A.J. (2016). Citizen science: a new approach to advance ecology, education, and conservation. *Ecological Research*, 31, 1–19.

Koh, L.P. and Wich, S.A. (2012). Dawn of drone ecology: Low-cost autonomous aerial vehicles for conservation. *Tropical Conservation Science,* 5, 121–32.

Kopf, R.K., Finlayson, C.M., Humphries, P., Sims, N.C., and Hladyz, S. (2015). Anthropocene baselines: assessing change and managing biodiversity in human-dominated aquatic ecosystems. *Bioscience*, 65, 798–811.

Le Quesne, T., Kendy, E., and Weston, D. (2010). The Implementation Challenge: Taking Stock of Government Policies to Protect and Restore Environmental Flows. WWF and The Nature Conservancy, Godalming, UK.

Leith, P., Haward, M., Rees, C., and Ogier, E. (2016). Success and evolution of a boundary organization. *Science, Technology and Human Values*, 41, 375–401.

Lewis, S.L. and Maslin, M.A. (2015). Defining the Anthropocene. *Nature*, 519, 171–80.

Lloyd, G.J., Korsgaard, L., Tharme, R., Boelee, E., Clement, F., Barron, J., and Eriyagama, N. (2013). Water management for ecosystem health and food production. In: Boelee, E. (ed.) Managing Water and Agroecosystems for Food Security. CA-CAB International Publishing, Oxford.

Lokgariwar, C., Chopra, V., Smakhtin, V., Bharati, L., and O'Keeffe, J. (2014). Including cultural water requirements in environmental flow assessment: an example from the upper Ganga River, India. *Water International*, 39, 81–96.

Lunn, Z., Conallin, J., Lwin, M.M., Baumgartner, L., Thorncraft, G., An, V.V., and Phonekhampeng, O. (2017). Determining Migration Corridors for Key Commercial, Subsistence and Conservation Freshwater Fish in Irrawaddy River Basin. Presentation at 2017 American Fisheries Society Annual Meeting, Tampa, Florida, USA.

Lynch, A., Cowx, I., Fluet-Chouinard, E., Glaser, S.M., Phang, S., Beard, T.D., Bower, S., Brooks, J., Bunnell, D., Claussen, J.E., Cooke, S., Kao, Y-C., Lorenzen, K., Myers, B., Reid, A., Taylor, J.J., and Youn, S-J. (2017). Inland fisheries—Invisible but integral to the UN Sustainable Development Agenda for ending poverty by 2030. *Global Environmental Change*, 47, 167–73.

Mason, C.F. and Macdonald, S.M. (2004). Growth in otter (*Lutra lutra*) populations in the UK as shown by long-term monitoring. AMBIO: A *Journal of the Human Environment*, 33, 148–52.

Matthaei, C.D. and Lange, K. (2016). Multiple stressor effects on freshwater fish: a review and meta-analysis. In: Closs, G.P., Krkosek, M., and Olden, J.D. (eds) Conservation of Freshwater Fishes. Cambridge University Press, Cambridge.

McGinnis, M.D. and Ostrom, E. (2014). Social-ecological system framework: initial changes and continuing challenges. *Ecology and Society*,19, 30.

Moore, J.W. and Olden, J.D. (2017). Response diversity, nonnative species, and disassembly rules buffer freshwater ecosystem processes from anthropogenic change. *Global Change Biology*, 23, 1871–80.

Moss, B.R. (2018). Ecology of Freshwaters: Earth's Bloodstream. 5th Edition, Wiley, UK.

Myers, N. (1993). Biodiversity and the precautionary principle. *Ambio*, 22, 74–9.

Naiman, R. and Dudgeon, D. (2011). Global alteration of freshwaters: influences on human and environmental well-being. *Ecological Resources*, 26, 865–3.

Newborne, P. and Dalton, J. (2016). Water Management and Stewardship: Taking Stock of Corporate Water Behaviour. IUCN, Gland, Switzerland and ODI, London, UK.

Nielsen, K.H. (2013). Scientific communication and the nature of science. *Science and Education*, 22, 2067–86.

Ogilvie, A., Belaud, G., Delenne, C., Bailly, J.S., Bader, J.C., Oleksiak, A., and Martin, D. (2015). Decadal monitoring of the Niger Inner Delta flood dynamics using MODIS optical data. *Journal of Hydrology*, 523, 368–83.

Olden, J.D., Konrad, C.P., Melis, T.S., Kennard, M.J., Freeman, M.C., Mims, M.C., Bray, E.N., Gido, K.B., Hemphill, N.P., Lytle, D.A., McMullen, L.E., Pyron, M., Robinson, C.T., Schmidt, J.C., and Williams, J.G. (2014). Are large-scale flow experiments informing the science and management of freshwater ecosystems? *Frontiers in Ecology and the Environment*, 12, 176–85.

Olden, J.D., Lawler, J.J., and Poff, N.L. (2008). Machine learning methods without tears: a primer for ecologist. *The Quarterly Review of Biology*, 83, 171–93.

Opperman, J., Grill, G., and Hartmann, J. (2015). The power of rivers: Finding balance between energy and conservation in hydropower development. Technical report, The Nature Conservancy, Washington, DC.

Organisation for Economic Cooperation and Development (OECD) (2017). Diffuse Pollution, Degraded Waters—Emerging Policy Solutions. OECD Studies on Water, OECD Publishing, Paris.

Pahl-Wostl, C., Arthington, A., Bogardi, J., Bunn, S.E., Hoff, H. Lebel, L., Nikitina, E., Palmer, M., Poff, L.N., Richards, K., Schluter, M., Schulze, R., St-Hilaire, A., Tharme, R., Tockner, K., and Tsegai, D. (2013). Environmental flows and water governance: managing sustainable water uses. *Current Opinion in Environmental Sustainability*, 5, 341–51.

Palmer, C.G., Muller, W.J., Gordon, A.K., Scherman, P-A., Davies-Coleman, H.D., Pakhomova, L., and de Kock, E. (2004). The development of a toxicity database using freshwater macroinvertebrates, and its application to the protection of South African water resources. *South African Journal of Science*, 100, 643–50.

Parker, H. and Oates, N. (2016). How do healthy rivers benefit society? A review of the evidence. Overseas Development Institute, Working Paper 430, ODI and WWF, London, UK.

Pegram, G. (2010). Global water scarcity: Risks and challenges for business. Lloyd's of London.

Peh, K.S.H., Balmford, A., Bradbury, R.B., Brown, C., Butchart, S.H., Hughes, F.M.R., and Gowing, D. (2013). TESSA: a toolkit for rapid assessment of ecosystem services at sites of biodiversity conservation importance. *Ecosystem Services*, 5, 51–7.

Pittock, J., Finlayson, M., Arthington, A.H., Roux, D., Matthews, J.H., Biggs, H., Harrison, I., Blom, E., Flitcroft, R., Froend, R., et al. (2015). Managing freshwater, river, wetland and estuarine protected areas. In: Worboys, G.L., Lockwood, M., Kothari, A., Feary, S., and Pulsford, I. (eds) Protected Area Governance and Management. ANU Press, Canberra.

Plavén-Sigray, P., Matheson, G.J., Schiffler, B.C., and Thompson, W.H. (2017). Research: The readability of scientific texts is decreasing over time. *eLife* 2017;6:e27725.

Pocock, M.J.O., Tweddle, J.C., Savage, J., Robinson, L.D., and Roy, H.E. (2017). The diversity and evolution of ecological and environmental citizen science. *PLoS One*, 12, 1–17.

Poff, N.L. (2014). Rivers of the Anthropocene? *Frontiers in Ecology and Environment*, 12, 427.

Poff, N.L., Brown, C.M., Grantham, T.E., Matthews, J.H., Palmer, M.A., Spence, C.M., Wilby, R.L., Haasnoot, M., Mendoza, G.F., Dominique, K.C., and Baeza, A. (2016). Sustainable water management under future uncertainty with eco-engineering decision scaling. *Nature Climate Change*, 6, 25–34.

Poff, N.L., Richter, B.D., Arthington, A.H., Bunn, S.E., Naiman, R.J., Kendy, E., Acreman, M., Apse, C., Bledsoe, B.P., Freeman, M.C., Henriksen, J., Jacobson, R.B., Kennen, J.G., Merritt, D.M., O'Keeffe, J.H., Olden, J.D., Rogers, K., Tharme, R.E., and Warner, A.

(2010). The ecological limits of hydrologic alteration (ELOHA): a new framework for developing regional environmental flow standards. *Freshwater Biology*, 55, 147–70.

Poff, N.L., Tharme, R.E., and Arthington, A.H. (2017). Evolution of environmental flows science, principles and methodologies. In: Horne, A., Webb, A., Stewardson, M., Richter, B., and M. Acreman (eds) Water for the Environment. Elsevier Press.

Pringle, R.M. (2017). Upgrading protected areas to conserve wild biodiversity. *Nature*, 546, 91–9.

Reed, M.S., Graves, A., Dandy, N., Posthumus, H., Hubacek, K., Morris, J., Prell, C., Quinn, C.H., and Stringer, L.C. (2009). Who's in and why? A typology of stakeholder analysis methods for natural resource management. *Journal of Environmental Management*, 90, 1933–49.

Rogers, K.H. (2006). The real river management challenge: integrating scientists, stakeholders and service agencies. *River Research and Applications*, 22, 269–80.

Roux, D.J. and Foxcroft, L.C. (2011). The development and application of strategic adaptive management within South African National Parks. *Koedoe*, 53, Art. #1049, 5 pages.

Roux, D.J., Nel, J.L., Ashton, P.J., Deacon, A.R., de Moor, F.C., Hardwick, D., Hill, L., Kleynhans, C.J., Maree, G.A., Moolman, J., and Scholes, R.J. (2008). Designing protected areas to conserve riverine biodiversity: Lessons from a hypothetical redesign of the Kruger National Park. *Biological Conservation*, 141, 100–17.

Sadoff, C., Hall, J., Grey, D., Aerts, J.C., Ait-Kadi, M., Brown, C., Cox, A., Dadson, S., Garrick, D., Kelman, J., McCornick, P., Ringler, C., Rosegrant, M., Whittington, D., and Wiberg, D. (2015). Securing Water, Sustaining Growth. Report of the GWP/OECD Task Force on Water Security and Sustainable Growth. University of Oxford, Oxford, UK.

Scholes, R., Briggs, R., Palm, C., and Duraiappah, A. (2010). Assessing state and trends in ecosystem services and human well-being. In: Ash, N., Blanco, H., Garcia, K., Tomich, T., Vira, B., Zurek, M., and Brown, C. (eds) Ecosystems and Human Well-Being: A Manual for Assessment Practitioners. Island Press, Washington.

Speed, R., Tickner, D., Naiman, R., Gang, L., Sayers, P., Yu, W., Yuanyuan, L., Houjian, H., Jianting, C., and Lili, Y. (2016). River restoration: A strategic approach to planning and management. UNESCO Publishing.

Stem, C., Margoluis, R., Salafsky, N., and Brown, M. (2005). Monitoring and evaluation in conservation: a review of trends and approaches. *Conservation Biology*, 19, 295–309.

Stern, M.J. (2018). Social science theory for environmental sustainability—A practical guide. Techniques in Ecology and Conservation Series, Oxford University Press, Oxford, UK.

Stewart, R.I.A., Dossena, M., Bohan, D.A., Jeppesen, E., Kordas, R.L., Ledger, M.E., Meerhoff, M., Moss, B., Mulder, C., Shurin, J.B., Suttle, B., Thompson, R., Trimmer, M., and Woodward, G. (2013). Mesocosm experiments as a tool for ecological climate-change research. *Advances in Ecological Research*, 48, 71–181.

Stewart-Koster, B., Bunn, S.E., Mackay, S.J., Poff, N.L., Naiman, P.J., and Lake, P.S. (2010). The use of Bayesian networks to guide investments in flow and catchment restoration for impaired river ecosystems. *Freshwater Biology*, 55, 243–60.

Stirzaker, R., Biggs, H., Roux, D., and Cilliers, P. (2010). Requisite simplicities to help negotiate complex problems. *AMBIO*, 39, 600–7.

Stoeckle, B.C., Kuehn, R., and Geist, J. (2016). Environmental DNA as a monitoring tool for the endangered freshwater pearl mussel (*Margaritifera margaritifera* L.): A substitute for classical monitoring approaches? *Aquatic Conservation: Marine and Freshwater Ecosystems*, 26, 1120–9.

Strayer, D.L. and Dudgeon, D. (2010). Freshwater biodiversity conservation: recent progress and future challenges. *Journal of the North American Benthological Society*, 29, 344–58.

Taberlet, P., Bonin, A., Zinger, L., and Coissac, E. (2018). Environmental DNA: for Biodiversity Research and Monitoring. Oxford University Press, Oxford.

The economics of ecosystems and biodiversity (TEEB) (2013). The economics of ecosystems and biodiversity (TEEB) for water and wetlands. TEEB report. IEEP, London and Brussels, and Ramsar Secretariat, Gland.

Thomsen, P.F. and Willerslev, E. (2015). Environmental DNA—An emerging tool in conservation for monitoring past and present biodiversity. *Biological Conservation*, 183, 4–18.

Tickner, D. and Acreman, M. (2013). Water security for ecosystems, ecosystems for water security. In: Lankford, B., Bakker, K., Zeitoun, M., and Conway, D. (eds) Water Security: Principles, Perspectives and Practices. Routledge, Oxford, UK.

Tickner, D., Parker, H., Oates, N.E., Moncrieff, C.R., Ludi, E., and Acreman, M. (2017). Managing rivers for multiple benefits—A coherent approach to research, policy and planning. *Frontiers in Environmental Science*, 5, 4.

Tomich, P.T., Argumedo, A., Baste, I., Camac, E., Filer, C., Garcia, K., Garbach, K., Geist, H., Izac, A-M., Lebel, L., Lee, M., Nishi, M., Olsson, L., Raudsepp-Hearne, C., Rawlins, M., Scholes, R., and van Noordwijk, M. (2005). Conceptual frameworks for ecosystem assessment: their development, ownership and use. In: Ash, N., Blanco, H., Brown, C., Garcia, K., Henrichs, T., Lucas, N., Raudsepp-Hearne, C., Simpson, R.D., Scholes, R., Tomich, T., Vira, B., and Zurek, M. (eds) (2005). Ecosystems and Human Well-Being: a Manual for Assessment Practitioners. Island Press, Washington DC.

United Nations Environment Program (UNEP) and United Nations University Institute for Environment and Human Security (UNU-EHS) (2016). International Water Quality Guidelines for Ecosystems (IWQGES), Scientific Background and Technical Guides, Draft for Regional Consultations. UNU, Bonn, Germany.

United States Agency for International Development (USAID) (2016). Environmental flows in Rufiji River Basin assessed from the perspective of planned development in Kilombero and Lower Rufiji Sub-basins. Technical Assistance to Support the Development of Irrigation and Rural Roads Infrastructure Project (IRRIP2), Final Technical Report. CDM International.

Vörösmarty, C.J., McIntyre, P.B., Gessner, M.O., Dudgeon, D., Prusevich, A., Green, P., Glidden, S., Bunn, S.E., Sullivan, C.A., Liermann, C.R., and Davies, P.M. (2010). Global threats to human water security and river biodiversity. *Nature*, 467, 555–61.

Walther, B.D., Limburg, K.E., Jones, C.M., and Schaffler, J.J. (2017). Frontiers in otolith chemistry: insights, advances and applications. *Journal of Fish Biology*, 90, 473–9

Wantzen, K.M., Ballouche, A., Longuet, I., Bao, I., Bocoum, H., Cisse, L., Chauhan, M., Girard, P., Gopal, B., Kane, A., Marchese, M.R., Nautiyal, P., Teixeira, P., and Zalewski, M. (2016). River Culture: an eco-social approach to mitigate the biological and cultural diversity crisis in riverscapes. *Ecohydrology and Hydrobiology*, 16, 7–18.

Ward, J.V. (1989). The four-dimensional nature of lotic ecosystems. *Journal of the North American Benthological Society*, 8, 1, 2–8.

Warner, A.T., Bach, L.B., and Hickey, J.T. (2014). Restoring environmental flows through adaptive reservoir management: planning, science, and implementation through the Sustainable Rivers Project. *Hydrological Sciences Journal*, 59, 770–85.

Wetlands International (2017). Water Shocks: Wetlands and Human Migration in the Sahel. Wetlands International, The Netherlands.

Winemiller, K.O., McIntyre, P.B., Castello, L., Fluet-Chouinard, E., Giarrizzo, T., Nam, S., Baird, I.G., Darwall, W., Lujan, N.K., Harrison, I., Stiassny, M.L.J., Silvano, R.A.M., Fitzgerald, D.B., Pelicice, F.M., Agostinho, A.A., Gomes, L.C., Albert, J.S., Baran, E., Petrere, Jr M., Zarfl, C., Mulligan, M., Sullivan, J.P., Arantes, C.C., Sousa, L.M., Koning, A.A., Hoeinghaus, D.J., Sabaj, M., Lundberg, J.G., Armbruster, J., Thieme, M.L., Petry, P., Zuanon, J., Torrente Vilara, G.T., Snoeks, J., Ou, C., Rainboth, W., Pavanelli, C.S., Akama, A., van Soesbergen, A., and Saenz, L. (2016). Balancing hydropower and biodiversity in the Amazon, Congo, and Mekong. *Science*, 351, 128–9.

Woodget, A.S., Austrums, R., Maddock, I.P., and Habit, E. (2017). Drones and digital photogrammetry: From classifications to continuums for monitoring river habitat and hydromorphology. *Wiley Interdisciplinary Reviews: Water*, 4, e1222.

World Bank (2016). High and Dry: Climate Change, Water, and the Economy. World Bank.

World Economic Forum (WEF) (2016). The Global Risks Report 2016. Insight report, 11th Edition, World Economic Forum.

World Economic Forum (WEF) (2017). The Global Risks Report 2017. Insight report, 12th Edition, World Economic Forum.

WWAP (United Nations World Water Assessment Programme) (2015). The United Nations World Water Development Report 2015: Water for a Sustainable World. UNESCO, Paris.

WWF (2013). Water Stewardship: Perspectives on Business Risks and Response to Water Challenges. WWF International, Gland, Switzerland.

WWF (2016). Living Planet Report: Risk and Resilience in a New Era. WWF International, Gland, Switzerland.

WWF-Greater Mekong (2016). Mekong River in the Economy. WWF-Greater Mekong.

WWF-Zambia (2016). Water in the Zambian Economy: Exploring Shared Risks and Opportunities in the Kafue Flats, 2nd edition. WWF-Zambia.

Zarfl, C., Lumsdon, A.E., Berlekamp, J., Tydecks, L., and Tockner, K. (2015). A global boom in hydropower dam construction. *Aquatic Sciences*, 77, 161–70.

Zielinski, L. (2016). Creating a Monitoring and Adaptive Management Framework for Reserve Flows in Kenyan River Basins. Unpublished Masters thesis. UNESCO-IHE, Delft, The Netherlands.

Ziv, G., Baran E., Nam, S., Rodríguez-Iturbe, I., and Levin, S.A. (2012). Trading-off fish biodiversity, food security, and hydropower in the Mekong River Basin. *PNAS*, 109, 5609–14.

3

Sampling Strategies and Protocols for Freshwater Ecology and Conservation

Leon A. Barmuta

Corresponding author: leon.barmuta@utas.edu.au

3.1 Introduction

Mention 'sampling' to a freshwater ecologist and we usually lunge for our sampling method of habit, be it a quadrat, transect, net, or electroshocker. Before we take any samples, it behoves us to step back and consider the questions our investigations seek to answer. It is rare for us to be able to census all the organisms or processes that we measure in wetlands or freshwaters: we are usually forced to work with incomplete information in the form of samples, which we hope are representative of the population, community, or process that we seek to measure and draw conclusions about. For this chapter, a *sample* consists of a number of *sampling units* which have been selected in some way from the parent *population*. It is important to note this difference from the vernacular use of 'sample' which refers to sample units, as in, for example, 'plankton sample': the individual net tow is simply one of many net tows (sample units) we have taken to estimate the abundance, diversity, or community composition of the target population (possibly for a single wetland in this example). This chapter considers the main issues that can affect the efficacy of sampling programmes in wetlands, both lentic and lotic, and draws attention to some of the less-used and recent developments in sampling regimes that show potential for wetlands and freshwaters.

3.1.1 Objectives, research questions, developing and testing hypotheses

Defining the objectives is crucial: it dictates when you need the information and what that should consist of (Green 1979; Downes et al. 2008; Anderson and Davis 2013a), but how should we elicit these objectives?

Many schemes have been proposed to describe how to conduct ecological research or monitoring. They vary in complexity from five (Cogălniceanu and Miaud 2009) to twelve (Morrison 2001) or even more steps (Ford 2000) when progressing from overall aim to drawing conclusions, and the process of planning and implementing a research or monitoring programme often involves iterations with stakeholders and researchers

Barmuta, L. A., *Sampling strategies and protocols for freshwater ecology and conservation.* In: *Freshwater Ecology and Conservation: Approaches and Techniques.* Edited by Jocelyne M. R. Hughes: Oxford University Press (2019).
© Oxford University Press 2019. DOI: 10.1093/oso/9780198766384.003.0003

to refine the objectives and study design (see Chapter 2). Sometimes this process is prescribed by an administrative body or procedure (Davis et al. 2001), especially when water quality is being assessed (e.g., Kuehne et al. 2017 briefly review the impact of the 1972 'Clean Water Act' in the USA). At the other extreme the initial aim may be vague and considerable effort needs to be expended to refine the questions and choose measurable, cost-effective, and relevant response variables (Barmuta et al. 2011; Anderson and Davis 2013a; b). All of these processes have sampling design at their heart.

Twenty-first-century investigations of wetlands have moved beyond the merely descriptive or inductive. Even exploratory studies aimed at generating hypotheses benefit from careful thought about the major processes that might generate the data collected. While the data 'should be allowed to speak', investigators should be able to articulate clearly why data need to be collected in a particular way, which additionally allows replicability of methods. Simply collecting data 'at random' or cleaving to a particular sampling design does not assure successful outcomes. It is better to identify several competing models or working hypotheses which will help target the variables to be measured and sampling designs to be used (Hilborn and Mangel 1997). Gotelli and Ellison (2004) briefly outline the different methods used to frame and test hypotheses in ecology, and both frequentist (Burnham and Anderson 2002) and Bayesian (Clark 2007) statisticians have argued strongly that ecologists need to articulate contrasting models *a priori* before collecting further data (Anderson 2008). This governs which variables are chosen and what sampling regimes are most appropriate.

As an example, Turner (1997) proposed four working hypotheses with contrasting predictions about the rates and patterns of loss of coastal wetlands in the northern Gulf of Mexico. The models were tested using historical maps of the size and extent of the wetlands, and the analysis supported dredging and formation of spoil banks as the most likely cause. That the subsequent debate queried this conclusion (Day et al. 2001; Turner 2001; Gosselink 2001) vindicates the strength of examining multiple working hypotheses: as knowledge improves, alternative models are proposed, and the variables and sampling regimes needed to test these models are identified.

However, that debate also serves to underscore the need to broaden stakeholder involvement to ensure that insights and information from a broader range of interests than those of a subset of wetland ecologists are captured. As well as developing a more comprehensive scientific basis for study design and data collection, broad-based stakeholder engagement at the planning stages of investigations increases the likelihood that the results of a sampling programme will be trusted and used (Turner et al. 2000; Barmuta et al. 2011; Chimner et al. 2017). Defining objectives in these ways will likely be iterative, and is potentially time-consuming, although there are tools and software (e.g., eWater's 'Concept' from https://toolkit.ewater.org.au) available to facilitate involvement from stakeholders with widely varying expertise. Burgman (2005) provides an excellent introduction to these processes, and many projects or investigations in wetlands and their catchments now explicitly incorporate stakeholder engagement in conservation planning and implementation (Ball 2008; Mfundisi 2008; Chapter 2 of this volume).

3.2 Sampling for ecology and sampling for conservation

Given the recent emphases on model-based inferences in ecology, there is probably very little difference between the sampling approaches used for 'pure' ecology and conservation. As a broad generalisation, conservation sampling typically addresses questions at broader spatial and temporal scales, and most management applications demand predictive rather than explanatory approaches (Mac Nally 2000). Nevertheless, the spatial and temporal scopes of the questions of interest need to be identified explicitly, and the choices of response variables will need to relate to these scales (Figure 3.1).

Accordingly, most real-world conservation questions need multiple lines of evidence (Linkov et al. 2009; Hope and Clarkson 2014; Chapter 2), and a (probably incomplete) survey of the lines of evidence that have been used in wetlands is presented in Table 3.1. Note that techniques normally associated with 'pure' ecology, such as field experiments and laboratory assays, can still be a component of conservation investigations because testing mechanisms that are thought to underlie the broader

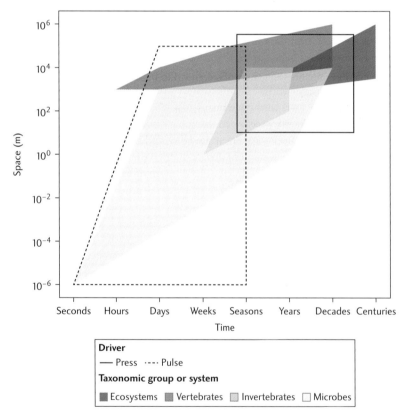

Figure 3.1 The spatial and temporal scopes over which different response variables and disturbance drivers can vary in wetlands. (Reproduced from Dafforn et al. (2016), with permission from CSIRO Publishing.)

Table 3.1 *Examples of lines of evidence that have been used in conservation-based studies in freshwater wetlands*

Line of evidence	Examples from freshwater conservation-based studies	Chapters in this book with examples of this line of evidence
Remotely sensed habitat and proxy data	Extent and intermittency of water bodies: where are the refuges during drought? Proxies of algal or macrophyte biomass: which wetlands appear eutrophic? Cover by floodplain and riparian vascular plants: have there been changes in vegetation cover over time?	4 *Water quantity and hydrology*; 9 *Plants, aquatic macrophytes and bryophytes*; 13 *Changes over time*; 9 *Remote sensing*.
Biodiversity surveys	Baseline and inventory: what's there? what's special? Conservation planning: what's conserved somewhere else already? Biosecurity: invasive or non-native species monitoring	14 *Secondary data*; 16 *Invasive aquatic species*; 17 *Freshwater ecosystem security and climate change*; 19 *Wetland landscapes and catchment management*.
Population surveys	Status surveys, including presence/absence and occupancy surveys and those using environmental DNA (eDNA): where are the rare, threatened or endangered species? Time series monitoring: which fish can be harvested? Is the population recovering? 'Place-based' surveys: is this migration route still being used?	9 *Plants, aquatic macrophytes and bryophytes*; 10 *Vertebrates*; 11 *Invertebrates*; 12 *Freshwater populations, interactions and networks*; 13 *Changes over time*; 9 *Remote sensing*; 19 *Wetland landscapes and catchment management*.
Field experiments and pseudo-experiments	Enclosures, exclosures, and mesocosms: what are the mechanisms? What are the effects of excluding a particular species? Restoration, habitat manipulation or construction: did the manipulation have the desired effect? Adaptive management interventions, including biomanipulation: has this intervention met the management goals? If not, how does management need to change to meet them?	12 *Freshwater populations, interactions and networks*; 16 *Invasive aquatic species*; 17 *Freshwater ecosystem security and climate change*; 18 *Restoration of freshwaters*.
Laboratory experiments and assays	Controlled laboratory and field experiments: what is the impact of an invasive species on the behaviour of native species? How does increasing a stressor affect an organism's physiology? Laboratory and field assays: which nutrients are limiting? Is this complex effluent toxic?	12 *Freshwater populations, interactions and networks*; 16 *Invasive aquatic species*; 17 *Freshwater ecosystem security and climate change*; 18 *Restoration of freshwaters*.

investigation can be important. For example, Stoffels et al. (2016) used laboratory tests of thermal performance for two species of freshwater crayfish, *Euastacus*, whose disjunct distributions within the same river have been attributed to water temperatures. They found that the species had identical aerobic scopes suggesting that this presumed mechanism was not important in explaining their well-documented spatial separation. Instead, differences in anaerobic locomotor performance were identified as a potential mechanism. Accordingly, attempting to predict range shifts in response to climate change based simply on records of the realised niche of organisms, even ectotherms, is fraught with difficulty unless important mechanisms that modify fitness in response to the thermal niche are better understood (Kearney et al. 2010).

The demand for multiple lines of evidence in conservation studies can mean that resources are spread more thinly than in classical 'pure' ecological studies. Accordingly, the taxonomic resolution for some types of conservation study or monitoring may need to be coarser, especially for 'difficult' taxa (e.g., phytoplankton, periphyton, many macroinvertebrate groups). Conversely, quantitative estimates of density may be overkill: semi-quantitative or qualitative data may suffice.

3.2.1 Classes of sampling types

Sampling programmes divide into four main types (Gilbert 1987) (Table 3.2), although there are many variants within probability and search sampling types (Thompson 2012). Probability-based sampling programmes are the most important class of sampling designs because sample units are selected or placed using methods that incorporate some form of randomisation. This ensures that the population being sampled is described explicitly and that every sample unit in the population can be sampled with known probability. As a result, the estimates derived from the sample

Table 3.2 *Classes of sampling type ranked from most to least desirable for drawing inferences*

Sampling type	Description	Disadvantages
Probability	Every sample unit in the population has a known probability of being selected in the sample. Randomisation of selection or placement of sample units is essential.	Randomisation may be difficult or expensive to implement. Investigation may be inherently model-based.
Search	Occurrence of response variable is known *a priori*. This (often historical information) is then used to select locations or times to sample.	Usually requires long-term data sets to establish basis for sample selection.
Judgement	Expert opinion or previous experience is used to inform the placement or selection of sample units.	Difficult to assess bias of expert opinion, or how the placement or selection of sample units can be used to extrapolate to population.
Haphazard or convenience	No probabilistic model used to select units. Frequently used because access, cost, ethical, or safety issues preclude probabilistic methods.	Extrapolation beyond the sampled units problematic.

(e.g., means, variances) are unbiased and the data can be used to infer properties about the population from which the sample is drawn using well-characterised statistical techniques.

An important feature of probability-based sampling is that analyses and conclusions drawn are *design based*: the validity of inferences is completely supported by the design and implementation of the sampling programme. The design and sampling protocol assures that any future reanalysis of the same data set will reach the same conclusion (Manly 2001; Theobald et al. 2007).

By contrast, *model-based* inference relies on explicit specification of the relationship between a sample and its population: it relies upon randomness that is inherent in the assumed model rather than the randomness of the sampling design (Manly 2001). The models can range from simple linear models (ANOVA or regression) to very complex models describing spatial or temporal processes (Dobbie et al. 2008). Opinions on its appropriateness range from pessimism ('real-world ecological systems are simply too complex to use model-based inference with much confidence': Theobald et al. 2007) to regarding them as inevitable, especially when faced with multiple working hypotheses (Manly 2001; Dobbie et al. 2008). A key disadvantage of model-based inference is that later investigators using the same data but different statistical models may reach different conclusions, and Manly (2001) describes two simple examples in which changes in the assumptions about the distribution used to model right-skewed data (the lognormal and gamma distributions) changes the estimated effects and conclusions reached. These problems are magnified as the complexity of the hypothesised processes increases, and professional statistical guidance is essential so that assumptions are clearly understood and advice given on critical aspects of the implementation of sampling (Dobbie et al. 2008).

Even if model-based inference is adopted at the outset, there is still a role for probability-based methods. Elements of the models being considered may require data where the sample units are selected or allocated 'at random' to ensure that models or hypotheses can be tested or compared validly (Manly 2001). There are, however, often logistical or other impediments that compromise this 'gold standard', and investigators need to be aware of the costs of using less formal sampling regimes such as professional judgement or haphazard sampling, both of which are probably very common in wetland studies (Haukos 2013) for a number of reasons (Table 3.2) (e.g., owners denying access, Hargiss and Dekeyser 2014). Such programmes are subjective and it is difficult to assess how representative the samples are of the populations. As a result, investigators are locked in to model-based inference, but nevertheless, the data collected may still be useful. For example Pierce et al. (2001) had to adopt sampling stations that had previously been established in the littoral zone of a lake to assess long-term temporal trends in fish populations; inclement weather prevented sampling some of the stations while still yielding useful information to quantify short-term variability. Judgement and haphazard sampling can also be useful for generating hypotheses or providing inputs to modelling studies (Haukos 2013).

Many texts on environmental assessment, ecology or wetland techniques introduce the principles behind simple random, stratified random, and systematic sampling

(Green 1979; Barnett 1997; Manly 2001; Quinn and Keough 2002; Gotelli and Ellison 2004; Downes et al. 2008; Anderson and Davis 2013a and c). Simple random sampling is, in theory, easy to implement, but it is also easy to introduce inadvertent bias in very small samples, especially if the population has strong underlying spatial or temporal structure (Green 1979; Hurlbert 1984). One response to these shortcomings has been to deploy systematic sampling designs to ensure more even representation over spatial or temporal patterns; however, substantial background information is needed to ensure that the chosen interval between sample units does not coincide with underlying periodic patterns in the population (Cochran 1977; Green 1979; Thompson 2012). A common recommendation that attempts to combine the strengths of simple random and systematic sampling is stratified random sampling (Green 1979; Manly 2001). If the population can be divided into well-defined strata that cover the population exhaustively and sample units within strata are more homogeneous within the strata than between them, then stratified random sampling can be much more efficient than either simple random or systematic sampling. Cochran (1977) and Thompson (2012) provide further details on the relative merits of allocating uneven numbers of sample units based on variance within strata or other criteria such as area or volume of each stratum, while Overton and Stehman (1996) and Manly (2001) describe examples of where stratification may become problematic, especially if strata are chosen after the data have been collected: so-called post-stratification (Overton and Stehman 1996).

There are many more types of probability-based sampling designs which are documented in comprehensive texts such as Cochran (1977) and Thompson (2012), while much recent progress has been made in cluster and adaptive sampling techniques, which can optimise cost-effectiveness (Thompson and Seber 1996), and strategies focussed on rare or cryptic species (Thompson 2004). All of these schemes seek to improve upon simple random sampling by increasing sampling efficiency and reducing sample costs. The next section of this chapter is a selective review of recent applications of these more advanced designs relevant to wetland investigations.

3.2.2 Logistic constraints

Practical issues can affect the planning or implementation of any sampling programme. Safety issues may prevent access to some habitats (e.g., soft muds and quick-sands, fast currents, waves), while hazardous fauna (e.g., hippopotamus, crocodiles) can also compromise sampling feasibility. Avoiding these hazards can change the population frame from which samples can be taken, and so inferences will need to be carefully qualified. Biosecurity issues (e.g., pathogens, invasive species) can also affect access, although hygiene protocols are often developed to mitigate these risks (e.g., for amphibians, see Dodd 2010).

Access to selected sites can be restricted because of social or political reasons. Wardrop et al. (2007) were denied access to 'a sizeable number of sites' in their survey of wetland condition, but they were able, however, to post-stratify their results based on wetland ownership since this was a strong correlate of non-response rates.

3.3 Habitat-centric strategies

Crudely we can distinguish between lentic (static) freshwaters, which typically mani-fest in the landscape as approximately 'round' habitats (e.g., ponds, lakes, marshes, peatlands) and running waters, which provide approximately linear, usually dendritic landscape features (e.g., perennial and ephemeral streams). Each presents challenges and opportunities for efficient sampling designs.

The conventional literature on probability-based surveys often assumes that an exhaustive, finite list of the sample members can be made from which randomised selection can be made. Many applications to regional lentic wetland surveys or sampling large bodies of water, however, have a potentially infinite population frame (i.e., coordinates of a map), but this presents opportunities to increase the efficiency of sampling by spatially balancing the allocation or selection of sampling units (Stevens and Jensen 2007). In these applications, space serves as an auxiliary variable because wetlands closer to each other tend to be more similar (i.e., are positively autocorrelated in space) than remote wetlands, and redundant information can be minimised by spatially balancing sample units across the population frame. If the population frame is a large body of water, then a spatially balanced design converges to a regular two-dimensional grid; for numerous, smaller wetlands, spatial balance can achieved via several strategies based on two-stage sampling and stratification (Stevens and Jensen 2007; Theobald et al. 2007). For example, Wardrop et al. (2007) used a Generalised Random Tessellation Stratified (GRTS) design to select wetlands for their sample. Because it is unknown *a priori* whether a given set of coordinates will be in the target population, this and related techniques allows sample sizes to be dynamically adjusted while still maintaining spatial balance (Stevens and Olsen 2004; Theobald et al. 2007) from which they could estimate wetland condition. GRTS has also been applied to sampling stream segments in south-eastern Brazil (Jiménez-Valencia et al. 2014) in which inclusion probabilities were assigned depending on stream size. In this instance, a number of factors meant that a given stream segment may be unavailable (e.g., inaccessibility; some streams were dry) so the GRTS allowed 'oversampling' from the GIS database so that extra sites were available to replace unavailable segments.

Sometimes there are extensive historical records, secondary databases and previous surveys that can be assimilated to address conservation questions and guide future monitoring and research (see Chapter 14). For example, migratory waterbirds are often used in surveys for conservation planning across regional and international scales. Xia et al. (2017) document how information from designed surveys carried out under different programmes can be combined with extensive citizen science data sets on waterbirds to identify conservation priorities in coastal wetlands in China, and they provide recommendations on how to address knowledge gaps in future surveys.

Recently, a number of techniques focussing on the network properties of lotic habitats has produced a flurry of novel, if challenging, opportunities for sampling designs (Peterson et al. 2013; Som et al. 2014; Saunders et al. 2016; Sály and Erős 2016) as well as spatially explicit models of population and meta population processes and community structure (Altermatt 2013; Mari et al. 2014; Lois et al. 2015; Paz-Vinas and Blanchet 2015).

As with more conventional geospatial designs, the data sets needed can be dauntingly large. McGuire et al. (2014) used over 600 water sample units collected every 100 m on 32 tributaries of a fifth order stream to separate the effects of fine-scale versus broad-scale processes and in-stream versus landscape controls on stream water chemistry. Optimising sampling programmes in spatially explicit dendritic networks also remains challenging (Sály and Erős 2016). Nevertheless, empirical studies show promise: Brennan et al. (2016) found dendritic modelling improved their ability to partition the influence of hydrologic transport from local geologic features on the biogeochemistry of a large river in Alaska; Frieden et al. (2014) found that autocovariance functions derived from networks generally increased the accuracy of predictions of macroinvertebrate diversity; and Isaak et al. (2017) demonstrate the use of these techniques to scale up point estimates of trout density from 108 sites to a 735 km river network in Canada.

3.4 Organism-centric strategies

Some sampling strategies focus on individual organisms or clusters of the target organism. Spatial point-pattern analysis (SPPA) has a long and productive history in ecology when the focus is on the spatial arrangement of individuals in a population (Velázquez et al. 2016), and the analytical techniques can provide powerful inferences about processes when combined with *a priori* hypotheses and ecological theory or knowledge (McIntire and Fajardo 2009). While SPPA has been used extensively for terrestrial plants, applications to freshwater systems remain scarce. Olden et al. (2001) regarded lakes as 'points' and then used Procrustes analysis to discriminate between different models of fish dispersal between the lakes, and Maheu-Giroux and de Blois (2007) used SPPA to determine the best predictors of the occurrence of the invasive *Phragmites australis* in networks of linear wetlands in Quebec. Advances in high-resolution remote sensing (e.g., Zhao et al. 2016) will likely make mapping of individuals of static organisms (e.g. wetland trees) more tractable. This should stimulate the use of these techniques, especially when combined with recent but under-used advances in SPPA (Velázquez et al. 2016). Li et al. (2015) illustrate the potential by adapting SPPA methods to detect clusters of mosquito larval sites when linear features such as riparian corridors influence the shape of such clusters.

Frequently, wetland organisms are strongly clustered in space (or time), and sampled data is often 'zero-rich'—a situation shared with many other ecological disciplines (Baeten et al. 2014; Warton et al. 2015)! A seemingly attractive route to increasing sampling efficiency for such species is that of adaptive sampling techniques (Thompson and Seber 1996). The basic structure of such programmes is that a conventional probability-based method is used to select sample units initially. If units in this initial sample meet some condition (e.g., the threshold abundance of a species is reached), then additional units are selected in the neighbourhood of these units until no further sample units in the neighbourhood meet the condition. At the end of the sampling process, there will be a number of variably-sized clusters of sample units, and well-developed theory provides the means to use the sample to estimate means and other quantities of interest (Smith et al. 2004).

While the approach is intuitively appealing to biologists, especially for rare or clustered species (i.e., we expend more effort in places where the species is likely to be found), Smith et al. (2004) caution that practicalities may limit its application. They identify four potentially problematic issues. First, increased efficiency is not guaranteed because it depends strongly on spatial distribution; Goldberg et al. (2007) found that adaptive sampling was no more efficient than simple random sampling in one study of rare benthic algae. Second, the final sample size is random and not known *a priori*, and so sampling costs could become prohibitive unless some form of stopping rule can be developed (Gattone and Di Battista 2011). Third, the results of the initial probability-based sample must be available in real time while the sampling is happening so that the additional quadrats or transects can be sampled. Sample units that require investigators to return to the laboratory to identify the organisms (e.g., plankton tows, benthic samples of algae or macroinvertebrates) are clearly not going to meet this criterion. Finally, mobile or sensitive species may need extensive modifications of adaptive sampling regimes. If detection of the species in the first sample is uncertain, then the adaptive clusters will not be taken rendering the estimate unreliable. Conversely, because adaptive sampling tends to target occupied habitat, disturbance to sensitive species may compromise its viability.

Mobile (or vagile) organisms have a wide variety of sampling procedures to estimate abundance ranging from distance-based (Buckland et al. 2016) to capture-mark-recapture methods (Pollock et al. 2002). Crucial to their validity is being able to estimate the detection probability of the target organism (Pollock et al. 2002; Kellner and Swihart 2014), and there is an extensive literature on these techniques, especially focused on individual species. More recently, wildlife programmes have become more focussed on multi-site and regional-scale investigations, with increasing emphasis on model-based inference (Buckland et al. 2016; Sollmann et al. 2016). Concurrently, there have been rapid improvements in technologies such as automated remote cameras, sound recordings, use of animal signs, and genetic techniques which make multiple samplings over short periods of time feasible, and increase the opportunities to estimate detectability reliably (Noon et al. 2012).

As a result, occupancy modelling (MacKenzie 2006) has become popular for broad-scale investigations. Instead of estimating abundance, the focus is on estimating the proportion of sample units that support (are 'occupied by') a given species or set of species (Noon et al. 2012). Mackenzie and Royle (2005) describe practical guidelines for designing occupancy surveys. Groff et al. (2017) provide a recent example of occupancy modelling for a salamander and a frog in montane ponds in Maine, USA. They concluded that including ponds associated with beaver activity and with proportionally large areas of shallow water (<1 m) would improve conservation outcomes for these species, and suggested that extrapolation of habitat information from outside their region was suboptimal. These occupancy methods may not suit all organisms, especially when the costs of repeat sampling visits to sites are high (e.g., fish). Gwinn et al. (2016) review the opportunities to estimate detection probabilities for fish using techniques other than occupancy modelling, and such techniques include well-tested methods such as capture-mark-recapture (CMR) and depletion sampling.

3.5 Temporal considerations

As with spatial patterns, attempts to infer trends in response variables over time need to identify diel, seasonal and other long-term cycles in behaviour that need to be accounted for in the sampling design. Temporal autocorrelation between sample times is to be expected, and designs need to ensure sufficient intensity of sampling to properly characterise the correlations (Overton and Stehman 1996).

The temporal behaviour of the organisms themselves and the conservation question also need to be considered in the design. These issues are highlighted by Goldberg et al. (2016) in their review of environmental DNA (eDNA) of aquatic species. This technique is potentially very useful for early detection of invasive species, as exemplified for *Oronectes* crayfish in the USA (Dougherty et al. 2016), whereas its use for monitoring declining species is more problematic because of difficulties with inferring abundance from eDNA data and the possibility of DNA fragments persisting after the species has become locally extinct. Moreover, changes in the activity patterns and life history of the organisms can affect the supply of eDNA, while transport processes and differential preservation between habitats can all affect detectability (Goldberg et al. 2016.).

Sampling over time also implies that there is a spatial rationale to the design (Green 1979), especially if the goal is to characterise long-term trends across regional or larger spatial scales (Overton and Stehman 1996). Simply revisiting exactly the same sites may result in biased estimates of regional trends (McDonald 2003); to mitigate these biases, Dobbie et al. (2008) introduce the basic principles of rotating panel designs where a different set of sample sites are visited on each occasion. These are more comprehensively explained by Fuller (1999) and McDonald (2003).

3.6 Conclusion

Sampling regimes in freshwaters depend on the spatial and temporal scope of the variables we seek to measure and, critically, on the questions or aims set by the investigation. Inferences reached solely from probability-based sampling designs are rare, but some examples relevant to regional or larger-scale surveys were outlined above. Such surveys are often oriented to assessing condition, in space, time, or both and considerable progress has been made in adapting these designs for both lentic and lotic wetlands. Model-based inferences are generally more common, and are becoming increasingly complex as remotely sensed and other sources of 'big data' reduce the costs of data acquisition. Simple sampling 'recipies' are no longer adequate, especially if larger spatial and temporal scales are the focus of the investigation. Involving professional statisticians to guide us through the sampling and analytical options available to us is, therefore, increasingly critical.

References

Altermatt, F. (2013). Diversity in riverine metacommunities: A network perspective. *Aquatic Ecology*, 47, 365–77.

Anderson, D. (2008). Model Based Inference in the Life Sciences: A Primer on Evidence. Springer, New York and London.

Anderson, J.T. and Davis, C.A. (eds) (2013a). Wetland Techniques. Volume 1. Foundations. Springer, Dordrecht, Netherlands.

Anderson, J.T. and Davis, C.A. (eds) (2013b). Wetland Techniques. Volume 3. Applications and Management. Springer, Dordrecht, Netherlands.

Anderson, J.T. and Davis, C.A. (eds) (2013c). Wetland Techniques. Volume 2. Organisms. Springer Dordrecht, Netherlands.

Baeten, L., Warton, D.I., Van Calster, H., et al. (2014). A model-based approach to studying changes in compositional heterogeneity. *Methods in Ecology and Evolution*, 5, 156–64.

Ball, T. (2008). Management approaches to floodplain restoration and stakeholder engagement in the UK: A survey. *Ecohydrology and Hydrobiology*, 8, 273–80.

Barmuta, L.A., Linke, S., and Turak, E. (2011). Bridging the gap between 'planning' and 'doing' for biodiversity conservation in freshwaters. *Freshwater Biology*, 56, 180–95.

Barnett, V. (1997). Statistical analyses of pollution problems. In V. Barnett and K.F. Turkman (eds) Statistics for the Environment 3. Pollution Assessment and Control, 3–41. John Wiley and Sons, Chichester.

Brennan, S.R., Torgersen, C.E., Hollenbeck, J.P., Fernandez, D.P., Jensen, C.K., and Schindler, D.E. (2016). Dendritic network models: Improving isoscapes and quantifying influence of landscape and in-stream processes on strontium isotopes in rivers. *Geophysical Research Letters*, 43, 5043–51.

Buckland, S.T., Oedekoven, C.S., and Borchers, D.L. (2016). Model-Based Distance Sampling. *Journal of Agricultural, Biological, and Environmental Statistics*, 21, 58–75.

Burgman, M.A. (2005). Risks and Decisions for Conservation and Environmental Management. Cambridge University Press, Cambridge.

Burnham, K.P. and Anderson, D.R. (2002). Model Selection and Multimodel Inference. A Practical Information-Theoretic Approach. Springer-Verlag, New York.

Chimner, R.A., Cooper, D.J., Wurster, F.C., and Rochefort, L. (2017). An overview of peatland restoration in North America: where are we after 25 years? *Restoration Ecology*, 25, 283–92.

Clark, J.S. (2007). Models for Ecological Data: An Introduction. Princeton University Press, Princeton.

Cochran, W.G. (1977). Sampling Techniques, 3rd edn. John Wiley and Sons, New York, NY, USA.

Cogălniceanu, D. and Miaud, C. (2009). Setting objectives in field studies. In C.K. DoddJr (ed.) Amphibian Ecology and Conservation: A Handbook of Techniques, 21–35, Oxford University Press, Oxford.

Dafforn, K.A., Johnston, E.L., Ferguson, A., Humphrey, C.L., et al. (2016). Big data opportunities and challenges for assessing multiple stressors across scales in aquatic ecosystems. *Marine and Freshwater Research*, 67, 393–413.

Davis, J.C., Minshall, G.W., Robinson, C.T., and Landres, P. (2001). Wilderness Stream Ecosystems. Gen. Tech. Rep. RMRS-GTR-70, U.S. Department of Agriculture, Forest Service, Rocky Mountain Research Station, Ogden, UT, USA.

Day J.W., J., Shaffer, G.P., Reed, D.J., Cahoon, D.R., Britsch, L.D., and Hawes, S.R. (2001). Patterns and processes of wetland loss in coastal Louisiana are complex: A reply to Turner 2001. Estimating the indirect effects of hydrologic change on wetland loss: If the earth is curved, then how would we know it? *Estuaries*, 24, 647–57.

Dobbie, M.J., Henderson, B.L., and Stevens, Jr, D.L. (2008). Sparse sampling: Spatial design for monitoring stream networks. *Statistics Surveys*, 2, 113–53.

Dodd, C.K. (ed.) (2010). Amphibian Ecology and Conservation—A Handbook of Techniques. Techniques in Ecology and Conservation Series. Oxford University Press, Oxford, UK.

Dougherty, M.M., Larson, E.R., Renshaw, M.A., Gantz, C.A., Egan, S.P., Erickson, D.M., and Lodge, D.M. (2016). Environmental DNA (eDNA) detects the invasive rusty crayfish *Orconectes rusticus* at low abundances. *Journal of Applied Ecology*, 53, 722–32.

Downes, B.J., Barmuta, L.A., Fairweather, P.G., et al. (2008). Monitoring Ecological Impacts: Concepts and Practice in Flowing Waters. Cambridge University Press, Cambridge.

Ford, E.D. (2000). Scientific Method for Ecological Research. Cambridge University Press, Cambridge.

Frieden, J.C., Peterson, E.E., Angus Webb, J., and Negus, P.M. (2014). Improving the predictive power of spatial statistical models of stream macroinvertebrates using weighted autocovariance functions. *Environmental Modelling and Software*, 60, 320–30.

Fuller, W.A. (1999). Environmental surveys over time. *Journal of Agricultural, Biological, and Environmental Statistics*, 4, 331–45.

Gattone, S.A. and Di Battista, T. (2011). Adaptive cluster sampling with a data driven stopping rule. *Statistical Methods and Applications*, 20, 1–21.

Gilbert, R.O. (1987). Statistical Methods for Environmental Pollution Monitoring. Van Nostrand Reinhold Company, New York.

Goldberg, N.A., Heine, J.N., and Brown, J.A. (2007). The application of adaptive cluster sampling for rare subtidal macroalgae. *Marine Biology*, 151, 1343–8.

Goldberg, C.S., Turner, C.R., Deiner, K., et al. (2016). Critical considerations for the application of environmental DNA methods to detect aquatic species. *Methods in Ecology and Evolution*, 7, 1299–307.

Gosselink, J.G. (2001). Comments on 'Wetland Loss in the Northern Gulf of Mexico: Multiple Working Hypotheses.' By R.E. Turner. 1997. Estuaries 20:1–13. *Estuaries*, 24, 636.

Gotelli, N.J. and Ellison, A.M. (2004). A Primer of Ecological Statistics. Sinauer Associates Inc. Publishers, Sunderland, Mass.

Green, R.H. (1979). Sampling Design and Statistical Methods for Environmental Biologists. John Wiley and Sons, New York, NY, USA.

Groff, L.A., Loftin, C.S., and Calhoun, A.J.K. (2017). Predictors of breeding site occupancy by amphibians in montane landscapes: Wetland Occupancy by Pool-Breeding Amphibians. *The Journal of Wildlife Management*, 81, 269–78.

Gwinn, D.C., Beesley, L.S., Close, P., Gawne, B., and Davies, P.M. (2016). Imperfect detection and the determination of environmental flows for fish: challenges, implications and solutions. *Freshwater Biology*, 61, 172–80.

Hargiss, C.L.M. and Dekeyser, E.S. (2014). The challenges of conducting environmental research on privately owned land. Environmental Monitoring and Assessment, 186, 979–85.

Haukos, D.A. (2013). Study design and logistics. In J.T. Anderson and C.A. Davis (eds) Wetland Techniques. Volume 1. Foundations, 1–47. Springer Netherlands, Dordrecht.

Hilborn, R. and Mangel, M. (1997). *The Ecological Detective: Confronting Models with Data*. Princeton University Press, Princeton, NJ, USA.

Hope, B.K. and Clarkson, J.R. (2014). A Strategy for Using Weight-of-Evidence Methods in Ecological Risk Assessments. *Human and Ecological Risk Assessment: An International Journal*, 20, 290–315.

Hurlbert, S.H. (1984). Pseudoreplication and the design of ecological field experiments. *Ecological Monographs*, 54, 187–211.

Isaak, D.J., Ver Hoef, J.M., Peterson, E.E., Horan, D.L., and Nagel, D.E. (2017). Scalable population estimates using spatial-stream-network (SSN) models, fish density surveys, and national geospatial database frameworks for streams. *Canadian Journal of Fisheries and Aquatic Sciences*, 74, 147–56.

Jiménez-Valencia, J., Kaufmann, P.R., Sattamini, A., Mugnai, R., and Baptista, D.F. (2014). Assessing the ecological condition of streams in a southeastern Brazilian basin using a probabilistic monitoring design. Environmental Monitoring and Assessment, 186, 4685–95.

Kearney, M., Simpson, S.J., Raubenheimer, D., and Helmuth, B. (2010). Modelling the ecological niche from functional traits. *Philosophical Transactions of the Royal Society B: Biological Sciences*, 365, 3469–83.

Kellner, K.F. and Swihart, R.K. (2014). Accounting for imperfect detection in ecology: A quantitative review. *PLoS ONE*, 9, e111436.

Kuehne, L.M., Olden, J.D., Strecker, A.L., Lawler, J.J., and Theobald, D.M. (2017). Past, present, and future of ecological integrity assessment for fresh waters. *Frontiers in Ecology and the Environment*, 15, 197–205.

Li, L., Bian, L., Rogerson, P., and Yan, G. (2015). Point pattern analysis for clusters influenced by linear features: an application for mosquito larval sites: point pattern analysis for clusters influenced by linear features. *Transactions in GIS*, 19, 835–47.

Linkov, I., Loney, D., Cormier, S., Satterstrom, F.K., and Bridges, T. (2009). Weight-of-evidence evaluation in environmental assessment: Review of qualitative and quantitative approaches. *Science of the Total Environment*, 407, 5199–205.

Lois, S., Cowley, D.E., Outeiro, A., San Miguel, E., Amaro, R., and Ondina, P. (2015). Spatial extent of biotic interactions affects species distribution and abundance in river networks: The freshwater pearl mussel and its hosts. *Journal of Biogeography*, 42, 229–40.

Mac Nally, R. (2000). Regression and model-building in conservation biology, biogeography and ecology: The distinction between and reconciliation of 'predictive' and 'explanatory' models. *Biodiversity and Conservation*, 9, 655–71.

MacKenzie, D.I. (2006). Occupancy Estimation and Modeling: Inferring Patterns and Dynamics of Species Occurrence. Elsevier, Burlington, MA.

Mackenzie, D.I. and Royle, J.A. (2005). Designing occupancy studies: General advice and allocating survey effort. *Journal of Applied Ecology*, 42, 1105–14.

Maheu-Giroux, M. and de Blois, S. (2007). Landscape ecology of *Phragmites australis* invasion in networks of linear wetlands. *Landscape Ecology*, 22, 285–301.

Manly, B.F.J. (2001). Chapman and Hall/CRC Press LLC, Boca Raton, Florida, USA.

Mari, L., Casagrandi, R., Bertuzzo, E., Rinaldo, A., and Gatto, M. (2014). Statistics for Environmental Science and Management Metapopulation persistence and species spread in river networks. *Ecology Letters*, 17, 426–34.

McDonald, T.L. (2003). Review of environmental monitoring methods: survey designs. *Environmental monitoring and assessment*, 85, 277–92.

McGuire, K.J., Torgersen, C.E., Likens, G.E., Buso, D.C., Lowe, W.H., and Bailey, S.W. (2014). Network analysis reveals multiscale controls on streamwater chemistry. *Proceedings of the National Academy of Sciences of the United States of America*, 111, 7030–5.

McIntire, E.J. and Fajardo, A. (2009). Beyond description: the active and effective way to infer processes from spatial patterns. *Ecology*, 90, 46–56.

Mfundisi, K.B. (2008). Overview of an integrated management plan for the Okavango Delta Ramsar site, Botswana. *Wetlands*, 28, 538–43.

Morrison, M.L. (2001). Wildlife Study Design. Springer, New York.

Noon, B.R., Bailey, L.L., Sisk, T.D., and Mckelvey, K.S. (2012). Efficient Species-Level Monitoring at the Landscape Scale. *Conservation Biology*, 26, 432–41.

Olden, J.D., Jackson, D.A., and Peres-Neto, P.R. (2001). Spatial isolation and fish communities in drainage lakes. *Oecologia*, 127, 572–85.

Overton, W.S. and Stehman, S.V. (1996). Desirable design characteristics for long-term monitoring of ecological variables. *Environmental and Ecological Statistics*, 3, 349–61.

Paz-Vinas, I. and Blanchet, S. (2015). Dendritic connectivity shapes spatial patterns of genetic diversity: A simulation-based study. *Journal of Evolutionary Biology*, 28, 986–94.

Peterson, E.E., Ver Hoef, J.M., Isaak, D.J., et al. (2013). Modelling dendritic ecological networks in space: An integrated network perspective. *Ecology Letters*, 16, 707–19.

Pierce, C.L., Sexton, M.D., Pelham, M.E., and Larscheid, J.G. (2001). Short-term variability and long-term change in the composition of the littoral zone fish community in Spirit Lake. *American Midland Naturalist*, 146, 290–9.

Pollock, K.H., Nichols, J.D., Simons, T.R., Farnsworth, G.L., Bailey, L.L., and Sauer, J.R. (2002). Large-scale wildlife monitoring studies: statistical methods for design and analysis. *Environmetrics*, 13, 105–19.

Quinn, G.P. and Keough, M.J. (2002). Experimental Design and Data Analysis for Biologists. Cambridge University Press, Cambridge.

Sály, P. and Erős, T. (2016). Effect of field sampling design on variation partitioning in a dendritic stream network. *Ecological Complexity*, 28, 187–99.

Saunders, M.I., Brown, C.J., Foley, M.M., et al. D.D. (2016). Human impacts on connectivity in marine and freshwater ecosystems assessed using graph theory: A review. *Marine and Freshwater Research*, 67, 277–90.

Smith, D.R., Brown, J.A., and Lo, N.C.H. (2004). Applications of adaptive sampling to biological populations. In W.L. Thompson (ed.) Sampling rare or elusive species: concepts, designs, and techniques for estimating population parameters, 77–122. Island Press, Washington.

Sollmann, R., Gardner, B., Williams, K.A., Gilbert, A.T., and Veit, R.R. (2016). A hierarchical distance sampling model to estimate abundance and covariate associations of species and communities. *Methods in Ecology and Evolution*, 7, 529–37.

Som, N.A., Monestiez, P., Ver Hoef, J.M., Zimmerman, D.L., and Peterson, E.E. (2014). Spatial sampling on streams: Principles for inference on aquatic networks. *Environmetrics*, 25, 306–23.

Stevens, D.L. and Jensen, S.F. (2007). Sample design, execution, and analysis for wetland assessment. *Wetlands*, 27, 515–23.

Stevens, D.L. and Olsen, A.R. (2004). Spatially Balanced Sampling of Natural Resources. *Journal of the American Statistical Association*, 99, 262–78.

Stoffels, R.J., Richardson, A.J., Vogel, M.T., Coates, S.P., and Müller, W.J. (2016). What do metabolic rates tell us about thermal niches? Mechanisms driving crayfish distributions along an altitudinal gradient. *Oecologia*, 180, 45–54.

Theobald, D.M., Stevens, D.L., White, D., Urquhart, N.S., Olsen, A.R., and Norman, J.B. (2007). Using GIS to generate spatially balanced random survey designs for natural resource applications. *Environmental Management*, 40, 134–46.

Thompson, W.L. (2004). Sampling Rare or Elusive Species: Concepts, Designs, and Techniques for Estimating Population Parameters. Island Press, Washington.

Thompson, S.K. (2012). Sampling, 3rd edn. Wiley, Hoboken, NJ.

Thompson, S.K. and Seber, G.A.F. (1996). Adaptive Sampling. New York: Wiley, 1996.

Turner, R.E. (1997). Wetland Loss in the Northern Gulf of Mexico: multiple working hypotheses. *Estuaries*, 20, 1–13.

Turner, R.E. (2001). Estimating the indirect effects of hydrologic change on wetland loss: If the earth is curved, then how would we know it? *Estuaries*, 24, 639–46.

Turner, R., van den Bergh, J.C.J.M., Söderqvist, T., Barendregt, A., van der Straaten, J., Maltby, E., and van Ierland, E.C. (2000). Ecological-economic analysis of wetlands: scientific integration for management and policy. *Ecological Economics*, 35, 7–23.

Velázquez, E., Martínez, I., Getzin, S., Moloney, K.A., and Wiegand, T. (2016). An evaluation of the state of spatial point pattern analysis in ecology. *Ecography*, 39, 1042–55.

Wardrop, D.H., Kentula, M.E., Jensen, S.F., Stevens Jr, D.L., Hychka, K.C., and Brooks, R.P. (2007). Assessment of wetlands in the Upper Juniata watershed in Pennsylvania, USA using the hydrogeomorphic approach. *Wetlands*, 27, 432–45.

Warton, D.I., Foster, S.D., De'ath, G., Stoklosa, J., and Dunstan, P.K. (2015). Model-based thinking for community ecology. *Plant Ecology*, 216, 669–82.

Xia, S., Yu, X., Millington, S., Liu, Y., Jia, Y., Wang, L., Hou, X., and Jiang, L. (2017). Identifying priority sites and gaps for the conservation of migratory waterbirds in China's coastal wetlands. *Biological Conservation*, 210, 72–82.

Zhao, T.-G., Yu, R.-H., Zhang, Z.-L., Bai, X.-S., and Zeng, Q.-A. (2016). Estimation of wetland vegetation aboveground biomass based on remote sensing data: A review. *Chinese Journal of Ecology*, 35, 1936–46.

Part II

Measuring the Component Parts

4

Water Quantity and Hydrology

Matthew McCartney

Corresponding author: m.mccartney@cgiar.org

4.1 Introduction

Water is, by definition, the key influence on the ecology of freshwater ecosystems. The source of water, discharge (i.e., volume and velocity), as well as turnover and residence times, all affect which organisms can live in different freshwater habitats and are key determinants of freshwater ecosystem structure and function. For example, fluxes of water, sediment, and nutrients between floodplains and rivers are key factors influencing river and floodplain ecology in many rivers.

Conversely, ecological processes in freshwater ecosystems can significantly influence the hydrological cycle. Typically located in the lowest part of landscapes, where both surface water and groundwater collect, they integrate outputs across all watershed scale processes, altering both water movement and water quality.

Freshwater ecosystems are naturally dynamic with water fluxes changing over time. The movement, distribution, and quality of water influence species richness, productivity, rates of organic matter accumulation, and nutrient cycling. Hydrology may restrict species richness and/or productivity in some places and enhance it in others. For example, both are typically lower in permanently flooded, stagnant wetlands, than in slow-flowing or seasonally flooded wetlands (Conner and Day 1982).

Changes to the volume and timing of freshwater flows, largely a consequence of economic development and increasing human demands, are a leading driver of global declines in freshwater biodiversity (World Wide Fund for Nature 2016) and are likely to be exacerbated by climate change (World Bank 2010). Knowledge of water fluxes and hydrological processes is essential to understand the current ecology, and the possible causes of both short- and long-term changes to freshwater ecosystems.

This chapter provides a brief introduction to hydrological monitoring for freshwater ecology and conservation. The next section deals with hydrological monitoring; the measurement of different hydrological variables. The following section briefly describes some ways in which hydrological data collected from monitoring networks can be used. Finally, a short case study and conclusions are presented.

McCartney, M., *Water quantity and hydrology*. In: *Freshwater Ecology and Conservation: Approaches and Techniques*. Edited by Jocelyne M. R. Hughes: Oxford University Press (2019). © Oxford University Press 2019. DOI: 10.1093/oso/9780198766384.003.0004

4.2 Hydrological monitoring

Freshwater ecosystems are adapted to the natural variability of water fluxes. Changes not only in magnitude but also in timing (e.g., seasonal patterns of precipitation or flow) can result in changes to freshwater habitat with consequences for productivity and/or biodiversity. Hydrological monitoring is required to understand basic physical processes, how water fluxes change both over the short-term (e.g., hours, days, seasons) and the long-term (e.g., years, decades), and the relationships between water and the biotic components of the ecosystems.

4.2.1 Data collection

The water budget

One of the basic scientific tenets of hydrology—and key to hydrological monitoring—is conservation of mass. This means that the inflows to any area over any period of time must equal the outflows plus any change in water storage. In general terms:

$$R + I = Eta + Q + \delta S + \delta G \qquad \text{(Equation 4.1)}$$

Where R is rainfall, I is inflow, Eta is actual evapotranspiration, Q is outflow, and δS and δG are changes in soil moisture and groundwater storage respectively. This may be applied to a whole basin, a sub-catchment, or may be adapted to describe the water balance of a subcomponent; for example, a river reach, a lake, a groundwater aquifer, or a wetland (Sutcliffe 2004). Each of the components of the water budget can be measured and changes over time determined.

Precipitation

Precipitation (rainfall, snow, and hail) is the simplest term of the water balance to measure. There are many different designs of rain gauge, but the basic principle is to 'catch', collect, and measure the depth of rain drops falling over a known area—normally a small circle of known diameter (typically 127 mm)—and collected through a funnel and collecting cylinder (Figure 4.1). So called rim and funnel or non-recording rain gauges are by far the most common, usually emptied and measured manually, at a specified time, every 24 hours, to provide an estimate of daily rainfall. More sophisticated rain gauges include automatic tipping buckets which comprise a see-saw like arrangement that tips when a pre-determined amount (typically 0.2 mm or 0.5 mm) of rain falls. These record not only the depth of rainfall but also when it fell. They can be used to estimate rainfall intensity (i.e., mm/hr) which can be useful to understand flow-generating processes.

One of the most important factors affecting rain gauge performance is exposure to wind. Studies have shown that for a given design the higher the rim above the ground the lower the catch efficiency (Rodda and Dixon 2012). For this reason some gauges are installed in pits with the rim at ground level, surrounded by an open grid or lattice to prevent rain drops splashing back into the gauge from the ground. However, relatively few of the worlds approximately 200,000 gauges are sited in this way (Rodda and Dixon 2012).

Figure 4.1 Rim and funnel rain gauge installed in Ethiopia (photo Matthew McCartney).

With a network of gauges spread over a geographic area and across the range of altitude, it is possible to determine the distribution of rainfall over the area (see Section 4.3.2). The density of rain gauges that should be included in a network depends on the types of rainfall likely to be encountered (i.e., frontal or convective) and the topography (i.e., position and elevation): more gauges are needed to give an adequate sample in mountainous areas (Putthividhya and Tanaka 2012). Despite being particularly important as water sources, mountainous areas tend to be under represented by rain gauges because of remoteness and difficulty of access.

Currently there are no international standards for rain gauges but the World Meteorological Organization (WMO) has conducted a number of studies comparing different types of gauges and observing methods (Vuerich et al. 2009) and has a guide to meteorological instruments and methods of observation (WMO 2012). In the UK, the national standard is BS 7843-3:2012 'Acquisition and management of meteorological precipitation from a gauge network' which provides a code of practice for the design and manufacture of storage and automatic rain gauges. Similarly standard BS 7843-4:2012 'Acquisition and management of meteorological precipitation data from a rain gauge network' provides guidance on the estimation of areal rainfall.

Rim and funnel rain gauges will collect hail, sleet, and snow, provided that the catch melts and runs into the gauge fast enough that the collector funnel doesn't overflow. The measured catch is then the liquid 'water equivalent' of the precipitation. It is important to note that snow is effectively water held in storage, and snowmelt is the water that enters the water budget. Snow depth is typically measured with a snow board, positioned carefully to avoid windblown snow accumulation, and read manually every 24 hours. Snowfall is then the amount of snow that has fallen since the previous observation (NWS 2013). The water content of a snow pack (i.e., total accumulated snow) can be determined by taking a core, melting, and measuring the depth of water in a standard rain gauge cylinder (NWS 2013).

Over the past 40 years, and with ever increasing sophistication, weather radars have been used to measure precipitation using pulses of microwave radiation. The amount of energy that bounces back from raindrops, ice particles, and snowflakes in clouds can be interpreted to provide estimates of precipitation over the area targeted by the radar. Networks of radars can be used to provide composite views covering large areas. Satellite-derived rainfall estimates use passive microwave sensors, in combination with infrared instruments aboard weather satellites, to indirectly measure rainfall based on the temperature of cloud tops (Mangadi et al. 2006). Both radar and satellite estimates must be calibrated with on-the-ground rain gauge data.

Measurements of precipitation should always follow guidance from national meteorological agencies since these will be based on best practice for the conditions experienced in that country.

Flow

Flow or discharge dominates the ecology of rivers (lotic habitat). In lakes and other wetlands (lentic habitat) flow may not be so directly important, but water level and how it changes can be critical and will be dependent on the water flowing in and out of the ecosystem. Records from gauging stations on rivers indicate the flow or discharge of water draining from a catchment or basin (i.e., the volume of water passing that point every second—m^3s^{-1}). The data derived can be used to compute average flows over a given time and describe the variability at different times of the year.

River flow is generally measured in terms of river water level (or stage) at a location where the level—measured relative to a fixed datum—can be related directly to the discharge (CEH 2017). The level is converted to flow using a 'rating equation', which is simply a mathematical relationship between the stage and discharge and is dependent on the cross-section of the channel at the site on the river where flow is being determined (WMO 2006). This site may be a structure built into the river channel (e.g., a weir or flume) (Clemmens et al. 2001) that provides a regular cross section for which a theoretical rating equation can be derived based simply on the channel geometry. In the UK there is a national standard for weir design BS ISO 1438:2017 'Hydrometry: Open channel flow measurement using thin-plate weirs' (BSI 2017).

Alternatively, flow may simply be determined at a site on a natural river channel where the relationship between level and flow is likely to be regular and stable; for instance, in the vicinity of a natural rock outcrop where there is little sedimentation.

In this case the relationship between stage and flow must be derived by a series of discharge measurements made—usually with a current meter—over the range of water level variation. Each discharge measurement comprises a number of velocity observations across the channel, usually at a fixed fraction of the depth. The product of velocity and channel depth is integrated to give the discharge linked to the water level at the time of observation. The rating equation is derived from repeated observations made across a range of river water levels (Gunston 1998).

Current meters measure the velocity of river water at a given location and depth. The traditional types are impellers (rotating on a horizontal axis) or cups (rotating on a vertical axis) which turn with a speed that is dependent on the velocity of the water. More modern electromagnetic sensors use the Faraday principle to measure water speed: as a conductor (water) moves through an electromagnetic field (produced by the sensor) it generates a voltage that is a function of the velocity. In small streams and rivers measurements are usually made manually, wading across the channel (Figure 4.2). On larger rivers measurements have to be made from bridges or from specially designed cableways (Buchanan and Somers 1976).

Stage is most usually measured using a graduated staff ideally fixed vertically to the river bank at a stable point in the river unaffected by turbulence or waves. Where there is a large range in the stage with a shelving river bank, a series of vertical staff gauges can be stepped up the bank to give continuity of readings (Figure 4.3). Depending on the flow regime of the river and the availability of observers, single readings of the stage at fixed times of the day may be adequate. More continuous measurements can

Figure 4.2 Determining river flow using a current meter in a shallow river in Ethiopia (photo Birhanu Zemadim).

Figure 4.3 (a) Graduated stage board (in Arabic) on the Nile River, Sudan; (b) a stage board on a small stream in Ethiopia (photos: Matthew McCartney).

be made with a chart recorder which records the level of a float in a stilling well connected by a horizontal pipe to the river, so that the level in the well is the same as that in the river. As the float moves up and down a pen traces water level on a slowly rotating chart. More modern designs rely on pressure transducers to measure the water depth and record the water level electronically every hour, or even more frequently, using data recorders (Freeman et al. 2004).

Advances in technology have allowed river flow measurements to be made using sound. The Acoustic Doppler Current Profiler (ADCP) uses the Doppler effect to determine water velocity by sending a sound pulse (transmitted from a transducer) into the water and measuring the change in frequency of that sound pulse reflected back by sediment or other particulates being transported in the water. The change in frequency, or Doppler Shift, that is measured by the ADCP is translated into water velocity. Mounted sideways on a wall or bridge piling in rivers or other open water channels, it can measure the speed of water and hence flow between banks (Laenen and Smith 1983). More recently ADCP tools that can be deployed from boats have been developed and are increasingly used (Mueller and Wagner 2009).

Advances in remote sensing technologies have facilitated the use of space-based instruments to determine changes in volumes of water stored and flowing in rivers, lakes, and wetlands (Alsdorf et al. 2007). Flow cannot be measured directly but must be inferred from the relevant hydraulic characteristics that can be measured from space-based platforms. Current methods rely on *in-situ* ground based observations to create regression relationships that link the variables observable from space to river discharge (Bjerklie et al. 2003). These include instruments that can determine surface water extent, water elevation and water slope. All have limitations associated with spatial and temporal resolution, and in some cases inability to penetrate clouds and the canopies of inundated vegetation. Nevertheless, considerable effort is going into improving the performance of space based measurements and because of the potential to fill gaps where there are insufficient ground data it is certain that advances will continue to be made (Fernandez-Prieto et al. 2012).

Evapotranspiration

Evapotranspiration can account for the large differences between incoming precipitation and the availability of water in freshwater ecosystems. However, it is the most difficult component of the hydrological cycle to quantify in part because it is determined by both the atmospheric demand (i.e., the ability of the air above a surface to absorb moisture) and by the supply of water to the surface (Shaw 1994). In this context it is important to distinguish between evaporation, transpiration, and evapotranspiration, and also between the potential (open water), reference, and actual quantities (Table 4.1). Factors that affect evapotranspiration from a landscape include biophysical variables, the extent of open water (i.e., in rivers, lakes, and reservoirs), the amount of soil

Table 4.1 *Terminology used to describe different aspects of evaporation/ evapotranspiration*

Term	Definition
Evaporation (E)	The physical process of water changing from a liquid to a gas or vapour. In the water cycle the liquid water is present in open water bodies (oceans, lakes, rivers), wet or moist substrates (damp soil, wet sand, etc.), and after rainfall on the branches and leaves of vegetation.
Transpiration (t)	The movement of water within a plant and the subsequent loss of water to the atmosphere as vapour through the stomata in its leaves.
Evapotranspiration (Et)	This is the sum of evaporation and transpiration and represents the total volume of water returned to the atmosphere from a land surface and its vegetation.
Open water evaporation (Eo)	Evaporation from an open water body.
Reference (Etr)	The amount of evapotranspiration from a short green crop (grass), completely shading the ground, of uniform height and with adequate water in the soil profile. It is a reflection of the atmospheric demand (i.e., the energy available to evaporate water, and the wind available to transport the water vapour into the lower atmosphere).
Actual (Eta)	The amount of evapotranspiration that actually takes place from a surface when both atmospheric demand and water supply constraints are taken into account.

moisture, land cover and land use, plant's growth stage or level of maturity (and hence depth of roots), and percentage of soil cover, as well as atmospheric variables, including solar radiation, humidity, temperature, and wind speed. Because it is dependent on so many factors determining actual evapotranspiration (Eta) accurately can be time-consuming and costly.

Based on the assumption that, when water supply is unlimited, evapotranspiration from a vegetated surface is a function only of atmospheric conditions, several methods have been developed to estimate reference evapotranspiration (Etr) conventionally for short grass, from meteorological data alone. For example, Thornwaite (1948) developed an empirical relationship from records in the eastern United States. The formula is based on mean monthly temperature and a heat index that is computed as a complex sum of monthly temperatures through the year with an adjustment made for the monthly hours of daylight corresponding to the latitude of the site. Because the relationship between temperature and transpiration is derived for a specific region, the formula is not necessarily applicable elsewhere. The Penman–Monteith equation approximates reference evapotranspiration, using daily mean temperature, wind speed, relative humidity, and solar radiation as inputs (Monteith 1965). Because it is based closely on the physical processes underlying evapotranspiration it is more generally applicable than the Thornwaite and other similar equations and so is recommended by FAO (Allen et al. 1998).

Open water evaporation can be estimated using an evaporation pan. This is simply a small open tank containing water (Figure 4.4). Allowing for addition of water to the tank by rainfall, the amount of water lost from the surface area of the tank in a given time (usually 24 hours) provides an estimate of open water evaporation. Although relatively simple to use, evaporation loss varies with dimensions of the tank. Furthermore, compared with a large body of water, they tend to overestimate evaporative losses for two reasons. First, because the tank receives heat supply from the sides of the tank and second, because the evaporating water surface is much moister than it surroundings it induces a so called 'oasis effect'. For these reasons, to estimate open water evaporation from a large body of water, it is common to apply empirical correction factors (typically about 0.7) to pan estimates (Allen et al. 1998).

Direct measurement of actual evapotranspiration is difficult and the instruments required to do this are expensive both to purchase and operate. One method is using a weighing lysimeter. The weight of a soil column is measured continuously and the change in storage of water in the soil is modelled by the change in weight. Evapotranspiration is then computed as the change in weight plus rainfall minus percolation (Fisher 2012). More modern techniques include eddy covariance which determines high frequency vertical fluxes (resulting from atmospheric turbulence) in the vicinity of a vegetated surface. The technique is mathematically complex but has proved to be useful in estimating water vapour fluxes in some circumstances, including from wetlands (Acreman et al. 2003; Kelvin 2011). Another approach to estimate actual evapotranspiration is the use of the energy balance. By measuring the components of the energy balance at a surface, using a bowen ratio instrument (Peacock and Hess, 2004) or a scintillometer (McJannet et al. 2011), the energy

Figure 4.4 An evaporation pan, Ethiopia (photo Matthew McCartney).

available for actual evapotranspiration can be determined and converted to a water vapour flux.

In recent years, estimating evapotranspiration has been improved by advances in remote sensing. Techniques such as the Surface Energy Balance Algorithm for Land (SEBAL) solve the energy balance at the Earth's surface and can be used to compute both potential and actual evapotranspiration at a range of spatial and temporal resolutions (Evans et al. 2009). They need to be validated but are an improvement over land-based measurements because they enable actual evapotranspiration to be estimated over large areas comprising mixed land cover.

Groundwater

For aquatic ecosystems the movement of water to and from groundwater can be critical, not just in terms of volumes but also water quality. In many cases aquatic ecosystems are groundwater 'dependent'. Since subsurface movements of water (both vertical and lateral) are not visible, they must be determined indirectly. The most common approach is to measure the difference in water level between two points, in conjunction with hydraulic conductivity, to calculate the potential flow velocity using 'Darcy's Law' (Darcy 1856). An estimate of the width and saturated thickness of the porous medium (i.e., aquifer or soil) is then used with the velocity to calculate the flow volume (Todd 1980; Brassington 2017).

$$Q = -\frac{KAdh}{dx}$$ (Equation 4.2)

Where: Q is the flow of water through an aquifer

A is the cross sectional area of the aquifer

K is the hydraulic conductivity (sometimes called the coefficient of permeability), which is a measure of a materials capacity to transmit water. K has high values for coarse sand and gravels (e.g. 10^{-3} to 10^{-1} ms^{-1}) and lower values for compact clays and consolidated rocks (e.g. 10^{-9} to 10^{-4} ms^{-1}) (Todd 1980)

dh/dx is the hydraulic gradient (i.e., the difference in the height of the water table above an arbitrary datum between two points in an unconfined aquifer or the difference in potentiometric water level (see below) between two points in a confined aquifer)

Observation wells (sometimes called dip wells), comprising an open pipe (often a plastic tube into which slots have been cut) inserted into the ground to a pre-defined depth, are the simplest way of measuring the level of the water table (WRAP 2000; ASTM 2004). A metal or plastic cap or rubber bung should be used to seal the top of the well when measurements are not being made and a concrete or clay plug should be installed around the top of the pipe to prevent the preferential flow of surface water down the sides of the pipe.

To gain an understanding of the seasonal fluctuations in water table and of the general shape of the water table across an ecosystem, weekly or monthly manual water level 'dip' measurements are usually sufficient. These can be collected with a 'dip meter', comprising an electrical sensor at the end of a measuring tape, which makes a sound (or lights up a bulb) when in contact with water (Figure 4.5). The level is recorded from the top of the monitoring pipe to the water level (the point at which, upon lowering it into the tube, the sensor first makes a sound). To make sense of the water levels across a site, it is necessary to accurately measure the elevations of the tops of the monitoring wells, and also ground level if the water level is to be described in relation to the ground surface.

For more detailed insights into the behaviour of the water table (e.g., how it fluctuates in response to rainfall events) it is necessary to monitor water levels with greater frequency. This can be achieved using pressure transducers, which, similar to those that are used to measure water levels in a river or stream (see above), measure the pressure of the water column above a point and store the results on a data logger. One important factor that must be considered when measuring groundwater levels with a pressure transducer is changes in atmospheric pressure, which can create non-trivial changes in the level of the water table in the absence of any water fluxes. Either vented cables must be used to eliminate responses to atmospheric pressure changes or barometric pressure must be measured in conjunction with water pressure and a correction

Figure 4.5 Measuring groundwater levels with a dip meter, Ethiopia (photo Birhanu Zemadim).

applied. Currently there are a multitude of commercially available transducers (commonly called 'divers') which vary considerably in cost and performance (USGS 2017). See the publications listed on *Techniques of Water-Resources Investigations Reports*, USGS website (https://pubs.usgs.gov).

In confined aquifers (i.e., aquifers sealed above by an aquiclude—a geologic unit that is incapable of transmitting significant quantities of water), groundwater may be under pressure and the potentiometric water level (i.e., the level coinciding with the hydrostatic pressure level of the water) is needed to determine fluxes using Darcy's equation (Todd 1980). In this case piezometers, which are wells that are open (i.e., slots cut in the well casing) at a specified depth and sealed (i.e., non-slotted) across the rest of their length, are used. Combinations of wells and piezometers installed across a range of depths, in different soil and geologic layers, and in transects down slopes, can be used to estimate fluxes laterally (within) and vertically (across) layers with different hydraulic properties.

Hydraulic conductivity may be determined in laboratory tests of rock material obtained through coring. There are also a number of methods of field measurement including 'falling head test', pumping tests, and Guelph Permeameter tests. All involve adding water or creating rapid changes in water level within an aquifer and determining the consequent changes in water level over time (Todd 1980; Brassington 2017).

Darcy's law has been shown to be applicable for saturated and unsaturated porous media but the volume of the porous medium must be large in comparison to the micro structure. In massively fissured rocks, such as limestones, or in rocks through which water moves predominantly in cracks (e.g., granites), more detailed consideration of flow patterns may be necessary. In this case experiments, using dyes or other forms of tracer, can be used to determine travel velocities and dispersion rates (Freeze and Cherry 1979).

Soil moisture

Soil moisture is extremely important for plants and at any given locality can, without replenishment, be quickly depleted through evapotranspiration. Insights into soil moisture changes can be important for understanding ecological processes in many wetlands (e.g., peatlands, seasonally inundated wetlands, and floodplains).

Determining the water content of unsaturated soil involves measuring the moisture content in the soil above the water table. Moisture content can be determined as a percentage of weight (gravimetric) or as a percentage of volume (volumetric). Methods can be destructive, involving field sample collection and laboratory analysis, or non-destructive, using sensors to detect the soil moisture content.

Field samples are collected using specially designed steel rings, which are inserted into the soil (at the surface or specific depth) and a sample of known volume of soil is extracted. The sample is put in an airtight bag and taken to a soil laboratory, where it is weighed, dried in an oven (normally at 105 °C), and then re-weighed. The difference in weight equates to the mass of water that was present in the sample. The limitation of this approach is that it requires considerable effort to repeat frequently.

Non-destructive detection methods generally make use of either the electrical properties or neutron scattering properties of the moist substrate. Electrically based methods include Electrical Resistance Tomography (ERT) (Brunet et al. 2010) and capacitance (e.g., Delta-T Theta probe) (Natural Resources Conservation service 2010) (Figure 4.6). Neutron scattering methods include the neutron probe (using an active neutron source) (Johnson et al. 2013) and COSMOS (using the naturally occurring cosmic ray source) (Zreda et al. 2012). All these methods should be calibrated with field samples.

4.2.2 Monitoring networks

A hydrological monitoring network comprises a set of instruments that are installed and operated to make observations with the aim of quantifying and better understanding water budgets and the different fluxes within a catchment or ecosystem or across a landscape.

Key elements of a successful hydrological monitoring program are:

- Clearly identified objectives for monitoring—what exactly is the purpose for monitoring? What questions are being answered or hypotheses tested?
- Designing the monitoring network to measure all hydrological variables required at locations which will provide useful information.
- Development of an optimal sampling protocol that ensures measurements are made and data collected with sufficient frequency.
- A data management framework that ensures that data are collected, quality controlled, analysed, and archived in ways that ensure they will be useful now and into the future.

Figure 4.6 Measuring soil moisture with a capacitance probe, Ethiopia (photo Birhanu Zemadim).

In many instances the priorities and questions to be answered are likely to change over time so that the monitoring network should be robust enough to maintain desired continuity and quality of data but flexible enough to adapt as required. There is a science to designing optimal networks in relation to the arrangement and distribution of monitoring instruments (Chang and Makkeasorn 2010) but in reality pragmatic opportunities and constraints (such as ease of access) often dominate site selection.

There is increasing awareness that engaging local people in establishing and operating hydrological monitoring networks can bring significant benefits in terms of: increased resilience and enhanced sustainability of the network; better cost effectiveness; increased awareness of water-related issues; and the importance of management of water within local communities (Zemadim et al. 2013).

If well designed, monitoring networks need not be complex to answer important questions about hydrological processes and the management of ecosystems. In South Africa, the GaMampa wetland was widely believed to make a significant contribution to dry season river flow of the Mohlapitsi River, a major tributary of the Olifants River. A simple network, comprising transects of manually observed dipwells, in conjunction with measurements of rainfall and river discharge, was used to gain better understanding of the seasonal wetland water balance. The results indicated that, the wetland contributed, at most, 12 per cent of the increase in dry season river flow observed over the river reach in which it was located. The remainder of the increase originated from groundwater flowing through the wetland. Hence, in terms of water management, preserving the natural state of the upper catchment was deemed just as important as protecting the wetland (McCartney et al. 2011).

4.3 Data analyses

4.3.1 Residence time

Residence time is the average time that a water molecule will spend in a water store (e.g., lake, aquifer, or wetland). At its simplest it is determined from the mass balance (Equation 4.1) by dividing the volume of the store by the total flow in or out of that store. In most aquatic ecosystems residence times vary from a few days to years (Table 4.2). Residence times can be an important driver of ecological dynamics. For example, in Mediterranean lakes droughts increased water residence times and through impacts on other variables, such as water temperature, water stability, nutrient concentrations, and turbidity, increased the total biomass of harmful cyanobacteria (Romo et al. 2013).

4.3.2 Areal rainfall

The transformation of point measurements of rainfall—as determined from rain gauges—into an estimate of the mean for an area can be done in several ways. Where gauges are evenly distributed over an area and the topography is relatively subdued the arithmetic mean may suffice. An advance on this is the method of 'Thiessen' polygons. In this case each gauge record is weighted by the size of the polygon around the gauge (Theissen 1911; Dartinguenave and Maidment 2017). Another method, for gauges spread over an elevation range, is to use regression analyses to determine the link between rainfall and altitude and sometimes other determinands, such as distance from the sea (Hayward and Clarke 1996). Isohyetal maps are maps with lines connecting points that receive equal rainfall. These are usually based on expert judgement to

Table 4.2 *Typical residence times for water in different types of terrestrial store (adapted from Freeze and Cherry 1979).*

Store	Typical residence times
Oceans	~4000 years
Lakes and reservoirs	Weeks to 10 years
Swamps	Weeks to 10 years
River channels	Days to weeks
Soil moisture	Days to weeks
Groundwater	Weeks to 10,000 years

interpolate rain gauge measurements and can be highly subjective (Sutcliffe 2004). More modern gridded rainfall data sets use a variety of statistical techniques to interpolate irregularly distributed rain gauge data into a regular array (Figure 4.7) (Rajeevan et al. 2006). In many cases they also integrate data from rain gauges with that from radar and satellites. Many such data sets (e.g., the Tropical Rainfall Measuring Mission (TRMM)) are available but may over- or underestimate rainfall in places with steep terrain or very localised rainfall and sparse ground networks for calibration (Abd Elbasit et al. 2014).

4.3.3 Hydrological modelling

Hydrological data are often collected to support hydrological modelling undertaken to better understand, and in some cases better manage, the complex non-linear processes occurring within aquatic environments. Computer models use mathematics to describe hydrological systems or processes. They are particularly useful for reproducing iterative calculations applied to large data sets, comprising long-term time series of data. For example, rainfall run-off models simulate the many physical processes in a catchment that transform precipitation into streamflow and/or evaporation. In so doing many simulate surface water and groundwater interactions (Beven 2001).

There are many types of model but in the simplest classification they can be divided into two broad categories: (i) deterministic models that aim to simulate the physical processes affecting flow in a catchment and/or ecosystem; and (ii) stochastic models that use statistical relationships between different characteristics to simulate elements of the hydrological cycle (Shaw 1994). Application of hydrological models enables researchers to 'test' theories of how complex aquatic systems function, and how changes (e.g., in climate) might affect the systems in the future. The types of data described above are needed both as input to these models and to calibrate and validate model performance in comparison with what is occurring in the 'real world'.

Over the years many hundreds of hydrological models have been developed. The choice of model to use in any particular circumstance is dependent to a large extent on which processes are deemed important, particularly in relation not just to the movement of water but also processes driven by water (e.g., hydrogeochemistry, erosion, and sediment transport) that are important for ecology (Beven 2001).

4.3.4 Environmental flows

There is now widespread awareness of the degradation of riverine and other aquatic ecosystems, resulting from human-induced changes in natural flow regimes. Both water abstractions and water storage behind dams modify natural river flow regimes with consequent impacts for downstream aquatic ecosystems. In this context, environmental flows are a flow regime that provides adequate patterns of flow quantity and quality to achieve or sustain a predefined condition of ecosystem functioning. These predefined objectives may be predicated on the conservation or rehabilitation of an entire river system and associated aquatic ecosystems, including wetlands, or they may

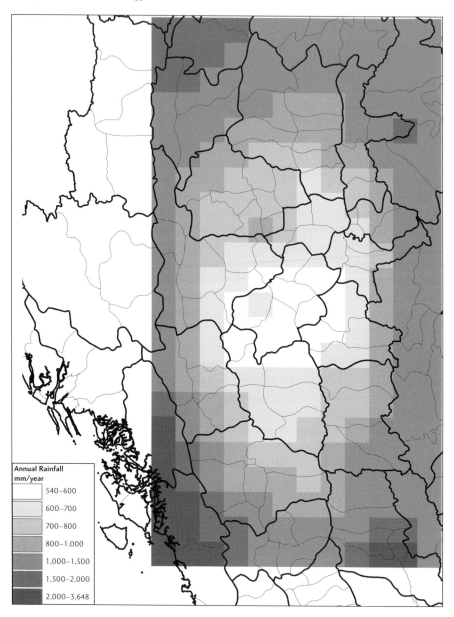

Figure 4.7 Example of gridded rainfall data. Annual rainfall across the 'Dry Zone' of Myanmar derived from the 'Aphrodite' database (Yatagai et al. 2012). The map shows the northern coastline of Myanmar and Bay of Bengal (lower left of the map); each grid cell measures 0.25° × 0.25°.

be targeted at conserving certain key species or elements of scientific, cultural, or recreational value (Tharme 2003).

In recent years, numerous methods have been developed for determining environmental flow requirements but all are dependent on having adequate data for analyses.
The simplest, so called *hydrological methods*, rely solely on historical hydrological records to determine natural flow patterns at specific locations. They assume (usually
based on empirical observation) that a certain, sometimes variable, proportion of the
flow must remain in the river to maintain critical species habitat and ecological functions (Reitberger and McCartney 2011). The most complex, so called *holistic methods*,
usually focus on the entirety of the riverine ecosystem, including floodplains, estuaries,
and coastal systems, and require significant amounts of hydrological and hydraulic
data combined with as much supporting ecological data as possible. In addition information on the needs of local people dependent on riverine resources may also be
included (King et al. 1999). The South African Building Block Methodology (King
and Louw 1998) and the Australian Holistic Approach (Arthington et al. 1992) were
the first attempts of this kind of holistic modelling and have been greatly improved
over recent years (O'Brien et al. 2017).

4.4 Case study: hydrological assessment for wetland conservation at Wicken Fen

Wicken Fen in East Anglia, UK is a unique habitat and a wetland of international
importance. The National Trust, a voluntary body founded to preserve the UK's
heritage, established the first UK nature reserve (just 0.06 ha) on the site in 1899. The
reserve, which has now expanded to 323 ha, is still owned and managed by the
National Trust and represents a last vestige of habitat for many species characteristic of
East Anglian Fen, which has been widely drained for agriculture since the seventeenth
century. The reserve has an exceptionally rich flora and fauna. Nearly 5,000 species
have been identified, including more than 121 that are documented as nationally rare
invertebrates (Colston and Broadbent-Yale 2000). It is classified as a Site of Special
Scientific Interest (SSSI) and has been designated as a Wetland of International
Importance under the terms of the Ramsar Convention.

Wicken Fen is criss-crossed by a network of ditches that divide it into a number of
management units. When this study was conducted a water level management plan
had been developed for the southern portion of the Fen but no similar plan existed for
the northern part, specifically two areas named Sedge Fen and Verall's Fen. Prior to the
study, the belief that Sedge Fen and Verall's Fen were drying had become prevalent,
based on both circumstantial and scientific evidence. The circumstantial evidence
arose from qualitative comparison of current and reported historic conditions that
suggested the Fen was drier than in the past (Lock et al. 1997). The scientific evidence
comprised two parts. First, the pH of the Fen soils was decreasing. This phenomenon
was attributed to reduced inflow of calcareous water from Wicken Lode, the principal
water course passing through the wetland (Gowing 1977). Second, a comparison of
two series of water table measurements made between 1928 and 1930 (Godwin 1931)

and between 1974 and 1976 indicated that, even allowing for differences in rainfall, summer minimum water levels in the 1970s were approximately 300 mm lower than they had been earlier in the century (Gowing 1977).

The study conducted in the early 2000s comprised three elements: (i) an examination of the evidence for drying based on contemporary and historic groundwater level measurements from a network of 17 observation wells located throughout Sedge and Verrall's Fens; (ii) the development of a hydrological model to investigate the water budget of the two Fens; and (iii) utilisation of the model to ascertain water requirements to sustain relatively rare M24 (*Molinia caerulea-Cirsium dissectum*) and S24 (*Phragmites australis-Peucedanum*) plant communities (McCartney and de la Hera 2004).

The study confirmed a clear seasonal pattern of winter wetting and summer drying and that, notwithstanding their importance for soil chemistry, the fluxes of water between the Fen and Wicken Lode were small in comparison to rainfall and evaporation. It also confirmed that summer minimum water levels between 1994 and 2000 were lower than those measured between 1928 and 1930. For summer rainfall of 200 mm and 350 mm, the divergence between minimum elevations was on average 450 mm and 200 mm respectively (McCartney and de la Hera 2004). Possible causes for the decline in the summer minimum were changes in climate, physical changes within the catchment in which Wicken Fen is located, and/or changes in the ditch network or vegetation within the wetland itself. The study concluded that a combination of factors was the most likely cause (McCartney and de la Hera 2004).

A number of possible remediation measures were considered including reconnecting the ditch network of Sedge and Verall's Fens directly to Wicken Lode, as they had been in the past. This would enable summer ditch water levels to be kept higher which in turn would enable recharge and maintenance of higher groundwater levels. However, it was noted that great care would need to be taken to ensure that there was no contamination of ditch water with nutrient-rich agricultural run-off that pollutes the river. An alternative was to induce flooding in the late spring or early summer (i.e., May or June). From a hydrological perspective this would top up groundwater and decrease the summer minimum. However, a major constraint to this idea was that the vegetation is not well adapted to spring/early summer inundation and large amounts of water would have to be pumped from Wicken Lode over a period of just a few days. Finally, the idea of increasing the density of ditches in the Fens with sluice gates to prevent summer drainage and/or pumping deeper groundwater into the Fen were also proposed for consideration (McCartney and de la Hera 2004).

This research was an early contribution to an ambitious 50-year project to recreate a large (3,700 ha) wetland area known as the Great Fen. The intention is to reconnect existing small areas of natural fen by purchasing farmland and recreating wetland habitat (EA et al. 2010). Key to this endeavour is detailed understanding the hydrology of the region, the hydro-ecological requirements of different wetland habitats (e.g., open water, swamp/reed fen, wet woodland, wet grassland) and the options for utilising existing surface and groundwater sources for rehabilitation. Good data obtained from a comprehensive monitoring network was a pre-requisite for quantifying water fluxes and facilitating computer simulations to determine what area could feasibly be restored under both existing and possible future climates (Mountford et al. 2004; see Great Fen project website).

4.5 Conclusion

Ecologists are increasingly focusing on freshwater flow regimes (both groundwater and surface water) as the determinant of freshwater ecosystem structure. Changes to the volume and regime of freshwater flows are already a leading driver of global declines in freshwater biodiversity, and the impacts of climate change are likely to accelerate this pressure. Changes to water timing as much as changes to total annual run-off are likely to have the most significant impact on freshwater ecosystems. As precipitation and evapotranspiration regimes continue to alter, they will have knock-on impacts into many aspects of ecology.

The case of Wicken Fen and the Great Fen project is not unique. The conservation and restoration of many wetlands around the world now requires detailed understanding and meticulous hydrological management. In many cases the preservation of unique flora and fauna is entirely dependent on continued human intervention and water management. In Spain, attempts to restore the Tablas de Daimiel National Park, severely degraded as a consequence of groundwater exploitation, have involved the drilling of emergency wells, inter-basin transfers of water, and the construction of dams (Llamas et al. 1996). Such interventions can only be undertaken and can only be successful if based on sound understanding of water budgets and hydrological processes. This understanding is reliant on accurate hydrological monitoring.

Hydrological data, whether it be precipitation measurements, water depth recordings, flow gaugings, or groundwater levels, are collected for many reasons but ultimately underpin decision-making in relation to water resources and aquatic ecosystems. Modern technologies, including novel methods of earth observation are improving in efficacy, but are still highly dependent on on-the-ground measurement for calibration and validation. Data collection, often over long periods of time, requires a lot of investment in financial and human resources and so is of high intrinsic value. In an era of change and uncertain future risks, hydrological monitoring is fundamental to identify emerging change and a prerequisite for the sustainable management of aquatic ecosystems.

References

Abd Elbasit, M.A.M., Adam, E.O., and Abu-Talib, K. (2014). The validation of satellite based rainfall measurements in arid and semi-arid regions of Sudan. Proceedings of the 10th International Conference of AARSE, October 2014, 93–101.

Acreman, M.C., Harding, R.J., Lloyd, C.R., and McNeil, D.D. (2003). Evaporation characteristics of wetlands: experience from a wet grassland and a reedbed using eddy correlation measurements. *Hydrology and Earth System Sciences*, 7, 11–21.

Allen, R.G., Pereira, L.S., Raes, D., and Smith, M. (1998). Crop evapotranspiration. Guidelines for computing crop water requirements. FAO Irrigation and Drainage Paper 56, Food and Agriculture Organization, Rome, Italy.

Alsdorf, D.E., Rodriguez, E., and Lettenmaier, D.P. (2007). Measuring surface water from space. *Reviews of Geophysics*, 45, RG 2002.

Arthington, A.H., King, J.M., O'Keeffe, J.H., Bunn, S.E., Day, J.A., Pusey, B.J., Bludhorn, D.R., and Tharme, R. (1992). Development of a holistic approach for assessing environmental flow requirements of riverine ecosystems In: Pilgrim, J.J. and Hooper, B.P. (eds) Proceedings of an international seminar and workshop on water allocations for the environment. The Centre for Water Policy Research, University of New England, Armidale.

ASTM D5092 (2004). Standard practice for design and installation of ground water monitoring wells. Available online.

Beven, K.J. (2001). Rainfall-runoff modelling. The primer. John Wiley and Sons Ltd, Chichester, UK.

Bjerklie, D.M., Dingman, S.L., Vorosmarty, C.J., Bolster, C.H., and Congalton, R.G. (2003). Evaluating the potential for measuring river discharge from space. *Journal of Hydrology*, 278, 17–38.

Brassington, R. (2017). Field Hydrogeology, 4th Edition. Wiley-Blackwell.

Brunet, P., Cement, R., and Bouvier, C. (2010). Monitoring soil water content and deficit using Electrical Resistivity Tomography (ERT)—a case study in the Cevennes area, France. *Journal of Hydrology*, 380, 146–53.

BSI (2017). BS ISO 1438:2017. Hydrometry. Open channel flow measurement using thin-plate weirs. Available online.

Buchanan, T.J. and Somers, W.P. (1976). Discharge measurements at gaging stations. Chapter A8 in Techniques of Water-Resources Investigations of the United States Geological Survey.

Centre for Ecology and Hydrology (CEH) (2017). How are flows measured? National River Flow Archive.

Chang, N.B. and Makkeasorn, A. (2010). Optimal site selection of watershed hydrological monitoring stations using remote sensing and grey integer programming. *Environmental Modeling Assessment*, 15, 469–86.

Clemmens, A.J., Wahl, T.L., Bos, M.G., and Replogle, J.A. (2001). Water Measurement with Flumes and Weirs. International Institute for Land Reclamation and Improvement, Wageningen, The Netherlands.

Colston, A. and Broadbent-Yale, P. (2000). Wicken Fen—the next 100 years. The National Trust.

Conner, W.H. and Day, J.W. (1982). The ecology of forested wetlands in the southeastern United States. In: Gopal, B., Turner, R.E, Wetzel, R.G., and Whigham, D.F. (eds) Wetlands: Ecology and Management. International Scientific Publishers, Jaipur, India.

Darcy, H. (1856). Les fountaines publiques de la ville de Dijon. Victor Dalmont, Paris.

Dartiguenave, C. and Maidment, D.R. (2017). Exercise 3. Computation of mean areal precipitation. Available online.

Environment Agency (EA), Huntingdonshire District council, Natural England and Wildlife Trusts (2010). Great Fen Masterplan. Available online.

Evans, R., Murrihy, E., Bastiaanssen, W., and Molloy, R. (2009). Using satellite imagery to measure evaporation from storages—solving the great unknown in water accounting. Available online.

Fernandez-Prieto, D., van Oevelen, P., Su, Z., and Wagner, W. (2012). Advances in earth observation for water cycle science. *Hydrology and Earth System Sciences*, 16, 543–9.

Fisher, D.L. (2012). Simple weighing lysimeters for measuring evapotranspiration and developing crop coefficients. *International Journal of Agricultural and Biological Engineering*, 5, 35–43.

Freeman, L.A., Carpenter, M.C., Rosenberry, D.O., Rosseau, J.P., Unger, R., and McLean, J.S. (2004). Use of submersible pressure transducers in water resource investigations. Book 8,

Section A, Instrumentation for measurement of water level. US State Department and USGS.

Freeze, R.A. and Cherry, J.A. (1979). Groundwater. Prentice Hall Inc., London.

Godwin, H. (1931). Studies in the ecology of Wicken Fen 1. The ground water level in the Fen. *Journal of Ecology*, 19, 449–73.

Gowing, J.W. (1977). The Hydrology of Wicken Fen and its Influence on the Acidity of the Soil. MSc thesis, Cranfield Institute of Technology, UK.

Gunston, H. (1998). Field Hydrology in tropical countries: a practical introduction. Intermediate Technology Publications, London, UK.

Hayward, D. and Clarke, R.T. (1996). Relationship between rainfall, altitude and distance from the sea in the Freetown Peninsula, Sierra Leone. *Hydrological Sciences Journal*, 41, 377–84.

Johnson, B., Thoma, M., and Barrash, W. (2013). Neutron probe installation, calibration and data treatment at the Boise Hydrogeological Research Site. Technical Report BSU CGISS 13–01. Available online.

Kelvin, J. (2011). Evaporation in Fen Wetlands. PhD Thesis. Cranfield University, UK.

King, J.M. and Louw, M.D. (1998). Instream flow assessment for regulated rivers in South Africa using the building block methodology. *Aquatic Ecosystem Health and Management*, 1, 109–24

King, J.M., Tharme, R.E., and Brown, C.A. (1999). Definition and implementation of instream flows. Thematic Report for the World Commission on Dams, Southern Waters Ecological Research and Consulting, Cape Town, South Africa.

Laenen, A. and Smith, W. (1983). Acoustic systems for the measurement of streamflow. US Geological Survey Water-Supply paper 2213. Available online.

Llamas, M.R., Garcia, M., and de la Hera, A. (1996). Landscape changes and ecological impacts caused by groundwater abstraction in the Upper Guardiana basin (Spain). Acts of Proceedings II Paesaggio Culturale Nelle Strategie Europee, Torino 16–17 May 1996.

Lock, J.M., Friday, L.E., and Bennett, T.J. (1997). The management of the Fen In: Friday, L.E. (ed.) Wicken Fen: the making of a wetland nature reserve. Harley Brooks, Colchester, UK.

Mangadi, P., Visser, P.J.M., and Ebert, E. (2006). Southern Africa satellite derived rainfall estimates validation. International Precipitation Working Group.

McCartney, M.P. and de la Hera, A. (2004). Hydrological assessment for wetland conservation at Wicken Fen. *Wetlands Ecology and Management*, 12, 189–204.

McCartney, M.P., Morardet, S., Rebelo, L-M., Finlayson, C.M., and Masiyandima, M. (2011). A study of wetland hydrology and ecosystem service provision: GaMampa wetland, South Africa. *Hydrological Sciences Journal*, 56, 1452–66.

McJannet, D.L., Cook, F.J., McGloin, R.P., McGowan, H.A., and Burn, S. (2011). Estimation of evaporation and sensible heat flux from open water using a large hyphen scintillometer. *Water Resources Research*, 47, W05545.

Monteith, J.L. (1965). Evaporation and the Environment. *Symposia of the Society for Experimental Biology*, 19, 205–34.

Mountford, J.O., Folwell, S.S., Manchester, S.J., Meigh, J.R., Wadsworth, R.A., and McCartney, M.P. (2004). Feasibility study for wetland restoration at Baston and Thurlby Fens. Report to the Baston and Thurlby Fens Project Steering Group. Centre for Ecology and Hydrology, Monks Wood, Cambridgeshire.

Mueller, D.S. and Wagner, C.R. (2009). Measuring discharge with acoustic doppler current profilers from a moving boat. U.S. Geological Survey Techniques and Methods 3A–22.

National Weather Service (2013). Snow Measurement Guidelines for National Weather Service's Surface Observing Programs. National Oceanic and Atmospheric Administration, Office of Climate, Water and Weather Services, Silver Spring, MD.

Natural Resources Conservation Service (2010). Capacitance-based soil moisture sensors guide sheet. Available online.

O'Brien, G.C., Dickens, C., Hines, E., Wepener, V., Stassen, R., and Landis, W.G. (2017). A regional scale ecological risk framework for environmental flow evaluations. *Hydrology and Earth System Sciences*, 22, 957–75.

Peacock, C.E. and Hess, T.M. (2004). Estimating evapotranspiration from a reed bed using the Bowen ratio energy balance method. *Hydrological Processes*, 18, 247–60.

Putthividhya, A. and Tanaka, K. (2012). Optimal rain gauge network design and spatial precipitation mapping based on geostatistical analysis from co-located elevation and humidity data. *International Journal of Environmental Science and Development*, 3, 124–9.

Rajaveen, M., Bhate, J., Kale, J.D., and Lal, B. (2006). High resolution daily gridded rainfall data for the Indian region: Analysis of break and active monsoon spells. *Current Science*, 91, 296–306.

Reitburger, B. and McCartney, M.P. (2011). Concepts of environmental flow assessment and challenges in the Blue Nile Basin, Ethiopia. In: Melesse, A. (ed.) Nile River Basin: Hydrology, Climate and Water Use. Springer Books, The Netherlands.

Rodda, J.C. and Dixon, H. (2012). Rainfall measurement revisited. *Weather*, 67, 131–6.

Romo, S., Soria, J., Fernandez, F., Ouahid, Y., and Baron-Sola, A. (2013). Water Residence time and the dynamics of toxic cyanobacteria. *Freshwater Biology*, 58, 515–22.

Shaw, E.M. (1994). Hydrology in Practice. Chapman & Hall, London.

Sutcliffe, J.V. (2004). Hydrology: A question of balance. International Association of Hydrological Sciences Proceedings and Reports.

Tharme, R.E. (2003). A global perspective on environmental flow assessment: emerging trends in the development and application of environmental flow methodologies for rivers. *River Research and Applications*, 19, 379–441.

Theissen, A.H. (1911). Precipitation for large areas. *Monthly Weather Review,* 39, 1082–4.

Thornwaite, C.W. (1948). An approach toward a rational classification of climate. *Geographical Review*, 38, 85–94.

Todd, D.K. (1980). Groundwater Hydrology. John Wiley and Sons Inc., New York.

United States Geological Survey (USGS) (2017). Measuring and mapping ground-water levels in wells. Available online.

Vuerich, E. Monesi, C., Lanza, L.G., Stagi, L., and Lanzinger, E. (2009). WMO field inter-comparison of rainfall intensity gauges, Instruments and Observing methods, Report No. 99. WMO/TD-No. 1504. World Meteorological Organization, Geneva, Switzerland.

Wetlands Regulatory Assistance Program (WRAP) (2000). Installing monitoring wells/piezometers in wetlands. Available online.

World Meteorological Organization (WMO) (2006). Technical Regulations. Volume III. Hydrology.

World Meteorological Organization (WMO) (2012). Guide to meteorological instruments and methods of observation. Available online.

World Wide Fund for Nature (WWF) (2016). Living Planet Report 2016. Risk and resilience in a new era. WWF International, Gland,Switzerland.

World Bank (2010). Flowing Forward: freshwater ecosystem adaptation to climate change in water resources management and biodiversity conservation. Water Working Note 10, Water Sector Board of the Sustainable Development Network of the World Bank Group.

Zemadim, B., McCartney, M., Langan, S., and Sharma, B. (2013). A participatory approach for hydrometeorological monitoring in the Blue Nile river Basin of Ethiopia, IWMI Research Report 155. International Water Management Institute (IWMI), Colombo, Sri Lanka.

Zreda, M., Shuttleworth, W.J., Zeng, X., Zweck, C. Desilets, D., Franz, T., Rosolem, R., and Ferre, T.P.A. (2012). COSMOS: The cosmic-ray soil moisture observing system. *Hydrology and Earth System Sciences Discussions*, 9, 4505–51.

5

Chemical Determinands of Freshwater Ecosystem Functioning

Nic Pacini, Libor Pechar, and David M. Harper

Corresponding author: nicopacini@gmail.com

5.1 The influence of geology, morphology, climate, and land use

Water contains so many dissolved compounds that almost every river system in the world is unique in its combination of them. The 'fingerprint' arises from the dissolution of gases and airborne particles into rain drops, added to flows over plant leaves and stems dissolving organics from the plant, and inorganics from deposited particles; then from the soil as it runs over the surface or seeps through it, before it converges into streams.

Proximity to the sea and altitude of the land where rain falls, influence the nature and the quantity of dissolved materials. Most soils the rainfall infiltrates through are derived from underlying rocks, with the major exception of the post-glacial soils of the northern half of the northern hemisphere, where soil origin might be rocks many kilometres to the north. In extremis, rain water might fall on and run off hard metamorphic rocks, such as granites close to the sea (e.g., Tipping et al. 2006), from which little is dissolved, but dominated by major ions such as magnesium (Mg^{2+}) and chloride (Cl^-). Rainwater that falls on calcareous, sedimentary rocks, such as chalk or limestone, readily dissolves calcium (Ca^{2+}) and carbonate/bicarbonate (CO_3^{2-}/HCO_3^-).

Soils are the product of long-term erosion and chemical weathering (usually millions of years), together with the shorter-term decomposition of vegetation by microbes and invertebrates (usually thousands of years). Regions at mid and high latitude that underwent glaciation are today characterised by rejuvenated soils and moderately mineralised surface waters; in contrast, tropical soils that were not scoured by retreating glaciers and that underwent long-term weathering under warmer climates, today give origin to highly dilute flows (Moss 2018). The nature of the vegetation and its decomposition depend upon latitude and altitude. Un-decomposed material (peat) dissolves into complex organic matter, staining the water brown (e.g., Worrall et al. 2003). Upland peat, dissolved from *Sphagnum* mosses is more acidic than lowland peat, derived from *Phragmites* reeds. Rain leaching through conifer forests dissolves different organic material and is more acidic than rain coming through deciduous vegetation.

Humans have altered the planet's atmosphere and every square centimetre of its surface; rainfall has been changed by dissolving greater quantities of sulphur and carbon

Pacini, N., Pechar, L., and Harper, D. M., *Chemical determinands of freshwater ecosystem functioning*. In: *Freshwater Ecology and Conservation: Approaches and Techniques*. Edited by Jocelyne M. R. Hughes: Oxford University Press (2019).
© Oxford University Press 2019. DOI: 10.1093/oso/9780198766384.003.0005

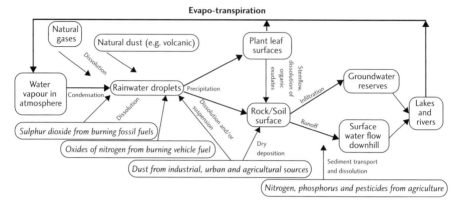

Figure 5.1 Physical processes driving the natural water cycle and the impact of anthropogenic forcing (italics) altering the chemical composition of natural waters.

dioxide and oxides of nitrogen even before taking account of the impact of vegetation changes and then industrialisation upon run-off (e.g., Muller et al. 2015). Figure 5.1 illustrates the major human influences upon water chemistry that need to be considered in freshwater assessment. Chemical elements are building blocks from which life depends, with the cycle of elements representing a permanent flux of matter linking the non-living components of ecosystems to their biotic structure and function (Weathers et al. 2013). Being a universal solvent, water is the main vector that mediates this exchange and makes this linkage possible. Aquatic chemistry is central to the understanding of element cycles, and basic ecosystem processes; this is true of aquatic ecosystems, but it includes terrestrial ecosystems as well, as all living creatures depend on water availability and composition.

Classical aquatic ecology (e.g., Hutchinson 1957) traditionally espoused a 'bottom-up' approach whereby chemical composition and its relation to geological background and climate history is regarded as a major control of ecosystem structure and processes (see also Moss 2018). Lovelock (2003) developed a complementary viewpoint, however, in which biotic structures and processes are at least as influential as physical ones in determining life on the planet. In parallel to this, the recently restated tenets of ecohydrology (Harper et al. 2016; Zalewski 2015) maintain that plants modify their physical and chemical habitat and that this property can be used in ecosystem restoration (see Chapter 18). Whether chemical characteristics are a 'major control' or rather the result of an intimate equilibrium between biotic and abiotic components, they provide ecologists with a fundamental tool for interpreting ecosystem processes. The primary objective of chemical monitoring is to follow the evolution and the product of this interaction.

5.2 pH and dissolved oxygen

5.2.1 Acidity

Natural fresh waters can be considered as a complex solution of weak concentrations of a large number of substances whose interactions include chemical dissociations (acid-base reactions) and act upon the H+/OH- balance. When the concentration of H+ ions (hereafter described by [H+]) prevails over the concentration of OH- ions (hereafter [OH-]) the solution is acid; when the latter prevails, the solution is alkaline. Acidity/alkalinity are master features that result from intricate chemical processes. They influence all physico-chemical and biological processes taking place in aqueous solutions (coagulation, absorption, precipitation, oxidation, reduction, hydrolysis, nitrification, denitrification, aerobic and anaerobic degradation of organic matter etc.). The pH value, defined as the negative logarithm of the activity of [H+][1] (or -log$_{10}$ [H+]), is scaled from 0 to 14. Aqueous solutions are said to have an acid reaction when pH <7, a neutral one when pH = 7, and an alkaline reaction when pH >7. The neutral value corresponds to a chemical balance between [H+] and [OH-] in distilled water at 25 °C and in the absence of CO_2 in the external atmosphere. Under natural conditions, the pH value of dilute waters is influenced by atmospheric CO_2 (*ca.* 0.04 per cent) invasion/evasion affecting the ionic balance and causing pH decrease/increase. Neutral pH values do not correspond to a biological ideal; in reality most freshwater organisms tend to prefer mildly basic waters (pH >7).

The pH results from the chemical form of diluted ions, their speciation[2] and the status of their chemical equilibria.[3] Within the pH range 4.5–9.5, the partition of $CO_2/HCO_3^-/CO_3^{2-}$ determines the pH of most natural waters, with rare exceptions (Pitter 2009). Uptake of free CO_2 (as gas dissolved in water) and of HCO_3^- (bicarbonate) by autotrophs during photosynthesis, and release of CO_2 by both autotrophs and heterotrophs during respiration, connect pH with key biological processes occurring within the aquatic environment. In productive waters, photosynthesis generates daily cycles that are closely coupled to changes in pH through successive periods of prevailing CO_2 uptake and O_2 release during the day and prevailing CO_2 release during the night (Figure 5.2a and Figure 5.2b). At the crossroad between photosynthesis and respiration, dissolved oxygen represents the net balance of production and consumption in aquatic ecosystems, and the determining factor controlling the liberation of reduced compounds at the water-sediment boundary. Depth distribution (stratification) and dissolved oxygen exhaustion in the hypolimnion, as well as daily changes in its distribution represent fundamental information on which it is possible to interpret major lake processes and classify lake type (see Wetzel 2001).

5.2.2 pH measurement

The measurement of pH is best performed directly on site rather than in the lab, because samples containing high plankton biomass or rich in organic matter are susceptible to pH changes. In such cases the pH difference between on site and laboratory measurements can reach up to one pH unit within a few hours. On site pH

---

[1] In dilute solutions, the activity coefficient can be ignored (i.e., $\alpha = 1$), thus pH can be simply defined as the negative log of the molar concentration of H+ ions.

[2] The prevailing chemical form under which a given ion is present in the solution.

[3] The ratio between the concentration of different coexisting ionic species of each single substance. Depending on the complexity of the chemical formula, several ionic species may coexist.

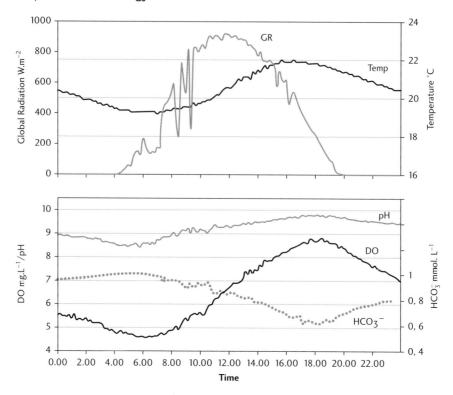

Figure 5.2a Diurnal courses of solar radiation and of water surface temperature measured at 20 cm depth (upper panel), dissolved oxygen, pH and bicarbonate ion (lower panel). Data collected on August 10, 2015 in Rod fishpond (20 ha, shallow human-made pond, southern Bohemia, Czech Republic). Temperature, solar radiation, pH and dissolved oxygen were recorded at 10-minute intervals and sent through GSM to a web server. The bicarbonate concentration was calculated according to standard equations.

measurements represent site-specific hydrochemical conditions and the pH balance generated by biotic processes. In the laboratory, pH measurement after bubbling the sample with air for 2 hours tells us about the hydrochemical water type. Such 'equilibrium pH' reflects acid-base ratios produced by dissolved chemical substances. Moreover, a positive value of the difference (on site pH – air equilibrated pH) indicates the influence of photosynthesis; a negative value shows the predominance of respiration.

Standard measurements are conducted electrometrically by means of a temperature compensated glass electrode probe. Colorimetric pH determination (less accurate) is performed with comparators that contrast standard coloured glass filters with the colour reaction induced by a pH indicator solution. Alternatively, pH indicator paper (e.g., litmus) can be used. pH measurement should be expressed with an accuracy of up to one decimal point and not more, because of the large variability of pH in time and space.

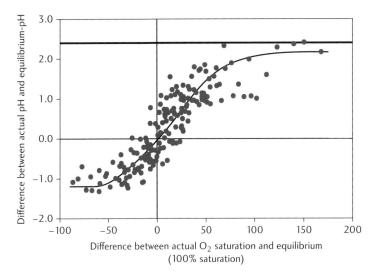

Figure 5.2b Relationship between pH and dissolved oxygen saturation in a productive (eutrophic) water body when photosynthesis and respiration are the main driving forces determining the pH/oxygen regime. The data were collected over several years (2000–2012) in southern Bohemian fishponds (10–100 ha). Intensive photosynthesis and respiration results in a shift of pH and oxygen saturation far from the equilibrium-pH value (pH 8.1) and well beyond 100% oxygen saturation. The upper limit of pH increase under such conditions is represented by the thick solid line at the top of the graph (pH 10.5).

5.2.3 Oxygen measurement

Methods of choice for dissolved oxygen assessment comprise polarographic electrochemical membrane probes, as well as the more recently developed fibre-optic sensors (see Wetzel and Likens 2000). Submersible probes allow measurement *in situ* without the need to collect samples and extract them from the water body; changes in temperature and pressure during these operations could rapidly affect oxygen solubility beside the risk of exposing the sample to the air. For instrument calibration, as well as for primary production studies, the classical chemical Winkler method is still the method of choice.

5.2.4 Significance of pH ranges

The pH level of most freshwaters lies between 6 and 9; however natural peat-bogs, laden in fulvic and humic acids, tend to reach pH 4.6–5.3; similar values can be measured in anthropogenically acidified lakes. Because of industrial emissions of nitric and sulphuric acid, leading to acid rain (pH 3–5), during the late twentieth-century Southern Bohemian lakes (Czech Republic) typically had a pH between 4 and 4.8. Acid rain has decreased during the last two decades in large parts of central and northern Europe (Kopáček et al. 2015), with natural recovery in Swedish streams, aided by liming (Holmgren et al. 2016). Streams and groundwater affected by mining activities may reach pH 3 or even less. Extreme acidity (pH <4) impacts on aquatic biota by favouring the solubility of metals (Al^{3+}, heavy metals), degrading chlorophyll, inhibiting photosynthesis and ionic (Ca^+ mediated) transport through cell walls.

The biogeochemical substrate and the pH can determine the distribution of organisms. The equilibrium of Al forms in mildly buffered waters is highly pH dependent. Increasing acidity below 4.5 leads to a release of mobile Al forms and to an increase of toxic Al^{3+} ions. Amphibian and fish damage because of Al hydroxide and low Ca^{2+} in acidic waters is common in insufficiently buffered Scandinavian streams. Some crustaceans and molluscs may suffer below pH 6.5 (Gunkel 1994), while cyprinid fishes survive with no apparent damage down to pH 5 and salmonids to 4.8. Enzymatic activities and photosynthesis can become inhibited by low pH, but also by strongly alkaline conditions (Sommer 1994). Planktonic autotrophs, which absorb dissolved free CO_2 directly become severely limited above pH 9, because above this value the carbonate equilibrium becomes entirely displaced and CO_2 is exhausted. Aquatic mosses, only able to utilise free CO_2, limit their distribution to cool and acidic mountain streams (so-called 'acidophile'); conditions that favour atmospheric CO_2 invasion (Wetzel 2001).

Above pH 11, no photosynthesis is possible (Sommer 1994); values above pH 10 may result in caustic burns to fish and to invertebrates. As pH increases above 9.5, under high temperature, ammonium (NH_4^-) turns into toxic ammonia ($NH_{3\ gas}$) causing seasonal fish kills and zooplankton poisoning in eutrophic freshwaters (common in temperate regions in summer and in the tropics).

5.2.5 Acidity/alkalinity buffering

The presence of weak acids and bases provides natural waters with chemical mechanisms that can oppose sudden changes in pH. This buffer effect, also called 'acid- (or base-) neutralising capacity', is directly proportional to the concentration of dissolved salts, in particular bicarbonate and carbonate, derived from the weathering of rocks. Freshwaters exposed to pH oscillations (low pH buffering capacity) that affect biological systems are often highly dilute, originating from harder, older, metamorphosed rocks (such as in Scandinavia). In such systems, summer photosynthesis can cause very significant pH increase in surface waters; a classical example is Lake Windermere, UK, that sometimes may reach pH 10 (Talling 2010). Conversely, in hard surface waters derived from softer, younger, sedimentary rocks, sudden increases of pH caused by intense photosynthesis are generally buffered by a deposition of $CaCO_3$, which re-establishes the CO_2/bicarbonate/carbonate chemical equilibrium (Wetzel 2001), visible as a thin ochre deposit of carbonate covering the surface of littoral vegetation.

Alkalinity, as well as 'free CO_2' ($CO_2+H_2CO_3$) can be determined by titration with acid (hydrochloric or sulphuric) using bromocresol green and phenolphthalein as indicators. A more complete and accurate result can be obtained by Gran titration as described by Mackereth et al. (1989). Interferences can be caused by dissolved organic matter, metal oxides, and any agent impacting on water colour.

5.3 Inorganic carbon, major ions, and conductivity

5.3.1 Inorganic carbon

Carbon is the major constituent of biomass. In most freshwaters, the concentrations of dissolved inorganic carbon (DIC), dissolved organic carbon (DOC) and particulate organic carbon (POC) tend to stand according to [DIC]>[DOC]>[POC] (Lampert and Sommer 2007). The DIC fraction varies from <20 mM in acidic soft waters to more than 5000 mM in alkaline hard waters and even higher in endorheic, saline lakes (Cole and Prarie 2009). The DIC pool tends to be regulated by CO_2 gas invasion/evasion according to the partial pressures present in the atmosphere and in the water body (Henry's law). A lake's pCO_2 controls the ambient pH by establishing the proportion between dissolved CO_2, HCO_3^-, and CO_3^-. Chemical equilibria tend to be overwhelmed by biological processes and by C advection. Hence, the surface layers of most lakes are supersaturated in CO_2 in respect to the atmosphere because of respiration excess in relation to photosynthesis, biotic and/or abiotic $CaCO_3$ precipitation, contribution from CO_2-rich groundwater and abiotic UV oxidation (Cole et al. 1994; Cole and Prairie 2009). Carbon efflux from wetlands attracts great interest because of its contribution to the atmospheric CO_2 pool; it tends to be higher (albeit more irregular) at lower latitudes and at higher temperatures because of higher biological process rates (Marotta et al. 2009), and in small-sized boreal lakes characterised by high total organic carbon (TOC; Raymond et al. 2013). Saline alkaline lakes (e.g., East African Rift Valley lakes) are the exception to this rule; their very high photosynthetic rates push pH above 9.0 displacing the carbonate equilibrium towards negligible CO_2 concentrations (Duarte et al. 2008). Given the large number of processes connected to DIC, spatial and temporal changes can vary widely, with daily depth profiles that are characteristic of different lake types (see Wetzel 2001). In stratified lakes, CO_2 accumulation in the hypolimnion is considered proportional to the intensity of respiration processes (Wetzel 2001). In running waters, the proportion of different carbon fractions is controlled by catchment processes rather than by internal biology (Nydahl et al. 2017). Catchment disturbance increases exogenous fluxes of terrestrial C towards inland waters; these represent a significant proportion of the C that ends up being released to the atmosphere from wetlands and lakes (Wilkinson et al. 2016).

5.3.2 Carbon measurement

Calculations of DIC from pCO_2 can be carried out from temperature, alkalinity, and pH data using equilibrium constants after correction for altitude and pressure (Stumm and Morgan 1996), but are uncertain in acidic (<pH 5.4) and in organic-rich waters (Nydahl et al. 2017). Experimentally, its concentration can be determined by acidifying a sample beyond the bicarbonate endpoint and then measuring gaseous CO_2. Direct measurements of pCO_2 can be achieved using either a gas-chromatograph or an infrared gas analyser. Accurate measurements can be carried out in the field using an *in-situ* gas equilibrator (Cole and Prairie 2009). Direct DIC measurements can also be performed by injecting water samples into gas-tight acidified vials. Filtration is not

recommended at pH <7.5 to avoid possible degassing. When alkalinity and DIC are constant or can be assumed to change monotonically, CO_2 change can be followed from continuously monitored temperature and pH data; problems arise above pH 8 because of potential $CaCO_3$ precipitation (Cole and Prairie 2009).

5.3.3 Major ions and conductivity

Major ions (Ca^{2+}, Mg^{2+}, Na^+, K^+ and HCO_3^-, Cl^-, SO_4^{2-}) represent on average >95 per cent of total dissolved solids (TDS) in freshwaters and their ionic charge enables aqueous solutions to convey electrical currents. Electrical conductivity (EC) is a function of the total dissolved concentration; it is the inverse of resistance that can be measured with a Wheatstone bridge. The relationship between TDS and EC shows a very good linear correlation and in most freshwaters r^2 >0.9 (slope 0.75 +/- 10%).

Modern conductivity meters use a standard cell with two platinum plates and an electronic unit for calculating temperature compensation; by convention, temperature corrected (20 °C) EC is the limnological standard parameter of choice, often referred to as specific conductance. EC meters are calibrated with standard KCl solutions. Field EC measurements are easy and reliable and EC is a very useful parameter for characterising changes in water chemistry in time/space (Figure 5.3). Underestimated EC measurements occur in alkaline waters; samples >10,000 μS should be diluted. The ionic composition of individual water bodies reflects local geochemical patterns and/ or local climatic conditions, such as spatial changes in the amount of precipitation. Ionic fingerprints are often best expressed as two triangular plots showing the proportions of Mg^{2+}/Ca^{2+} and Na^+/K^+ together with Cl^-, SO_4^{2-}, and CO_3^{2-}/HCO_3^- (e.g., for Triassic Sandstones; Kimblin 1995).

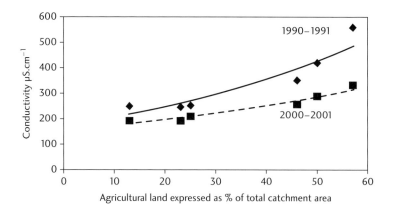

Figure 5.3 The political changes that followed the demise of the communist regime in Czechoslovakia in 1989 translated into significant changes in agricultural land use caused by a switch from intensive state-managed farming to private ownership and less fertilizer application. The impact of agricultural run-off on fishponds in southern Bohemia decreased, with a remarkable change in the relationship between the proportion of agricultural land and fishpond conductivity.

Major cations tend to be present in most freshwaters in the following order of importance: $[Ca^{2+}] > [Mg^{2+}] > [Na^{+}] > [K^{+}]$. Ca and Mg contribute to water hardness. Ca is a universal component of supporting structures from molluscs (shells) to vertebrates (bones) and has a relevant role in cell membranes. Many clades of aquatic organisms are distributed according to gradients of soft-water (Ca deficient) and hard-water (Ca abundant) species (Wetzel 2001). Despite this, biological uptake has little effect on concentrations, and in hard water lakes Ca carbonate precipitation can lead to massive seasonal change in $[Ca^{2+}]$. $CaCO_3$ precipitation implies that at high concentrations, Ca may pose a human health problem as well as impacts on hydraulic infrastructure.

Mg is an essential element constitutive of chlorophyll. It is widespread and its natural concentration depends on the weathering of ferromagnesic rocks as well as carbonates. It is a conservative element as it precipitates only at high pH (>10) and undergoes minor biological uptake. Na is highly soluble and available worldwide; it can become the dominant cation in soft-water regions characterised by intense precipitation. Excess Na concentrations can prevent irrigation as Na substitutes for Ca^{2+} and Mg^{2+} and affects soil structure. Salinisation risk is addressed by monitoring the Sodium Absorption Ratio (the proportion of Na^{+} that can be retained in the soil). Potassium, commonly available at low concentrations, is an essential element for plant growth and is frequently added to fertiliser mixtures. Atomic absorption spectrometry is the method of choice for the determination of major cations. Ca and Mg can be also determined by titration with ethylene-diammine tetracetic acid (EDTA).

Major anions comprise mainly Cl^{-}, which is highly soluble, common, and abundant, and SO_4^{2-} which is generally available but in lesser amounts, rarely limiting, while high $[Cl^{-}]$ is often associated with sewage and its dispersion in water bodies (Chapman 1992). During the late twentieth century, in the northern hemisphere, acid rains caused by industrial emissions contributed significant amounts of SO_4^{2-} to freshwaters. Cl^{-} and SO_4^{2-} can be determined by titrimetric methods and Cl^{-} by specific selective electrodes. In labs, both are commonly assessed by ion exchange chromatography.

5.4 Organic carbon and nutrients

5.4.1 Organic carbon

Dissolved Organic Carbon (DOC) comprises a complex mixture of humic and fulvic acids, amines, urea, as well as a variety of organic substances of anthropogenic origin; these represent collectively an overwhelming proportion of the Total Organic Carbon (TOC) present in freshwaters, of which living organic matter is usually just a minor fraction (Wetzel 2001). Particulate Organic Carbon (POC) becomes relevant only during floods and in highly turbid rivers.

In lakes, the balance between endogenous (primary production-derived) and exogenous TOC can vary depending on catchment C exports. Accumulated TOC can be converted to CO_2 and CH_4 in lake sediments and released to the atmosphere. Many lakes are autotrophic, and more so when anthropogenic N and P fluxes push primary production. Anthropogenic disturbance tends to increase the advected coloured terrestrial organic matter (Weyhenmeyer et al. 2014), which may cause a reduction in phytoplankton growth through light shading (Seekell et al. 2015), potential impacts on aquatic bacterial communities (Findlay and Sinsabaugh 2003; Tranvik and Bertilsson 2001), and enhanced CO_2 emissions (Lapierre et al. 2013). Running waters are heterotrophic; they typically export less C than they import because primary production is greatly inferior to respiration (Weathers et al. 2013).

TOC is measured on unfiltered samples following inorganic carbon removal by acidification and expulsion of the resulting carbonic acid with CO_2-free gas (air or N_2); then the sample is treated by a hot catalyst (e.g., platinum at 680 °C) to combust the remaining organic C to CO_2. Alternatives include photo-oxidation in the presence of persulphate. In lakes, both biochemical and chemical oxygen demand (BOD and COD) are sometimes assessed as proxies of TOC, keeping in mind that [COD]>[BOD]>[TOC], because of the potential presence of non-C oxygen demand in freshwaters (Chapman 1992). In rivers, COD and BOD are standard indicator parameters of anthropogenic pollution. Low [COD], [BOD], [TOC] indicate low anthropogenic impact and good drinking water potential. The assessment of humic acids can be relevant to monitor the toxicity of organic pollutants and metals that can become part of organic complexes. Humic and fulvic acids generate a low oxygen demand (relatively stable, low BOD) but oxidise with potassium dichromate during COD assessment (Chapman 1992). Spectral properties of DOC characterise similarities between lentic water bodies and reveal seasonal changes in C cycling (Loiselle et al. 2009).

5.4.2 Nutrients

Elements that are constituents of aquatic autotrophic biomass include the *macronutrients:* N, P, S, K, Mg, Ca, Na, Cl (approximately in this order of magnitude), that are > 0.1 per cent of living organic matter, and *micronutrients* (Fe, Mn, Cu, Zn, B, Si, Mo, V, Co, as well as others) present in trace amounts. In the case of diatoms and chrysophytes, Si should be considered a macronutrient (Lampert and Sommer 2007). Macronutrients are released through bedrock weathering and are available to autotrophic organisms, unlike N and P which are scarce and limit primary production in ecosystems ranging from terrestrial to freshwaters, to marine (Elser et al. 2007); co-limitation by both occurs in undisturbed catchments where human impact has not affected nutrient ratios (Moss 2018). While N can be fixed from the (virtually infinite) atmospheric compartment by means of natural bacterial processes, P availability is dependent on rock weathering, Fe/Mn oxo-hydroxide dissolution, and from adsorbed-P release from accumulated fine sediments (sources that are limited and finite). Cycles of P, and far more so N, have been greatly enhanced by fertiliser inputs and this is the most significant biogeochemical global change characterising our era (Steffen et al. 2015), causing profound detrimental effects on biodiversity, trophic structure and biogeochemical cycles (Jeppesen et al. 2010).

Monitoring the bioavailable N and P forms is an important tool in aquatic ecosystem management and helps prevent eutrophication (excessive bacterial, algal and/or macrophytic growth; Harper 1992, Chislock et al. 2013). The atomic ratio of dissolved (or total) C, N, and P can be compared to the ratio indicated by Redfield (106 C: 16 N: 1 P; Redfield et al. 1963), representing average algal nutrient content in the marine environment. An unbalanced N:P proportion may indicate more or less strong N or P limitation. An alternative ratio (116 C: 9.9 N: 1 P) was described for algae collected in freshwaters by Boyd and Lawrence (1967).

Bioavailable N and P are conservatively assessed by measuring the concentrations of dissolved N and P forms, defined as the fractions that are not retained by a 0.45 µ pore size filter. All the P present in a filtered sample aliquot is converted to free PO_4^{3-} within a strongly acid medium and complexed with Mo to yield a distinctive blue colour whose intensity is proportional to PO_4^{3-} concentration. With a spectrophotometer, this reaction is able to detect down to 3 µg $P-PO_4$ L^{-1}, a level strongly limiting for most autotrophs (Lampert and Sommer 2007). Regulatory agencies often require the measurement of 'Total P', defined as the phosphate concentration that can be measured in an unfiltered sample after digestion, with digestion procedures dependent on sample matrix. Lake surface samples, where most P is embedded in plankton biomass, are digested in an autoclave in the presence of concentrated potassium persulphate. Samples including mineral matrices, such as stream water or lake benthos, require stronger extraction including boiling acid mixtures, or better, *aqua regia* in a pressurised microwave.

Particulate P (i.e., P fractions retained by the 0.45 µ filter) consists of the P embedded in organic biomass, P adsorbed onto organic and/or Fe/Mn oxo-hydroxides, and P contained within more or less refractory minerals (e.g., apatites, vivianite). Under low redox conditions, Fe/Mn-bound phosphate can become released; this tends to occur at compartment interfaces, such as at the sediment water interface within the shallow lake littoral, or in the deep pelagic, as well as within the water column at the thermocline boundary. P release is linked to photosynthesis/respiration cycles and tends to occur typically in summer during early morning hours when oxygen levels are lowest because of night-time bacterial respiration. The bound-P that may become released can be assessed by sequential extraction (Psenner et al. 1988; Hupfer et al. 2009).

Unlike P, that has negligible gas forms and only one effective oxidation state, nitrogen biogeochemistry is dominated by atmospheric N_2 fixation, by transformations between dissolved N species of different oxidation state, and by denitrification producing N_2 gas. Under pristine conditions, bacterial processes dominate N inputs; in contrast to N releases from catchments generated by N leaching from fertilised agricultural plots. Ammonia is immediately bioavailable for most autotrophs, and its determination is of primary importance for establishing N limitation. This should be done with minimal time delay after sampling to avoid transformations and ammonia volatilisation (predominantly at pH >9). Highly sensitive spectrophotometric methods are available, but vulnerable to contamination if samples are not stored and analysed correctly in clean containers. Nitrate is more stable; its determination requires a laborious catalytic reduction process, whereby nitrate is transformed into NO_2 that can be readily measured. This last N species is highly unstable and is present exclusively under reduced conditions. Alternatively, the nitrate ion concentration can be estimated directly by ion chromatographic techniques. Organic dissolved and particulate N forms can be adequately oxidised using potassium persulphate in an autoclave (the same digestion procedure can be used for TN and TP for surface lake water), or digested with strong acids with a heating block or in a microwave for the determination of N in refractory matrices.

5.5 Heavy metals and xenobiotic compounds

In small amounts, Fe, Cu, Co, Mn, Mo, and Zn are essential for the physiology and biochemical regulation of living tissues (Chapman 1992); at higher concentrations they are the cause of generalised toxicity. The so-called 'heavy metals' (As, Cu, Hg, Pb, Cr, Cd, Ni, Zn) defined as elements characterised by high density (>5 g cm^{-3}), high atomic number, and low solubility, are the elements that cause the greatest toxic risk to individual organisms as well as to communities. Other, less common dangerous metals include Be, Tl, V, Sb, and Mo (Chapman 1992).

Depending on specific chemical properties, metals can become more or less water soluble under different pH or redox conditions (metal solubility increases with decreasing pH for most metals except V and Hg). Toxicity effects are greatly dependent on their chemical form, on the environmental context (Di Toro et al. 1990) and on their environmental fate (Newman and Unger 2002). Environmental conditions can modify the bioavailability and toxicity of contaminants, and thus have indirect effects on food-web components affected by chemicals (Gessner and Tlili 2016).

Monitoring should be oriented towards the assessment of the concentration of the chemical species most likely to be present that can be deduced from the solubility diagrams characterising single contaminants. Soluble metal compounds (e.g., hydroxides) and, in particular, free metal ions are closest to the bioavailable forms that can affect biochemical processes (methylmercury represents a notable exception being more toxic than free Hg), although some regulatory agencies may require the determination of total metal concentrations for conservative assessments. Potential metal toxicity is determined during ecological risk assessment procedures (USEPA 1998), by comparing environmental concentrations with natural background to establish whether extant levels are natural or caused by anthropogenic activities (Suter et al. 2000). Potential metal toxicity is also determined by comparison with local and/or international water quality standards; these are generally elaborated for the protection of human health rather than for the protection of aquatic organisms. In a few cases, such as with Cu, a significant discrepancy exists, because moderate concentrations are tolerated by humans while free Cu^{2+} is lethal to aquatic life already at very low levels unless bound to humic acids (Newman and Unger 2002). In general, metals do not cause biomagnification through trophic transfer, apart from organic species of Hg, for As, Rb, and Cs (Newman 2014). Hardness can decrease metal availability because of competition in the uptake of Ca^{2+} and Mg^{2+} and metals at biological interfaces (Campbell and Tessier 1996). The presence of fine organic sediments can represent an important sink that effectively reduces dissolved metal concentrations through adsorption on particle surfaces; fine sediments may constitute important secondary sources of successive metal release. The measurement of dissolved metals requires immediate filtration through 0.45 µS cm^{-1} membrane filters, then acidification to pH 2 and cool storage to prevent sorption and processes caused by bacterial growth. Samples for total metal analysis should be kept frozen.

Synthetic compounds, most of which can be termed 'xenobiotic' as they do not occur naturally in ecosystems, have become pervasive, distributed worldwide across biomes, and are critical markers of our era—the Anthropocene (Bernhardt et al. 2017). The economic growth of the synthetic chemical industry surpasses the rate of change of other drivers of global change. Pesticides and pharmaceuticals are the most diverse, most widely distributed, and most potentially active on biological systems because they are purposely designed to produce biological effects. Despite their generalised impact on ecosystems, they are little studied by ecologists and are regulated according to eco-toxicological tests conducted on single-species incubated in a laboratory. The potential toxicity of synthetic organic contaminants can be consulted at the US Environmental Protection Agency's ECOTOX database (https://cfpub.epa.gov/ecotox/index.html).

5.6 Chemical analysis

Chemical analysis is expensive, time-consuming, and requires the development of specialised skills. It should be carefully assessed how much chemical detail is necessary during an ecological investigation and how many resources should be invested in to reach the project objectives.

Selecting representative sampling sites and strategies represents the first most critical step, followed by collection, storage, analysis, and data recording (see Nollet and De Gelder 2014). Field ecologists may not have sufficient skills to perform accurate chemical analysis; while in the field, surveyors should be prepared to gather essential descriptive information, to measure basic physical parameters, as well as to collect, handle, and store samples of acceptable quality for processing later.

A number of variable physico-chemical parameters such as temperature, conductivity, pH, redox, and dissolved oxygen saturation must be measured directly in the field using portable probes (also see Chapter 6). Probes must be regularly checked and maintained according to the specific instructions provided by the supplier (this includes consulting online resources for greater clarity), and duly calibrated before recording field data.

Field samples are collected using pre-washed sampling devices (selected according to water body type) in clean 1 L polyethylene (PE) bottles and stored in portable cool boxes. Borosilicate glass is a suitable substitute for PE, but more expensive and breakable; high-density PE is preferred. When not in use, reusable 1 L PE bottles are best kept filled with 50 ml of 5 per cent hydrochloric acid, rinsed with distilled water, and then rinsed once with the sample itself before filling to the brim (not so if samples are rich in particles; Nollet and De Gelder 2014). With the shortest possible delay (hours), a sample aliquot should be filtered through 0.45 µ cellulose nitrate (or polycarbonate) filters; this filtrate is virtually sterile (free of microorganisms). It should be stored in pre-washed and filtrate-rinsed PE bottles and kept refrigerated/frozen for later analysis. Filtration in the field can be performed using acid-washed portable filter holders, such as Swinnex Micropore connected to hand-pumps or PE syringes. Cool storage minimises microbial respiration, which is highly temperature-dependent, and absence of light prevents UV-oxidation and photosynthesis. Chemical preservatives may have to be added if analysis is delayed by more than 24 hours. Samples for nutrient analysis can be preserved using a few drops of P-free chloroform and samples for major cations or metals are best fixed by acidification.

Each sample should provide three fractions: the total unfiltered, the filtrate, and the particles collected on the filter. Concentrations measured in a 0.45 μ filtrate are considered equivalent to the bioavailable pool, that is, the portion that can be directly assimilated by aquatic organisms. The original unfiltered sample can be used for total concentrations, while the portion retained by the 0.45 μ filter can be used for estimating particulate concentrations (although these can also be deduced from the difference between the former two fractions).

Chemical analysis should be conducted in a clean dust-free environment; a portable spectrophotometer can be used in the field to confirm the lack of sample alteration during transport. A general characterisation of the water chemistry can be achieved using one of three increasingly sophisticated methods: (i) chemical test kits; (ii) traditional colourimetry (e.g., Mackereth et al. 1989), or (iii) ion chromatography + atomic absorption spectroscopy (AAS); further, more sophisticated techniques also exist. The right choice will depend on the aim of the survey together with considerations of the time and cost, taking into account the number of samples and the accuracy required.

Chemical test kits are inexpensive, easy to transport, very easy to learn, and quick to use when the number of samples is very low (up to 10). Accuracy varies depending on the chemical species; test kits are not reliable for measuring low concentrations and are not recommended for producing precise results. Traditional colourimetric methods require moderate training and are time-consuming, but may achieve high accuracy. The cost per sample decreases rapidly with samples >10 and replicates are required (at least three are recommended). Some constituents require the preparation of complex reagents and include laborious multi-step procedures (e.g., total N and P, nitrate).

Ion chromatography, AAS, inductively coupled plasma mass spectrometry (ICP-MS), and other techniques require equipment operated by dedicated laboratory technicians, are often ISO certified and provide highly accurate results within a short time. The cost of an individual sample is variable and it will need to be negotiated with the laboratory; if equipment purchase and maintenance are not considered, the actual cost to the operator of a single sample is very low. These techniques are able to analyse several chemical constituents at once, whereas colourimetric methods require a separate procedure to be set up and calibrated for each chemical species. These advanced techniques should be the option of choice every time a laboratory is within reach; chemical assessment kits can be reserved to provide indicative ranges of concentration for specific chemical species of interest. The results obtained from using even the most sophisticated and expensive technique will depend on correctly collected, pre-treated, fixed, and stored samples—a top priority for any freshwater ecologist.

References

Bernhardt, E.S., Rosi, E.J., and Gessner, M.O. (2017). Synthetic chemicals as agents of global change. *Frontiers in Ecology and the Environment*, 15, 84–90.

Boyd, C.E. and Lawrence, J.M. (1967). The mineral composition of several freshwater algae. In J.W. Webb (ed.) Proceedings of the 20th Annual Conference, South-Eastern Association of Game and Fish Commissioners, Columbia, South Carolina, USA, 413–24.

Campbell, P.G.C. and Tessier, A. (1996). Ecotoxicology of metals in the aquatic environment: geochemical aspects. In M.C. Newman and C.H. Jagoe (eds) Ecotoxicology: a hierarchical treatment, CRC Press, Boca Raton FL, USA, 11–58.

Chapman, D. (1992). Water Quality Assessments. A Guide to the Use of Biota, Sediments and Water in Environmental Monitoring. UNESCO, WHO, UNEP, Chapman & Hall, London.

Chislock, M.F., Doster, E., Zitomer, R.A., and Wilson, A.E. (2013). Eutrophication: causes, consequences, and controls in aquatic ecosystems. *Nature Education Knowledge*, 4, 10.

Cole, J.J., Caraco, N.F., Kling, G.W., and Kratz, T.K. (1994). Carbon dioxide supersaturation in the surface waters of lakes. *Science*, 265, 1568–70.

Cole, J.J. and Prairie, Y.T. (2009). Dissolved CO_2. In: Encyclopedia of Inland Waters, Vol. 2. Elsevier, Oxford, 30–4.

Di Toro, D.M., Mahony, J.D., Hansen, D.J., Scott, K.J., Hicks, M.B., Mayr, S.M., and Redmond, M.S. (1990). Toxicity of cadmium in sediments: the role of acid volatile sulfide. *Environmental Toxicology and Chemistry*, 9, 1487–502.

Duarte, C.M., Prairie, Y.T., Montes, C., Cole, J.J., Striegl, R., Melack, J., and Downing, J.A. (2008). CO_2 emissions from saline lakes: A global estimate of a surprisingly large flux. *Journal of Geophysical Research*, 113, G04041.

Elser, J.J., Bracken, M.E., Cleland, E.E., Gruner, D.S., Harpole, W.S., Hillebrand, H., Ngai, J.T., Seabloom, E.W., Shurin, J.B., and Smith, J.E. (2007). Global analysis of nitrogen and phosphorus limitation of primary producers in freshwater, marine and terrestrial ecosystems. *Ecology Letters*, 10, 1135–42.

Findlay, S. and Sinsabaugh, R.L. (2003). Aquatic ecosystems: interactivity of dissolved organic matter. Academic Press, Boston.

Gessner, M.O. and Tlili, A. (2016). Fostering integration of freshwater ecology with ecotoxicology. *Freshwater Biology*, 61, 1991–2001.

Gunkel, G. (ed.) (1994). Bioindikation in Aquatischen Ökosystemen. Bioindikation in limnischen und küstennahen Ökosystemen-Grundlagen, Verfahren und Methoden. Gustav Fischer Verlag, Jena.

Harper, D.M. (1992). Eutrophication of Freshwaters—Principles, Problems and Restoration. Chapman and Hall, London.

Harper, D.M., Pacini, N., and Zalewski, M. (2016). Ecohydrology is a fundamental component of integrated water management. *Ecohydrology and Hydrobiology*, 16, 201–2.

Holmgren, K., Degerman, E., Petersson, E., and Bergquist, B. (2016). Long-term trends of fish after liming of Swedish streams and lakes. *Atmospheric Environment*, 146, 245–51.

Hupfer, M., Zak, D., Roßberg, R., Herzog, Ch., and Pöthig, R. (2009). Evaluation of a well-established sequential phosphorus fractionation technique for use in calcite-rich lake sediments: identification and prevention of artifacts due to apatite formation. *Limnology and Oceanography: Methods*, 7, 399–410.

Hutchinson, G.E. (1957). A Treatise on Limnology, Volume 1, Geography, Physics and Chemistry. John Wiley, Chapman and Hall, London.

Jeppesen, E., Moss, B.R., Bennion, H., Carvalho, L., De Meester, L., Feuchtmayr, H., Friberg, N., et al. (2010). Interaction of climate change and eutrophication. In: M. Kernan,

R.W. Battarbee, and B.R. Moss (eds) Climate Change Impacts on Freshwater Ecosystems. Wiley-Blackwell, Chichester, 119–51.

Kimblin, R.T. (1995). The chemistry and origin of groundwater in Triassic sandstone and Quaternary deposits, northwest England and some UK comparisons. *Journal of Hydrology*, 172, 293–311.

Kopáček, J., Bičárová, S., Hejzlar, J., Hynštová, M.M., Kaňa, J., Mitošinková, M., Porcal, P., Stuchlík, E., and Turek, J. (2015). Catchment biogeochemistry modifies long-term effects of acidic deposition on chemistry of mountain lakes. *Biogeochemistry*, 125, 315–35.

Lampert, W. and Sommer, U. (2007). Limnoecology: the ecology of lakes and streams. 2nd edition. Oxford University Press, Oxford.

Lapierre, J.-F., Guillemette, F., Berggren, M. and del Giorgio, P.A. (2013). Increases in terrestrially derived carbon stimulate organic carbon processing and CO2 emissions in boreal aquatic ecosystems. *Nature Communications*, 4, 2972.

Loiselle, S., Bracchini, L., Arduino, M.D., Ricci, M., Tognazzi, A., Cozar, A., and Rossi, C. (2009). Optical characterization of chromophoric dissolved organic matter using wavelength distribution of absorption spectral slopes. *Limnology and Oceanography*, 54, 590–7.

Lovelock, J. (2003). The living Earth. *Nature*, 426, 769–70.

Mackereth, F.J.H., Heron, J., and Talling, J. F. (1989). Water analysis: Some Revised Methods for Limnologists. Scientific Publication no. 36. Freshwater Biological Association, Ambleside, Cumbria.

Marotta, H., Duarte, C.M., Sobek, S., and Enrich-Prast, A. (2009). Large CO_2 disequilibria in tropical lakes. Global Biogeochemical Cycles, 23, GB4022.

Moss, B. (2018) Ecology of Freshwaters: Earth's bloodstream. 5th edition, Wiley.)

Muller, F., Chang, K-C., Lee, C-L., and Chapman, S. (2015). Effects of temperature, rainfall and conifer felling practices on the surface water chemistry of northern peatlands. *Biogeochemistry*, 126, 343–62.

Newman, M.C. (2014). Fundamentals of Ecotoxicology: the science of pollution. 4th edition, CRC Press, Boca Raton FL., USA.

Newman. M.C. and Unger M.A. (2002). Fundamentals of Ecotoxicology. Lewis Publishers for CRC Press, Boca Raton FL., USA.

Nollet, L.M.L. and De Gelder, L.S.P. (2014). (eds) Handbook of Water Analysis. 3rd edition, CRC Press. Boca Raton, FL., USA.

Nydahl, A.C., Wallin, M.B., and Weyhenmeyer, G.A. (2017). No long-term trends in pCO_2 despite increasing organic carbon concentrations in boreal lakes, streams, and rivers. *Global Biogeochemical Cycles* 31, 985–95.

Pitter, P. (2015). Hydrochemie. 5th edition. VŠCHT Publisher, Prague.

Psenner, R., Dinka, M., Pettersson, K., Pucsko, R., and Sager, M. (1988). Fractionation of phosphorus in suspended matter and sediment. *Archivium Hydrobiologiae Beihefte Ergebnisse der Limnologie*, 30, 98–103.

Raymond P.A., Hartmann, J., Lauerwald, R., Sobek, S., McDonald, C., Hoover, M., Butman, D., Striegl, R., Mayorga, E., Humborg, C., Kortelainen, P., Dürr, H., Meybeck, M., Ciais, P., and Guth, P. (2013). Global carbon dioxide emissions from inland waters. *Nature*, 503, 355–9.

Redfield, A.C., Ketchum, B.H., and Richards, F.A. (1963). The influence of organisms on the composition of sea water. In: M.H. Hill (ed.) The Sea, Vol. 2. Wiley Interscience, New York, 25–79.

Seekell, D.A., Lapierre, J.-F., Ask, J., Bergström, A.-K., Deininger, A., Rodríguez, P., and Karlsson, J. (2015). The influence of dissolved organic carbon on primary production in northern lakes. *Limnology and Oceanography*, 60, 1276–85.

Sommer, U. (1994). Planktologie. Springer-Verlag, Berlin.

Steffen, W., Richardson, K., Rockström, J., Cornell, S.E., Fetzer, I., Bennett, E.M., Biggs, R., Carpenter, S.R., de Vries, W., de Wit, C.A., Folke, C., Gerten, D., Heinke, J., Mace, G.M., Persson, L.M., Ramanathan, V., Reyers, B., and Sörlin, S. (2015). Planetary boundaries: guiding human development on a changing planet. *Science*, 347, 6223.

Stumm, W. and Morgan, J.J. (1996). Aquatic Chemistry, Chemical Equilibria and Rates in Natural Waters. 3rd edition. John Wiley & Sons, New York.

Suter, G.W. II, Efroymson, R.A., Sample, B.E., and Jones, D.S. (2000). Ecological risk assessment for contaminated sites. Lewis Publishers for CRC Press, Boca Raton FL., USA.

Talling, J.F. (2010). pH, the CO_2 system and freshwater science. *Freshwater Reviews*, 3, 133–46.

Tipping, E., Lawlor, A.J., and Lofts, S. (2006). Simulating the long-term chemistry of an upland UK catchment: Major solutes and acidification. *Environmental Pollution*, 141, 151–66.

Tranvik, L.J. and Bertilsson, S. (2001). Contrasting effects of solar UV radiation on dissolved organic sources for bacterial growth. *Ecology Letters*, 4, 458–63.

USEPA (1998). Guidelines for ecological risk assessment. EPA/630/R-95/002F. Risk Assessment Forum, Washington, USA.

Weathers, K.C., Strayer. D.L., and Likens, G.E. (2013). Fundamentals of Ecosystem Science. Academic Press, Waltham, MA, USA.

Wetzel, R.G. (2001). Limnology: Lake and River Ecosystems. 3rd edition. Academic Press, San Diego, CA, USA.

Wetzel, R.G. and Likens, G. (2000). Limnological Analyses. 3rd edition. Springer, New York.

Weyhenmeyer, G.A., Prairie, Y.T., and Tranvik, L.J. (2014). Browning of Boreal freshwaters coupled to carbon–iron interactions along the aquatic continuum. *PLoS ONE*, 9, e88104.

Wilkinson, G.M., Buelo, C.D., Cole, J.J., and Pace M.L. (2016). Exogenously produced CO2 doubles the CO2 efflux from three north temperate lakes. *Geophysical Research Letters*, 43, 1996–2003.

Worrall, F., Burt, T., and Adamson, J. (2003). Controls on the chemistry of runoff from an upland peat catchment. *Hydrological Processes*, 17, 2063–83.

Zalewski, M. (2015). Ecohydrology and hydrologic engineering: Regulation of hydrology–biota interactions for sustainability. *Journal of Hydrologic Engineering*, 20, 1, A4014012

6

Physical Variables in Freshwater Ecosystems

Curt Lamberth and Jocelyne M.R. Hughes

Corresponding author: curt@oxfordenvironment.co.uk

6.1 Introduction

Understanding the processes driving freshwater ecosystems requires not only insight into the dynamics of the organisms under study, but how those organisms interact with their physical environment. It is a fundamental observation in ecology that plants and animals interact with, and are influenced by the soils, atmosphere, and water that they inhabit. For freshwater ecosystems it is water, and in particular the hydrochemistry, that drives the distribution and composition of aquatic plants and animals (Moss 2018; Finlayson et al. 2018; Patten 1990) but the quantity and quality of water in a water body is in turn affected by the catchment setting, including size and topography, the geology, climate, and human impacts (Frid and Dobson 2013). The previous two chapters in this volume have considered hydrology and water chemistry and the techniques required to measure the spatial and temporal variation of each. In this chapter we will consider three categories of physical variables that can be measured for different freshwater ecosystems: (1) variables measured or described at the catchment or sub-catchment scale (e.g., bathymetry, depth, topography, geology); (2) those in or near to the water (e.g., temperature, turbidity, solar radiation); and (3) variables used to describe the substrate (e.g., particle size, mineral vs. peat).

The techniques we will review and evaluate could contribute to answering questions such as: What effect does a change in climate have on the morphology of freshwater ecosystems? (see contributions in Kernan et al. 2010); How does light influence phytoplankton presence/absence and abundance in lakes (Moss 2018)? Does substrate depth or coarseness affect refuge-potential of amphibians (Babbitt et al. 2010) or fish (Ligon et al. 2016)? Do changing water temperatures affect the distribution of fish (Van Zuiden et al. 2016) or the diversity of freshwater molluscs (Kemp et al. 2016)? Why do freshwater ecologists need to understand the physical environment when managing and conserving aquatic or wetland environments (Ausden 2007; Frid and Dobson 2013; Brooks et al. 2014; McBride et al. 2011)? What role does geology play in understanding the long-term stability of a fen mire ecosystem (Jablonska et al. 2014)? What are the functional impacts of changing sediment loads on stream invertebrates (Buendia et al. 2013) or changing land use on riverine fish (Leitão et al. 2018)?

Lamberth, C. and Hughes, J. M. R., *Physical variables in freshwater ecosystems*. In: *Freshwater Ecology and Conservation: Approaches and Techniques*. Edited by Jocelyne M. R. Hughes: Oxford University Press (2019).
© Oxford University Press 2019. DOI: 10.1093/oso/9780198766384.003.0006

6.2 Approaches to collecting physical variables

6.2.1 Habitat and sampling considerations

There are different approaches to the collection of physical variable data which vary according to freshwater type (e.g., wetland, river, lake, peatland); scale of investigation (e.g., studying individual organisms, communities of organisms, or entire freshwater landscapes); and sampling design (e.g., one-off survey, or monitoring project to generate a time series). Temporal (e.g., diurnal, weekly, monthly, yearly) or spatial changes (e.g., within the water column, along the length of a river, across a transect) in physical variables such as light, temperature, and humidity may have a significant effect on the biological system and are fundamental to understanding biotic processes in freshwaters, but there is often poor consideration of the appropriate method of collection and the ultimate use of the data: the most appropriate survey design will always be the one that produces the right data to answer the research questions. This 'reverse planning' process in ecological surveys is reviewed by Sutherland (2006) and Wheater et al. (2011) and is related to whether the surveyor is conducting a deductive survey (where the investigation is question- or hypothesis-driven) or an inductive survey (for description, classification, or recording). Examples of deductive surveys using physical variables are given in Section 6.1 above, and examples of inductive surveys include statutory/standard protocols (e.g., River Habitat Surveys (RHS)), or as part of a national database recording scheme (e.g., UK Environmental Change Network (ECN), Great Lakes Network (GLKN), Lake Victoria monitoring (Sichangi and Makokha 2017)).

A general approach to the measurement of physical variables should include a clear distinction between sampling spatially, either at single or multiple points taken at a similar time (e.g., a grid of points across a mire); or as part of a time series (e.g., monitoring the water levels in a pond over a year). This leads to the fundamental question of how to record the data and, therefore, the complexity of methods that should be used (Box 6.1). For example, if a time series of light data is required under conditions of extreme cold over a month at a remote location, then some form of data logger or memory will be required. The electronics will require power from a battery and the type of battery is dictated by the voltage and power requirements of the equipment, the recharge method if used (e.g., solar power and controllers), and the temperature extremes to which the equipment is going to be exposed (see Section 6.2.2).

6.2.2 Battery considerations

Surveys that require field data loggers or powered equipment require batteries, and these may influence the sampling strategy (e.g., position of the instrument and data logger; length of time monitoring can take place; time of the year instrument and data logger can be used; number of instruments or data loggers that are deployed). Primary (non-rechargeable) batteries can have better high- and low-temperature performance. For example, AA/FR6/LR6 lithium cells specify a lower operating temperature of –40°C, while lithium and alkaline batteries operate up to around 60°C but will need replacing more often. Common secondary or rechargeable batteries based on nickel-metal hydride do not charge well below 0°C, and cease to function below around –10°C.

Box 6.1 Choosing the correct method or technique for measuring physical variables—ten questions to answer when planning the survey

1. What physical variable needs to be measured?
2. Does the survey need to be conducted in the field; e.g., installing sondes in a lake to monitor temperature, or using secondary sources; e.g., measuring the surface area of a water body using lidar?
3. Over what time duration will the study take place and how often will sampling occur within that time?
4. Should it be a single point location or a spatial distribution of points, and at what scale?
5. How are the data going to be recorded?
6. What are the electrical power requirements?
7. What accuracy and resolution are required to answer the research questions?
8. What equipment calibration is required and how often?
9. What is the cost of the equipment and the time/financial resources required to take the measurements?
10. What are the project constraints in terms of: health and safety of the researcher; remoteness of location and access; size or weight of equipment; loss, theft, or vandalism; harshness of the environment; e.g., water logging, flows, extreme events from temperature and flows, physical abrasion; or interference from animals or biofouling.

Heavier lead acid batteries loose capacity to such an extent that at –20°C they have 50 per cent of rated capacity and older batteries often fail at temperatures just below 0°C (Hutchinson 2004). Modern lithium-ion batteries have the best power density, are lightweight, made in a variety of shapes and voltages, and are supplied with in-built short-circuit protection. They must be charged and discharged between voltage limits of 4.2V maximum and 2.4V minimum and require specialised charging circuitry. Lithium-ion batteries are currently limited to an effective operating temperature of down to –10°C with only a 20 per cent capacity loss, but below this temperature lithium-ion batteries cease to work. New rechargeable lithium-based batteries are under development capable of operation down to –50°C with only a 50 per cent capacity loss.

Solar or wind recharging of batteries may be required for sites where access is restricted or to enable long periods of unattended operation (Box 6.2). A recharging system consists of three components: a method of generating the power; a charge controller; and a rechargeable battery. Small wind turbines are available but are harder to set up than solar panels that are effective and easy to use. The size of the solar panel depends on the power required to keep the system in operation, the voltage required to drive the controller and charging process, the duration and amount of light the solar panel receives, and the efficiency of the solar panel. Charge controllers are essential to prevent overcharging, over-discharging or short-circuiting the battery. Finally, there are a variety of commercially available sealed-for-life button-sized loggers, each

Box 6.2 Selecting a solar panel and lithium-ion battery system—a worked example

In this hypothetical project, a piece of equipment is operating continuously in the UK in winter and may need to run for one week without solar recharge. It is sited away from any trees or buildings and will be orientated at 45 degrees. The equipment requires 0.099 W of power at 3.3 V. Per day the equipment will use 0.099 W × 24 hours = 2.376 W-hours or 720 mAh at 3.3 V. Thus, to run the equipment for seven days without recharge will require a battery of capacity 720mAh × 7 = 5040mAh plus an extra 10 per cent for losses, necessitating a lithium-ion battery (3.3 V) of capacity 5.5 Ah. An efficient charge controller converts the solar panel energy normally at 12 V into the required charging voltage for the battery. Here the charging voltage will be 4.2 V as set by the type of battery. The charging process is only 80 per cent efficient so the system should be over-rated by at least 20 per cent. It is recommended to use an online solar panel size calculator designed to work for small solar installations. The above example assumes a maximum of only 1.6 hours of sunlight per day in winter, and therefore a 12 V solar panel providing at least 15 W *actual* output power is required. Solar panels are rated at 1,000 Wm^{-2} so in the UK where typical sunlight power is less by around 30 per cent, a 12 V solar panel rated at 20 W by the manufacturer should be used.

one with a dedicated sensor including temperature and humidity. Temperature range and resolution vary between the different models and should be chosen to suit the application or sampling strategy, and most have a programmable time interval. Some button loggers are not designed for total immersion and so will require a waterproof capsule or extra protection.

6.3 Catchment setting

6.3.1 Bathymetry, topography, area, and length

There are a number of practical reviews and resources that deal with the measurement and mapping of bathymetry for freshwater bodies (Huertos and Smith 2013; Gao 2009; Wilcox and Los Huertos 2005; Haag et al. 2005; Wetzel and Likens 2000). Bathymetry is the study of underwater topography and is highly relevant to understanding the ecology and spatial dynamics of wetlands and water bodies. Bathymetric surveys (z coordinates) are usually conducted in parallel with mapping x and y coordinates, so that a complete picture of the underwater shape, water surface area, and length of the water body is produced. The bathymetry of a water body is usually the result of the interplay of water depth, water storage, organic matter storage, and substrate dynamics, as well as the geology, catchment setting, and geomorphological history. The range of methods that can be used to describe and quantify the bathymetry is varied in sophistication and use of technology and depends on the research questions, available resources, the accuracy or resolution required, and the depth and area of the water body.

Huertos and Smith (2013) provide a practical overview of the topic and a summary of commonly used field survey equipment, ranging from the (low-tech) use of a meter rule to measure the water depth across a grid in (e.g., a small pond (see Section 6.3.2)), to ground-based lidar for surveying an expansive tidal marsh. They provide details on the use of field survey equipment, such as a total station and laser level, for measuring x, y, and z coordinates. A vertical benchmark is the starting-off point of all topographic surveys, including bathymetry, and the survey will need to be connected to either a local vertical datum point which will produce a relative survey, or to a published datum/benchmark which allows the survey to be compared to others, or referenced to sea level. The ultimate end product from a bathymetric survey is a 3-D map or digital model, and there is a range of commercial or open-source GIS software that can be used to visualise the data. The data can be used to model stage to area to volume relationships (Lei et al. 2014); or to produce, for example, models of habitat/topography, trophic status, and autotrophic structure (Althouse et al. 2014). Surveys that only require 2-D variables on area or length of the water body under study can use maps, aerial images or remotely sensed images to obtain the data required (see Chapter 9 and Chapter 14).

Bathymetric surveys that operate over large areas, or are undertaken regularly as part of a monitoring programme, or require high resolution, may use a choice of boat-based technologies requiring different skills, expertise, and resourcing. Popielarczyk et al. (2015) review and use geodetic and hydroacoustic methods to survey Lake Hancza in Poland (as well as temperature, water depth, and bottom sediment); and a useful evaluation of hydroacoustic and laser-optic methods to conduct a bathymetric survey of transboundary Lake Constance was undertaken by Wessels et al. (2015). Hydraulics Research (HR) Wallingford together with the UK Environment Agency have developed a remote controlled boat (ARC boat) that collects data on flow, depth, and suspended sediment concentrations using acoustic Doppler current profiling, sub-bottom profiling, and echo sounding. The resulting river cross sections display detailed bathymetric data as well as flow, discharge, and other physical parameters by width and depth. Bathymetric data such as these may be stored in open-access databases, including the UK Lakes Portal website and NOAA's National Geophysical Data Center website for the Great Lakes.

6.3.2 Water depths and levels

Variations in water level (stage or gauge height) or water depth for a water body provide essential eco-hydrological information for the measurement of flow, discharge, or water surface profile. Seasonal changes in water level can dictate the type of freshwater habitat that develops. For example, shallow seasonal water level fluctuations over a large area favour the development of marsh or swamp habitat. Small seasonal variations of water, especially if water flows are low, will promote the preservation of organic material; if buffered by base-rich (carbonate) geology will form fen type habitats; while unbuffered water supplied by rainwater will form acidic peat-based bog habitats. Some freshwaters experience extreme seasonal water-level variations (e.g., Irish turloughs where water levels drop suddenly in spring by up to several metres (Skeffington 2006) and support a unique ecology including the fen violet (*Viola stagnina*)). A recent study showed that the fen violet is opportunistic and takes advantage of the lack of competition for light by occupying bank-side draw-down zones of rivers and lakes. Under more hydrologically stable conditions or where grazing or haymaking are absent the fen violet is out competed by other plants and must survive as seed until the next disturbance event (Lamberth, unpublished). Measurement of water levels can be made using a number of techniques with the choice dependent on: type of water body and remoteness of location; sampling frequency and accuracy in relation to frequency and range of water level change; equipment costs; and survey time.

The simplest method uses a ruler or staff installed in the water body and firmly attached to the bed substrate or immobile structure. The staff should have a known 'zero' or reference point such as a local map datum. If there is more than one measuring location within a site the relative elevations of all the locations should be known so that water levels can be compared. The staff should be installed where it can be easily read, and in a location where the full range of water levels can be measured: deep enough to measure low summer water levels and high enough to measure flood water levels. Freshwaters influenced by tidal ranges can be particularly difficult due to the large diurnal ranges. If the precise location is known, it is possible to estimate the local coordinate reference elevation using a contour map (error around 2 to 5 m vertical), a hand held GPS (very variable), using lidar data of a nearby flat area such as a road (0.15 m or better), a differential GPS (10 to 20 mm error), or a nearby mapping agency elevation reference point (e.g., triangulation point). For further details on GPS units refer to US Geological Survey 'Global Positioning Application and Practice' web page, and US Army Corps of Engineers (1998).

The manual measurement of water levels within a borehole or well can be done using a dip-system or well dipper. At its simplest, a weighted length of thin rope or a stick can be used to measure the distance from a known point (e.g., ground level or well top), down to the water surface. More reliable methods use a graduated steel tape, an electric tape which sounds when it touches the water surface, or water displacement from an air line using a pressure gauge (Cunningham and Schalk 2011). Automated systems normally consist of a sensor, a memory or storage device with power supply, or a device to transmit data. The water-level measurement can be automated using different technologies such as a float-wire-shaft encoder, using capacitance, resistive, piezoelectric pressure; or acoustic (ultrasonic sound), radar, or optical methods (Guaraglia and Pousa 2014). Float and wire, acoustic, and radar methods are expensive and are generally used for permanent monitoring stations or bathymetric surveys (see Section 6.3.1).

There are a number of excellent water-level logging products for the measurement of water depth or water level (e.g., see 'Divers', and loggers supplied by Eijkelkamp Soil & Water, Global Water Instrumentation, Schlumberger Water Services, Solinst Canada Ltd, Van Essen Instruments), with good battery life and memory capacity, and available for a range of depths and resolutions. These devices generally use a piezoelectric pressure sensor measuring absolute pressure and must be calibrated for temperature and compensated for changes in atmospheric pressure. A complete system has high initial costs and must include loggers for every monitoring site, a dedicated logger for atmospheric pressure measurement for every site or area, a reader to transfer the data to a computer, thin stainless steel cables to suspend and fix the loggers in place, and computer software. Products to measure gauge pressure (the difference between atmospheric pressure and water pressure) use vented systems with a breather tube that extends above the water surface and do not require atmospheric correction. These devices measure the depth of water from the surface to the sensor and must be secured or suspended in place to a fixed object. The distance of a logger above a lake or stream-bed must be known to be able to calculate the total depth of the water body.

6.3.3 Geology

The geology of a wetland, river, or lake catchment is a key influence on the hydrochemistry of the water body. Much has been published on this hydrogeochemical relationship; the extent to which storages and fluxes in hydrogeochemical cycles influence different freshwater types, and how the relationship is altered or exacerbated through anthropogenic influences (see Moss 2018; Mitsch and Gosselink 2015; Mitsch et al. 2009; Berner and Berner 2012). Gaining an overview of catchment geology is fundamental to the interpretation and understanding of freshwater ecosystems and processes and should be a first step in any freshwater investigation. Some of the geological variables to consider in a survey are: whether there are easily weathered rocks in the catchment (leading to waters with high nutrient loads) or resistant rocks (where water may be similar in composition to precipitation); the extent of aquifers, areas of recharge or discharge, and the groundwater flow direction; the type of rock—whether igneous, metamorphic, sedimentary, surficial deposit or base rock; the nature of the rock—e.g., porosity, hydraulic properties, grain size; and, therefore, the influence of rocks on the hydrochemistry reflected by, for example, pH, total dissolved solids/salts, individual anions and cations in the water body (see Chapter 5).

Hiscock (2006) provides an excellent introduction to the theory and practice of hydrogeology, and Brassington (2017) explains the practicalities of field hydrogeological surveys. It is one of a number of practical guides in the 'Geological Field Guides Series' published by John Wiley & Sons which are useful to ecologists wishing to survey and observe geological features. For freshwater studies that include geological variables or need to obtain a catchment view of the geology, it is recommended to begin with a desk study where geological and topographic maps (producing a geological cross section), aerial and remotely sensed images are consulted before going into the field. Further geological information may be obtained from mining companies, or from unpublished borehole surveys (providing stratigraphy) and geological surveys from country specific Geological Surveys or government departments. From these it is possible to gain data on dominant base rocks, which rocks will behave as possible aquifers, thickness and boundaries of aquifers, distribution of outcrops, surficial deposits, springs, wells, or boreholes (Brassington 2017). These are all clues to inform where there might be hydrogeochemical storages and fluxes influencing the water body under study. Brassington (2017) describes how to assess the nature of rocks (e.g., porosity, specific yield, permeability, and dispersion), both in the field and as indicative values.

The catchment scale influence of geology on the hydrochemistry of water bodies is undisputed, but it is worth remarking on the role of climate and human influences in this relationship. Moss (2018) discusses the interplay of rock, evaporation, and precipitation dominated waters where waters are either exorheic (waters flow from atmosphere to ocean with only a small part evaporated back to the atmosphere) or endorheic where water entering the catchment is mostly evaporated back to the atmosphere due to arid or anthropogenic conditions. This interplay is reflected in the ionic content of water bodies where, broadly, high calcium ratios and high total dissolved loads are found in waters of endorheic basins where evaporation is high; high sodium ratios and low dissolved loads are found in precipitation dominated waters that have passed over oceans and are depositing sea spray; and, between these two extremes, a range of high calcium ratios and moderate total dissolved loads are found in exhoreic waters where precipitation is influenced by geology (rock weathering). Significantly, catchments with igneous or hard rocks are low in dissolved salts, have low pH of around 3.5 to 5.5 and have low buffering capacity; in contrast, the buffering effect of dolomite, limestone, or carbonate-rich catchments is far greater due to hard water and a pH of around 7.5 to 8.5.

Human influences may have a more profound impact on the ecology of freshwaters than the influence of the natural geological setting with notable effects due to agriculture (e.g., compaction and nitrate-rich fertilisers), deforestation (e.g., decrease in ionic concentrations, increase in runoff), and urbanisation (e.g., run-off rich in hydrocarbons and other pollutants, wastewater and sewage, reduced infiltration (Frid and Dobson 2013)). The well-documented adaptations of organisms to their different hydrogeochemical environments adds further weight to the importance of quantifying or documenting physical variables in freshwaters to explain ecological processes. Further resources on hydrogeological methods and techniques can be found on the websites of national geological institutions (e.g., US Geological Survey, British Geological Survey, Geological Survey of India, Geoscience Australia, China Geological Survey), and the Ramsar Convention website hosts the 4th edition of its 'Handbooks for the Wise Use of Wetlands' published in 2010, including Handbook 11 on 'Managing Groundwater'.

6.4 In and near to the water

6.4.1 Temperature and humidity

The measurement of temperature, in or near to the water, is essential for the study of freshwater ecosystems as it controls the rate of chemical reactions; (therefore) the rate of biological processes; the equilibria of many physio-chemical processes such as the amount of dissolved oxygen (see Chapter 5); and other variables which are affected by temperature (e.g., pH and electrical conductivity), and thus need temperature correction. One of the simplest methods is the use of conventional liquid-in-glass thermometers. This method requires no power, can be used for liquid or air measurements, can be calibrated and have a resolution (precision) of better than 0.2°C and is low cost. They are, however, fragile, do require care in reading the scale and in use, as liquid-in-glass thermometers are manufactured either for partial immersion (to a specified depth), total immersion, for air measurements, or for a specific orientation. Excellent organic-based alternatives to the standard liquid mercury-filled thermometers are reviewed in 'Mercury Thermometer Alternatives: Advantages of Changing' on the US National Institute of Standards and Technology website. Electrical conductivity, dissolved oxygen, REDOX, and pH meters should all have temperature measurement inbuilt to allow for automatic temperature compensation.

Thermistor- and thermocouple-based instruments are widely available for accurate manual temperature measurements, and both methods require power, an amplifier, and a display. The probes have a lower thermal inertia than conventional glass thermometers so respond to temperature variations quickly, but they must be made waterproof and the electrical circuitry kept dry. Thermistor-based units are generally simpler, lower cost and least accurate with resolution of 1°C and accuracy generally ±1°C to ±2°C. Thermistors are based on the variation of resistance with temperature. It is recommended that thermistor based instruments are calibrated at least at three temperature points; for example, 0°C, 10°C, and 35°C or points covering the expected temperature range to be measured. Thermocouple-based units are more complex but have the advantage of being accurate and cover a wide temperature range. They are based on the voltage generated at the junction between two differing metals (e.g., copper-constantin). There are a variety of thermocouple probe types available commercially, based on specific metal pairs. Each meter will specify the type of thermocouple probe to be used, with type K and T thermocouples being the most common. Type K has a resolution typically of ±2.2°C while type T thermocouples can be calibrated to an accuracy of ±1.0°C or better. Thermistor- and thermocouple-based instruments can be connected to a datalogger, interface, or to the Internet, allowing temperatures to be measured remotely either continuously and real-time or logged at a specified interval for later retrieval. Furthermore, there is a range of commercially-produced automatic transmitters (called sondes) for measuring temperature and other physical (as well as chemical) water parameters, including turbidity, solar radiation and water levels (see 'Sonde and CTD' web page on the US National Oceanographic and Atmospheric Administration website).

Electronic temperature dataloggers are widely available, the most useful being those that are waterproof, designed for total immersion, or usable in all weather conditions. Ideally, the logger should have an inbuilt real-time-clock, synchronised when interfaced with a computer so that the time of measurement is recorded. Logging intervals of 1 or more seconds are common, but limitation of the total memory capacity or number of possible readings dictates the frequency of measurement and time period over which the unit can be used. Temperature measurement standards that are both precise and accurate tend to be set for applications where data will be compared nationally or internationally. For example, there are defined rules for air and ground temperature measurement for international weather stations including time of measurement, location of station, and type of equipment. However, generally there are no standard protocols for taking temperature measurement in water bodies and so measurements should follow best practice where available and measurements should be made and recorded so that they can be repeated by another observer at a later date. 'Best practice' recommendations exist for wadeable streams (US Environmental Protection Agency 2014) and there are many good examples for larger lakes (Lakes Environmental Association 2015).

Where temperature measurements are taken depends on the research questions. For faster streams and rivers, measurements should be taken in the main flow with the thermometer or sensor held beneath the water for sufficient time to allow the temperature to reach equilibrium (around 30 seconds to 1 minute). Temperature sensors in larger meters used for measuring pH and electrical conductivity may take minutes to reach equilibrium especially if the waters are cold. If access to the centre of a steam is difficult a sample of about 0.5 to 1 L can be taken with a suitable container and the temperature measured immediately. Smaller ponds and lakes have a marked temperature gradient at the edges, and a standard sampling protocol should be developed to reduce measurement errors. For example, measuring pond water temperature at 1 m from the edge (or less for easier safe access), at a depth of 50 mm, and in or out of emergent vegetation.

Humidity is the amount of gaseous water (water vapour) in the atmosphere. In freshwater ecosystems it affects plant evapotranspiration; temperature by evaporative cooling; and is one of the controlling factors in the evaporation of water from water surfaces, influencing catchment water budgets. Humidity can be reported either as absolute humidity (mass per volume), specific humidity (dimensionless ratio), or relative humidity (percentage). Relative humidity is the most common measurement and should always be reported with temperature so that the dewpoint temperature (temperature at which air is fully saturated and liquid water starts to condense) can be calculated. Field measurements of humidity are straightforward and a range of hand held electronic meters are widely available and almost always include an accurate measurement of temperature. For an introduction and practical guide to humidity measurements see the 'How to measure humidity' web page on the UK National Physical Laboratory website and Box 6.3.

Box 6.3 Guidelines for measuring relative humidity

(a) Check the calibration of the humidity and temperature probe periodically against a known standard; e.g., calibration salts, choosing two values covering the range of interest.

(b) Check to see if probes are rated for either non-condensing use (<95 per cent relative humidity) where the sensor will not come into contact with liquid water, or for condensing use where the sensor will be exposed to 100 per cent relative humidity or condensing water. Non-condensing probes will be damaged irreversibly by contact with water, whilst condensing probes will take time to recover after saturation before an accurate reading is observed.

(c) Provide additional protection from water contact by wrapping the probe carefully in a thin water vapour-permeable polypropylene, geotextile fabric with edges sewn or heat fused together; e.g., roofing underlay.

(d) Decide between low cost humidity meters that generally have a low accuracy of +/- 5 per cent relative humidity, and more expensive, higher quality meters accurate to within +/- 2 per cent relative humidity or better, and supplied with a calibration certificate.

6.4.2 Colour, clarity, turbidity, and suspended sediment

Colour, clarity, turbidity, and suspended sediments are variables associated with light transmission and particulate transport that reflect a range of environmental factors (Davies-Colley and Smith 2001). Colour is one of the standard tests applied to potable drinking water (see World Health Organisation website) but could be applied to natural waters where this is of ecological interest (e.g., for peaty, manganese, or iron-rich waters). Colour can be expressed using spectrophotometric methods or by visual comparison using hazen units (Hongve and Akesson 1996). A simple note of the colour at the time of sampling is often all that is needed.

Water clarity is a useful measure of the ability for light to penetrate or transmit through water and is often used as a surrogate or replacement for turbidity or suspended sediment load. Water clarity is measured easily using a Secchi or black disk and the methods give similar results (Anderson 2005; Wheater et al. 2011; Bidwell 2013). A Secchi disk is a weighted round, flat disk of alternating black and white quadrants, generally 0.2 m in diameter but for greater depths can be up to 0.5 m, or for shallow streams it can be made smaller. The disk can be made inexpensively from a variety of resistant materials and the pattern painted on. The Secchi disk measurement is done by lowering the disk on the end of a tape measure into the water and the depth at which the disk pattern is no longer visible is recorded. The disk is then raised until the pattern is just visible again and a second depth measurement is made. The Secchi depth is the average of these two depth measurements and is reported in metres (Becker Nevers and Whitman 2010). It is best to take the Secchi depth measurements consistently either in or out of shade and at a particular time of day, preferably noon, to minimise variations in light levels. Frequency of measurement is dictated by the temporal variation of the event under study (e.g., hourly in relation to short-term runoff events, or monthly for long term seasonal variations). There are a number of studies with well-documented methods comparing Secchi disk with other physical variables and relating Secchi depth with the amount of photosynthetic active radiation (PAR) available for plant photosynthesis in different water bodies (e.g., Alaskan lakes (Koenings and Edmundson 1991), Polish lakes (Borowiak 2014), and a Turkish reservoir (Bayram and Kenanoğlu 2016)).

Water turbidity is caused by fine suspended particles of clays and silts, plant material or microscopic organisms, and dissolved materials such as plant acids or coloured compounds. Turbidity differs from clarity in that turbidity is an optical property of a liquid that causes light to be scattered or absorbed rather than being transmitted directly. Clarity as measured by a Secchi disk is an optical measure of the amount of light that is transmitted directly without scattering (Anderson 2005). High loads of fine suspended particulate matter causing turbidity prevents the penetration of light through water and may result in deposits of silt on plant surfaces reducing the overall productivity rate by interfering with photosynthesis. Turbidity can be an indication of potential pollution or water quality of a water body: fine sediment particles affect the respiration of fish, can be a transport mechanism for adsorbed pollutants, and can provide shelter for water borne pathogens. The simplest method of measurement uses a turbidity tube that measures the 'cloudiness' of a water sample. The design is based on a miniature Secchi disk and construction details are described in Myre and Shaw (2006), while other methods for measuring turbidity are reviewed by Bidwell (2013).

Sediment transport within the freshwater environment plays an important role in the processes of erosion, assists in the transport of pollutants, is a major driver for the morphological evolution of rivers, and affects ecosystems (Bartram and Ballance 1996). The definition of suspended sediment is inexact and the method of determination should always be specified. Suspended sediment is not the same as turbidity: suspended sediment is the solid particulate matter that is transported by water and is measured in units of mass per volume. Gray et al. (2000) compare two methods for the determination of solids in water: suspended-sediment concentration (SSC) and total suspended solids (TSS). The scope of the SSC method groups samples into three categories: samples with sediment that settles within an allocated storage time (days to weeks) and measured by evaporation with a correction factor for dissolved solids; samples with low concentrations of sand or clay which must be filtered; and samples which can be wet-sieved to allow for the determination of coarse and fine material. The TSS method measures the total suspended solids filtered from a specified volume of water and dried at 105°C.

Bedload is sediment that is transported by water by rolling, lifting, sliding, or bouncing along the bed of a stream or river. Bedload is related to the stability of river and stream channels and in turn affects habitat suitability for many aquatic organisms. Rivers and streams with occasional or continuous high energy flows will have a higher bedload. There are a number of measurement methods available (US Geological Survey 2003) of which the Helley-Smith sampler is well documented (US Geological Survey 1999; US Geological Survey 1980; Bunte et al. 2008). Of recent interest, due to their effects on aquatic food webs, is the accumulation of microplastics in coastal and lake sediments and the effect of microplastic pollution on fish and marine invertebrates (Horton et al. 2017). Microplastics have been sampled in sediment cores from Lake Ontario where their accumulation was found to have occurred over the past 38 years or more (Corcoran et al. 2015). To date there are no standard methods for sampling and identifying microplastics in freshwaters although progress is being made for large and medium rivers (Liedermann et al. 2018); for identification using Raman spectroscopy (Araujo et al. 2018); and for riverine fish (McNeish et al. 2018).

6.4.3 Solar radiation and PAR

The sun is the primary source of energy for the biosphere. Part of the incident solar radiation is reflected back into space or absorbed or scattered by the atmosphere and the remaining radiation which reaches earth's surface is terrestrial radiation. Terrestrial radiation comprises a range of wavelengths (or frequencies) grouped into the high-energy ultraviolet (UV) band, the visible band (which includes photosynthetically active radiation, PAR or PhAR, as used by plants to drive photosynthesis), and the infrared band (IR) associated with heat. Much of the energy in terrestrial radiation is re-radiated back into space as long wavelength IR or thermal radiation. The measurement of photosynthetically active radiation may be required for freshwater ecosystem assessment as it provides an excellent gauge of how much light is available to unicellular and multicellular phototrophic organisms such as algae, aquatic plants, protists, and phytoplankton. The measurement of PAR is complicated (Box 6.4) because, firstly, light levels vary considerably over time, and secondly, visible light is a range of wavelengths (colours or

Box 6.4 Choosing a light measurement method—four important points to consider

(a) Which wavelength range or band is of interest and of greatest importance? It is important that the sensors used for the measurement of PAR have a filter to cut out UV and IR light so that the sensor responds only to the frequencies of interest; e.g., from 400 to 700 nm as many sensors have an increased sensitivity to far red and IR radiation. The inter-calibration of sensors is of particular interest to allow for data comparison between studies but is often overlooked (Long et al. 2012).

(b) Within what medium (air or water) is the light going to be measured? There are a number of methods available to measure PAR (Jewson 1984) and PAR is reported in units of Wm^{-2} or photon flux density as mol photons m^{-2} s^{-1}. Electronic meters are now available at moderate cost including full spectrum underwater quantum meters specifically designed for hobby aquarists with a ±5 per cent calibration uncertainty and a measurement range up to around 4,000 $umol$ m^{-2} s^{-1}.

(c) Meters should have in-built electronic compensation for use either in air or in water where light attenuation is greater especially for longer (red) wavelengths. Light meters measuring in Lux are designed with a greater sensitivity to green light, similar to that of the human eye. Green light is not absorbed by plants for photosynthesis but is either transmitted through or reflected from plant leaf surfaces so Lux meters can only be used for approximate measurements of PAR and data conversion to PAR units depends on the light spectra being measured.

(d) What method will be used to compensate for variations in incident light levels or how can such variations be minimised so that the results are comparable; e.g., time of day, sunny or shade, weather conditions, angle of observation? Compensation for changes in ambient light could be accomplished by dual-sensor measurements where one sensor measures incident PAR and the second sensor measures PAR at the location of interest (Davies-Colley 1984; Rørslett 1997). Changes in water turbidity or colour affects PAR penetration in freshwaters and an interesting method relating PAR to the Secchi disk depth has been proposed by Man'kovsky (2001).

frequencies) and the absorption of each wavelength of light differs as it passes through the atmosphere, water, or reflects off a surface.

6.4.4 Evaporation and evapotranspiration

In any wetland system water loss by evaporation and evapotranspiration can be significant and must be considered when calculating the water budget or water balance model. Evaporation is driven by energy taken from the air, from land and vegetation surfaces, and from sunlight. Both evaporation and evapotranspiration are normally expressed in mm per day (or inches per day) or other time units, similar to the measurement of rainfall. Evapotranspiration is the combination or sum of the amount of water that has evaporated from the land surface and transpired from plants. Evapotranspiration is an important factor in the hydrological cycle because it can cause a significant loss of water from wetlands especially under hot and windy conditions and when the relative humidity is low.

It is not necessary to know the detailed theoretical basis behind the measurement or calculation of evaporative processes (see classic studies by Penman; Shaw et al. 2010) but a basic understanding of evaporation processes combined with a simple method of calculation of evaporative loss are essential in studies requiring data for this variable. Hydrologists refer to two types of evapotranspiration; potential evaporation or potential evapotranspiration and actual evaporation or actual evapotranspiration. These terms, confusingly, are used interchangeably in the literature but evapotranspiration includes the loss of water from the ground by transpiration as well as by direct evaporation so relates to the nature of the vegetation cover (Irmak and Haman 2017; Lhommel 1997). Potential evaporation is the amount of evaporation that could occur from a surface if the supply of water to that surface was unrestricted (e.g., just after heavy rainfall). High rates of soil evaporation with little or no rainfall can lead to a concentration of soluble salts at or near the soil surface leading to soil salinity. Evaporation from a soil surface is restricted by the availability of water present in the soil, the vegetation cover and the rate of replacement of soil water by upward capillary action. Thus the amount of water evaporated from a surface where the supply of water to that surface is restricted is called the actual evaporation or actual evapotranspiration.

Evapotranspiration rates differ for each type of vegetation (tall, short, species, structure, rooting depth, etc.) and this is related to the crop coefficient. The crop coefficient is relative to the reference standard evapotranspiration rate for a crop of a well-watered grass area. The total evaporative loss for a catchment is calculated from the sum of the contributing land areas for each component of the evaporative pathway. For example, in a small catchment made up of open water, dry rocky scrub, and saturated marshland with tall swamp vegetation, the total evaporative loss will be the sum of: loss from open water (supply of water unrestricted) plus loss from the marshland (tall vegetation with unrestricted water supply so a high evaporative loss) plus the small evaporative loss from the dry scrub but over a larger contributing area (restricted by soil water availability). The amount of water evaporated directly from a water surface is termed 'open water evaporation' and can be used to estimate potential evapotranspiration. Open water evaporation can be calculated from meteorological data using a number of methods, such as the Penman Method (Finch and Hall 2001), or estimated using the open pan method (Allen et al. 1998).

There are several other ways to obtain evapotranspiration data. National weather centres offer country-specific evapotranspiration data for 5 × 5 or 10 × 10 km grid squares (e.g., MORECS in the UK) but may charge for this service. Evapotranspiration can be calculated from meteorological data and appropriate crop coefficients using published data. The calculation of evapotranspiration for a vegetated surface requires an estimate of the dominant crop type or vegetation and the following weather data: radiation or sunlight, temperature, humidity, and wind speed. There are evapotranspiration calculation methods detailed for sites where meteorological data are limited and where one or more variables are unavailable. The standard Penman-Monteith method for calculating reference evapotranspiration (see Chapter 4) is available online from the Food and Agricultural Organization of the United Nations (Allen et al. 1998) and an evaluation of different methods for different regions is provided by Gao et al. (2017).

6.5 Substrate

The underlying substrate in any wetland will reflect storages and fluxes in hydrogeochemical processes operating both at the catchment and site scae. For enclosed wetlands the substrate may consist of organic soils that have built up over hundreds of years, and for riverine floodplains the substrate may consist of fine sand and coarse gravel deposited over the seasons and reworked during flood events. A defining feature of many wetlands is the presence of distinctive soil types. These are described and defined by Mitsch and Gosselink (2015) for a variety of wetlands in different climatic zones. In general, wetland or hydric soils store more organic matter, contain microorganisms that emit gases such as methane, ammonia, hydrogen sulphide, and experience less leaching than terrestrial soils—they sustain reduced rather than oxidised conditions (Keddy 2010; Batzer and Sharitz 2006). When quantifying and describing the substrates of freshwater ecosystems, the key factors to consider are: type of substrate or soil; how it varies across the site and over time; rate of change—accretion and erosion; texture, porosity; and characterisation of the substrate. The variables that are chosen will, of course, depend on the aims of the survey, the research questions, and the management implications. With all these methods, a suitable sampling strategy should be deployed to assess spatial variation in substrate depth and characteristics, as well as for monitoring or repeat samples.

6.5.1 Sampling

Sampling techniques will vary depending on the depth of water, accessibility of the substrate, and whether it is a terrestrial wetland or aquatic freshwater body. In terrestrial freshwaters it is possible to use extendible hand augers to assess the depth of the substrate and normally take samples up to 2 m depth (but some can go to 5 m). Beyond 2 (to 5) m, for compacted or solid substrates it becomes necessary to use machinery (e.g., a drill rig) to obtain deep samples or cores down to the bedrock. In peat-based wetlands, a peat probe can be used to assess the depth of peat and optimal sampling points; a gouge corer to (easily) extract cores of up to 1 to 2 m; and a Russian corer for obtaining undisturbed cores of up to several metres depth (see review by Frew 2014; Low et al. 2018; Bogology website; Chapter 13). Winton et al. (2017) used a stainless steel box corer to sample tropical peatlands in Madre de Dios, Peru, as part of a study to investigate methane emissions from tropical peatlands; box corers range from a simple metal box that is inserted into the substrate and the undisturbed monolith is cut out manually, to box corers that are deployed from a research vessel to extract undisturbed substrate from the bottom of a lake using a tripwire guillotine to cut free the monolith. A key problem when collecting substrate samples is compressibility and the best results are obtained if the sample or core is kept saturated. If only shallow samples are required for the survey, gardeners' bulb planters can be bought cheaply to extract substrate samples to a few centimetres depth across the site, or simply digging a hole to a constant depth with a trowel or spade may be all that is required.

For deep aquatic or lacustrine environments, a piston corer or Mackereth sampler is more suitable to extract saturated sediments without disturbing the sample. Alternatively, for soft sediment, a steel push-pipe with PVC liner may be sunk into the substrate by gravity and the core extracted. These can be combined with a stilling pipe, through which the sample can be taken to avoid loss of substrate and minimise disturbance to the core. Once the core is extracted it is advisable to keep it out of direct sunlight so as not to contaminate the sediments. It can then be removed from the pipe or cylinder or the cylinder cut in half, making sure that the core is labelled carefully ready for further analysis in the laboratory, or alternatively, the Troels-Smith scheme (Troels-Smith 1955) can be used to describe the physical components of the substrate from each core in the field. These sampling methods may be used together with surface substrate samplers (e.g., grab samplers) or dredgers. Grab samplers (e.g., Petersen, Ponar) can be used from the edge of a small boat but care needs to be taken with the combined weight of many grab samples to avoid capsizing the vessel (Frew 2014). The USGS 'Guidelines for Collecting Bottom Sediments from Streams' can be consulted via the Internet and provides instructions for collecting and processing riverine sediment samples including samplers and sieves. Ultimately, the choice of sampler will depend on the type of freshwater and the research questions, and whether surface samples or a stratigraphic sample are required.

Geophysical methods have been used in ecological surveys of wetland substrate (e.g., ground penetrating radar (GPR), electrical resistivity) to characterise tropical peatlands (Comas et al. 2015) and the distribution of a burrowing crayfish (Sweat et al. 2017). A study by Holden et al. (2002) on blanket peat used GPR to identify sub-surface piping and depth of peat to the mineral transition. The drawback with using GPR in freshwaters or saturated sediments such as clay, is its limited penetration (1 to 1.5 m) because water is a dielectric; lower-frequency GPR has better penetration but the resolution is lower (see review by Robinson et al. 2013). Finally, it is worth mentioning standard soil pits. These may be useful to describe soils or substrate surrounding a wetland or freshwater body, or can be dug within terrestrial wetlands if the conditions are sufficiently dry. Pits can be dug with a spade to produce clean sides and thus enable the depth of each layer to be measured or samples to be taken with a trowel.

6.5.2 Type of substrate and soil characterisation

It is beyond the scope of this chapter to review the methods for describing and characterising freshwater substrates or soils, or the wider catchment soils, in any great detail. In this section, however, we will provide guidance to some of the resources that deal with the practical techniques needed to interpret and characterise substrate and soil samples from field surveys.

When a substrate becomes saturated with water (even for a few days) and there is active microbial activity, the substrate becomes depleted in oxygen (anaerobic). Anaerobic conditions encourage the accumulation of organic matter, produces reduced elements such as iron and manganese, and the accumulation of methane, hydrogen sulphide, and ammonia. Over time, hydric soils develop and will be present whether or not saturated conditions exist at the time of the survey, and they can be readily identified by their characteristic grey colour (blue-grey or greyish waterlogged soils sometimes with orange mottles are called gleys), with mottles or redoximorphic features (colour patterns caused by oxidation or reduction of iron or manganese), or their high organic accumulation at the surface (peat). There are other overarching factors at the catchment scale that influence hydric soil formation such as climate (and therefore hydrological storages and fluxes), geology (whether carbonate-rich, igneous or hard rocks, and therefore pH conditions or buffering capacity), topography, as well as vegetation type and land use. Locally within the water body, the structure, function, and productivity of *in situ* plants (macrophytes) and microbial organisms will affect soil development. Characterising hydric soils and modelling the complex relationship (catchment and local scale) between reduced conditions, microbial activity, plant productivity, water depth and time, is the evidence-base for the management of wetland biogeochemical cycles and the conservation of freshwater substrates (Reddy and Delaune 2008; Mitsch and Gosselink 2015; excellent review by Bodelier and Dedysh 2013).

Overviews of all aspects of the properties of soils, soil formation, and global soil groups can be found in the online 'Encyclopedia of Soils in the Environment' published by Elsevier, Weil and Brady (2017), Schoonover and Crim (2015), European Soil Data Centre (ESDAC) website, International Soil Reference and Data Centre (ISRIC) website including the World Data Centre for Soils (see Procedures and Standards report), and the Global Soil Biodiversity Atlas which can be freely downloaded from the ESDAC website. Resources that focus on the practicalities of delineating and describing types of waterlogged substrates, hydric soils, and organic-rich wetland soils or histosols, and their analysis and classification, include Vepraskas and Craft (2016), USDA (2017), Vasilas and Vasilas (2013), Vasilas et al. (2013), Delaune et al. (2013), Okalebo et al. (2002), and Chapter 5 (Wetland delineation) in 'Wetlands: Characteristics and Boundaries' downloadable from the National Academy Press website.

Parameters that are routinely described or measured when characterising hydric soils are soil horizons (described, measured or sampled with depth, see Section 6.5.1); colour (e.g., to distinguish between organic and mineral soils, or to infer mineral composition); texture (see Section 6.4.4); structure (aggregation and resistance to physical change); density (soil weight per unit volume); and organic matter content or C content (Vasilas and Vasilas 2013). The value of characterising wetlands using soil data (among other variables) for organic C, total N, available P, pH, and particle size distribution, is demonstrated by Sakane et al. (2011), who use their results to provide guidelines for the future protection and agricultural use of wetlands in Kenya and Tanzania. Hydrology, soil redox potential, pH, and temperature dynamics were used by Seybold et al. (2002) to understand the capacity of tidal freshwater wetlands to attenuate agrichemicals along the James River, Virginia, USA. They monitored the soil for one year at 20 cm and 50 cm depths at three locations in the wetland and found continuous sub-surface reduced conditions, high organic matter accumulation, and continuous biological activity. Furthermore, the time taken for the restoration of hydric soils should not be underestimated and is illustrated in a study by Moreno-Mateos et al.

(2012) of 621 global wetland sites. They demonstrate that even after 100 years since restoration, biological structure and biogeochemical functions (carbon storage in soils) were, respectively, 23 per cent to 26 per cent lower than in reference sites.

6.5.3 Sedimentation, accretion, and erosion

Assessing whether freshwater habitats are naturally accreting or eroding is important for understanding plant community dynamics, changes in ecosystem productivity, water quality, impacts of catchment land use practices, and, therefore, for making evidence-based management decisions and monitoring the success of restoration. Section 6.3.3 describes the techniques for measuring suspended sediment in water, but if the material is deposited, it is possible in the short-term to examine rates of accumulation (or removal) within the freshwater body, both spatially and temporally. For example, Ibáñez et al. (2010) in a study of sediment dynamics in the Ebro River delta (marshes and wetlands) in Spain used Surface Elevation Tables (SET), marker horizons and lead-210 dating to demonstrate surface marsh subsidence and accretion dynamics as a result of reduced freshwater sediment loadings from the Ebro River and sea level rise. Methods that can be used to measure and monitor sedimentation rates, accretion, or erosion are reviewed by Vasilas et al. (2013), Thomas and Ridd (2004), the British Society for Geomorphology's online 'Geomorphological Techniques', and Goudie (1990). The most commonly used techniques to measure sedimentation in freshwater ecosystems can be divided into continuous (e.g., sediment accumulation sensors, video cameras) or discontinuous (e.g., sediment traps, marker horizon), and those that measure accumulation (e.g., anchored tiles—discontinuous) or elevation (e.g., SET—discontinuous).

For organic substrates such as those found in peatlands there are valuable publications and resources which explain how to describe and quantify accretion or erosion, in response to degradation or management. For example, the online Briefing Notes of the IUCN UK Committee Peatland Programme, Bonnett et al. (2009), Gell and Finlayson (2016), Schumann and Joosten (2008), Oldfield and Richardson (1990), the websites of the International Mire Conservation Group, the International Peat Society, and Bogology. Blais et al. (2015) collate a range of research chapters that use 'natural archives', including peat and other freshwater substrates, for monitoring environmental contaminants in space and time; Turetsky et al. (2004) review dating techniques suitable for recent peat deposits of a few hundred years; while Chapter 13 in this volume considers a range of techniques, including palaeotechniques, to investigate freshwater changes over time.

6.5.4 Texture, particle size, porosity, penetrability, and soil moisture

There are many physical attributes that could be used to describe wetland substrate and which are key to understanding different freshwater ecological processes. For example, quantifying the penetrability of substrate to explain the distribution of freshwater mussels (Pandolfo et al. 2016); measuring substrate size in relation to macroinvertebrate diversity and abundance for tropical streams (Graca et al. 2015); describing soil texture or measuring particle size to explain the distribution of submersed macrophytes (Li et al. 2012); and examples in Moss (2018), Finlayson et al. (2018), and

Keddy (2010). Soil moisture, soil water deficit, and soil compaction measurements at the local or catchment scale may provide data for modelling ecological flows; estimating run-off from surrounding land into water bodies; understanding infiltration rates, flow pathways, and recharge processes of groundwater-fed wetlands; or for understanding upstream hydrological processes for the management of downstream riverine, floodplain, or wetland plant communities.

A range of techniques, of varying accuracy and resolution, for estimating or quantifying texture, particle size, porosity, soil moisture, and penetrability are reviewed or described by Jones et al. (2006), Chapter 2 of Wheater et al. (2011), Chapter 5 of Tiner (2017), DeLaune et al. (2013), Vasilas and Vasilas (2013), and Vasilas et al. (2013). Guy (1977) reviews these various methods including the use of sieves for estimating particle size. There are some excellent online open-access resources which cover the theory and practice of soil and substrate physical attributes, including 'Geomorphological Techniques' on the British Society for Geomorphology website; 'Methods of Soil Analysis' (see Part 4 for Physical Methods) on the Alliance of Crop, Soil and Environmental Science Societies (ACSESS) website; FAO Soils Portal; UK Soil Observatory; and Natural England Technical Information Note TIN037 on 'Soil Texture'.

For all these methods it is important to choose a technique that is appropriate to the scale of investigation or resolution required. For example, if it is only estimates of substrate texture that are required to generate hypotheses on the distribution of organisms in a wetland, then choose simple descriptors that can be undertaken in the field rather than taking precise measurements or samples back to the laboratory; for an investigation that requires quantitative data on sediment size and distribution, for example, the size ratio of movable to immovable riverbed particles in relation to redd building by spawning salmon (Riebe et al. 2014), it would be necessary to deploy a suitable sampling strategy with accurate measurement of sediment.

6.6 Examples

The following examples have been chosen to illustrate the importance of measuring physical variables to explain freshwater ecological processes. In each case, there is an explanation of why the physical variable has been measured and how it relates to management or conservation decisions.

Lake Baikal, Russia, is the deepest (>1,600 m depth from water surface), oldest (*circa* 25 million years), and most diverse (2,500 animal and 1,000 plant species) lake on earth; it contains 20 per cent of global surface (liquid) freshwater; and is listed as a UNESCO World Heritage site. It has been the focus of a huge body of research over many decades, to document, quantify, and monitor physical and biological variables, and, in particular, in relation to cultural eutrophication, pollution, global warming, and changing seasonal cover of surface ice (Izmest'eva et al. 2016). Diverse and endemic communities of diatoms (primary producers) support pelagic food webs, including crustaceans, fish, and, a top predator, the endemic Baikal seal (Moore et al. 2009). Measurements of the deepest sections of the lake were taken using hydrographic echosounders, published online by the INTAS team in 2002, and consolidated with previous databases to produce a digital bathymetric map (Sherstyankin et al. 2006) which highlights the immense stratigraphical and topographical variation with depth. The dynamic physical environment of Lake Baikal is inextricably linked to energy transfer across trophic levels, affecting food web structure and function. Key physical variables that have been monitored, among others, are water temperature, transparency by Secchi disc (depth increases as plankton density decreases), and extent of surface water ice. As water and air temperatures increase, surface ice decreases in duration and thickness, affecting cycles of cladocerans and favouring autotrophic picoplankton (chlorophyll a levels have increased) over cold water endemic diatoms. These physical changes create challenges in the long-term management of Lake Baikal, but it is vital to identify the physico-ecological linkages that drive these unique assemblages of aquatic species in order to make future policy decisions.

The second example focuses on the relationship between sediment load and fish productivity. The Mekong is a transboundary river flowing from China across Laos, Thailand, Cambodia, and into the Mekong Delta in southern Vietnam. This vast river system is important for fish (diversity, productivity, and endemism) and sediment production and discharges the highest sediment load for any river system in the world (160 million tonnes p.a.). Most of the sediments are fine sand, clay, and silt: 99 per cent of sediment grain size is smaller than small gravel (>4.75 mm) and 41 per cent is finer than coarse sand (>0.45 mm). Baran et al. (2015) review the types of sediment (including particulate organic matter), size of sediment, and the dependence of fish productivity (18 per cent of world's freshwater fish yield) on the different sediment types. The importance of sediment for transporting nutrients downstream and maintaining the fertility of the Mekong floodplains and Delta has been demonstrated (Piman and Shrestha 2017; Baran et al. 2015); and Kondolf et al. (2014) use physical and biological variables to model the impacts of the construction of 140 proposed hydropower dams on sediment supply, fish, and food security for the river basin.

Sampling the accumulation of sediment or organic material in a freshwater wetland can provide valuable data on historic catchment land use changes or long-term climate change. Wetlands where suitable substratum may be found include mires and peat-based wetlands (high organic content); floodplains with deep sedimentary environments (high mineral content); lakes, ponds, and enclosed water bodies with limited water turnover or erosion (organic or mineral content). The next two examples use riverine sediment accumulation for (a) assessing the success of a created riverine wetland, and (b) quantifying past catchment land uses and nutrient loadings. Mitsch et al. (2014) reviewed and measured gross sedimentation (bottle sediment trap), sediment accretion (horizon marker), and net sedimentation (SET, coring), to assess sediment accumulation rates in created riverine wetlands at the Olentangy River Wetland Research Park, Ohio. They calculated an accumulation rate of 30 cm over two decades, questioning the lifetime and future management of these created habitats. Floodplain and depressional freshwater wetlands in Georgia, USA, were investigated for soil accretion, sediment deposition, and nutrient accumulation (Craft and Casey 2000). Soil cores were extracted and analysed to quantify changes over 30 and 100 years. Wetland sediment accumulation for the 100-year rates (1036 g m^2 yr^{-1}) were higher than for the 30-year rates (118 g m^2 yr^{-1}) reflecting changes in land use and wetland vegetation type, productivity, and hydrology, with overstocking 100 years previously and improved catchment management in the last 30 years.

The final example is for a blanket peat wetland (long-term accumulation of organic substrate; ombrotrophic) in Keighley Moor Reservoir catchment in northern England, where the underlying geology is impermeable Carboniferous Millstone Grit series (Blundell and Holden 2015). They used simple stratigraphic techniques (gouge corer, Troels-Smith scheme) to produce a catchment-wide overview of peatland development, and together with macrofossil reconstruction from a central core (monolith tin for 0–0.5 m depth; narrow gauge Russian corer for depths to 1.9 m), provided the historical context for peatland restoration targets. They found that the current lack of *Sphagnum* moss was atypical of past vegetation assemblages and was due to wildfire in the early 1900s and subsequent controlled burning. This study of peat substrate provided the evidence for the management of the wetland by rewetting the site, restricting further controlled burns to encourage *Sphagnum* growth, and therefore decreasing levels of dissolved organic carbon leaving the site and reducing flood peaks by slowing run-off.

References

Allen, R.G., Pereira, L.S., Raes, D., and Smith, M. (1998). Crop evapotranspiration-Guidelines for computing crop water requirements. FAO Irrigation and drainage paper 56, FAO, Rome. Available online.

Althouse, B., Higgins, S., and Vander Zanden, M.J. (2014). Benthic and planktonic primary production along a nutrient gradient in Green Bay, Lake Michigan, USA. *Freshwater Science*, 33, 487–98.

Anderson, C.W. (2005). National Field Manual for the Collection of Water-Quality Data. (TWRI Book 9), Vol. 6.7 Turbidity. Available from US Geological Survey website.

Araujo, C.F., Nolasco, M.M., Ribeiro, A.M.P. and Ribeiro-Claro, P.J.A. (2018). Identification of microplastics using Raman spectroscopy: Latest developments and future prospects. *Water Research*, 142, 426–440

Ausden, M. (2007). Habitat Management for Conservation: A Handbook of Techniques. Oxford University Press.

Babbitt, K.J., Veysey, J.S., and Tanner, G.W. (2010). Measuring habitat. In: C.K. Dodd (ed.) Amphibian Ecology and Conservation: A Handbook of Techniques. Oxford University Press.

Baran, E., Guerin, E., and Nasielski, J. (2015). Fish, Sediment and Dams in the Mekong River. CGIAR Research Program on Water, Land and Ecosystems (WLE), International Water Management Institute (IWMI) and WorldFish, available online.

Bartram, J. and Ballance, R. (eds.) (1996). Water Quality Monitoring—A Practical Guide to the Design and Implementation of Freshwater Quality Studies and Monitoring Programmes. United Nations Environment Programme and the World Health Organization (UNEP/WHO). Available online on WHO website.

Batzer, D.P. and Sharitz, R.R. (eds) (2006). Ecology of Freshwater and Estuarine Wetlands. University of California Press, Berkeley.

Bayram, A. and Kenanoğlu, M. (2016). Variation of total suspended solids versus turbidity and Secchi disk depth in the Borçka Dam Reservoir, Çoruh River Basin, Turkey. *Lake and Reservoir Management*, 32, 209–24.

Becker Nevers, M. and Whitman, R.L. (2010). Lake Monitoring Field Manual. Lake Michigan Ecological Research Station. Available on US Geological Survey website.

Berner, E.K. and Berner, R.A. (2012). Global Environments: Water, Air and Geochemical Cycles. 2nd edition, Princeton University Press, New Jersey.

Bidwell, J.R. (2013) Physical and chemical monitoring of wetland water. In: Anderson, J.T. and Davis, C.A. (eds) Wetland Techniques, Volume 1, Foundations. Springer, Dortrecht.

Blais, J.M., Rosen, M.R., and Smol, J.P. (eds) (2015). Environmental Contaminants—Using Natural Archives to Track Sources and Long-Term Trends of Pollution. Springer, Dortrecht.

Blundell, A. and Holden, J. (2015). Using palaeoecology to support blanket peatland management. *Ecological Indicators*, 49, 110–20.

Bodelier, P.L.E. and Dedysh, S.N. (2013). Microbiology of wetlands. *Frontiers in Microbiology*, 4, 79. PMC3617397.

Bonnett, S.A.F., Ross, S., Linstead, C., and Maltby, E. (2009). A Review of Techniques for Monitoring the Success of Peatland Restoration. University of Liverpool, Natural England Commissioned Reports, Number 086.

Borowiak, D. (2014). Optical properties of Polish lakes: The Secchi Disc transparency. *Limnological Review*, 14, 131–44.

Brassington, R. (2017). Field Hydrogeology. The Geological Field Guide Series. 4th edition, John Wiley & Sons, New Jersey and West Sussex.

Brooks, S., Stoneman, R., Hanlon, A., and Thom, T. (2014). Conserving Bogs—The Management Handbook. Yorkshire Peat Partnership. Book available online.

Buendia, C., Gibbins, C.N., Vericat, D., Batalla, R.J., and Douglas, A. (2013). Detecting the structural and functional impacts of fine sediment on stream invertebrates. *Ecological Indicators*, 25, 184–96.

Bunte, K., Abt, S.R., Potyondy, J.P., and Swingle, K.W. (2008). A comparison of coarse bedload transport measured with bedload traps and Helley-Smith samplers. *Geodinamica Acta*, 21, 53–66.

Comas, X., Terry, N., Slater, L., Warren, M., and Kolka, R. (2015). Imaging tropical peatlands in Indonesia using ground-penetrating radar (GPR) and electrical resistivity imaging (ERI): implications for carbon stock estimates and peat soil characterization. *Biogeosciences*, 12, 2995.

Corcoran, P.L., Norris, T., Ceccanese, T., Walzak, M. J., Helm, P.A., and Marvin, C.H. (2015). Hidden plastics of Lake Ontario, Canada and their potential preservation in the sediment record. *Environmental Pollution*, September 2015, 204, n17–25.

Craft, C.B. and Casey, W.P. (2000). Sediment and nutrient accumulation in floodplain and depressional freshwater wetlands of Georgia, U.S.A. *Wetlands*, 20, 323–32.

Cunningham, W.L. and Schalk, C.W. (2011). Groundwater technical procedures of the US Geological Survey (No. 1-A1). Available online from US Geological Survey website.

Davies-Colley, R.J., Vant, W.N., and Latimer, G.J. (1984). Optical characterisation of natural waters by PAR measurement under changeable light conditions. *New Zealand Journal of Marine and Freshwater Research*, 18, 455–60.

Davies-Colley, R.J. and Smith, D.G. (2001). Turbidity, suspended sediment, and water clarity: a review. *Journal of the American Water Resources Association*, 37, 1085–101.

Delaune, R.D., Reddy, K.R., Richardson, C.J., and Megonigal, J.P. (eds) (2013). Methods in Biogeochemistry of Wetlands. Number 10 in the Soil Science Society of America Book Series, Soil Science Society of America, Wisconsin.

Finch, J.W. and Hall, R.L. (2001). Estimation of Open Water Evaporation: A Review of Methods. Environment Agency R&D Handbook W6-043/HB. Available online from the UK GOV website.

Finlayson, C.M., Everard, M., Irvine, K., McInnes, R.J., Middleton, B.A., van Dam, A.A., and Davidson, N.C. (eds) (2018). The Wetland Book. Volume 1, Structure and Function, Management, and Methods. Springer.

Frid, C. and Dobson, M. (2013). Ecology of Aquatic Management. Oxford University Press.

Frew, C. (2014). Coring methods, section 4.1.1. In: Cook, S.J., Clarke, L.E., and Nield, J.M. (eds) Geomorphological Techniques (Online Edition). British Society for Geomorphology, London, UK.

Gao, F, Feng, G., Ouyang, Y., Wang, H., Fisher, D., Adeli, A. and Jenkins, J. (2017). Evaluation of reference evapotranspiration methods in arid, semiarid, and humid regions. Journal of the American *Water Resources Association*, 53, 791–808.

Gao, J. (2009). Bathymetric mapping by means of remote sensing: methods, accuracy and limitations. *Progress in Physical Geography*, 33, 103–16.

Gell, P.A. and Finlayson, C.M. (2016). Understanding change in the ecological character of internationally important wetlands. *Marine and Freshwater Research*, 67, entire volume.

Goudie, A.S. (ed.) (1990). Geomorphological Techniques. 2nd edition, Routledge, Taylor and Francis Group, London and New York.

Graca, M.A.S., Ferreira, W.R., Firmiano, K., Franca, J., and Callisto, M. (2015). Macroinvertebrate identity, not diversity, differed across patches differing in substrate particle size and leaf litter packs in low order, tropical Atlantic forest streams. *Limnetica*, 34, 29–40.

Gray, J.R., Glysson, G.D., Turcois, L.M., and Schwarz, G.E. (2000). Comparability of Suspended-Sediment Concentration and Total Suspended Solids Data. US Geological Survey, Water Resources Investigations Report 00–4191. Available online on US Geological Survey website.

Guaraglia, D.O. and Pousa, J.L. (2014). Introduction to Modern Instrumentation: For Hydraulics and Environmental Sciences. De Gruyter Open Ltd, Berlin and Warsaw.

Guy, H.P. (1977). Laboratory Theory and Methods for Sediment Analysis. Techniques of Water-Resources Investigations, Book 5, US Geological Survey. Available from the USGS website.

Haag, K.H., Lee, T.M., and Herndon, D.C. (2005). Bathymetry and Vegetation in Isolated Marsh and Cypress Wetlands in the Northern Tampa Bay Area, 2000–2004. U.S. Geological Survey Scientific Investigations Report 2005–5109. Available online.

Hiscock, K.M. (2006). Hydrogeology: Principles and Practice. Blackwell, Oxford.

Holden, J., Burt, T.P., and Vilas, M. (2002). Application of ground penetrating radar to the identification of subsurface piping in blanket peat. *Earth Surface Processes and Landforms*, 27, 235–49.

Hongve, D. and Akesson, G. (1996). Spectrophotometric determination of water colour in hazen units. *Water Research*, 30, 2771–5.

Horton, A.A., Walton, A., Spurgeon, D.J., Lahive, E., and Svendsen, C. (2017). Microplastics in freshwater and terrestrial environments: Evaluating the current understanding to identify the knowledge gaps and future research priorities. *Science of The Total Environment*, 586, 127–41.

Hutchinson, R. (2004). Temperature effects on sealed lead acid batteries and charging technologies to prolong cycle life. Sandia Report SAND2004-3149. Sandia National Laboratories, Albuquerque, New Mexico. Available online.

Ibáñez, C., Sharpe, P.J., Day, J.W., Day, J.N., and Prat, N. (2010). Vertical Accretion and Relative Sea Level Rise in the Ebro Delta Wetlands (Catalonia, Spain). *Wetlands*, 30, 979–88.

Irmak, S. and Haman, D.Z. (2017). Evapotranspiration: Potential or Reference? University of Florida IFAS Extension Publication #ABE 343. Available online from the University of Florida Extension website.

Izmest'eva, L.R., Moore, M.V., Hampton, S.E., Ferwerda, C.J., Gray, D.K., Woo, K.H., Pislegina, H.V., Krashchuk, L.S., Shimaraeva, S.V., and Silow, E.A. (2016). Lake-wide physical and biological trends associated with warming in Lake Baikal. *Journal of Great Lakes Research*, 42, 6–17.

Jablonska, E., Falkowski, T., Chormański, J., Jarzombkowski, F., Kłosowski, S., Okruszko, T., Pawlikowski, P., Theuerkauf, M., Wassen, M.J., and Kotowski, W. (2014). Understanding the long-term ecosystem stability of a fen mire by analyzing subsurface geology, eco-hydrology and nutrient stoichiometry—Case study of the Rospuda Valley (NE Poland). *Wetlands*, 34, 815–28.

Jewson, D.H., Talling, J.F., Dring, M.J., Tilzer, M.M., Heaney, S.I., and Cunningham, C. (1984). Measurement of photosynthetically available radiation in freshwater: comparative tests of some current instruments used in studies of primary production. *Journal of Plankton Research*, 6, 259–73.

Jones, J.C., Reynolds, J.D., and Raffaelli, D. (2006). Environmental variables. In: W. J. Sutherland (ed.) Ecological Census Techniques. 2nd edition, Cambridge University Press.

Keddy, P.A. (2010) Wetland Ecology: Principles and Conservation. 2nd edition, Cambridge University Press.

Kemp, M., de Kock, K.N., Zaayman, J.L., and Wolmarans, C.T. (2016). A comparison of mollusc diversity between the relatively pristine Marico River and the impacted Crocodile River, two major tributaries of the Limpopo River, South Africa. *Water SA*, 42, 253–60.

Kernan, M., Battarbee, R.W., and Moss, B. (eds) (2010). Climate Change Impacts on Freshwater Ecosystems. Wiley-Blackwell.

Koenings, J.P. and Edmundson, J.A. (1991). Secchi disk and photometer estimates of light regimes in Alaskan lakes: Effects of yellow color and turbidity. *Limnology and Oceanography*, 36, 91–105.

Kondolf, G.M., Rubin, Z.K., and Minear, J.T. (2014). Dams on the Mekong: Cumulative sediment starvation. *Water Resources Research*, 50, 5158–69.

Lakes Environmental Association (2015). 2014 Temperature Monitoring Project Report. Available from Maine Lakes website.

Lei, Y., Yang, K., Wang, B., Sheng, Y., et al. (2014). Response of inland lake dynamics over the Tibetan Plateau to climate change. *Climate Change*, 125, 281–90.

Leitão, R.P., Zuanon, J., Mouillot, D., Leal, C.G., Hughes, R.M., Kaufmann, P.R., Villéger, S., Pompeu, P.S., Kasper, D., de Paula, F.R., Ferraz, S.F.B., and Gardner, T.A. (2018). Disentangling the pathways of land use impacts on the functional structure of fish assemblages in Amazon streams. *Ecography*, 41, 219–32.

Lhommel, J.P. (1997). Towards a rational definition of potential evaporation. *Hydrology and Earth System Sciences Discussions*, 1, 257–64.

Li, Z.Q., Kong, L.Y., Zhang, M., Cao, T., Ju, J., Wang, Z.X., and Lei, Y. (2012). Effect of substrate grain size on the growth and morphology of the submersed macrophyte *Vallisneria natans* L. *Limnologica- Ecology and Management of Inland Waters*, 42, 81–5.

Liedermann, M., Gmeiner, P., Pessenlehner, S., Haimann, M., Hohenblum, P. and Habersack, H. (2018). A methodology for measuring microplastic transport in large and medium rivers. *Water*, 10, 414

Ligon, F.K., Nakamoto, R.J., Harvey, B.C., and Baker, P.F. (2016). Use of streambed substrate as refuge by steelhead or rainbow trout *Oncorhynchus mykiss* during simulated freshets. *Journal of Fish Biology*, 88, 1475–85.

Long, M.H., Rheuban, J.E., Berg, P., and Zieman, J.C. (2012). A comparison and correction of light intensity loggers to photosynthetically active radiation sensors. *Limnology and Oceanography: Methods*, 10, 416–24.

Los Huertos, M. and Smith, D. (2013). Wetland bathymetry and mapping. In Anderson, J.T. and Davis, C.A. (eds) Wetland Techniques, Volume 1, Foundations. Springer, Dortrecht, 49–86.

Low, R., Farr, G., Clarke, D., and Mould, D. (2018). Hydrological Assessment and Monitoring of Wetlands. In: Finlayson, C.M. et al. (eds) The Wetland Book. Springer.

Man'kovsky, V.I. (2001). Method for the evaluation of the spectral values of underwater quantum irradiance within the band of photosynthetically active radiation according to the depth of visibility of a Secchi disk. *Physical Oceanography*, 11, 299–304.

McBride, A., Diack, I., Droy, N., Hamill, B., Jones, P., Schutten, J., Skinner, A., and Street, M. (eds) (2011). The Fen Management Handbook. Scottish Natural Heritage, Perth. Book available online.

McNeish, R.E., Kim, L.H., Barrett, H.A., Mason, S.A., Kelly, J.J. and Hoellein, T.J. (2018). Microplastic in riverine fish is connected to species traits. *Nature Scientific Reports*, 8, 11639.

Mitsch, W.J., Gosselink, J.G., Anderson, C.J., and Zhang, L. (2009). Wetland Ecosystems. John Wiley & Sons, New Jersey.

Mitsch, W.J., Nedrich, S.M., Harter, S.K., Anderson, C., Nahlik, A.M., and Bernal, B. (2014). Sedimentation in created freshwater riverine wetlands: 15 years of succession and contrast of methods. *Ecological Engineering*, 72, 25–34.

Mitsch, W.J. and Gosselink, J.G. (2015). Wetlands. 5th edition, John Wiley & Sons, New Jersey.

Moore, M.V., Hampton, S.E., Izmest'eva, L.R., Silow, E.A., Peshkova, E.V., and Pavlov, B.K. (2009). Climate Change and the World's 'Sacred Sea'—Lake Baikal, Siberia. *BioScience*, 59, 405–17.

Moreno-Mateos, D., Power, M.E., Comin, F.A., and Yockteng, R. (2012). Structural and functional loss in restored wetland ecosystems. *PLOS Biology*, 10, e1001247.

Moss, B. (2018). Ecology of Freshwaters: Earth's Bloodstream. Wiley-Blackwell.

Myre E. and Shaw, R. (2006). The turbidity tube: simple and accurate measurement of turbidity in the field. Michigan Technological University. Available online.

Okalebo, J.R., Gathua, K.W., and Woomer, P.L. (2002) Laboratory Methods of Soil and Plant Analysis—A Working Manual. TSBF—CIAT and SACRED Africa, Kenya. Available online.

Oldfield, F. and Richardson, N. (1990) Peats and lake sediments: formation, stratigraphy, description and nomenclature. In A.S. Goudie (ed.) Geomorphological Techniques. 2nd edition, Routledge, Taylor and Francis Group. pp. 496–528.

Pandolfo, T.J., Kwak, T.J., and Cope, W.G. (2016). Microhabitat suitability and niche breadth of common and imperilled Atlantic slope freshwater mussels. *Freshwater Mollusk Biology and Conservation*, 19, 27–50.

Patten, B.C. (ed.) (1990). Wetlands and Shallow Continental Water Bodies. Volume 1: Natural and Human Relationships. SPB Academic Publishing bv The Hague.

Piman, T. and Shrestha, M. (2017). Case Study on Sediment in the Mekong River Basin: Current State and Future Trends. UNESCO and Stockholm Environment Institute (SEI), available online.

Popielarczyk, D., Templin, T., and Łopata, M. (2015). Using the geodetic and hydroacoustic measurements to investigate the bathymetric and morphometric parameters of Lake Hańcza (Poland). *Open Geosciences*, 7, 854–69.

Reddy, K.R. and Delaune, R.D. (2008). Biogeochemistry of Wetlands: Science and Applications. CRC Press, Taylor and Francis Group.

Riebe, C.S., Sklar, L.S., Overstreet, B.T., and Wooster, J.K. (2014). Optimal reproduction in salmon spawning substrates linked to grain size and fish length. *Water Resources Research*, 50, 898–918.

Robinson, M., Bristow, C., McKinley, J., and Ruffell, A. (2013). Ground Penetrating Radar, section 1.5.5. In: Cook, S.J., Clarke, L.E., and Nield, J.M. (eds) Geomorphological Techniques (Online Edition). British Society for Geomorphology, London, UK.

Rørslett, B., Hawes, I., and Schwarz, A. (1997). Features of the underwater light climate just below the surface in some New Zealand inland waters. *Freshwater Biology*, 37, 441–54.

Sakané, N., Alvarez, M., Becker, M., Böhme, B., Handa, C., Kamiri, H.W., Langensiepen, M., Menz, G., Misana, S., Mogha, N.G., Möseler, B.M., Mwita, E.J., Oyieke, H.A., and Wijk, M.T. (2011). Classification, characterisation, and use of small wetlands in East Africa. *Wetlands*, 31, 1103.

Schoonover, J.E. and Crim, J.F. (2015). An introduction to soil concepts and the role of soils in watershed management. *Journal of Contemporary Water Research and Education*, 154, 21–47.

Schumann, M. and Joosten, H. (2008). Global Peatland Restoration Manual. Institute of Botany and Landscape Ecology, Greifswald University, Germany. International Mires Conservation Group website.

Seybold, C.A., Mersie, W., Huang, J., and McNamee, C. (2002). Soil redox, pH, temperature, and water-table patterns of a freshwater tidal wetland. *Wetlands*, 22, 149–58.

Shaw, E.M., Beven, K.J., Chappell, N.A. and Lamb, R. (2010). Hydrology in practice. CRC Press.

Sherstyankin, P.P., Alekseev, S.P., Abramov, A.M., Stavrov, K.G., De Batist, M., Hus, R., Canals, M., and Casamor, J.L. (2006). Computer-based bathymetric map of Lake Baikal. *Doklady Earth Sciences*, 408, 564–9.

Sichangi, A.W. and Makokha, G.O. (2017). Monitoring water depth, surface area and volume changes in Lake Victoria: integrating the bathymetry map and remote sensing data during 1993–2016. Model. Earth Syst. Environ., 3, 533.

Skeffington, M.S., Moran, J., Connor, Á.O., Regan, E., Coxon, C.E., Scott, N.E., and Gormally, M. (2006). Turloughs–Ireland's unique wetland habitat. *Biological Conservation*, 133(3), 265–90.

Sutherland, W.J. (ed.) (2006). Ecological Census Techniques. 2nd edition, Cambridge University Press.

Sweat, S.C., Mutiti, S., and Skelton, C.E. (2017). Application of hydrogeophysical techniques to study the distribution of a burrowing crayfish in a wetland. *Wetlands Ecology and Management*, 25, 149–58.

Thomas, S. and Ridd, P.V. (2004). Review of methods to measure short time scale sediment accumulation. *Marine Geology*, 207, 95–114.

Tiner, R.W. (2017). Wetland Indicators—A Guide to Wetland Formation, Identification, Delineation, Classification, and Mapping. 2nd edition, CRC Press, Taylor and Francis Group.

Troels-Smith, J. (1955). Characterisation of unconsolidated sediments. *Danm. geol. Unders.* Ser. IV, 3(10), 73.

Turetsky, M.R., Manning, S.W., and Wieder, R.K. (2004). Dating recent peat deposits. *Wetlands*, 24, 324–56.

US Army Corps of Engineers (1998). Engineering and Design: Using Differential GPS Positioning for Elevation Determination. Available from the Defence Technical Information Centre website.

US Department of Agriculture (2017). Field Indicators of Hydric Soils- A Guide for Identifying and Delineating Hydric Soils. Version 8.1. Available from USDA website.

US Geological Survey (2003). Proceedings of the Federal Interagency Sediment Monitoring Instrument and Analysis Research Workshop, 9–11 September 2003, Flagstaff, Arizona. US Geological Survey Circular 1276. Available from USGS website.

US Geological Survey (1999). Techniques of Water-Resources Investigations. Field methods for measurement of fluvial sediment, 03-C2. Information Services, US Geological Survey. Available from USGS website.

US Geological Survey (1980). A field calibration of the sediment-trapping characteristics of the Helley-Smith bedload sampler. Geological Survey Professional Paper 1139. Available from USGS website.

US Environmental Protection Agency (2014). Best Practices for Continuous Monitoring of Temperature and Flow in Wadeable Streams. Global Change Research Program, National Center for Environmental Assessment, Washington, DC, EPA/600/R-13/170F. Available from the National Technical Information Service website.

Van Zuiden, T. M., Chen, M. M., Stefanoff, S., Lopez, L., and Sharma, S. (2016). Projected impacts of climate change on three freshwater fishes and potential novel competitive interactions. *Diversity and Distributions*, 22, 603–14.

Vasilas, L.M. and Vasilas, B.L. (2013). Hydric Soil Identification Techniques. In Anderson, J.T. and Davis, C.A. (eds) Wetland Techniques, Volume 1, Foundations. Springer, Dortrecht.

Vasilas, B.L., Rabenhorst, M., Fuhrmann, J., Chirnside, A., and Inamdar, S. (2013). Wetland Biogeochemistry Techniques. In: Anderson, J.T. and Davis, C.A. (eds) Wetland Techniques, Volume 1, Foundations. Springer, Dortrecht.

Vepraskas, M.J. and Craft, C.B. (eds) (2016). Wetland Soils—Genesis, Hydrology, Landscapes and Classification. 2nd edition, CRC Press, Taylor and Francis Group.

Weil, R.R. and Brady, N.C. (2017). The Nature and Properties of Soils. 15th edition, Global edition, Pearson Education Ltd.

Wessels, M., Anselmetti, F., Artuso, R., Baran, R., Daut, G., Gaide, S., Geiger, A., Groeneveld, J.D., Hilbe, M., Möst, K., Klauser, B., Niemann, S., Roschlaub, R., Steinbacher, F., Wintersteller, P., and Zahn, E. (2015). Bathymetry of Lake Constance. State-of-the-art in surveying a large lake. *Hydrographische Nachrichten*, 2, 6–11.

Wetzel, R.G. and Likens, G.E. (2000). Limnological Analyses, 3rd edition. Springer, New York.

Wheater, C.P., Bell, J.R., and Cook, P.A. (2011). Practical Field Ecology—A Project Guide. Wiley-Blackwell, Oxford.

Wilcox, C. and Los Huertos, M. (2005). A simple, rapid method for mapping bathymetry of small wetland basins. *Journal of Hydrology*, 301, 29–36.

Winton, R.S., Flanagan, N., and Richardson, C.J. (2017). Neotropical peatland methane emissions along a vegetation and biogeochemical gradient. *PLoS ONE*, 12, e0187019.

Microorganisms 1

Phytoplankton, attached algae, and biofilms

David C. Sigee

Corresponding author: david.sigee@manchester.ac.uk

7.1 Introduction

Freshwater environments can be divided into two main types—standing (lentic) waters (lakes and wetlands) and flowing (lotic) waters (rivers and estuaries). In both cases algae are a major component of the ecosystem and can be grouped into three main categories—phytoplankton (free floating; Figure 7.1), benthic algae (substratum-attached; Figure 7.5), and biofilms (mainly present in the gelatinous surface film of rocks and stones; Figures 7.6 and 7.7). The environmental context of freshwater algae has direct relevance to the techniques used to sample and study them. This chapter should be read together with Chapter 8 in this volume, which deals with zooplankton and other microorganisms.

Algae are important for the study and management of freshwater bodies in a number of ways:

Ecosystem dynamics. They are an integral part of the freshwater ecosystem, acting as primary producers. Along with photosynthetic higher plants, they are the key source of biomass within the system. Phytoplankton are a major part of the food chain in the open water of lakes (pelagic food chain), while benthic algae and biofilms are more important at the edges of lakes, in wetlands and in running waters.

Index of chemical status. Their population density and species composition closely relate to the chemical status of the water body—including the loading of soluble inorganic (e.g., phosphates and nitrates) and organic nutrients, level of heavy metals, and acidity. Analysis of freshwater algae in terms of biomass and species composition can thus be used to monitor general trophic status, with some adapted species being particularly useful as bioindicators.

Analysis of human impacts and restoration. Excessive growth of freshwater algae can generate major problems in freshwater environments. This is particularly the case where human activities have led to increased soluble nutrient concentrations

Sigee, D. C., *Microorganisms 1: Phytoplankton, attached algae, and biofilms*. In: *Freshwater Ecology and Conservation: Approaches and Techniques*. Edited by Jocelyne M. R. Hughes: Oxford University Press (2019).
© Oxford University Press 2019. DOI: 10.1093/oso/9780198766384.003.0007

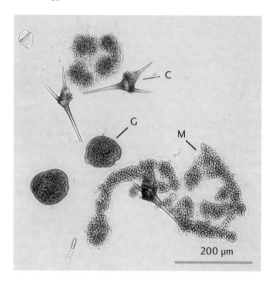

Figure 7.1 Fresh phytoplankton sample from the late summer bloom in a eutrophic lake, showing the three dominant algal genera:

C—*Ceratium* (dinoflagellate)
G—*Gomphosphaeria* (colonial blue-green)
M—*Microcystis* (colonial blue-green)
Under intense bloom conditions, these colonial blue-greens can be regarded as 'nuisance algae'—able to produce toxins and cause environmental problems.

(eutrophication) and nuisance algae such as colonial blue-greens form dense blooms (Figure 7.1). These have the capacity to bring about irreversible breakdown of the natural ecosystem and the release of toxins make the water body unusable for human activities such as extraction of drinking water and recreation.

In view of the intrinsic importance of freshwater algae, they are a major aspect of international legislation of freshwater systems—including the USA Clean Water Act (1987: amended 2002) and the European Union Water Framework Directive (2000).

7.2 Diversity of freshwater algae

Light and electron microscope examination of freshwater algae shows them to be a very diverse group of organisms in relation to size, colour and morphology (Table 7.1). The most important distinction is between prokaryotes (blue-green algae) and eukaryotes (all other algae). Prokaryotes (bacteria-like organisms) have small individual cell size, characteristic simple cell structure, unique molecular characteristics, distinctive biochemical systems, and the ability to tolerate extreme environmental conditions. Eukaryotic algae, with greater cell size and complexity, have evolved the ability to form different cell types and complex multicellular structures.

Table 7.1 *Range of freshwater algae: Taxonomic diversity, visible characteristics, and typical habitats*

Algal division (Phylum)	Taxonomic diversity[a]	Typical colour	General morphology and presence of flagella	Location within major aquatic habitats	Typical environmental examples[b]
1. Blue-green algae **Cyanophyta**	124	Blue-green	Non-flagellate unicells, filaments, or globular (may be very large) colonies.	Lakes and streams: Planktonic, benthic, biofilm	Synechocystis (u) Microcystis (gc)
2. Green algae **Chlorophyta**	302	Grass-green	Unicells, filaments, or spherical colonies. Some with flagella.	Lakes, rivers, estuaries: Planktonic, benthic, biofilm	Chlorella (u) Cladophora (f)
3. Euglenoids **Euglenophyta**	10	Various colours	Mostly flagellate, unicellular.	Lakes and ponds: Planktonic	Euglena (u) Colacium (s)
4. Yellow-green algae **Xanthophyta**	90	Yellow-green	Unicells, filaments, or colonies. Flagellate zoospores and gametes.	Wide habitat range: Planktonic, benthic, epiphytic	Botrydium (u) Tribonema (f)
5. Dinoflagellates **Dinophyta**	37	Red-brown	Unicells. All with flagella.	Lakes and estuaries: Planktonic	Ceratium(u) Peridinium (u)
6. Cryptomonads **Cryptophyta**	12	Various colours	Unicells. Most with flagella.	Lakes: Planktonic	Rhodomonas (u) Cryptomonas (u)
7. Chrysophytes **Chrysophyta**	72	Golden brown	Unicells or colonies. Some with flagella.	Lakes and streams: Planktonic	Mallomonas (u) Dinobryon (bc)
8. Diatoms **Bacillariophyta**	118	Golden brown	Centric or pennate symmetry. Mostly unicells, no flagella.	Lakes, rivers, estuaries: Planktonic, benthic, biofilm	Stephanodiscus (u) Aulacoseira (f)
9. Red algae **Rhodophyta**	25	Red	Often large filamentous or flat structures. No flagella.	Mainly streams, some lakes: Benthic	Batrachospermum(f) Hildenbrandia (ft)
10. Brown algae **Phaeophyta**	4	Brown	Filamentous, forming multicellular cushions. No flagella.	Lakes and streams: Benthic	Pleurocladia (f) Heribaudiella (f)

[a] Taxonomic diversity: number of genera within the group in the USA (Wehr and Sheath, 2003)

[b] Example morphologies u-unicell, gc—globular colony, f-filament, s-stalked, bc—branched colony, ft—flat thallus

7.2.1 Algal size

Within the phytoplankton there are four major recognised size categories:

Picoplankton Maximum diameter 0.2–2 μm. Unicellular blue-green algae.
Nanoplankton 2–20 μm. Unicellular eukaryotic algae.
Microplankton 20–200 μm. Unicellular and colonial algae.
Macroplankton 200 μm–2 mm. Large colonial blue-green algae

Free-floating algae thus have a size range of about 10^4, equivalent to the size range seen in a tropical rain forest. Benthic algae range from single celled organisms such as diatoms (able to move across the substratum) to extended morphologically complex structures such as *Cladophora*, made even more complex by the attachment of epiphytic diatoms such as *Tabellaria*. The biofilm community typically has unicellular algae, living in association with other unicells such as bacteria and protozoa (Figure 7.7)—but may develop to form multicellular growths (periphyton) extending from the surface of rocks and other surfaces (Figures 7.5 and 7.6).

7.2.2 Taxonomy

Freshwater algae can be divided into ten major groups (phyla) on the basis of microscopical appearance (Table 7.1) and biochemical/cytological characteristics. The most obvious criteria, when observed microscopically in the living state, are colour, overall morphology (unicellular/multicellular), and motility (presence or absence of flagella). Biochemical features provide an important diagnostic tool, with pigmentation—specific chlorophylls and associated carotenoids being particularly relevant. Increasingly, molecular technology is becoming important in defining major algal groups and as a diagnostic tool.

7.3 Standing (lentic) and flowing (lotic) waters

Algal diversity also extends to their location within a particular ecosystem and whether that is a standing or flowing system. Although standing and flowing aquatic systems appear to comprise well defined ecosystems, the distinction is not absolute—slow flowing rivers may show some lentic characteristics (e.g., extensive phytoplankton populations), while rapidly flooded wetlands may have lotic features such as growths of benthic algae. Some general distinctions between lentic and lotic systems in relation to their algal populations are shown in Table 7.2, including major differences in the occurrence of planktonic and benthic populations.

7.3.1 The balance between planktonic and benthic algae

Within a particular ecosystem, freshwater algae occur both within the main water body (free floating: planktonic), or associated with substratum (benthic)—which includes mud and stones at the bottom of lakes and rivers. Substratum-associated algae may be permanently attached at a particular site, lie loose on solid surfaces (e.g., overwintering blue-green algal spores) or may actively move across the surface (e.g., pennate diatoms). Algae may also be epiphytically attached to phytoplankton (Sigee 2004) and to the submerged surfaces of higher plants such as pondweeds, water lilies, and rushes (Bellinger and Sigee 2015).

Table 7.2 *Freshwater algae in standing and flowing aquatic systems. Major algal communities are shown for typical standing (lentic) and flowing (lotic) aquatic systems, along with the main physical and biological characteristics that influence the algal populations.*

Ecosystem	Major algal community	Major physical characteristics	Major biological characteristics
LENTIC SYSTEMS			
Lake	Phytoplankton	Stratification Wind-generated turbulence	Nutrient competition Zooplankton grazing Parasitism
Wetlands			
(a) Flood plain	Phytoplankton Epiphytic algae	Periodic desiccation	Competition between algae and
(b) Permanent	Benthic algae	Periodic flooding	macrophytes
LOTIC SYSTEMS			
River	Benthic algae and biofilms	Linear flow Flow-generated turbulence	Colonisation Competition Biofilm grazing
Estuary			
(a) Mudflats	Epipelic biofilms	Desiccation High light Exposure to salt water	Competition and grazing at mud surface
(b) Outflow channels	Phytoplankton	Mixing with saltwater Turbidity	Zooplankton grazing in water column

In many systems a single species may occur both in planktonic and substrate-associated form and its presence in a particular state represents a balance within the ecosystem. Examples of this include:

Standing waters

Colonial blue-green algae, typical of eutrophic lakes, have a major planktonic phase as a bloom-forming population, but form resistant spores at the end of the growing season—sedimenting to the mud at the bottom of the lake and overwintering there before regenerating to form new planktonic populations in early spring (Misson and Latour 2012) and early summer.

Running waters

Algae such as *Cladophora* are able to maintain their position within a fast flowing current due to their basal attachment, but extensive growths may break off to form plankton filaments which drift with water flow and can colonise new areas.

7.4 Phytoplankton sampling and assessment

The appearance of a typical phytoplankton sample from a eutrophic lake is shown in Figure 7.1. Analysis of phytoplankton populations typically involves initial sampling (collection strategy, method of collection, sample concentration) followed by assessment (total biomass, species counts, data analysis)—as summarised in Figure 7.2.

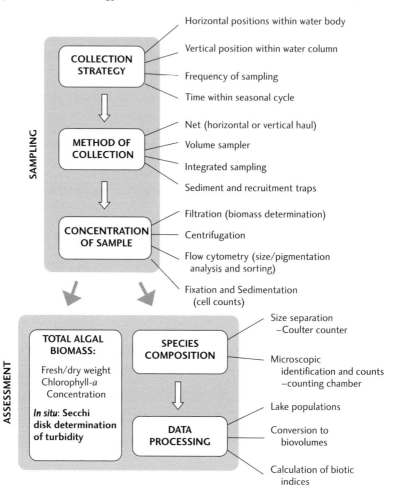

Figure 7.2 Standard approaches to phytoplankton sampling and assessment.

7.4.1 Phytoplankton sampling

Collection strategy

Collection of phytoplankton samples is labour-intensive, and it is important to match the collection strategy with the objectives of the study. Schneider (2009), for example, details the calculations required to estimate the scope of research programmes in space and time. Where a particular water body is being intensively investigated, sampling might be carried out at monthly (or more frequent) intervals throughout the entire seasonal cycle, on each occasion taking samples from at least three sites on the lake to allow calculation of mean values with confidence limits. Details of vertical positioning within the water column may also be important. A less intensive approach might involve simply collecting a surface sample at bloom periods within the seasonal cycle to get information on maximum biomass and species of algae present.

Method of collection

Collection of phytoplankton typically occurs from the top stratified layer (epilimnion) of the standing water body using a trawl net, volume sampler, or integrated collection tube at a fixed point in time (Figure 7.3). A well-established sampling procedure involves the use of a phytoplankton net, which is dragged behind the boat at steady uniform speed over a set period of time. The mesh size of the net (typically about 50 μm) results in retention of much of the phytoplankton, but smaller sized unicellular algae may pass through. Accumulation of zooplankton with the algal sample is reduced by placing a zooplankton net in front of the phytoplankton net, removing species such as *Daphnia*—which may alter the algal species composition by continued ingestion and may also affect determination of phytoplankton biomass by adding gut contents to the estimate. Net sampling is particularly useful where phytoplankton populations are low and a relatively large volume of surface water can be trawled to obtain enough sample for analysis.

One drawback of net sampling is that it is difficult to get an accurate value of the volume of water that has been trawled and thus the algal count per unit volume of water. This is best obtained from volume samplers or by integrated sampling at a particular site (Figure 7.3). In the case of volume samplers, a standard collecting device such as a Van Dorn Sampler (Bellinger and Sigee 2015) can be lowered to the required depth, where it fills with water. A weight is then dropped down the suspension cord to close to close the upper and lower stoppers, and the sample brought to the surface. Integrated sampling involves lowering an open flexible weighted tube from the water surface (boat), collecting a vertical sample within the epilimnion to a depth of 5 m. The weighted end can then be raised to the surface via a cord, and the entire contents

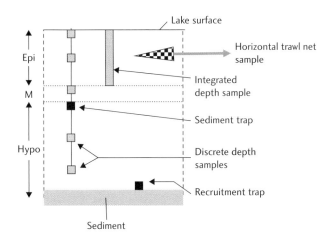

Figure 7.3 Methods for sampling lake phytoplankton.

Epilimnion (Epi)—integrated and trawl net samples
Hypolimnion (Hypo)—sediment and recruitment traps
Entire water column, including the metalimnion (M)—discrete depth samples
Adapted from Bellinger and Sigee (2015).

of the tube poured into a collecting vessel. The integrated sample thus represents a phytoplankton mixture within the top 5 m of the water column that can be counted and expressed per unit volume of water.

In addition to the above procedures, phytoplankton samples can also be obtained over a period of time in an open sediment trap (collecting live and dead cells sedimenting through the water column) or a recruitment trap (collecting benthic algae that are rising from the sediment). For example, Sigee et al. (2007) used sediment traps over a relatively short (nine days) period in their studies on cell death during late season decline in a eutrophic lake.

Sample concentration

Phytoplankton cells and colonies within the environmental samples may need to be concentrated for further analysis. An integrated sample, for example, can be subdivided into separate aliquots which can be concentrated in different ways:

Filtration. Samples for dry weight or total pigment analysis are normally brought back to the laboratory and filtered as rapidly as possible using a cellulose acetate filter membrane (0.45 µm) or glass fibre filter (Whatman GF/C, 1.2 µm).

Centrifugation. Microscopy of live cells, observing colours and motility in the fresh state, provides a useful preliminary to more detailed observation and counts of chemically fixed cells. Samples of fresh phytoplankton can be rapidly concentrated, if necessary, by gentle centrifugation.

Sedimentation. Cell/colony counts are normally carried out on samples that have been chemically fixed and then concentrated by sedimentation. Chemical fixation is normally carried out immediately upon collection by the addition of a few drops of iodine solution (kills and stains cells for light microscopy) or by addition of buffered glutaraldehyde (kills cells and maintains structure for light/scanning electron microscopy). Iodine is routinely used for standard light microscope counts, but has the disadvantage that original colour is completely lost. The iodine sample is normally concentrated back in the laboratory by placing a well-mixed aliquot in a 100 ml measuring cylinder, allowing the cells to sediment over a 24-hour period, then removing the top 90 ml. Fixation increases the density of cells and aids the sedimentation process.

7.4.2 Phytoplankton assessment

Assessment of the phytoplankton population within concentrated environmental samples involves determination of total biomass, analysis of species composition and data processing.

Total phytoplankton population: estimation of algal biomass

Although phytoplankton biomass can be directly obtained from filtered lake water samples as dry weight, it is normally expressed as total biovolume (from microscope population counts) or as chlorophyll-a concentration. The latter can be determined by filtration of integrated water samples, extraction of chlorophyll using an organic solvent such as ethanol, and then carrying out a spectroscopic analysis to derive the pigment concentration (Bellinger and Sigee 2015). Changes in lake surface water chlorophyll-a concentration provide a useful index of temporal shifts in phytoplankton biomass and can be inversely related to changes in lake transparency, as determined by surface visualisation of a sectored marker (Secchi disk) within the water column.

Species composition

Identification and counts of algal species within trawl net or integrated water samples is routinely carried out under a light microscope (× 40 objective)—see Figure 7.1.

Identification. Identification of most algae to genus and species level is normally achieved using a light microscope (× 40 or × 100 objective)—carried out on the basis of size, colour (fresh sample), and morphology in reference to a standard key (e.g., Bellinger and Sigee 2015). More detailed analysis may be useful with certain groups of algae, including diatoms (scanning electron microscopy) and unicellular blue-greens (molecular analysis). See Box 8.1 in Chapter 8.

Algal counts. These are normally carried out by pipetting 1 ml of a well-mixed sample into the well of a graduated counting slide (Sedgewick Rafter Slide: Figure 7.4) and counting individual cells and colonies within individual squares according to standard procedures (Bellinger and Sigee 2015). The resulting population counts provide direct quantitative information on algal numbers per unit water volume in lake surface waters in the case of the integrated water samples, but only relative proportions within the overall phytoplankton population in the case of the trawl net samples, where the volume of sampled lake water is difficult to determine.

Data processing

Information obtained from microscopical analysis of the environmental samples can be processed in various ways—including conversion to biovolumes and calculation of biotic indices. The software 'Opticount' can be used to aid the counting process for final count and biovolume estimation (Hepperle 2004).

Algal biovolumes. The average volume of individual algae (cells or colonies) within the sample is referred to as the 'mean unit volume' and can be estimated (as μm^3) by measurement of size and approximating the alga to a standard shape (Reynolds and Bellinger 1992). Hillebrand et al. (1999) described and used the geometrical shape method for algal biovolume estimation. The 'species population biovolume' is the product of mean unit volume and count for a particular species, and provides a measure of the volume occupied by a single species population ($\mu m^3 L^{-1}$). The cumulative volume of all algal species within the sample gives the 'total phytoplankton biovolume' ($\mu m^3 L^{-1}$), providing an estimate of phytoplankton biomass that can be related to chlorophyll-a concentration and Secchi disk measurement.

Biotic indices. Species counts obtained from environmental samples can be used to assess phytoplankton diversity. This can be assessed in various ways (Bellinger and Sigee 2015), including species richness (total number of species, Margalef index), species dominance (Simpson index), and as a combined index of diversity (Shannon-Wiener index). Phytoplankton diversity is also linked to overall growth (productivity), as indicated by total biomass. A recent study by Skácelová and Leps (2014), for example, on standing

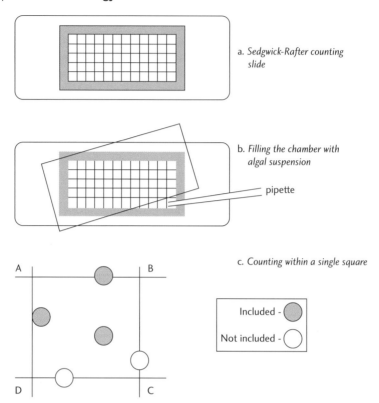

a. *Sedgwick-Rafter counting slide*

b. *Filling the chamber with algal suspension*

pipette

c. *Counting within a single square*

A   B

D   C

Included -
Not included -

Figure 7.4 Phytoplankton Counts: The Sedgwick-Rafter slide.
a. Slide with counting chamber, showing raised rim and central squared area.
b. Coverslip displaced to allow chamber infill with phytoplankton suspension.
c. Hypothetical square, showing protocol for counting. Single cells, filaments, or colonies lying inside the square or touching two sides (AB, AD) are counted, but those touching BC, DC are not. Full details of counting procedure are given in Bellinger and Sigee (2015). Adapted from Bellinger and Sigee (2015).

waters in the Czech Republic showed that diversity was minimal under conditions of low and high productivity, but elevated at intermediate growth levels.

7.4.3 Non-standard procedures for monitoring phytoplankton

In addition to the standard procedures outlined in Figure 7.2, where lake samples are obtained, concentrated and analysed by light microscopy, other approaches can also be adopted to monitor phytoplankton populations in standing waters. These include:

Chromatography: pigment analysis

Rather than making detailed and time-consuming cell counts, simple spectral analysis of phytoplankton samples using high performance liquid chromatography (HPLC) can provide a rapid assessment of total biomass (Chl-a concentration) and diagnostic carotenoid pigments. These allow quantification of major phytoplankton groups and include fucoxanthin (diatoms), zeaxanthin (blue-greens) alloxanthin (cryptomonads), and peridinin (dinoflagellates). This information may be all that is required if the researcher is looking at, for example, broad changes in eutrophication within a large study area. Paerl et al. (2005) used this approach, coupled with a diagnostic computer program, in a phytoplankton community analysis of various estuary systems in the Southern United States.

Flow cytometry: pigment analysis

This is a rapid technology that allows the analysis of millions of single-cells from aqueous samples in 1-5 minutes. It measures the scattered light and the fluorescence emitted by the cells when passing one by one though a zone of intense illumination (laser beam) in a fast flow medium (Adan et al. 2017). It can be used to characterise and identify phytoplankton communities based on the composition of their fluorescent pigments. The instrument is equipped with blue (488 nm) and red (638 nm) excitation lasers and two to eight emission filters. Flow cytometry is a viable approach for monitoring the changing seasonal patterns of abundance, composition, and biovolume of phytoplankton in both lotic (Read et al. 2014) and lentic (Pomati et al. 2013) waters.

Monitoring buoys

The use of remote sensing buoys, equipped with a depth probe to determine changes in chlorophyll-a concentration (total algal biomass), turbidity (total planktonic biomass), specific conductivity (total ion concentration), and water temperature can be particularly useful in looking at lake dynamics. The monitoring buoy illustrated in Bellinger and Sigee (2015), for example, was permanently stationed in the lake and allowed data to be obtained without making a direct visit to the site. The buoy provided a detailed daily and seasonal time sequence that would not otherwise have been available, with information obtained both from surface waters and different depths within the water column.

Molecular analysis

Information obtained from extracted DNA can give precise data on species composition of environmental samples and is being increasingly used in environmental analysis. This approach avoids time-consuming microscopical identification and counting procedures. It also has the advantages that it can identify algae occurring at low numbers (due to DNA amplification), can be used in environments where sampling and subsequent cell counts may be difficult (e.g., radioactive sites) or where the algae are relatively inaccessible (e.g., biofilms). It is also particularly useful in identifying algal species where morphological differences are minimal (e.g., unicellular blue-green algae). Molecular probes have been used to detect key nuisance organisms such as the potentially toxic dinoflagellate *Pfiesteria* (Litaker et al. 2003) and bioindicators such as N_2-fixing blue-green algae (Stancheva et al. 2013). Potential drawbacks with molecular analysis are that identification may by tentative (sometimes only to genus level) and taxonomic quantitation (relative numbers of different algae) is difficult. It may thus be difficult to extrapolate molecular data to species populations and total biomass.

7.5 Substratum-associated algae: from biofilm to benthic communities

This section focuses particularly on algae that are associated with the inorganic substratum, dealing initially with biofilms (containing unicellular organisms) then leading on to more extended benthic communities (mainly multicellular algae; Figure 7.5). Such bottom-living organisms are particularly prominent in shallow flowing waters, where adhesion to the substratum is essential to maintain position within the water body and light is able to penetrate to the depth at which the algae are growing.

7.5.1 Biofilms

Freshly exposed surfaces such as rocks and stones are rapidly colonised by pioneer bacteria and algae, with a sequence of development which leads to biofilm formation. The developing biofilm can vary from a simple monolayer of algae to a complex mixture of microorganisms (unicellular algae plus bacteria, fungi, and protozoa) within a layer of polysaccharide (the biofilm matrix)—as illustrated in Figure 7.6. In rapidly flowing streams and on unstable sediments the biofilm community represents the conclusion of the colonisation sequence—seen, for example, as the diatom biofilm on estuarine sediments (Sutherland et al. 1998) In more stable conditions the biofilm can

Figure 7.5 River periphyton. View of fresh samples obtained by scraping stone surfaces.

Main picture: Filamentous mat of *Phormidium* (P—blue-green) attached to a dense layer of underlying organic debris (D).
Inset: Stellate cluster of diatoms (possibly *Synedra*) radiating out from an original point of attachment—scale bar: 200 μm.

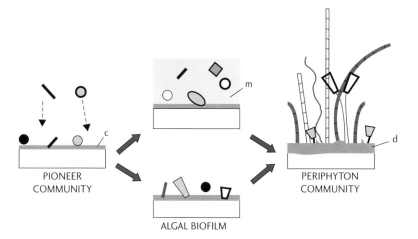

Figure 7.6. Algal colonisation of inorganic surfaces: Formation of biofilm and periphyton communities.

The initial pioneer community typically develops into a biofilm, which can be a simple monolayer of algae (lower figure) or a mixture of organisms within a mucilaginous matrix (top figure). A dense mat of filamentous growths (periphyton) can develop from either of these. Each community has its own region of associated organic material: conditioning layer (c), mucilage matrix (m), and organic debris (d). Diagrammatic representation (not to scale) with algae shown as solid objects or elongate filaments.
Adapted from Bellinger and Sigee (2015).

develop into an extensive algal community—the periphyton. This is no longer enclosed within a gelatinous matrix and typically includes both filamentous blue-green algae and diatoms (Figure 7.5), some attached by elongate stalks (Figure 7.6).

The different components of a mucilaginous biofilm (Figure 7.7) can be visualised by a range of light and electron microscope techniques (Sigee 2004; Bellinger and Sigee 2015) and comprise a structured polysaccharide matrix, water channels, and a range of microorganisms—bacteria, unicellular algae, protozoa, and fungi. The biofilm is in dynamic equilibrium with the overlying water—involving microorganism exchange, loss of biofilm matrix floccules, and gaseous/nutrient exchange. Aspects of study include determination of total biofilm biomass, algal biomass, microbial community species composition, and biofilm structure.

Total biofilm biomass

Total biofilm biomass includes the organic matrix plus inorganic constituents and associated microorganisms. The organic matrix is the main component and is normally used as the general index of biomass—measured as carbohydrate. This can be expressed as monosaccharide units per unit area (μg glucose cm^{-2}) or more simply as total carbon (μg C cm^{-2}). As an example of natural systems, Sutherland et al. (1998) used a modified phenol sulphuric acid procedure with spectrophotometry to monitor carbohydrate levels in samples of estuarine sediment. Using this approach they were able to establish that stabilisation of estuarine sediments related to levels of surface biofilm.

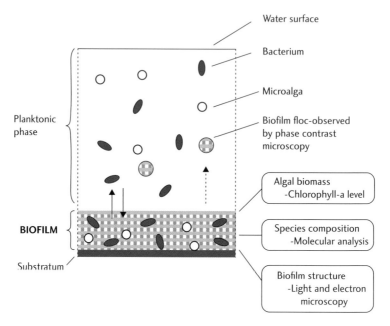

Figure 7.7 Biofilm analysis

Hypothetical community composed of bacteria and algae embedded in extracellular polymeric matrix (shaded). The biofilm is in equilibrium with a planktonic phase, which includes bacteria, microalgae, and globules of detached matrix (flocs).
Adapted from Bellinger and Sigee (2015).

Measurement of biofilm biomass has also been useful in experimental systems. In a study on the impact of wave action, Droppo et al. (2007) showed that biofilm matrix increased exponentially (after a seven-day lag period) in undisturbed conditions, but was limited by erosion and biomass loss during increased wave action.

Algal biomass

The biomass of eukaryotic algae within natural or experimental biofilms can be assessed in terms of phospholipid fatty acid (PLFA) analysis—(e.g., Droppo et al. (2007)). More usually, however, the combined biomasses of eukaryotic and prokaryotic algae are determined as chlorophyll-a concentration using similar organic solvent extraction and spectrophotometric analytical techniques similar to those noted previously (Section 7.4) for phytoplankton. Chlorophyll-a determinations can either be expressed per unit area of biofilm (μg chlorophyll-a cm^{-2}) or per unit biofilm biomass (μg chlorophyll-a mgC^{-1}).

In natural biofilms, Sutherland et al. (1998) demonstrated a clear correlation in estuarine diatom biofilms between algal and total biofilm biomass, expressed per unit area. In other systems (as with phytoplankton populations) algal biomass has been shown to relate to aquatic nutrient status. Lyon and Ziegler (2009), for example, obtained chlorophyll-a/carbon values of 2.8-19.9 μg mg^{-1} for epilithic biofilm communities of four headwater streams (Arkansas, USA)—with significant correlation to trophic status and biofilm nitrogen content (0.7–2.9 per cent total mass).

Algal growth in biofilms can also be studied on artificial substrates (e.g., plastic plates) placed in the aquatic environment. Using this approach, Jarvi et al. (2002) studied riverine biofilm growth at various points in the water column and on the riverbed, noting a biofilm algal bloom in early spring. This preceded the spring phytoplankton bloom that occurred in the water column.

Microbial species composition

Identification and enumeration of biofilm microorganisms can be carried out in two main ways—microscopically or using molecular techniques. For microscopic analysis, biofilms can be scraped off a defined area of substrate into filter-sterilised distilled water, fixed in formaldehyde to stabilise the cells, then released into the added liquid by sonication or vigorous shaking. Cell suspensions can subsequently be transferred to a sedimentation chamber or to a counting slide (e.g., Figure 7.4), identified, and counted. Augspurger et al. (2008) used this approach to make cell counts and study the stream biofilm food web, using tiles placed for set times within the lotic environment.

Information on the taxonomic composition of biofilms can also be obtained using a range of molecular and biochemical techniques—including pigment, phospholipid, and nucleotide analyses. Droppo et al. (2007) used phospholipid fatty acid (PLFA) gas chromatographic analysis to look at broad temporal shifts in the main taxonomic groups—including eukaryotes (determination of polyenoics), Gram-negative biota (monosaturated fatty acids), and Gram-positive bacteria (tertiary and branched fatty acids). More specific taxonomic information can be obtained by molecular analysis of genomic and chloroplast DNA, and enabled Droppo et al. (2007) to demonstrate a transition from bacteria to algae in the five-day biofilm, with subsequent domination by the green alga *Scenedesmus* (9d) followed by *Phormidium* (15d).

7.5.2 Benthic algae

The development of periphyton illustrated in Figure 7.6 can lead to dense communities of attached algae which are often found attached to solid inorganic substrates such as rocks and stones (epilithic algae). Mature attached communities (Figure 7.5) are frequently dominated by mats of filamentous blue-green algae such as *Phormidium*, interspersed with diatoms which may be either attached at their base (e.g., *Synedra*) or project out on stalks (e.g., *Gomphonema*). Attached algae are important ecologically in rivers and shallow lakes as a major aspect of the benthic community, and are also important experimentally for the determination of bioindices and water quality—particularly in flowing waters (Bellinger and Sigee 2015).

Sampling and processing

Epilithic algae can be removed from the stone surface at site by scraping with a sharp knife/stiff brush, or the entire stone taken back to the laboratory for subsequent analysis. A standard protocol needs to be adopted where algae are being sampled for environmental monitoring to ensure comparability between different sites. The system adopted by Kelly et al. (1995) was designed to standardise collection of diatoms for river environmental assessment and involved selection of particular boulders, washing to remove river invertebrates, removal of algae (with a stiff brush) and collection as a suspension in stream water, 24-hour sedimentation, then final chemical processing to remove surface organic material to reveal details of the diatom frustule for identification. Other protocols have been adopted for sampling attached algae in shallow lakes (e.g., King et al. 2006).

Species counts and biomass

Attached algae, particularly diatoms, are normally identified by light microscopy since the distinctive frustule ornamentation of particular species is readily detectable under phase contrast optics. Scanning electron microscopy may also be used for higher resolution. Species counts can be made using a counting chamber (Figure 7.4) and expressed per unit area of substrate where a known area of surface was sampled. Algal biomass can be estimated as chlorophyll-a concentration from chemically untreated samples using extraction/quantitation techniques noted earlier.

7.6 Assessment of eutrophication and control of nuisance algae in standing waters

Nutrient enrichment (eutrophication) of freshwater environments can lead to the growth of 'nuisance' algae (Figure 7.1) and a range of adverse consequences. The major nutrients involved, phosphates and nitrates, can enter the freshwater system internally (e.g., from phosphate-rich bedrock) or externally, by inflow of domestic effluent (including sewage), industrial waste, or from agricultural practices (e.g., application of fertilisers). The nutrient status of a lake or river depends to a large extent on its catchment area, and the occurrence of high nutrient levels with associated algal growth, may simply represent part of a natural range in lake status rather than the direct result of human activity. In this section the algal response to eutrophication is considered specifically in relation to standing waters—looking at variation in natural systems, the effects of human activities, monitoring techniques, and remedial measures to control the growth of colonial blue-green algae.

7.6.1 Variation in natural systems

According to established criteria (OECD 1982; Technical Standard Publication 1982) standing waters can be subdivided into three major categories on the combined basis of nutrient concentrations and the standing algal population (chl-a concentration)—Table 7.3. Oligotrophic lakes are typically deep mountain water bodies surrounded by low-nutrient moorland, with little inflow of phosphates and nitrates into the system. They normally have clear water, little growth of algal populations and a relatively uniform annual cycle—dominated for much of the year by diatoms. At the other end of the scale, eutrophic lakes are typically lowland water bodies, with high nutrient levels derived internally or externally from the surrounding catchment. Major algal blooms in spring (diatoms) and late summer (colonial blue-greens, dinoflagellates; Figure 7.1)

Table 7.3 *Trophic Classification of Temperate Freshwater Lakes, based on a Fixed Boundary System. Standing waters are classified according to both inorganic nutrient concentrations (non-shaded area) and phytoplankton productivity (shaded area). In addition to the three major trophic categories shown here, two further groups occur at either end of the sequence: ultra-oligotrophic and hypertrophic. Boundary values are mainly from the OECD classification system (OECD 1982) with the exception of orthophosphate and DIN, which are from a Technical Standard Publication (1982)—measured as the mean surface water concentrations during the summer growth phase.*

	Trophic category		
	Oligotrophic	Mesotrophic	Eutrophic
Nutrient concentration ($\mu g\ l^{-1}$)			
Total phosphorus (mean annual value)	4–10	10–35	35–100
Orthophosphate	<2	2–5	5–100
Dissolved inorganic nitrogen (DIN)	<10	10–30	30–100
Chlorophyll-a concentration ($\mu g\ l^{-1}$)			
Mean concentration in surface waters	1–2.5	2.5–8	8–25
Maximum concentration in surface waters	2.5–8	8–25	25–75
Turbidity (Secchi depth—m)			
Mean annual value	12–6	6–3	3–1.5
Minimum annual value	>3.0	3–1.5	1.5–0.7

lead to marked fluctuations in nutrient concentrations and also impact on populations of other microorganisms (bacteria, protozoa, fungi) and zooplankton. Under natural conditions, where dense algal blooms are a normal annual occurrence, the lake is able to return to its original state (homeostasis) on a yearly basis and no long-term ecological damage has occurred.

7.6.2 Effects of human activities

Where eutrophication occurs, it is often difficult to exclude human contribution, particularly where the surrounding land has agricultural use. In some cases, the adverse impact of human activity is very clear. Wetland areas are particularly sensitive and a good example is the wetland system (Norfolk Broads) in East Anglia, UK. During the last century this region was particularly exposed to increased nutrient input from agriculture (farming intensification) and domestic effluent (increased residential human population plus seasonal tourist influx). The resulting blooms of colonial blue-green algae were so intense that normal homeostatic control ceased to operate—zooplankton populations became reduced, fish populations became imbalanced (increased zooplankivorous fish) and the regeneration of pondweeds and other water plants (e.g., water lilies) was prevented (Madgwick 1999). In addition to the effect of external nutrients on algal growth, destruction of water lily and reed beds by boating activities removed a major zooplankton refuge, further limiting the natural control of blue-green algae. In this situation the summer algal bloom becomes very intense (peak Chlorophyll-a levels >100 µg L^{-1}). With planktonic algae concentrated at the water surface, light intensity rapidly attenuates with depth—reducing the growth of other planktonic algae and benthic macrophytes. Loss of other planktonic algae (outcompeted) means that the algal population becomes very uniform.

7.6.3 Monitoring environmental change

Why monitor?

Monitoring the process of eutrophication, long-term changes in algal populations, and deterioration of the freshwater environment is particularly important where the water body is:

- A site for recreation (e.g., boating activities) or can be accessed within a general recreational area (e.g., public parkland).
- Used as a source of drinking water.
- Part of a conservation area, where freshwater sites are being maintained in their original state or restored to an earlier condition.

In addition to these specific situations, there is a general concern that eutrophication and its adverse consequences should be minimised wherever possible. Legislation introduced by different administrations around the world tends to adopt this approach. The European Union, for example, in its Water Framework Directive (2000) adopted procedures for monitoring water quality and had a long-term aim to restore standing and running waters to an 'original state' that existed before human intervention.

Procedures

A wide range of procedures are available to record the occurrence and effects of nutrient enrichment, both in terms of past changes and current trends.

Past changes in trophic status. Where no records are available, historic changes in trophic status are normally assessed by sediment analysis. This typically involves taking a vertical core of the lake sediment and relating the occurrence of key algal species to depth within the sampled core—which can itself be dated by isotope analysis. Within the core the majority of algal remains are lost, but the silica frustules of diatoms are not degraded and allow key species to be identified and used in the assessment of environmental change (see, for example, Bennion and Battarbee 2007).

Current changes in trophic status. Techniques to monitor these are wide-ranging (Bellinger and Sigee 2015) and include:

- Direct measurement of soluble phosphate and nitrate concentrations. These are normally determined in surface waters (epilimnion) and at different times of year to relate to the OECD classification (Table 7.3).
- Determination of algal biomass at key periods of growth. Typically measured as chlorophyll-a concentration to relate to the OECD classification (Table 7.3).

- Analysis of phytoplankton taxonomic composition, looking at the occurrence of key bio-indicator species—that is, algae adapted to particular environmental parameters. Single species populations, expressed as cells/colonies ml^{-1}, can be combined in the calculation of 'bioindices'—which give a numerical assessment of trophic status (Bellinger and Sigee 2015).
- Measurement of bacterial and viral populations. Bacterial counts, determined by microscopy (fluorescent counts), plating (colonies on nutrient agar), or molecular analysis rise dramatically with increased aquatic nutrient status. Counts of bacteria such as *Escherichia coli* and *Salmonella typhi* have particular significance as indicators of sewage contamination as also do viruses such as *Coronavirus* and *Norovirus*.
- Use of invertebrate bioindicator species, with determination of bioindices. These are particularly useful in running waters, but may also be used in standing waters—particularly in shallow areas of the littoral zone.
- Observations on the macrophyte population. The occurrence, or lack, of higher plants may provide a useful indicator of trophic status. Mention has already been made of the suppression of macrophyte (particularly waterlily) beds in the Norfolk Broads (wetlands), where visible loss of macrophytes tied in with the very obvious occurrence of blooms of blue-green algae. At the other end of the trophic spectrum, lack of reed beds at the edge of the lake is typical of oligotrophic mountain lakes. Transition to mesotrophic lakes normally involves the introduction of littoral macrophytes as well as changes in the phytoplankton population.

7.6.4 Practical measures to control eutrophication and the growth of nuisance algae

Measures to limit eutrophication in standing waters tie in with control of colonial blue-green algae, blooms of which often triggered the remedial action in the first place. Control of colonial blue-greens in standing waters can be related to the general pelagic food chain (Figure 7.8) and consists of four major approaches:

Bottom-up control. This involves reduction in availability of organic and inorganic nutrients—particularly soluble phosphates and nitrates, directly limiting algal growth. Reversal of the eutrophication process, with restoration to an earlier situation, typically entails a reduction in the external loading of nutrients, including—limitation in the release of domestic effluent (e.g., sewage treatment), agricultural changes (limitation in spread of fertiliser, introduction of barrier strips around freshwater sites), and reduction in industrial discharges. Procedures to limit external loading vary with the particular freshwater system involved, as summarised in Sigee (2004).

Reduction in the internal loading of nutrients may also be important. Long-term agricultural and domestic inflow into the Norfolk Broads (wetlands), for example, has resulted in the accumulation of high-phosphate sediments. In parts of the system where these are particularly high, removal of contaminated sediment by dredging may be necessary to prevent continued release into the water column (Pitt et al. 1996).

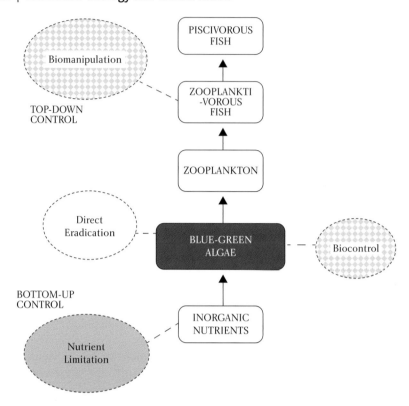

Figure 7.8 Major approaches to the control of blue-green algae in standing freshwaters.

The diagram shows four major types of control procedure in relation to the main pelagic food chain of standing waters—involving changes to ecological balance (biomanipulation, biocontrol), elimination of biota (direct eradication), and chemical composition of the aquatic medium (nutrient limitation). Solid arrows indicate direction of nutrient and carbon flow. Adapted from Sigee (2004).

Top-down control: promotion of algal removal by zooplankton. In the natural ecosystem, the growth of planktonic algae is largely controlled by zooplankton. Grazing activity of the larger Cladoceran species such as *Daphnia magna* is particularly important in limiting populations of colonial blue-green algae. In situations of excessive eutrophication (e.g., Norfolk Broads), it is the loss of this predation that results in uncontrolled blue-green algal growth and irreversible environmental deterioration (see previously). Promotion of zooplankton populations can be achieved by:

- Removal of zooplanktivorous fish (also referred to as biomanipulation) but not piscivorous fish such as Perch and Pike.
- Provision of refuges where zooplankton populations can increase. This may involve restoration of waterlily and reed beds, with limitation in the future destruction of these higher plant communities by restriction on boating activities. Floating biofilters could be an alternative in water bodies with fluctuating water levels (Castro-Castellon et al. 2016).

Direct limitation in blue-green algal growth. Direct limitation of algal growth, using non-biological procedures outside the food chain, may be achieved in various ways. These include:

- Destruction of algal populations by algicide. This is particularly effective in small water bodies such as swimming pools and parkland ponds, where there is a short-term need to eradicate algae and prevent human or domestic animal exposure. The main problem with this approach is that a wide range of aquatic organisms, in addition to algae, are destroyed—so there is major adverse environmental impact.
- Changes to water column dynamics. Algal blooms typically occur as dense standing daytime populations at the water surface under conditions of high light. Artificial circulation of water within the column means that dispersed cells receive less overall light and growth becomes limited (Visser et al. 2016). This can be achieved by injecting air into the base of the water column, referred to as hypolimnetic aeration.
- Trawling surface waters. In some situations, removal of surface algae by trawling is a simple option.
- Removal by flushing. In most standing waters there is an inflow and outflow, with gradual replacement of water. The 'aquatic residence time' is normally so long that phytoplankton populations are not flushed out and are able to increase. In some situations, however, particularly in man-made lakes, it is possible to promote a massive through-flow of water—leading to large-scale removal of algae by flushing out or purging. This simple approach can be particularly useful in situations such as spent nuclear fuel storage ponds, where access to the highly radioactive pond may be very limited.

Biological control. Biological control involves the introduction of biological agents or their products to control the growth of blue-green algae. Since naturally occurring agents are normally used, this technique (unlike the use of algicides) has minimal impact on other freshwater organisms. Biological control can be carried out in two main ways:

- Use of leaf litter or Barley straw. Application of either of these agents leads to long-term decomposition and the release of degradation products that can limit the growth of algae (Ridge et al. 1999; Murray et al. 2010). The technique is best used in relatively small water bodies such as ponds, where decomposition products can build up to high concentrations, with application early in the growth season to prevent the initial build-up of algal populations.
- Isolation and use of live biocontrol agents. This aspect is very much in its infancy and involves the use of naturally occurring algal antagonists that are able to destroy blue-green algae by secretion of antibiotics (fungi, bacteria, actinomycetes), predation (Zooplankton, protozoa), or parasitism (fungi, viruses). This approach has the potential advantage that particular algal species can be targeted, but has the disadvantages that complete eradication of blue-green algae rarely occurs, the

antagonist may be expensive to isolate and produce in large quantity and may itself be rapidly diluted and removed within the ecosystem (Sigee et al. 1999).

Use of the above techniques for reversal of eutrophication, elimination of harmful algal blooms and restoration of the original ecosystem are best carried out in concert—with different situations requiring different management combinations. With a particular water body, a fully integrated management policy (Costanza and Voinov 2000; Sigee 2004) might involve:

- Development of a computer model to predict how external climatic factors such as rainfall and temperature may affect nutrient levels and algal development.
- Regular monitoring of the water body to detect early development of harmful algal populations, with designation of a trigger point at which control measures will be implemented.
- Adoption of a clear strategy for control of algal blooms, using a range of techniques (see Figure 7.8).
- Issue of notices to limit public access during bloom periods, when algal toxins reach unacceptable levels.
- Making sure that the measures carried out and the objectives of the remedial procedures comply both with local byelaws and international legislation—such as the USA Clean water Act (1987: amended 2002) and the European Union Water Framework Directive (2000).

7.7 General overview

This chapter gives a broad overview of freshwater algae, providing information on their taxonomic diversity, methods of enumeration, and occurrence in different freshwater environments. The relevance of these algae to human activities is considered, particularly in relation to eutrophication and the occurrence of harmful algal blooms. References to more detailed accounts are given throughout, and the reader is pointed particularly to more extensive texts, such as Reynolds (1990), Wetzel (2001), Wehr and Sheath (2003), Sigee (2004), and Bellinger and Sigee (2015).

References

Adan, A., Alizada, G., Kiraz, Y., Baran, Y., and Nalbant, A. (2017). Flow cytometry: Basic principles and applications. *Critical Reviews in Biotechnology*, 37, 163–76.

Augspurger, C., Gleixner, G., Kramer, C., and Küsel, K. (2008). Tracking carbon flow in a 2-week-old and 6-week-old stream biofilm food web. *Limnology and Oceanography*, 53, 642–50.

Bellinger, E.G. and Sigee, D.C. (2015). Freshwater Algae: Identification, Enumeration and Use as Bioindicators, 2nd Edition. John Wiley and Sons, Chichester.

Bennion, H. and Battarbee, R, (2007). The European Union Framework Directive: opportunities for palaeolimnology. *Journal of Palaeolimnology*, 38, 285–95.

Castro-Castellon, A.T., Chipps, M.J., Hankins, N.P., and Hughes, J.M.R. (2016). Lessons from the 'Living-Filter': An in-reservoir floating treatment wetland for phytoplankton reduction prior to a water treatment works intake. *Ecological Engineering*, 95, 839–51.

Costanza, R. and Voinov, A. (2000). Integrated ecological economic regional modelling. In: J. Hobbie (ed.) Estuarine Science. Island Press, Washington.

Droppo, I.G., Ross, N., Skafel, M., and Liss, S.N. (2007). Biostabilisation of cohesive sediment beds in a freshwater wave-dominated environment. *Limnology and Oceanography*, 52, 577–89.

European Union (2000). Directive 2000/60/EEC of the European Parliament and of the Council of 23 October 2000 on establishing a framework for community action in the field of water policy. *Journal of the European Community*, 327, 1–72,

Hepperle, D. (2004). Opticount software. Available online from Sequentix website.

Hillebrand, H., Dürselen, C.D., Kirschtel, D., Pollingher, U., and Zohary, T. (1999). Biovolume calculation for pelagic and benthic microalgae. *Journal of Phycology*, 35, 403–24.

Jarvi, H.P., Neal, C., and Warwick, A. (2002). Phosphorus uptake into algal biofilms in a lowland chalk river. *Science of the Total Environment*, 282/283, 353–73.

Kelly, M.G., Penny, C.J., and Whitton, B.A. (1995). Comparative performance of benthic diatom indices used to assess river quality. *Hydrobiologia*, 302, 179–88.

King, L.C., Clarke, G., Bennion, H., Kelly, M., and Yallop, M. (2006). Recommendations for sampling littoral diatoms in lakes for ecological status assessments. *Journal of Applied Phycology*, 18, 15–25.

Litaker, R., Vandersea, M.W., Kibler, S.R., Reece, K.S., Stokes, N.A., Steidinger, K.A., Millie, D.F., Bendis, B.J., Pigg, R.J., and Tester, P.A. (2003). Identification of Pfeisteria piscida and Pfeisteria-like organisms using internal transcribed spacer-specific PCR assays. *Journal of Phycology*, 39, 754–61.

Lyon, D.R. and Ziegler, S.E. (2009). Carbon cycling within epilithic biofilm communities across a nutrient gradient of headwater streams. *Limnology and Oceanography*, 54, 439–49.

Madgwick, F. (1999). Strategies for conservation management of lakes. In D. Harper (ed.) The Ecological Bases for Lake and Reservoir Management. Kluwer Academic Publishers, Dordrecht.

Misson, B. and Latour, D. (2012). Influence of light, sediment mixing, temperature and duration of the benthic life phase on the benthic recruitment of *Microcystis*. *Journal of Plankton Research*, 34, 113–19.

Murray, D., Jefferson, B., Jarvis, P., and Parsons, S.A. (2010). Inhibition of three algae species using chemicals released from barley straw. *Environmental Technology*, 31, 455–66.

OECD (1982). Eutrophication of waters, assessment and control. OECD Cooperative Programme on monitoring inland waters (eutrophication control). Organisation for Economic Cooperation and Development, Paris, France.

Paerl, H.W., Dyble, J., and Pinkney, J.L. (2005). Using microalgal indicators to assess human- and climate-induced ecological change in estuaries. In: S.A. Bortone (ed.) Estaurine Indicators. CRC Press, Boca Raton.

Pitt, J., Kelly, A., and Phillips, G. (1996). Control of nutrient release from sediment. In: F. Madgwick and G. Phillips (eds) Restoration of the Norfolk Broads. Broads Authority and Environment Agency, Norwich.

Pomati, F., Kraft, N.J., Posch, T., Eugster, B., Jokela, J., and Ibelings, B.W. (2013). Individual cell based traits obtained by scanning flow-cytometry show selection by biotic and abiotic environmental factors during a phytoplankton spring bloom. *PloS one*, 8, e71677.

Read, D.S., Bowes, M.J., Newbold, L.K., and Whiteley, A.S. (2014). Weekly flow cytometric analysis of riverine phytoplankton to determine seasonal bloom dynamics. *Environmental Science: Processes and Impacts*, 16, 594–603.

Reynolds, C.S. (1990). The Ecology of Freshwater Phytoplankton. Cambridge University Press, Cambridge, UK.

Reynolds, C.S. and Bellinger, E.G. (1992). Patterns of abundance and dominance of the phytoplankton of Rostherne Mere, England: evidence from an 18-year data set. *Aquatic Sciences*, 54, 10–36.

Ridge, I., Walters, J., and Street, M. (1999). Algal growth control by terrestrial leaf litter: a realistic tool? In D. Harper (ed.) The Ecological Bases for Lake and Reservoir Management. Kluwer Academic Publishers, Dordrecht.

Schneider, D.C. (2009). Quantitative ecology: measurement, models and scaling. Academic Press/Elsevier, London, UK.

Sigee, D.C. (2004). Freshwater Microbiology: Diversity and Dynamic Interactions of Microorganisms in the Aquatic Environment. John Wiley and Sons, Chichester.

Sigee, D.C., Glenn, R., Andrews, M.J., Bellinger, E.G., Butler, R.D., Epton, H.A.S., and Hendry, R.D. (1999). Biological control of cyanobacteria: principles and possibilities. In: D. Harper (ed.) The Ecological Bases for Lake and Reservoir Management. Kluwer Academic Publishers, Dordrecht.

Sigee, D.C., Selwyn. A., Gallois, P., and Dean, A.P. (2007). Patterns of cell death in freshwater colonial cyanobacteria during the late summer bloom. *Phycologia*, 46, 284–92.

Skácelová, O. and Leps, J. (2014). The relationship of diversity and biomass in phytoplankton communities weakens when accounting for species proportions. *Hydrobiologia,* 724, 67–77.

Stancheva, R., Sheath, R.G., Read, B.A., McArthur, K.D., Schroepfer, C., Kociolek, J.P., and Fetscher A.E. (2013*).* Nitrogen-fixing cyanobacteria (free-living and diatom endosymbionts): their use in southern California stream bioassessment. *Hydrobiologia*, 720, 111–27.

Sutherland, T.F., Amos, C.L., and Grant, J. (1998). The effect of buoyant biofilms on the erodibility of sublittoral sediments of a temperate microtidal estuary. *Limnology and Oceanography*, 43, 225–35.

Technical Standard Publication (1982). Utilisation and Protection of Waterbodies. Standing Inland Waters. Classification. Technical standard 27885/01, Berlin. Germany.

USA (1987; amended 2002). The Clean Water Act. US Government Printing Office.

Visser, P.M., Ibelings, B.W., Bormans, M., and Huisman, J. (2016). Artificial mixing to control cyanobacterial blooms: a review. *Aquatic Ecology*, 50, 423–41.

Wehr, J. and Sheath, R. (2003). Freshwater Algae of North America. Academic Press, Amsterdam.

Wetzel, R. (2001). Limnology: Lake and River Systems. Academic Press, San Diego.

8

Microorganisms 2

Viruses, prokaryotes, fungi, protozoans, and microscopic metazoans

Julia Reiss

Corresponding author: julia.reiss@roehampton.ac.uk

8.1 Introduction

Organisms that are invisible to the naked eye are the most abundant component of any freshwater community. These tiny organisms span domains and phyla and include viruses, prokaryotes (archae and bacteria), and protists (single-celled eukaryotes such as single-celled fungi, algae—see Chapter 7, and protozoans), as well as multicellular fungi and microscopic metazoans (such as nematodes and rotifers). Microscopically small organisms do not only exceed macroscopic eukaryotes in terms of their numbers by far, they are also extremely bio-diverse (Green and Bohannan 2006) and contribute substantially to energy flows in freshwater ecosystems. For example, it is estimated that, globally, heterotrophic biota in inland waters respire 1.2 Pg of terrestrial derived carbon each year and release it to the atmosphere (Battin et al. 2009) and as a very rough approximation more than two-thirds of this carbon is respired by heterotrophic microbes, if we assume that production (the turnover of matter over time) is scale invariant with body mass and consider a typical size distribution in fresh waters.

Microscopically small organisms drive the bulk of ecosystem processes on this planet and techniques for estimation of their biodiversity, their sampling, and calculation of community processes, such as production, are vital if we want to assess ecosystem health and functioning of fresh waters. Here I give an overview of microscopically small organisms and techniques used by biologists studying these organisms in fresh waters (other than algae and biofilm, see Chapter 7). Because the organisms discussed here span domains of life and a plethora of phyla and functional groups (i.e., extensive differences exist when it comes to techniques for assessing them) I have highlighted literature that has many relevant references cited within and that will help with further reading (see also Tables 8.1 and 8.2). It is helpful to read this chapter in conjunction with Chapter 7, where a number of generic techniques have been described for surveying and sampling microorganisms and to consult the textbooks 'Freshwater Microbiology'

Reiss, J., *Microorganisms 2: Viruses, prokaryotes, fungi, protozoans, and microscopic metazoans*. In: *Freshwater Ecology and Conservation: Approaches and Techniques*. Edited by Jocelyne M. R. Hughes: Oxford University Press (2019).
© Oxford University Press 2019. DOI: 10.1093/oso/9780198766384.003.0008

Table 8.1 *Taxonomic guides or references to the main groups of microorganisms reviewed in this chapter*

Group	Sub-group	Reference
Viruses	Operational taxonomic units	Weinbauer 2004
Prokaryotes	Operational taxonomic units	Duarte et al. 2010
Fungi	Operational taxonomic units	Duarte et al. 2010
	Fungal species	Gulis et al. 2005
Protozoans	Ciliates	Foissner and Berger 1996; Foissner 1991
	Heterotrophic flagellates	Auer and Arndt 2001
	Amoebae	Page 1976
Microscopic metazoans	Meiofauna (sediment metazoans)	Rundle et al. 2002; Reiss and Schmid-Araya 2008
	Zooplankton (open water metazoans)	Thorp and Covich 2009; Ruttner-Kolisko 1974

(Sigee 2004), 'Wetland Techniques' (Anderson and Davis 2013), and 'Methods in Stream Ecology' (Hauer and Lamberti 2011).

8.2 Biodiversity

The term 'biodiversity' refers to the extent of genetic, taxonomic, and ecological diversity over all spatial and temporal scales (Harper and Hawksworth 1994). Most commonly, 'species richness' is used as a synonym for 'biodiversity' but 'biodiversity' in fresh waters can certainly be measured and assessed in different ways. After a short overview of size ranges within microscopically small fresh water organisms, I here discuss non-taxonomic groupings, operational taxonomic units, and species.

8.2.1 Size ranges

Viruses are biological entities and consist of a single- or double-stranded DNA or RNA, surrounded by a protein (some viruses have a lipid coat), and are typically smaller than 100 nm, with a range of about 10–300 nm, as observed by transmission electron microscopes. However, there are viruses considerably larger than this and an example is a group of viruses found in the protozoan *Acanthamoeba*, which can be seen with a light microscope and which are 1 μm in length (Philippe et al. 2013).

If we exclude viruses, the body mass range in a typical freshwater community spans more than 16 orders of magnitude from bacteria to fish. One-third of this range is occupied by bacteria and archae, which are typically 0.2–20 μm in length. However, there are exceptions, such as *Achromatium oxaliferum*, a large sediment bacterium commonly found in freshwater and brackish environments, with cell sizes around 40 μm (Rhodes et al. 2012).

Another third of the body mass range in fresh waters is occupied by protists and microscopically small metazoans with body lengths ranging from ~20 μm to 2,000 μm (and six orders of magnitude differences in body weight). These organisms constitute

Table 8.2 *Studies or reviews that explain either sampling method, sample processing, or counting of microscopically small organisms. Examples are given for particular freshwater systems.*

Group	System	Reference
Viruses	Lake	Bronner et al. 2016
Prokaryotes	River	Trimmer et al. 2003
	Groundwater	Griebler and Lueders 2009
Fungi	Stream	Gulis and Suberkropp 2007
	Lake	Wurzbacher et al. 2016
Protozoans	Sediment	Reiss and Schmid-Araya 2008
	Open water	Scherwass et al. 2010; Kammerlander et al. 2016
	Artificial substrates	Weitere et al. 2003
Microscopic metazoans	Sediment	Reiss and Schmid-Araya 2008
	Open water	Kammerlander et al. 2016

an important part of biodiversity and energy flow in aquatic ecosystems (e.g., Reiss and Schmid-Araya 2010). Because of their small body sizes and short generation times, these organisms can respond particularly quickly to changes in environmental conditions (Finlay 2002). For instance, in planktonic communities, 'spring blooms' can result in marked changes in community size and structure as light and temperature levels rise (Gaedke 1992; Rojo and Rodríguez 1994).

Representatives of all phyla of true fungi can be found in fresh water and very common are the so-called aquatic hyphomycetes (ascomycetes and basidiomycetes). These fungi have microscopically small spores (conidia) but their hyphae can grow to substantial networks (e.g., Gulis and Suberkropp 2007) and it is difficult to assign sizes to these species.

8.2.2 Non-taxonomic groupings

Freshwater biologists often group microscopic organisms according to their habitat (e.g., 'biofilm' (see Chapter 7), 'benthos', 'plankton', 'epibiont assemblage'). Within a habitat, these groups are further assigned to size classes and, confusingly, the range of these size bins depends on the tradition of sampling for that particular group (often determined by mesh nets used or counting techniques used).

For example, small benthic fauna are divided into the artificial groups of micro- and meiofauna, terms based on size classes for which the definitions are not unanimously agreed. Microfauna are defined as benthic organisms that are 20–200 μm in length. Meiofauna have been defined as those animals which pass a 1,000 or 500 μm sieve and are retained on a 42 μm sieve or on a 63 μm sieve (Robertson et al. 2000). Therefore, this group contains metazoans small enough to pass the upper mesh class and protozoans large enough to be retained on the denser mesh and many of them are around 200 μm in size. Some metazoan species, especially their juvenile stages, can pass the lower mesh net. Hence, many benthic species are enclosed in both the definition of meiofauna and microfauna. I suggest using the terms 'protozoans' and 'microscopically small metazoans' (micro-metazoans) when studying microscopically small animals.

Other non-taxonomic groupings include grouping organisms by their traits ('how they look'—e.g., viruses, or 'what they do'—e.g., bacteria)—such as their feeding strategy, diet, or role in the energy cycle (e.g., 'pathogen', 'symbiont', 'herbivore', 'predator', 'filter feeder', or 'shredder'). Excellent synopses exist on macroscopic freshwater invertebrates regarding their traits (e.g., Usseglio-Polatera et al. 2000); however, for microscopically small species it is necessary to consult specific taxonomic keys to find information on their traits. For example, Foissner and Berger (1996) list salinity tolerance and diet for many freshwater ciliates.

8.2.3 Taxonomy—operational taxonomic units

Genome sequencing is now a standard method to estimate biodiversity of viruses, prokaryotes, fungi, and other organismal groups and while often a species name cannot be assigned, the number of 'operational taxonomic units' (OTUs) can give an estimate of biodiversity in a sample or habitat.

Virus taxonomy is clearly an important scientific discipline for freshwater science and conservation because viruses are generally regarded as pathogens that will shape food webs (Weinbauer 2004). The majority of the viruses found in the freshwater environment are typically prokaryotic viruses (Weinbauer 2004). One way to classify viruses is by genome sequence alone (i.e., by ignoring other biological data—including host) and this approach has to date yielded thousands of species (Thompson et al. 2015) but it is not clear how many different viruses occur in particular freshwater habitats (but see Bronner et al. 2016).

Species within the bacteria and archaea are also distinguished according to their genetic similarity and the most popular metric is 16S rRNA gene sequence (deposited in GenBank and the Ribosomal Database Project). Schloss et al. (2016) point out that OTUs from aquatic environments only represent 16.5 per cent of all described, due to sampling bias towards zoonotic environments. With recent estimates of global diversity of 1 trillion species of bacteria, archae, and micro-fungi (Locey and Lennon 2016) it seems obvious that most freshwater bacteria and archaea have yet to be described. Equally, genetics (18S rDNA) are a common approach to estimate fungal biodiversity in fresh waters (Duarte et al. 2010) and OTU biodiversity estimates, depending on the method, have been shown to be higher than estimates based on microscope techniques (Duarte et al. 2010).

8.2.4 Taxonomy—species identification via microscopy

Viruses, bacteria, and protists are often classified according to their traits or morphology because consensus on how to assign species to a phylogeny or taxonomy is ongoing. It is important to know that these 'artificial' groups are in use when searching for relevant taxonomic literature and identification keys and this is especially true for protozoan species where phylogenetic classification is ever changing. Box 8.1 includes practical advice on using a microscope and the enumeration of microscopically small organisms.

Protozoans are heterotrophic, single-celled eukaryotes and are typically assigned to either: heterotrophic flagellates (they possess flagella), ciliates (they possess cilia and at

Box 8.1 Microscope and enumeration techniques

Biologists studying microscopically small organisms rely on microscopes when it comes to studying external or internal features of microbes (e.g., for taxonomy or enumeration; see Chapter 7). While viruses and many bacteria are so small that transmission electron microscopes (TEM; e.g., Šimek et al. 2001) are used for their identification and enumeration (see Chapter 7); it is possible to identify and count many prokaryotes and microscopic eukaryotes with light microscopy.

Light microscope types used include: the inverted microscope, the compound microscope (this includes the epifluorescence microscope) and the stereo-microscope. The latter is useful for enumeration of the 'larger' eukaryotes such as larger meiofauna or zooplankton but for taxonomy the maximum magnification (up to × 400) is generally too low for species identification. Species that are over 40 micrometres in length can usually be identified with × 400 and × 600 magnification using a compound- or inverted microscope (using the × 10 ocular and the × 40 or × 60 objective, see Chapter 7). Differential Interference Contrast (DIC, also called Normarski) and Phase Contrast can be very useful microscope features—especially for the unicellular eukaryotes.

Microscopic pro- and eukaryotes are so small and abundant that it will be necessary to bring samples back to the laboratory to identify taxa and count their numbers. USB microscopes, however, are useful tools in the field when it comes to counting macroscopic fauna.

Fixing-, sedimentation-, filter mounting-, and staining techniques are described or referenced in the examples in Table 8.1; however it is worth noting that many microscopic species either cannot be fixed or not identified when they are dead. For example, bdelloid rotifers are best observed alive so it is possible to count the number of toes and then fix them to view their jaws (called trophi).

To study living protozoans and metazoans, individuals have to be transferred individually from a sample into a small drop of water on a microscope slide using a pipette or an eyelash glued to a pipette (the latter is used for nematodes) so they can be 'squeezed' with a microscope slip that has Vaseline 'feet' (see Foissner and Berger 1996) which both slows the individual down and gives a much better microscope image, allowing to use the higher objectives (up to × 1,000 magnification); for example, ciliates (Foissner and Berger 1996).

If the aim of the study does not call for detailed taxonomy, enumeration can be automated; for example, by cell counters. Enumeration techniques of bacteria include flow cytometry (see Chapter 7). Automated image analysis is very common for plankton samples but not so much for benthic samples where the organisms have to be separated from biofilm or sediment particles (with a pipette, a needle, or by centrifugation techniques with a medium denser than water—such as Lugol).

When using a light microscope, simple microscope slides can be used for enumeration purposes but often specially designed counting chambers are used to estimate abundances in a water sample. The Sedgewick Rafter Counting Chamber (gridded and holds 1 ml of water; see Chapter 7) is an excellent example for the latter

---

Box 8.1 *Continued*

---

and useful for counting organisms as small as flagellates up to the larger micro-metazoans. For larger meiofauna and zooplankton, petri dishes that are used with a stereomicroscope can be more useful because they hold a larger volume of water.

---

least one macro and micro-nucleus), or amoebae (they have pseudopodia). Essential taxonomic keys for these three groups are (and see references within): Foissner and Berger (1996), and Foissner (1991) for ciliates; Auer and Arndt (2001) for heterotrophic flagellates; and Page (1976 for amoebae.

When it comes to microscopic metazoans that dwell in sediments, there is rich taxonomic literature for these species (see references in Rundle et al. 2002), which is impressive as they belong to more than ten metazoan phyla in fresh water (Robertson et al. 2000). These include freshwater cnidarians and Platyhelminthes; nematodes, rotifers, gastrotrichs, and tardigrades; and tiny annelids and arthropods (such as cladocerans and copepods). Taxonomic keys to freshwater micro-metazoans that live in, and on, sediments (meiofauna) are further listed in Reiss and Schmid-Araya (2008). An online taxonomic key for a meiofaunal group is the 'Illustrated Key to Nematodes Found in Fresh Water' by the Nematology Laboratory of the University of Nebraska (2017). The website 'plingfactory' run by Michael Plewka, is not an official key to microscopic species but gives a very impressive and accurate overview of microscopic groups in freshwater (e.g., the rotifers) and this includes prokaryotes, protists, and zooplankton.

Thorp and Covich (2009) also provide a good overview of freshwater micro-metazoans including those from the open water (i.e., small-sized zooplankton), comprising planktonic rotifers (see also Ruttner-Kolisko 1974), the planktonic cladocerans, and the open-water copepods. Many fungal species within leaf samples can be 'encouraged' to sporulate by aerating the samples. The spores can then be stained with trypan blue and their species- characteristic shape observed under the microscope and identified to species level using the key by Gulis et al. (2005).

8.3 Freshwater environments

Field- and laboratory techniques will not only depend on the types of organisms targeted but also on the freshwater system and micro-habitats sampled. In Chapter 7, lotic versus lentic environments and biofilms are discussed. Large reservoirs of freshwater biota are also found in environments such as the hyporheic zone of streams and rivers and groundwater (almost all fresh water that is not bound in ice on this planet is groundwater; see Chapter 1).

Organisms themselves are habitats and hosts for symbionts, parasites, and other hitchhikers such as epibionts. One example is the water-hoglouse *Asellus aquaticus*, whose carapax can be covered with a thick carpet of ciliates and other epibionts (Cook et al. 1998). Another obvious example are hosts of viruses as viruses all need their intracellular machinery. Lytic viruses infect cells, replicate, and then destroy cells by lysis, setting free viral progeny and cellular lysis products (Weinbauer 2004).

8.4 Sampling and assessment

A spoonful of stream sediment will sample up to 1,000 individuals of microscopically small metazoans (and many more protozoans, bacteria, and viruses) and dunking a 1 L Octoberfest beer-glass into a pond will easily sample 100 waterfleas. Still, for a scientific sampling campaign, the total sample volume might have to be much larger if densities vary greatly in micro-habitats and if variables such as species richness are measured.

Sampling devices include Hess samplers (for sediment; see Fig. 5.3 in Anderson et al. 2013), the Bou-Rouche pump (to sample the hyporheos in the hyporheic zone), plankton nets, or the Schindler-Patalas trap and bailers (all for plankton or groundwater boreholes; see Fig. 5.3 in Anderson et al. 2013). All of these sampling techniques have to use fine mesh sizes or sample the water/sediment without any loss of water. Because microscopically small organisms are very fast colonisers of new habitats, artificial substrates, such as microscope slides (e.g., Weitere et al. 2003) or stone tiles, can be a good way to sample micro-communities.

The aims of a study and the statistical analysis that can best address the questions will determine the sampling design and sampling technique. For example, which subhabitats are sampled, and the sample volume and sample (pseudo-) replication will depend on the questions and purpose of the study. However, all scientific studies of freshwater microorganisms will aim for replication and for the replication to yield a reliable estimate of the response variable measured (often 'mean density'). In its simplest form, the minimum number of replicates can be estimated by performing a pilot sampling and calculating which number of replicates will yield less than 20 per cent standard error of the mean (Elliott 1977). In the case of species richness, sample number can be regressed against cumulative species richness, and when species richness plateaus the optimal replicate number has been reached.

8.4.1 Sampling viruses and bacteria

Biodiversity studies on viruses and bacteria in freshwater samples call for analysis of the genomes found, but the use of metagenetics, targeted metagenomics, and viral metagenomic are not without controversy when it comes to the sequencing approaches. Weinbauer (2004) gives an excellent overview of methods associated with prokaryote viruses. Further examples of sampling for freshwater viruses and bacteria and their evaluation (e.g., preparation for transmission electron microscopy, epifluorescence microscopy, and molecular analyses) include Šimek et al. (2001) who sampled the mesoeutrophic Římov Reservoir in South Bohemia and Bronner et al. (2016) and Sime-Ngando et al. (2016a) who sampled viruses and prokaryotes in Lake Pavin (the latter are contributions to the book 'Lake Pavin' that includes chapters on viruses, prokaryotes, flagellates, ciliates, and fungi (Sime-Ngando et al. 2016b)). Depending on the study aim, bacteria sampling can be performed to test for the presence of certain functional groups and in this case, the samples are not treated with fixative. For example, samples can be incubated with substances utilised by the organisms and their transformation is measured. One example for this is the study by Trimmer et al. (2003) who demonstrated that anaerobic ammonium oxidation (Anammox) bacteria are present in sediments of the River Thames, UK, by incubating samples with labelled $^{15}NH_4^+$ and either $^{14}NO_2^-$ or $^{14}NO_3^-$ (or both). In this vein, Biolog-ECO microplates are a very useful tool for freshwater ecologists, where the number of functional groups present in a sample can be estimated by incubating bacteria on different substrates (in a multiwell dish) which they can either utilise or not, giving a colour reaction (Sala et al. 2006).

8.4.2 Sampling protozoans, metazoans, and fungi

Protozoans and microscopically small metazoans are as omnipresent as other microbes. Their sampling can be tricky, especially when samples have to be sieved (e.g., necessary for hyporheic samples), or when plankton nets are used, because many species can pass even very small mesh sizes and sampling with nets is only really suitable for the hard-bodied species, such as monogont rotifers and micro-crustaceans. Often a combination of sieved and un-sieved samples is the most appropriate method (e.g., for stream benthos (Reiss and Schmid-Araya 2008)).

Freshwater hyphomycetes are abundant on decomposing leaves in running waters but spores can also be sampled from foam on the water and other micro-habitats. Sampling, culturing, sporulation, molecular approaches to sampling, and assessment of fungi in freshwater environments are described in Gulis and Suberkropp (2007); see also Wurzbacher et al. (2016).

8.4.3 Conversions and calculations: population characteristics such as production

Microscopically small organisms have vast population sizes but they are also important in terms of their biodiversity, their role in the energy cycle, their biomass, and turnover of that biomass over time (secondary production). Biodiversity estimates from different environments can be compared as taxon richness or functional richness and often it makes sense to calculate the Shannon-Wiener or the Simpson index (e.g., Reiss and Schmid-Araya 2008).

For those species, where individuals can be counted and measured, it is straightforward to estimate biomass by multiplying their average body mass by their abundance. In most cases, mass will be calculated from body dimensions and published equations (e.g., Fuhrman and Azam 1980; Reiss and Schmid-Araya 2010) and assuming a sphere shape (or the most appropriate geometric shape that resembles their body form the most). Assuming a density of 1 (or 1.1 for metazoans) will give a rough estimate of body fresh weight which can then be converted to carbon units (e.g., Reiss and Schmid-Araya 2008; Reiss and Schmid-Araya 2010). When generation times (or intrinsic rate of population increase) are known (they can be estimated from allometric principles), then production can be calculated from biomass and temperature data (Reiss and Schmid-Araya 2010).

Fungi do not have a set body size but their biomass can be calculated by measuring the ergosterol content in a sample. Gulis and Suberkropp (2007) give a detailed overview of how fungal biomass (i.e., the amount of fungal mycelium in carbon units) can be estimated via the ergosterol extraction method (also described in Duarte et al. 2010). Pascoal and Cássio (2004) explain how to calculate spore mass.

Virus production is a very different concept compared to the secondary production of other microscopically small organisms because the amount of viruses released into the water depends on lysis, in freshwaters mostly of bacterial cells (Weinbauer 2004; also see Chapter 7 in Sigee 2004).

8.5 Bioindicators

Microorganism loading in fresh waters can indicate pollution or contamination (e.g., from soil run-off or sewage influx including pathogens), and the composition of fresh-water 'micro-communities' can be used as indicators of water quality. Many microscopic species are present in fresh waters as spores (i.e., there is a 'seedbank' of non-active microbes) and due to their short generation times respond quickly to changes in their environments, such as temperature, nutrient loading, oxygen availability, or pollution. Microscopic species are present in freshwater bodies throughout the year (except for temporary meiofauna), which means they do not leave the system like insect larvae do. In addition, many species occur worldwide, meaning that indices based on these small-sized assemblages can find application across countries; which is not feasible with macrofauna.

For instance, within Europe, it is only possible to compare the presence and dominance of macrofaunal groups as opposed to individual species because these can be absent from regions within Europe (e.g., the water-hoglouse *Asellus* which dominates British streams is absent in Portugal). Comparisons between countries are important when it comes to biomonitoring and management of fresh water on a larger scale; for example, within the European Union. For instance, the EU's 'Water Framework Directive' sought to conserve or restore freshwater bodies throughout Europe to at least 'good' ecological status by 2015 (a goal that was not reached). The EU funded the assessment of rivers and other freshwater bodies, including 'biological quality' based on macrofauna and fish (e.g., metric 'Biological Monitoring Working Party' (Hawkes 1997)). However, microbes are an omnipresent component of all freshwaters and their use in biomonitoring is explained in the following paragraphs.

A common functional measure of ecosystem heath in fresh waters is to measure the biochemical oxygen demand (BOD) as a proxy for the trophic status of the system. The BOD is essentially the 'biological response' of microscopically small organisms (often it is assumed these are mainly bacteria) to organic loading. It is the amount of dissolved oxygen used to respire organic substances in a sample. This is a semi-standardised method (defined water volume, temperature at 20°C and darkness) used by water treatment works and governmental agencies (e.g., Environment Agency in the UK; United States Environmental Protection Agency) and often the five-day Biochemical Oxygen Demand: BOD5 is used (Delzer and McKenzie 1999).

Ciliate community composition can be used to determine the trophic status of surface freshwaters (based on the system outlined by Foissner and Berger 1996) but this approach is only used in Germany and Austria despite the fact that protozoans in general make excellent bioindicators because of their ubiquity and ease of analysis (Payne 2013). Foissner and Berger (1996) give a brief overview of the four volumes of Foissner's ciliate taxonomy, in which they descibe typical assemblages in oligotrophic up to hypereutrophic environments. The Foissner tables include a score for a ciliate species. The scores of all ciliate species found in a system can be added up and will indicate its trophic status, similarly to other biomonitoring metrics that are based on macrofauna (e.g., Biological Monitoring Working Party (Hawkes 1997)).

Many micro-metazoans such as *Hydra* or *Daphnia* are model organisms in toxicity tests (e.g., LD_{50} to toxins). The most famous of all is the terrestrial nematode *Caenorhabditis elegans* and (unsurprisingly) evaluating the nematode community of fresh waters can indicate eutrophication. The ratio of the orders Secernentea and Adenophorea, as well as the so-called Maturity Index, are both indicators of nutrient loading in freshwaters (see Beier and Traunspurger 2001 and references therein). Similarly, freshwater hyphomycetes species are sensitive to heavy metals and nutrient loading in streams and rivers (Pascoal et al. 2010) and have equally been suggested to make excellent bioindicators of fresh water quality (Solé et al. 2008).

8.6 Case studies

In this section, case studies are discussed to demonstrate the use of techniques and approaches associated with a particular aim: to demonstrate the abundance and activity (e.g., secondary production) of different organismal groups in the system. These case studies were chosen because they involve sampling many different microbial groups but they were mostly conduced in one or two systems only. I want to stress that, to answer general ecological questions, it is often necessary to sample many similar systems because within-system replication is essentially pseudo-replication (see Chapter 2). In other words, rather than sampling a single stream throughout a whole year, a sampling design that involves sampling 20 streams on one sampling occasion in, for example, July only, will produce data that can answer the question. Patterns of data can then be compared between systems, generalisations can be made, and new hypotheses or models can be generated. The conclusions obtained in such a study reach beyond an observation towards estimating abundance and occupancy, testing hypotheses, formulating theory, and applying the evidence to freshwater conservation issues.

8.6.1 Ashdown Forest streams

Streams in the Ashdown Forest (South-East England), such as the Broadstone Stream or the Lone Oak, are all low-order, nutrient poor, and slightly acidic streams. The Broadstone Stream has been studied since the 1970s and is famous among freshwater ecologists. The macrofaunal community in Broadstone and Lone Oak has been described in numerous papers but two separate studies did set out to sample microbes in these systems. The overarching aim of both studies was to demonstrate that microbes are an important component of the stream community.

Jenkins et al. (2013) exposed cotton strips in 31 streams in the Ashdown Forest over seven days in summer 2011 and 49 days in winter 2012. They compared this with data from an identical study conducted in 1978 and 1979. One of their objectives was to demonstrate that the pH level in a stream determines microbial activity (i.e., to show the potential to break down organic matter in the stream). They secured 93 cotton strips to metal rods in shallow riffle sections on each occasion, harvested these, dried them, and estimated tensile strength. Those strips that easily ripped apart were decomposed the most and the value for tensile strength proofed to give a sufficient range of decomposition levels. This is not only an affordable way to demonstrate decomposition but also a very intuitive proxy for microbial activity (in this case decomposition). Jenkins and colleagues found that decomposition increased with pH—but this pattern could only be observed in winter, not summer.

In one of these 31 streams, Lone Oak, Reiss and Schmid-Araya (2008) sampled benthic ciliates and micro-metazoans over the course of a year in monthly intervals, to show that these organisms are a diverse and abundant component of the stream community. A main focus was accurate estimates of their abundance which meant a pilot sampling was performed to estimate the number of replicates needed to get a representative mean of total abundance in a particular month (eight samples). They sampled the stream sediment with a Hess sampler, which did not have a mesh net attached but a 5 L plastic bag. The entire sample was transferred to a large bucket, the sediment was stirred, and pseudo-replicates of 50 ml were taken for ciliates. While one of these samples was simply stored on ice, the others were fixed with glutharaldeyde. The rest of the bucket sample was poured over 500 μm and 40 μm sieves to select for meiofauna (Reiss and Schmid-Araya 2008).

The ciliates were extracted from the sediment by centrifugation. They were stained and mounted on slides using the Quantitative Protargol staining technique. Micro-metazoans were counted and identified alive. All ciliate and micro-metazoan individuals found on a sampling occasion were identified to species or the closest taxonomic level and their body length and width were measured until a total of 50 to 100 individuals was reached for each group.

Body dimensions (length and width) were converted to species-specific biovolume using published regression models (references in Reiss and Schmid-Araya 2008) and species-specific biovolumes of fixed ciliate cells were then converted into carbon content by assuming 0.14 pg C μm³. In the case of meiofauna, body volume was converted into individual fresh weight assuming species specific gravities. The individual carbon content was estimated by assuming a dry/wet weight ratio of 0.25 and a dry weight carbon content of 40 per cent (i.e., carbon is assumed to be 10 percent of the wet weight). These conversions were necessary to show patterns of both density and biomass over the course of a year. The estimates were then used, in combination with equations for the relationship between generation time and body mass and stream temperature, to calculate secondary production (Reiss and Schmid-Araya 2010).

The Lone Oak case study demonstrates that ciliates and meiofauna are hugely abundant (e.g., ciliates reached up to almost 1 million individuals m⁻²) and despite of 'low' standing stock of biomass, this biomass was turned over frequently; that is, secondary production was up to ~ 1 g carbon m⁻² yr⁻¹ because annual P/B ratios (production divided by biomass) for the whole assemblage exceeded 11.

These results show that ignoring microscopic metazoans and protozoans in freshwater ecology or conservation will lead to wrong conclusions about the productivity of the system and the ability of the system to attenuate and recycle nutrients, including carbon.

8.6.2 Lake Constance

While simultaneous sampling of different microbial groups is still uncommon for benthic environments (and especially running waters), there is a tradition of sampling a larger body mass range, from bacteria to fish, in the open water of lakes. A 'classic' example is the study of Lake Constance, Germany, which is known among protistologists because the ciliate community, and its role in the food web, is very well described.

Gaedke (1992) was able to demonstrate a macro-ecological pattern in Lake Constance: she found temporal variation in the biomass of small versus large plankton in Lake Constance, with shallower biomass spectra slopes in winter than in summer. She suggested this occurs because larger organisms have an intrinsic lag-time when it comes to maximising growth rates, depending on resource supply.

These observations were possible because of an extensive sampling regime that required the fixing of samples, and a focus on a large range of body sizes and groups. Plankton abundance was measured weekly (larger phytoplankton twice a week) over the course of seven months, at the site of maximum depth and a large array of organisms, belonging to seven groups, were counted: free-living bacteria, autotrophic picoplankton, larger eukaryotic phytoplankton, heterotrophic flagellates, ciliates, rotifers, and crustaceans. The water column, from the surface to 20 m deep, was sampled with a 2-m-long tube sampler (4 L volume). Ten sequential samples were taken. Small plankton was fixed with formalin and filtered in nuclepore membranes and counted by epifluorescence microscopy with DAPI (4',6-diamidino-2-phenylindole is a fluorescent stain that binds to DNA). Larger phytoplankton including *Gymnodinium*, ciliates, and rotifers were fixed with formalin and enumerated by the Utermoehl technique (this involves letting a sample stand so the organisms can sink to the bottom of a cylinder—a microscope slide); crustaceans were counted in petri-dishes under a stereo microscope. Gaedke also describes how she estimated individual body mass from transforming body volume (which had been determined from length and width measurements and the most similar geometric shape) to carbon units, using published equations for all seven groups.

8.6.3 Římov reservoir

Šimek et al. (2001) sampled the meso-eutrophic Římov reservoir in South Bohemia (at a depth of 0.5 m) and conduced an experiment with these samples. They designed this experiment to track changes in the bacterial community due to viral lysis and flagellate grazing. This experiment involved sieving the samples into size-fractions but one of their experimental treatments included samples that were not manipulated (not sieved). These latter samples represent an excellent case study when it comes to describing abundance and production of viruses, bacteria, and protozoans in the open water of the reservoir.

The authors used a range of techniques to target each of the three groups. Bacteria were stained with DAPI and counted and measured under an epifluorescence microscope. The size and shape of of the bacteria was used to estimate carbon content of the cells (derived from the literature), which was multiplied by abundance to get biomass estimates. Bacterial production was measured by a thymidine incorporation method and to describe the community composition, 16S rDNA sequences were analysed and *in-situ* hybridisation with fluorescent oligonucleotide probes was performed. Protozoans were fixed with Lugol and formaldehyde but instead of counting them, previous estimates on their abundance were used. Viruses were stained and counted with both epifluoresence microscopy and TEM, the latter was also used to deteremine visibly infected bacteria and burst size. Using these methods, the authors show that viral abundaunce was as high as 5×10^7 viruses ml^{-1}; almost ten times more than the abundance of bacteria (12×10^6 bacteria ml^{-1}); and 1,000 times more than the protozoans (roughly 10^4 cells ml^{-1}). Although it was not the primary aim of this study, these estimates of abundances are important as they highlight the capacity of these organisms to flux energy (carbon) within the microbial food web (which in turn fuels higher tropic levels). For example, doubling time of this standing stock of bacteria abundance (and biomass) was estimated to be only 24 hours (reservoir water temperature was 18–20 degrees Celsius) which explains the high abundances of protozoans feeding on them.

Aquatic ecologists need these type of estimates to compare systems. The microbial community in the Římov reservoir displays a classic pyramid of numbers where the small are more abundant than the large organisms. Theoretical ecology is using the relationship between the range of organismal sizes and abundance (imagine a classic abundance pyramid flipped to the right by 90 degrees) to compare 'healthy' systems with those that are impacted by human activity (see Petchey and Belgrano 2010).

8.7 Conclusion

Both the scientific community and conservation bodies are now aware of the fact that, in order to judge ecosystem health, we need the tools to measure abiotic factors as well as the biota present in these systems. Molecular techniques, as well as a long tradition of publications by dedicated taxonomists and freshwater ecologists, make it possible to accurately assess microscopically small organisms and to compare these communities across systems—and to therefore judge ecosystem health. Microbes have the largest size range within fresh waters, the highest abundances, the highest species richness, and are the main drivers of important ecosystem processes such as whole-system respiration: they truly rule fresh waters.

References

Anderson, J.T. and Davis, C.A. (eds) (2013). Wetland Techniques: Volume 2: Organisms. Springer Science and Business Media.

Anderson, J.T., Zilli, F.L., Montalto, L., Marchese, M.R., McKinney, M., and Park, Y.L. (2013). Sampling and processing aquatic and terrestrial invertebrates in wetlands. In: Anderson, J.T. and Davis, C.A. (eds) Wetland Techniques. Springer, Netherlands.

Auer, B. and Arndt, H. (2001). Taxonomic composition and biomass of heterotrophic flagellates in relation to lake trophy and season. *Freshwater Biology*, 46, 959–72.

Battin, T.J., Luyssaert, S., Kaplan, L.A., Aufdenkampe, A.K., Richter, A., and Tranvik, L.J. (2009). The boundless carbon cycle. *Nature Geoscience*, 2(9), 598.

Beier, S. and Traunspurger, W. (2001). The meiofauna community of two small German streams as indicator of pollution. *Journal of Aquatic Ecosystem Stress and Recovery*, 8, 387–405.

Bronner, G. et al. (2016). Study of Prokaryotes and Viruses in Aquatic Ecosystems by Metagenetic and Metagenomic Approaches. In: Lake Pavin. Springer.

Cook, J.A., Chubb, J.C., and Veltkamp, C.J. (1998). Epibionts of Asellus aquaticus (L.) (Crustacea, Isopoda): An SEM study. *Freshwater Biology*, 39, 423–38.

Delzer, G.C. and Mckenzie, S.W. (1999). Five-day biochemical oxygen demand. *TWRI Book 9*, 7, 1–20.

Duarte, S. et al. (2010). Assessing the dynamic of microbial communities during leaf decomposition in a low-order stream by microscopic and molecular techniques. *Microbiological Research*, 165, 351–62.

Elliott, J.M. (1977). Some Methods for the Statistical Analysis of Samples of Benthic Invertebrates. Freshwater Biological Association, Kendall, UK.

Finlay, B.J. (2002). Global dispersal of free-living microbial eukaryote species. *Science*, 296(5570), 1061–3.

Foissner, W. (1991). Basic light and scanning electron microscopic methods for taxonomic studies of ciliated Protozoa. *European Journal of Protistology*, 27, 313–30.

Foissner, W. and Berger, H. (1996). A user-friendly guide to the ciliates (Protozoa, Ciliophora) commonly used by hydrobiologists as bioindicators in rivers, lakes, and waste waters, with notes on their ecology. *Freshwater Biology*, 35, 375–482.

Fuhrman, J.A. and Azam, F. (1980). Bacterio plankton secondary production estimates for coastal waters of British-Columbia Canada Antarctica and California USA. *Applied and Environmental Microbiology*, 39, 1085–95.

Gaedke, U. (1992). The size distribution of plankton biomass in a large lake and its seasonal variability. *Limnology and Oceanography*, 37, 1202–20.

Green, J. and Bohannan, B.J. (2006). Spatial scaling of microbial biodiversity. *Trends in Ecology and Evolution*, 21, 501–7.

Griebler, C. and Lueders, T. (2009). Microbial biodiversity in groundwater ecosystems. *Freshwater Biology*, 54, 649–77.

Gulis, V., Marvanová, L. and Descals, E. (2005). An illustrated key to the common temperate species of aquatic hyphomycetes. In Methods to study litter decomposition (pp. 153–167). Springer, Dordrecht.

Gulis, V. and Suberkropp, K. (2007). Fungi: biomass, production, and sporulation of aquatic hyphomycetes. In: Hauer, F.R. and Lamberti, G.A. (eds) Methods in Stream Ecology. Academic Press.

Harper, J.L. and Hawksworth, D.L. (1994). Biodiversity: measurement and estimation. *Philosophical Transactions: Biological Sciences*, 345, 5–12.

Hauer, F.R. and Lamberti, G.A. (eds) (2011). Methods in stream ecology. Academic Press.

Hawkes, H. (1997). Origin and development of the biological monitoring working party score system. *Water Research*, 32, 964–8.

Jenkins, G.B., Woodward, G., and Hildrew, A.G. (2013). Long-term amelioration of acidity accelerates decomposition in headwater streams. *Global Change Biology*, 19, 1100–6.

Kammerlander, B. et al. (2016). Ciliate community structure and interactions within the planktonic food web in two alpine lakes of contrasting transparency. *Freshwater Biology*, 61, 1950–65.

Locey, K.J. and Lennon, J.T. (2016). Scaling laws predict global microbial diversity. *Proceedings of the National Academy of Sciences*, 113, 5970–5.

Nematology Laboratory of the University of Nebraska (2017). Illustrated Key to Nematodes Found in Fresh Water. See 'nematodes' on the University of Nebraska-Lincoln website. (nematode.unl.edu)

Page, F.C. (1976). An Illustrated Key to Freshwater and Soil Amoebae. Freshwater Biological Association, Scientific Publication No. 34, Freshwater Biological Association, Ambleside, Cumbria.

Pascoal, C. et al. (2010). Realized fungal diversity increases functional stability of leaf litter decomposition under zinc stress. *Microbial Ecology*, 59, 84–93.

Pascoal, C. and Cássio, F. (2004). Contribution of fungi and bacteria to leaf litter decomposition in a polluted river. *Applied and Environmental Microbiology*, 70(9), 5266–73.

Payne, R.J. (2013). Seven reasons why protists make useful bioindicators. *Acta Protozoologica*, 52, 105–13.

Petchey, O.L. and Belgrano, A. (2010). Body-size distributions and size-spectra: universal indicators of ecological status? *Biology Letters*, 6(4), 434–7.

Philippe, N. et al. (2013). Pandoraviruses: amoeba viruses with genomes up to 2.5 Mb reaching that of parasitic eukaryotes. *Science*, 341, 281–6.

Plewka, M. (2017). Pling—Life in Water. PlingFactory website (http://www.plingfactory.de/) accessed 05/07/18.

Reiss, J. and Schmid-Araya, J.M. (2008). Existing in plenty: Abundance, biomass and diversity of ciliates and meiofauna in small streams. *Freshwater Biology*, 53, 652–68.

Reiss, J. and Schmid-Araya, J.M. (2010). Life history allometries and production of small fauna. *Ecology*, 91, 497–507.

Rhodes, G., Porter, J., and Pickup, R.W. (2012). The bacteriology of Windermere and its catchment: Insights from 70 years of study. *Freshwater Biology*, 57, 305–20.

Robertson, A.L., Rundle, S.D., and Schmid-Araya, J.M. (2000). An introduction to a special issue on lotic meiofauna. *Freshwater Biology*, 44, 1–3.

Rojo, C. and Rodríguez, J. (1994). Seasonal variability of phytoplankton size structure in a hypertrophic lake. *Journal of Plankton Research*, 16, 317–35.

Rundle, S., Robertson, A.L., and Schmid-Araya, J.M. (2002). Freshwater Meiofauna: Biology and Ecology. Blackhuys.

Ruttner-Kolisko, A. (1974). Plankton Rotifers: Biology and Taxonomy. Schweizerbart.

Sala, M.M., Pinhassi, J., and Gasol, J.M. (2006). Estimation of bacterial use of dissolved organic nitrogen compounds in aquatic ecosystems using Biolog plates. *Aquatic Microbial Ecology*, 42, 1–5.

Scherwass, A. et al. (2010). Changes in the plankton community along the length of the River Rhine: Lagrangian sampling during a spring situation. *Journal of Plankton Research*, 32, 491–502.

Schloss, P.D. et al. (2016). Status of the archaeal and bacterial census: An update. *mBio*, 7.

Sigee, D.C. (2004). Freshwater Microbiology: Diversity and Dynamic Interactions of Microorganisms in the Aquatic Environment. John Wiley and Sons, Chichester.

Sime-Ngando, T., Bettarel, Y., Colombet, J., Palesse, S., Ram, A.S.P., Charpin, M., and Amblard, C. (2016a). Lake Pavin: A Pioneer Site for Ecological Studies of Freshwater Viruses. In: Sime-Ngando, T. et al. (eds) Lake Pavin. Springer International, Switzerland.

Sime-Ngando, T., Boivin, P., Chapron, E., Jezequel, D., and Meybeck (2016b). Lake Pavin: History, Geology, Biogeochemistry, and Sedimentology of a deep Meromictic Maar Lake. Springer International, Switzerland.

Šimek, K., Pernthaler, J., Weinbauer, M.G., Hornák, K., Dolan, J.R., Nedoma, J., Mašín, M., and Amann, R. (2001). Changes in bacterial community composition and dynamics and viral mortality rates associated with enhanced flagellate grazing in a mesoeutrophic reservoir. *Applied and Environmental Microbiology*, 67, 2723–33.

Solé, M., Fetzer, I., Wennrich, R., Sridhar, K.R., Harms, H., and Krauss, G. (2008). Aquatic hyphomycete communities as potential bioindicators for assessing anthropogenic stress. *Science of the Total Environment*, 389, 557–65.

Thompson, C.C. et al. (2015). Microbial taxonomy in the post-genomic era: Rebuilding from scratch? *Archives of Microbiology*, 197, 359–70.

Thorp, J.H. and Covich, A.P. (2009). Ecology and Classification of North American Freshwater Invertebrates, 3rd edition. Academic Press, San Diego.

Trimmer, M., Nicholls, J.C., and Deflandre, B. (2003). Anaerobic Ammonium Oxidation Measured in Sediments along the Thames Estuary, United Kingdom. *Applied and Environmental Microbiology*, 69, 6447–54.

Usseglio-Polatera, P. et al. (2000). Biological and ecological traits of benthic freshwater macroinvertebrates: relationships and definition of groups with similar traits. *Freshwater Biology*, 43, 175–205.

Weinbauer, M.G. (2004). Ecology of prokaryotic viruses. *FEMS Microbiology Reviews*, 28, 127–81.

Weitere, M., Schmidt-Denter, K., and Arndt, H. (2003). Laboratory experiments on the impact of biofilms on the plankton of a large river. *Freshwater Biology*, 48, 1983–92.

Wurzbacher, C. et al. (2016). High habitat-specificity in fungal communities in oligo-mesotrophic, temperate Lake Stechlin (North-East Germany). *MycoKeys*, 16, 17–44.

9

Wetland Plants and Aquatic Macrophytes

Jocelyne M.R. Hughes, Beverley R. Clarkson,
Ana T. Castro-Castellon, and Laura L. Hess
Corresponding author: jocelyne.hughes@ouce.ox.ac.uk

9.1 Introduction—Why survey vegetation in freshwaters?

Ecologists, conservationists, and managers frequently need to recognise and survey different aquatic plant species, vegetation types, plant communities, or habitat. It is, after all, the vegetation that defines the extent of a freshwater wetland—whether a 'terrestrial' wetland, dominated by water-tolerant bryophytes, grasses, sedges, and herbs; an 'aquatic' one, dominated by aquatic macrophytes; or a mosaic of both (Chapter 1). The scale of the study is key to the surveyor's approach, which depends on the questions the surveyor wishes to answer: evaluate the relationship between rare or endemic plant distributions and environment (water chemistry, hydrology, or substrate) to make conservation decisions at the site or catchment scale (Schneider et al. 2015); or model vegetation as habitat (for, e.g., aquatic vertebrates, macroinvertebrates, wetland birds, or other fauna (Webb et al. 2010)) to make evidence-based decisions at the landscape scale. Evaluating the effects of freshwater plants in wetland food webs, trophic structure, or as nutrient sinks or sources involves intricate field and experimental research, the results of which are vital in assessing their role in biogeochemical cycles (Fu et al. 2014; Warfe and Barmuta 2006). Monitoring vegetation over time generates data that enable models to be constructed, predictions of future impacts or threats to be made, best practice for wetland restoration to be implemented, or resilience strategies formulated to counter climate change impacts.

The morphological and biochemical traits of autotrophs in freshwater ecosystems explain small-scale interactions between and within species and are fundamental to understanding ecosystem functions, modelling ecological networks, and assessing the impacts of non-native invasive plant species (Schultz and Dibble 2012; Moor et al. 2017). Surveying aquatic vegetation may also be part of using a fixed protocol to assess water quality, or common standards monitoring (e.g., River LEAFPACS2 (WFD-UKTAG 2014), US Environment Protection Agency (2002), Interagency Freshwater Group (2015)), and may involve using metrics or routine survey procedures. At the landscape scale, monitoring changes in the spatial distribution of freshwater vegetation or habitat can quantify the impacts of climate change or environmental flows on wetland area

Hughes, J. M. R., Clarkson, B. R., Castro-Castellon, A. T., and Hess L. L., *Wetland plants and aquatic macrophytes.*
In: *Freshwater Ecology and Conservation: Approaches and Techniques.* Edited by Jocelyne M. R. Hughes: Oxford
University Press (2019). © Oxford University Press 2019. DOI: 10.1093/oso/9780198766384.003.0009

(Gallant 2015). This chapter reviews the methods used to survey both 'terrestrial' and 'aquatic' freshwater plants and considers the approaches taken and some of the specialised equipment and technologies used.

9.2 Planning a survey

Sutherland's classic text on Ecological Census Techniques devotes the first chapter to the steps needed to plan a research programme. He uses 'reverse planning' (or designing the programme backwards) by starting with the question and then working out how to answer it. This process applies to all freshwater plant surveys where the following questions need to be answered (after Sutherland 2006a): What is the specific question? What results are necessary to answer the question? What data are needed to complete these results? What protocol is required to obtain these data? Can the data be collected in the time available? Is modification of the planning required in response to time available? How should the data sheets be organised? Having started the survey and encountered reality, what modifications to the survey plan need to be done?

In addition to using reverse planning to organise a survey, it's vital to have clear aims, justification, and research questions—this drives everything, including the methodology (Karban et al. 2014). The scale of the survey will inform the approach taken, whether the survey is investigating a species, a habitat, or a landscape. The differing scales may make a significant difference between choosing to use field experiments and mesocosms to study a species, structural methods and transects to study a habitat, or remote-sensing and photographic techniques for studies at the landscape scale. For freshwater vegetation, the overall habitat type will influence the methods used with (broadly) wetlands and peatlands falling under 'terrestrial' habitats, and rivers and lakes under 'aquatic'. The former requires boots and the latter requires a boat or a wet suit. Finally, an important consideration for the planning of a freshwater plant survey is the resources available, including time, money, and technical expertise. If you have little of these, then the survey needs to be conducted as efficiently as possible, whereas having access to adequate resources should enable the surveyor to conduct a monitoring programme to generate data for modelling and to answer complex ecological questions.

9.2.1 Approaches to freshwater plant surveys

A scientific approach to vegetation surveys can take an inductive or deductive approach: surveys that are question or hypothesis driven will be deductive, and those involving data collection for description or classification or general recording of species are inductive. Many freshwater vegetation surveys involve a mixture of inductive surveys (especially at the reconnaissance stage) followed by deductive surveys that are delimited by hypotheses or research questions (these may emerge from the initial inductive survey). Examples of inductive approaches are surveys for their own sake, national database recording, baseline surveys, descriptive presence/absence or statutory protocols. Deductive approaches include question-or hypothesis-driven surveys or field experiments—the 'why?' and 'what?' of freshwater plant ecology or conservation research. See Chapters 2 and 3 for further discussion of these different approaches.

Other ways of categorising freshwater vegetation survey methods are influenced by the location of the survey, the type of equipment needed, and the technical expertise required. There are three general categories of methods: field methods (e.g., presence/absence surveys, abundance surveys, structural surveys, monitoring programmes, combined with surveys to collect geochemical and hydrological data); methods involving controlled experiments that may take place in a laboratory, in a glasshouse, or in the field (e.g., mesocosms, field experiments, field or laboratory trials); and remote or secondary data methods, which involve data collection in the field, at a computer, or via the Internet (e.g., remote sensing, aerial photos, lidar data, drones, databases, and meta analyses—see Chapter 14).

There are a number of books offering practical advice on conducting terrestrial and freshwater vegetation surveys including Kent (2012), Little (2013), Mueller-Dombois and Ellenberg (1974), and Kershaw (1973). Barker (2001) can be downloaded from the Internet and is an excellent practical manual on vegetation survey techniques, including monitoring and photographic techniques. Cronk and Fennessy's (2001) monograph on the biology and ecology of wetland plants includes sections on methods for measuring primary productivity. Specific texts dealing with remote sensing for conservation, and delimiting or monitoring wetlands and freshwater plant communities include Horning et al. (2010) and Tiner et al. (2015) respectively.

9.2.2 Survey impacts on plants

An important consideration in any vegetation survey is the impact of the surveyor on the habitat or target species (Malakoff 2004). This is especially true in fragile freshwater habitats such as calcareous fens, some upland peat habitats, or meromictic lakes, where the field survey itself could create significant disturbance. Before embarking on a freshwater vegetation survey it's important to conduct an ethical review of the research, follow any country guidelines for sampling rare aquatic plants (e.g., Natural England 2009), assess how destructive the methods are going to be and/or consider alternative techniques, find out if a licence is required to collect freshwater plants from the wild or to survey them, and, if the survey is being conducted overseas, make sure you comply with the country's permit and licence requirements.

9.3 Techniques and equipment specific to freshwater plant surveys

If the wetland under survey is shallow or the water table is just below the surface, and it can be walked in normal walking shoes or knee-high rubber boots, it should be possible to undertake the survey in much the same way as a terrestrial vegetation survey. For example, transects can be set up and sampled systematically using quadrats, or samples can be taken randomly, and plants can be recorded in the same way as in terrestrial vegetation surveys. The season/time of year may also be an important consideration in your decision to use 'terrestrial' or 'aquatic' techniques (e.g., sampling temperate wetlands when water levels are low in summer and high in winter, or tropical floodplains during the wet and dry season, or conducting surveys when flowers/fruits are present for species identification, which may be when water levels are high).

The terrestrial approach to surveying aquatic vegetation doesn't work if the water is any deeper because it becomes impossible looking through the water column to see what plants are growing at the bottom. Furthermore, walking around in a wetland or lake can hugely disturb the sediment, making it difficult to observe anything at all. If this is the case, the survey has to be conducted either from a boat, or by snorkelling, and may require the use of specialised equipment. Observing, sampling, or surveying underwater vegetation may require (a) dedicated sampling equipment, (b) particular methods of observation, or (c) techniques for detecting aquatic plants from the surface of the water. Some of these techniques are destructive and the plants are uprooted or cut away, while others rely on observation or remote techniques and have little impact on the habitat.

9.3.1 Equipment—grab and grapnel

Grapnels hook aquatic plants around their spikes, enabling the surveyor to pull up a sample (Figure 9.1a). There are two types of grapnel—standard (2–3 spiral spikes) and rake (a line of spikes), that can be fixed either onto a rope or a rake handle, depending on the depth of water. 'Weed weasels' consist of a double line of spikes, and some

Figure 9.1a Equipment for conducting an aquatic plant survey in shallow water, including a bathyscope for viewing underwater plants, an extendable grapnel in three sections in the foreground, a standard spike grapnel on a length of rope in the tray, and white bucket and tray (photo by J. Hughes).

commercial ones have cutters included (see website of UK Upland Waters Monitoring Network, Aquatic Macrophyte Sampling Methodology).

A grab has two spring-loaded shutters that 'grab' a bottom sample. These are usually used to sample benthic sediments but are also useful for bottom-growing aquatic plants in deep waters. There are three standard types of grab sampler: Eckman, Petersen, and Ponar. They come in different sizes, materials, and weights and should be chosen to suit the type of substrate and water depth in which the plants grow (Sliger et al. 1990; Dromgoole and Brown 1976; Spears et al. 2009).

9.3.2 Observation—bathyscope, snorkelling, and underwater camera

A bathyscope is an opaque/darkened viewing tube with a clear acrylic glass bottom (Figure 9.1a). When positioned on the surface of the water, it allows plants to be viewed much more clearly than by looking into the water. It can be used from the side of a boat to observe within the water column, or when surveying shallow water bodies by foot or in waders to make it easier to observe presence/absence or percentage cover of submerged species (Interagency Freshwater Group 2015).

Snorkelling techniques for surveying aquatic plants require a certain amount of expertise and experience but underwater surveys can be conducted in much the same way as a marine survey. It is possible to carry out transects, presence/absence surveys, and underwater quadrats in this way (e.g., surveying for invasive aquatic plants by Maine VLMP (2015), and a survey of Scottish rivers by Keruzore et al. (2013).

A purpose-built underwater camera or a camera in a waterproof housing (Bergshoeff et al. 2017) can be particularly useful to observe and record presence/absence of plant species and the survey can be carried out from a boat or a floating platform. In situations where the water is turbid or there is insufficient light, this technique may not provide suitable resolution for the photos or videos.

9.3.3 Detection—hydroacoustics and eDNA

Hydroacoustic surveys can be used to estimate aquatic plant biomass and are used for mapping submerged species in a variety of freshwater habitats. The technique is efficient on time and can generate data more cheaply than conventional manual monitoring techniques for submerged macrophytes. A recent evaluation of hydroacoustic systems by Radomski and Holbrook (2015) found significant differences in results for plant abundance and height due to differing signal processing approaches and advised on standardisation of data collection equipment as well as signal processing. While such remotely sensed data can be used in conjunction with satellite images to investigate changes over time, it is not a method to use for studies needing data on abundance of individual species (Madsen and Wersal 2017; Bucas et al. 2016).

Using traditional PCR techniques, environmental DNA (eDNA) in water samples can be analysed to detect both early traces of non-native invasive aquatic plants, and rare or endangered species. These techniques were trialled successfully for early detection of *Myriophyllum spicatum* in water bodies in Montana and Michigan (Newton et al. 2016), and to survey *Egeria densa* in Japan (Fujiwara et al. 2016). eDNA methods are potentially a powerful tool for the efficient monitoring and detection of aquatic plants but the techniques are still being tested and refined for aquatic habitats (Goldberg et al. 2016; Taberlet et al. 2018). See Chapter 10 for use of eDNA techniques for surveying freshwater vertebrates.

9.4 Sampling strategies for freshwater plant surveys

A range of methods has been used for sampling vegetation in wetlands and freshwaters (Tiner 2016). In the case of terrestrial wetlands (bog, fen, swamp, and marsh) they have usually been adapted from vegetation-sampling techniques developed for other terrestrial ecosystems (e.g., Mueller-Dombois and Ellenberg 1974; Barbour et al. 1980). For aquatic habitats, such as deep lakes and rivers, a three-dimensional approach may be needed, where the methods chosen will reflect the shape of the water body as well as water depth, and density and structure of aquatic macrophytes within the water column. In such situations, sampling at predetermined depths within the water column at the chosen sampling site using appropriate equipment (see Section 9.3) may be the only way to fulfil the aims of the study (Jeppesen et al. 1998). The selection of appropriate sampling techniques depends on the type of data needed, the size and inherent variation of the survey area, available budget, timing, and investigator expertise. This section will focus on field sampling strategies that have been commonly used to characterise vegetation composition and pattern, and monitor change in wetlands and freshwaters; sampling using unmanned aerial vehicles (UAVs) is discussed in Section 9.6.1.

9.4.1 Sampling approaches

Placement of representative samples can be subjective, random, or systematic. Subjective sampling, where the investigator chooses the location, can sometimes lead to erroneous conclusions, and is considered less scientific than other approaches (see Chapter 3). Random sampling is statistically desirable because it removes all bias; however, it can be inefficient and time-consuming, and can under-sample distinctive or rare vegetation patches (Barbour et al. 1980). Similarly, systematic sampling (i.e., regular placing of samples along a grid or transect) may also be inflexible, having some of the same limitations as a complete random design.

Several modifications have been developed to address the disadvantages, including a stratified random design (Barbour et al. 1980). Here, the investigator stratifies the survey area into several homogeneous units, and locates the samples randomly within each unit, the number of samples established usually being proportional to the unit's extent. Systematic sampling within the homogeneous units may also be used. The stratified design ensures a representative example of the range of physical and floristic features of the target plant community, whether for a terrestrial wetland or aquatic habitat (Croft and Chow-Fraser 2009).

The stratified random (or stratified systematic) approach has been applied in wetlands to measure and quantify the vegetation composition, and to help understand functional processes such as species–environment relationships, successional change, and impacts of disturbance. The first step involves mapping or stratifying the wetland based on species cover dominance, vegetation structure, landform, and other physical features. This can be done using good-quality aerial images, remotely sensed images, maps, and/or field knowledge to determine relatively homogeneous representative units (e.g., vegetation types) in which samples will be located. Recent technologies and tools such as Geographic Information Systems (GIS) or hand-held Global Positioning System (GPS) devices including apps on mobile devices greatly increase accuracy and time efficiency in field sampling. Methods to measure vegetation will depend on the vegetation structure and budget available. In some situations, using a series of plots (e.g., quadrats) may be more cost-effective than plot-less techniques (e.g., points, transect lines), as it provides quantitative data on an areal basis suitable for analysing a range of metrics (e.g., cover, composition, diversity, structure, species richness, density) (Rochefort et al. 2013 and references therein).

Where changes over a physical or hydrochemical gradient are being investigated, it may, however, be desirable to use systematic transect sampling (e.g., to investigate plant community structure in relation to frequency of inundation across a floodplain; or changes in plant associations across a gradient of height, water depth and pH for an ombrotrophic bog). A relatively fast and efficient approach to monitoring across a wetland gradient uses the line-intercept method, where a transect line is set up across the gradient and plants are recorded when they touch the line at fixed distances, either using presence/absence or abundance counts within contiguous or spaced sections of the same length. The measurements can be repeated by setting up the line at fixed time intervals, using permanent markers at each end of the transect. This is a particularly useful approach for monitoring habitats that experience rapid change along a gradient or where the effects of management need to be assessed in a time- and cost-efficient way.

Transects are also used for macrophyte surveys in deep lakes, rivers, and aquatic environments (Titus 1993). Sampling, for example, along radial lines from the centre, or along the length of the water body, enables the surveyor to conduct a plant census, a visual survey, a hydroacoustic survey, or to examine the structure of plant communities in relation to water depth, turbidity, bathymetry, or substrate (Chapter 6). The same transect lines can be used to survey for fish, plant–animal relationships, or food-web dynamics. Transects can be surveyed from a boat or by snorkelling, in conjunction with the equipment described in Section 9.3.

The advantages and disadvantages of the various sampling approaches and their fitness for purpose can be explored further in classic texts such as Kershaw (1973), Mueller-Dombois and Ellenberg (1974), Barbour et al. (1980), Kent (2012), Sutherland (2006b), Henderson and Southwood (2016), Jeppesen et al. (1998), Hurford et al. (2010), and Barker (2001). Once the sampling approach has been chosen, the next step is to decide on the nature of the sampling unit or plot (i.e., what shape, what size, how many, and what data to collect).

9.4.2 Plot shape and size

Plots are usually rectangular, square, or circular, and of any practical size. Square or rectangular plots are favoured in many quantitative vegetation studies as they are easy to set out using quadrat frames (fixed or floating), or poles and measuring tapes, and delimit distinct edges and boundaries for measurement and relocation (Figure 9.1b;

Figure 9.1b Sampling *Empodisma minus*-dominated vegetation at Danby Hill mire, Southland, New Zealand, using 2 × 2 m plots delineated by bamboo corner poles and tape measure (photo by B. Clarkson).

Figure 9.1c; Table 9.1). The plot size depends on what plant species and parameters are being measured, and should increase with vegetation height. A minimal area for sampling within a vegetation type can be determined using a nested plot technique (Mueller-Dombois and Ellenberg 1974; Barker 2001), in which plot areas are doubled sequentially and species numbers recorded. The minimal area is achieved where a 10 per cent increase in area yields only 5 per cent (10 per cent is sometimes used) more species on a species–area curve. Minimal areas should be established in the range of vegetation types in the wetland or water body and to ensure adequate representation of the target plant community, the selected plot size should not be less than the single largest minimal area. When using a fixed protocol or national vegetation classification protocol for wetlands and aquatic habitats, if plots are used their size is often predetermined (e.g., in the UK there are fixed plot sizes for short or tall herbs and grasses, for submerged or emergent macrophytes, or for bryophyte-dominated plant communities).

9.4.3 Sample size

An adequate sample size (number of replicates) that reflects the particular characteristics of a vegetation community can be inferred by constructing species–area curves using equal-sized plots randomly selected from the vegetation type (or homogeneous unit/wetland) being surveyed in a manner similar to that described above. Another

Figure 9.1c Surveying 25 permanent monitoring plots at Marley Fen, Wytham Woods, UK, using 2 × 2 m plots. Each plot has two permanent plastic marker pegs inserted into the peat in two corners and they are marked with a 2.5 m bamboo pole with a numbered flag. In such tall, dense, calcareous fen vegetation it is difficult to relocate a plot without the tall bamboo pole. Each plot has a dipwell tube to measure the depth of the water table and a GPS coordinate (photo by J. Hughes).

method is to calculate and plot the running mean for the more important (or abundant) species from randomly selected plots. An adequate sample size is obtained when this line first lies between 5 per cent (or 10 per cent) of the mean of the maximum sample size (Kershaw 1973; Mueller-Dombois and Ellenberg 1974). Because wetlands often comprise complex mosaics of different plant communities, having multiple smaller plots from the wider area rather than one or two large plots is better for ecological studies requiring statistical analysis. Linear riverine habitats can create additional sampling challenges, but ultimately the sample size and location of samples for linear freshwaters will be dictated by the research questions and knowledge of plant community structure and function in a lateral or linear dimension.

Choosing the correct sample size for aquatic plant habitats may be logistically more complex, requiring samples to be taken spatially across the water body as well as with depth within the water column. Samples may consist of dragging a rake grapnel across the lake bed along several transects (distance or timed), observing macrophytes through a bathyscope along systematic transects, or locating a survey boat at randomly chosen sample sites in a lake from which grab samples can be taken. For example, Alahuhta et al. (2017) conducted a cross-continental survey of aquatic macrophytes to investigate niche conservatism using different length transects and sample plots of varying sizes; Nagasaka (2004) used a combination of five grab samples taken at each of 12 sites, and 37 transects to examine changes in *Elodea nuttallii* populations in Lake Kizaki, Japan; and De Souza et al. (2015) used floating PVC 1 × 1 m quadrats to sample 60 sites in a reservoir in Brazil for macrophyte cover, periphyton biomass, and environmental variables. Unfortunately, there is no completely objective criterion for determining an adequate sample size (or minimal sample area) for quantitative analysis, and ultimately the decision needs to be made by the surveyor based on knowledge, experience, and the type of statistical test to be used (Barker 2001; Little 2013).

9.5 Survey methods and approaches used in freshwater plant ecology and conservation

This section provides a general overview of methods and approaches that can be used in freshwater plant surveys with a focus on field surveys, experimental approaches, and remote sensing, and the range of data that can be collected from a sample unit. Key texts on vegetation survey techniques are cited in Section 9.2.1, and the methods are summarised in Table 9.1.

9.5.1 Field surveys

Field surveys can be divided into structural and floristic. Structural surveys focus on what the plant looks like, traits, life form, morphology, size, stratification, and its function in the ecosystem. An overview of structural survey methods can be found in Kent (2012), including classic methods devised by Raunkier, Fosberg, Elton, and others. Such methods can be applied to freshwater habitats and can be used to answer questions on, for example, the influence of habitat structure (depth, density of plants, leaf shape) on macroinvertebrate abundance; the distribution of fish spawning areas in vegetated channels (ratio of floating to submerged leaves); the effects of wetland vegetation height on light penetration and seedling success; the impact of non-native species grazing on plant community structure and ecosystem function; or quantifying the habitat structure of individual leaves of aquatic plants using fractal techniques. Such descriptive methods are very useful for surveying large areas, if time is a constraint, for conducting an initial reconnaissance of a study area, or when structure is an important variable explaining the distribution of animal or other plant species. This is particularly the case for freshwater habitat classifications, which, if implemented creatively, can create a baseline, leading to hypotheses that can then be tested using floristic techniques. For example, see Freshwater Habitat Classifications for the UK on the JNCC website, and for the USA on the Nature Conservancy website. Researchers may alternatively devise their own freshwater structural methods or classifications specific to a site, region, or research question (e.g., Petr 2000; Thomaz and Cunha 2010).

In contrast, floristic surveys are undertaken at the species level and therefore require expertise in plant identification. There are many excellent resources for identifying plants on the Internet including interactive guides, although it is likely that many countries or regions will have their own country floras and plant guides, including

Table 9.1 *Factors to consider when creating a freshwater vegetation survey, experimental design, or using remote-sensing images*

Research approach	Habitat type	Temporal scale	Spatial scale	Experimental or survey design		Other considerations
				Treatment or predictor variable	Response variable or changes in:	
Mesocosm and field experiment	• Planktonic and/or benthic. • Submerged and/or floating vegetation. • Naturally colonised or planted vegetation.	• Acclimation period. • Duration of the experiment: hours, days, months, years. • Monitoring and surveillance e.g., diurnal, seasonal, annual.	• Length and width (or radius) of container or experimental set-up. • Depth e.g., position of experiment in water column. • Volume or discharge of freshwater. • Horizontal and vertical surface areas.	• Chemical e.g., toxin addition, nutrient addition or removal. • Physical e.g., light (natural or artificial), temperature, substrate, suspended sediment. • Hydrology and fluid dynamics e.g., water exchange, mixing, batch, continuous flow, simulation of flood and drought conditions. • Biological e.g., number of organisms, target species, trophic levels.	• Target organism(s) e.g., behaviour, physiology, adaptation. • Population e.g., growth, productivity, mortality, intra-species interactions. • Community e.g., grazing, predation, inter-species interactions. • Ecosystem processes e.g., biogeochemistry, water security.	• Number of trophic levels assessed e.g., primary, secondary, tertiary. • Number and frequency of treatments. • Species richness, abundance and diversity of target species or community. • Number of controls. • Number of mesocosm or experimental replicates per treatment.

(Continued)

Table 9.1 Continued

Research approach	Habitat type	Temporal scale	Spatial scale	Experimental or survey design		Other considerations
				Treatment or predictor variable	Response variable or changes in:	
Field survey	• Terrestrial wetland vegetation: marsh, fen, bog, natural, restored or constructed. • Aquatic vegetation: shallow, deep, emergent, submerged, natural, restored or constructed. • Ombrotrophic, minerotrophic, lotic, or lentic.	• Duration of survey: time it takes to collect samples or conduct visual observations. • One-off survey or continuous survey. • Repeated observations: monitoring interval, total duration of monitoring effort in weeks, years, decades. • Repeated observations of vegetation combined with continuous monitoring of environmental variables.	• Political or economic area e.g., Ramsar wetlands in a country; floodplain wetlands in a water company region. • Hydrogeological landscape, region, catchment, habitat, or feature. • Protected area e.g., based on conservation status or rarity of landscape, habitat, species. • Sample site e.g., whole wetland, vegetation type, structural class, plant population or community, substrate type, flow regime.	• Biological: vegetation/plant species e.g., cover, count, height, native vs alien, vegetation structure, phenology. • Physico-chemical: soil or plant nutrients, water storage or flux, pollutants and water quality e.g., heavy metals, phosphorus. • Disturbance e.g., grazing, fire, flood, drought. • Environmental change e.g., temperature, recharge and discharge, nutrient loading, increase in acidity.	• Plant species richness and abundance, cover, count, height, vegetation structure, density etc. • Community composition, ecosystem processes, ecological condition: spatially and temporally • Physico-chemical parameters affecting vegetation variables. • Disturbance e.g., invertebrate/ vertebrate grazing; wetland extent from drainage or rewetting.	• Number of samples e.g., plots, transects, photo points. • Spatial extent of sampling; what areal coverage? Random, stratified or systematic? • Frequency of sampling (in space or time) or duration of a timed aquatic sample. • Number of replicates, paired samples, nested samples. • Number of benthic samples or water column samples or both.

| Remote sensing from satellite, manned aircraft, or drone (UAV) platforms using optical or microwave sensors (Table 9.2) | • All wetland types including lotic, lentic, terrestrial wetlands and aquatic vegetation. Understory vegetation may not be detectable.
• Sensors detect spatial and temporal patterns of vegetation structure, phenology, and inundation state, which are associated with habitat type.
• Presence of surface water on at least one imaging date aids in discriminating between wetland and non-wetland vegetation. | • One-off survey or repeated observations and monitoring.
• For satellite sensors, temporal scale is dictated by imaging repeat cycle and operational period of sensor (Table 9.2).
• For optical satellite sensors, clouds limit availability of useable (cloud-free) imagery in some seasons or regions. | • Political or economic area e.g., Ramsar wetlands in a country; floodplain wetlands in a water company region.
• Hydrogeological landscape, region, catchment, habitat, or feature.
• Protected area e.g., based on conservation status or rarity of landscape or habitat. | • Biological: extent, patch size, per cent cover, phenology for vegetation types.
• Physico-chemical: inundation regime, sediment load, nutrient load, water chemistry.
• Disturbance: land cover and land use change, fire, mining, pollution, altered hydrologic regime (water withdrawals, damming, ditching, draining), hurricanes. | • Plant species richness and abundance, carbon stocks (based on scaling up from field surveys).
• Plant community composition, spatial extent, structure, productivity.
• Changes in plant community composition, spatial extent, productivity, carbon stocks, mortality; extent of algal blooms; altered inundation period, inundation depth, hydrologic connectivity. | • Appropriate sensor and platform, or combination of multiple sensors and platforms.
• Identification of critical seasons based on annual cycles of streamflow, precipitation, and phenology, or critical years (drought or flood years).
• Adequate temporal span for capturing change or for characterising seasonal and interannual variability.
• Appropriate validation data set for accuracy assessment. |

(Continued)

Table 9.1 *Continued*

Research approach	Habitat type	Temporal scale	Spatial scale	Treatment or predictor variable	Response variable or changes in:	Other considerations
		• For aerial surveys including with UAVs, time scale is dictated by sampling needs and project budget.	• Site e.g., whole wetland, vegetation/habitat type, structural class. For some sites it may be possible to work at the scale of a plant population, community, substrate type or flow regime. • Trade-offs occur between spatial extent of study area, spatial resolution, and temporal resolution.	• Environmental change: altered hydrologic or freeze/thaw regime related to climate change; salt water intrusion; natural floodplain processes (erosion and deposition; channel migration and avulsion).	• Shifts in local and regional distributions of wetland types and plant communities; changes in vegetation productivity and carbon stocks; successional changes e.g., in oxbow lakes.	• Robust sampling scheme for training and validation samples. • Coordination in timing of field sampling and image acquisitions.
Secondary data	• See Chapter 14.					

freshwater species and bryophytes. If a floristic survey needs to be undertaken, what do surveyors record in each sampling unit and do the research questions require qualitative or quantitative data? Presence/absence data are qualitative and provide data on species occurrence only. Such data can be combined with structural surveys and/or habitat mapping in wetlands to provide the evidence for management decisions; or collected along a line-intercept transect to monitor changes in a floodplain wetland. Presence/absence data are rapid to collect and therefore cheap compared with quantitative surveys and are often used in international conservation methodologies (e.g., Biodiversity Hotspots, Global Strategy for Plant Conservation, Important Plant Areas).

Quantitative data include species cover and height, stratum height, abundance counts, biomass, and yield, and are often measured in quadrats or other sample units both in terrestrial wetlands and aquatic habitats (Barker 2001). In aquatic habitats, floating PVC quadrat frames can be made stationary by pinning them in place using vertical bamboo stakes pushed into the substrate. In this way, the quadrat floats on the surface and the bamboo poles mark out the base of the quadrat at the bottom of the water column. The frame can be used to estimate percentage cover and height of individual species, and habitat structure can be measured at different depths in much the same way estimates can be made of canopy height and tiers within a forest. In short terrestrial wetland habitats a cover pin frame can be used to obtain objective data, but this is logistically difficult in deep water. Small-scale biomass measurements are objective but destructive and time consuming, so the surveyor would need to assess whether objective cover estimates or modelling biomass were appropriate alternatives.

Finally, the techniques described above can be used to monitor freshwater habitats over time. Regulatory, conservation, and allied organisations globally are increasingly required to assess and report on the health and ecological condition of wetlands. Monitoring is important for establishing baselines in extent and condition, detecting changes, and identifying trends over time (Barker 2001; Environmental Change Network website selecting 'Measurements' and 'Aquatic Sites'). Management and restoration outcomes can only be assessed if an adequate monitoring programme is in place (Rochefort et al. 2013). A number of bespoke monitoring protocols have been developed, including rapid assessment methods that involve considerably smaller investments of time and effort than more intensive monitoring (Dorney et al. 2018). It is worth noting that much of the effort in creating robust protocols for monitoring wetland vegetation is a result of assessing and monitoring carbon sequestration in peatlands and other wetland carbon stores for climate change mitigation (Malmer et al. 2005).

9.5.2 Experimental approaches

Experimental approaches in wetland plant ecology enable key parameters to be manipulated to better understand the controls and influences on the ecosystem or species. The parameters that are manipulated are usually identified from initial field surveys, and therefore experimental approaches are often a natural progression in the research. Experimental studies carried out in parallel with field surveys can provide a holistic perspective to the problem being investigated.

The ecological questions that can be answered using experimental approaches include: identifying environmental stressors on wetland plants (e.g., floods, droughts and temperature extremes via field trials or creating artificial wet/dry regimes in a glasshouse (Bucak et al. 2012)); assessing the range of hydrochemical conditions a species can tolerate (e.g., in relation to acidity or eutrophication, or as nutrient sinks (Olsen et al. 2015)); quantifying species interactions (e.g., effects of allelochemicals from one plant species on another, or impacts of an invasive non-native macrophyte species on greenhouse gas emissions (Attermeyer et al. 2016) or functional feeding groups); impacts of removing/adding a grazing species on plant productivity and diversity via removal experiments or exclosures (e.g., effects of cattle on floodplain wetlands or planktivorous fish on submerged macrophytes (Jeppesen et al. 1998)); identification of functional and evolutionary traits of wetland plants (e.g., leaf area to leaf mass ratio in relation to leaf gas exchange); and evaluating the effects of remediation or management on freshwater plant species richness and abundance via controlled experiments. Semlitsch and Boone (2010) review the use of mesocosms for amphibian research but include useful sections on general considerations and experimental design.

Experiments can take place in the field, laboratory, glasshouse, flume, or artificial ponds (Figure 9.2a; Figure 9.2b). The main experimental approaches vary with scale

Figure 9.2a Nine mesocosm units (10 cm width × 30 cm length × 10 cm depth) containing the macrophyte *Phalaris arundinacea* to test the effects of flow and time on *Microcystis aeruginosa* cell filtration and root allelochemical production. The experiment was conducted on a continuous flow setting for 5, 7 and 11 days (photo by A. Castro-Castellon).

Figure 9.2b Macrocosm (21 m width × 10 m length × 1 m depth) containing three species of aquatic macrophyte, Farmoor Reservoir, UK. *Phragmites australis, Carex acutiformis,* and *Phalaris arundinacea* were grown hydroponically in a floating macrocosm in continuous-flow conditions for several months, to quantify their effectiveness as a biofilter (photo by A. Castro-Castellon).

from unenclosed field experiments at one end of the continuum to single-species experiments in batches or continuous culture at the other end (Sommer 2012). In the middle are mesocosms, which are defined as a contained subset of a larger ecological system where at least three or more species interact (Odum 1984). An example of a natural mesocosm is 'phytotelmata', which is a naturally formed contained aquatic habitat populated by aquatic organisms (e.g., natural small ponds, tree holes, and bromeliads) (Srivastava et al. 2004).

The research questions should define the experimental design and type of mesocosm used. An artificial aquatic mesocosm is defined as a contained system between 1 and 100 L in volume; a microcosm is <1 L in volume; and a macrocosm has a volume >100 L. The artificial containers can be made of glass, ceramic, metal, concrete, or plastic, and are either placed in a natural water body, or water is supplied to the container by pipe or by diverting river water. Some meso- and/or macrocosms are engineered designs with a combination of materials. For example, floating living or artificial plant bed systems (or treatment wetlands) may be considered as a mix of natural to artificial mesocosm/macrocosm. The mixed materials used to construct these experimental systems and embed the plants are fixed to a buoyant metal frame providing continuous flow and biochemical interchange in the aquatic medium. Table 9.1 presents different types of experimental methods using mesocosms based on habitat, temporal and spatial scales, and experimental design.

9.5.3 Remote sensing approaches

Remote sensing—the use of data collected by satellite and airborne sensors to characterise the physical properties of an area on the ground—is an indispensable tool for answering many types of research and management questions in freshwater ecology and conservation (Table 9.1). For all but the smallest wetlands, ponds, or stream reaches, remote sensing provides the most practical means of estimating the areal extents of freshwater habitats and how those change through time, and of scaling up field measurements to larger areas. Remote sensing provides a synoptic view and allows systematic sampling through time of areas that are difficult to access. The technology and analytic techniques of remote sensing have evolved rapidly from the mid-twentieth century, when the first modern wetland maps were created using labour-intensive manual interpretation of analogue aerial photographs, to the modern era of digital imagery acquired by satellites, piloted aircraft, or drones and analysed rapidly using sophisticated software and machine-learning algorithms. Remote sensing analyses can provide information on wetland extent (present and historic), hydrology (dynamics of surface water extent and water level height), water properties (dynamics of suspended sediments, chlorophyll a, CDOM) and vegetation properties (type, structure, phenology, biomass), allowing researchers or managers to answer key questions related to freshwater ecosystem function and conservation.

Wetlands remote sensing employs a wide range of sensor types, platforms, spatial scales, and analytic approaches. A comprehensive treatment of wetlands remote sensing can be found in Tiner et al. (2015), and Guo et al. (2017) review more than 250 remote-sensing studies of freshwater and coastal wetlands. Recent review articles have also addressed topics such as remote sensing for wetland classification (Mahdavi et al. 2017), multispectral and hyperspectral mapping of wetland vegetation (Adam et al. 2014), radar remote sensing of flooded vegetation (Henderson and Lewis 2008; Tsyganskaya et al. 2018), and remote sensing of lakes (Dörnhöfer and Oppelt 2016), rivers (Palmer and Ruhi 2018), aquatic vegetation (Silva et al. 2008), and water quality (Gholizadeh et al. 2016; Matthews 2011).

9.6 Remote sensing methods

9.6.1 Sensor types and platforms

Sensors used in remote sensing can be grouped according to which range of the electromagnetic spectrum they employ, their spatial resolution, on which type of platform (satellite or airborne) they are deployed, and whether they are passive or active. Passive sensors collect reflected sunlight or naturally emitted thermal or microwave radiation, while active sensors such as radar or lidar generate their own signal, a portion of which is then scattered back to the sensor. Table 9.2 lists some of the satellite systems most frequently used for freshwater ecology applications; a complete, continuously updated list of Earth Observation satellites can be found on the 'Satellite Missions Database' web page of the 'Sharing Earth Observation Resources' website (directory.eoportal.org).

Passive optical instruments include satellite sensors such as Landsat-8 OLI or Sentinel-2 MSI (Table 9.2), as well as digital cameras or multispectral scanners

Table 9.2 *Optical and microwave satellite sensors frequently used for remote sensing of lakes, rivers, and wetlands*

Satellite	Sensor	Spatial resolution (m)	Bands* and Modes**	Repeat cycle (days)	Years operational
Optical, Moderate Resolution					
NOAA 1-16	AVHRR	1,000	1 VIS, 1 NIR, 1 SWIR	1	since 1981
Terra, Aqua	MODIS*	250–1,000	10 VIS, 3 NIR, 3 SWIR	3	since 1999
Envisat	MERIS*	300–1,200	9 VIS, 2 NIR	3	2002–2012
Sentinel-3A	OLCI	300–1,200	7 VIS, 3 RE, 3 NIR	2	since 2016
Optical, Fine Resolution					
Landsat 1–3	Multispectral Scanner	60	2 VIS, 2-3 NIR	18	1972–1983
Landsat 4–5	Thematic Mapper	30	3 VIS, 1 NIR, 2 SWIR	16	1982–2011
Landsat 7	Enhanced Thematic Mapper Plus	15–30	3 VIS, 1 NIR, 2 SWIR, 1 Pan	16	1999–2003
Landsat 8	Operational Land Imager*	15–30	4 VIS, 1 NIR, 2 SWIR, 1 Pan	16	since 2013
SPOT 1-3	HRV	20	2 VIS, 1 NIR	26	1986–2009
SPOT 4	HRVIR	10–20	2 VIS, 1 NIR, 1 SWIR, 1 Pan	26	1998–2013
SPOT 5	HRG	2.5–20	2 VIS, 1 NIR, 1 SWIR, 1 Pan	26	2002–2015
SPOT 6–7	NAOMI	2–8	3 VIS, 1 NIR, 1 Pan	≤26	since 2012
Sentinel-2A, 2B	MSI*	10–20	4 VIS, 3 RE, 3 NIR, 2 SWIR	5	since 2015
Optical, Very Fine Resolution					
IKONOS	OSA	1–4	3 VIS, 1 NIR, 1 Pan	≤11	1999–2015

(Continued)

Table 9.2 *Continued*

QuickBird	MS Scanner; Pan Camera	0.6–2.4	3 VIS, 1 NIR, 1 Pan	≤3	2001–2015
WorldView*	WV60; SpaceView	0.3–2	5 VIS, 2 NIR, 1 RE, 1 Pan	1–3	since 2007
RapidEye	REIS	5	3 VIS, 1 NIR, 1 RE	1–6	since 2009
Microwave (Synthetic Aperture Radar; SAR), Very Fine to Moderate Resolution					
JERS-1	SAR	12.5	L-band; HH pol	45	1993–1997
RADARSAT-1,2	SAR	1–100	C-band; single-, dual-, or quad-pol	24	1995–2007
ALOS; ALOS-2	PALSAR; PALSAR-2	1–100	L-band; single-, dual-, or quad-pol	46; 14	2006–2011
COSMO-SkyMed	SAR-2000	1–100	X-band; single-pol	16	since 2007
TerraSAR-X	TSX-SAR	1–16	X-band; single-, dual-, or quad-pol	11	since 2007
Sentinel-1	C-SAR	3.5–40	C-band; dual-pol	12	since 2014

* *Optical band abbreviations: VIS—Visible; NIR—Near Infrared; SWIR—Shortwave Infrared; RE—Red Edge; Pan—Panchromatic. Optical bands used primarily for aerosols, water vapour, and cirrus are not listed. Bands shown are for WorldView-2 and WorldView-3.*

***Microwave polarisation modes: single-, dual-, and quad-pol modes use one, two, or all of the four possible transmit-receive configurations: HH—horizontal transmit, horizontal receive; HV—horizontal transmit, vertical receive; VV—vertical transmit, vertical receive; VH -vertical transmit, horizontal receive.*

deployed on piloted aircraft or unmanned aerial vehicles (UAVs; also called drones). Optical sensors operate in the visible, near-infrared, and shortwave infrared regions of the electromagnetic spectrum at wavelengths between 0.3 and 3.0 μm (Figure 9.3). Hand-held spectrometers used in the field sample over very narrow bands, allowing continuous spectra to be recorded for water, soil, or vegetation samples. Hyperspectral sensors, or imaging spectrometers, similarly collect across hundreds of narrow bands; to date they have been deployed primarily on aircraft. Satellite multispectral sensors for land observation, such as OLI, sample fewer and broader bands, located within spectral regions (atmospheric 'windows') within which absorption by H_2O, CO_2, O_3, and other atmospheric constituents is minimised.

The contrasting reflectances of materials at different bands are the basis for classifying land cover or estimating physical variables from optical imagery. Green leaf spectra (Figure 9.3) are characterised by low reflectance in the visible range, where sunlight is

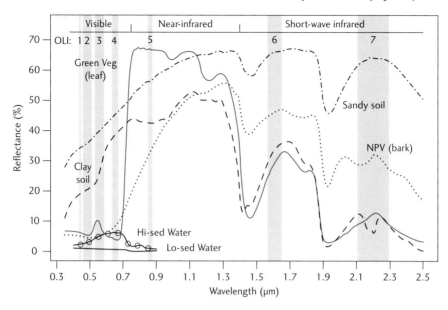

Figure 9.3 Spectral responses collected with field spectrometers in the optical portion of the electromagnetic spectrum for green vegetation (single leaf), non-photosynthetic vegetation (NPV; bark), sandy soil, clay soil, and lake water with total suspended solids (TSS) concentrations of 1.9 mg/L and 20.8 mg/L. Grey bars show bands of Landsat 8 Operational Land Imager (OLI) in the visible (ultra blue, blue, green, red; bands 1 to 4), near-infrared (NIR; band 5), and shortwave infrared (SWIR1, SWIR2; bands 6 and 7) portions of the spectrum. Water spectra from Novo et al. (2013). Land spectra courtesy of VIPER Lab, UC Santa Barbara.

strongly absorbed by chlorophylls and other leaf pigments; high reflectance in the near-infrared, owing to strong scattering from cell walls within the mesophyll layer, with little absorption; and a drop-off in reflectance in the shortwave infrared from strong absorption by water (Homolová et al. 2013). Several optical sensors have 'red edge' bands targeted in the region between red and near-infrared, which is sensitive to chlorophyll status and leaf structure. Soil reflectance is affected by soil mineral composition, texture, organic matter content, and soil moisture. Non-photosynthetic vegetation (NPV) refers to senescent or dormant vegetation, as well as woody trunks, branches, and twigs; it displays higher shortwave infrared reflectance relative to green vegetation. Water is, of course, a prime target in wetlands remote sensing. The very low reflectance of water in the near- and shortwave-infrared range generally allows it to be readily distinguished from other land cover types using optical imagery. Many optical sensors have a broad 'panchromatic' band, which is essentially a black and white band with higher spatial resolution than is possible for the narrow multi-spectral bands.

Lidar systems are active optical sensors that record the range and position of returns from laser pulses, building a 'point cloud' from which digital elevation models (DEMs) or forest canopy models can be derived. Near-infrared lasers are used for terrestrial applications, and green light lasers for bathymetric studies of coastal or inland underwater topography. The vertical information provided by lidar systems makes them highly complementary to two-dimensional satellite or aerial imagery, and airborne laser scanning has become the method of choice for many forestry applications (Vauhkonen et al. 2014).

Synthetic aperture radar (SAR) sensors have been widely used for wetland studies owing to their ability to penetrate cloud cover and to detect flooding beneath vegetation canopies (Hess et al. 1995). SARs are active sensors operating in the microwave portion of the electromagnetic spectrum; currently operational satellite SARs operate at X-band (~3 cm), C-band, (~5.5 cm), or L-band (~23 cm) wavelengths (Table 9.2). SAR instruments measure the amplitude and polarisation of the signal backscattered by ground targets from transmitted pulses. Operating at much longer wavelengths than optical sensors, they respond to quite different scene properties, and are sensitive mainly to surface roughness and to the size and orientation of vegetation canopy elements such as leaves, branches, and trunks. Because of their longer wavelength, L-band SARs are better able to penetrate forest canopies (and to detect sub-canopy flooding) than C- or X-band SARs (Figure 9.4).

Although satellite-based remote sensing is the most common approach for larger study areas, in recent years numerous studies have demonstrated the utility of unmanned aircraft system (UAS) surveys for assessing wetland vegetation at very fine resolutions (Marcaccio et al. 2015). Comprising an unmanned aerial vehicle (UAV or drone) and its sensor payload and control systems, a UAS is able to fly below cloud cover, permitting optical image acquisition during seasons when cloud-free satellite

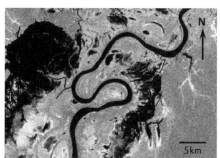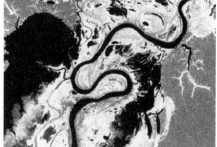

Figure 9.4 ALOS PALSAR scenes, lower Purus River floodplain, central Amazon, illustrating capability of L-band SAR to detect flooding beneath forest canopy. Scenes were acquired on 24 December 2008 (left; low water stage) and 29 March 2010 (right; high water stage). Bright areas indicate flooded forest, which extends to the floodplain boundary at high water. At low water, only low-lying portions of forested floodplain are inundated. Successional shrub communities within the large lake at the west margin of the scene are emergent and visible at low water but completely submerged at high water. Medium-toned areas at north-west and south-east borders of the scene are upland rainforest. Images courtesy of Japan Aerospace Exploration Agency and Alaska Satellite Facility.

imagery is not available. Through use of 'structure from motion' processing of image sequences along flight tracks, aerial mapping and digital elevation models can be produced with accuracies comparable to those from photogrammetric stereo mapping and lidar methods, and at much lower cost (Boon et al. 2016; Kalacska et al. 2017). When considering a UAS survey, it is critical to be aware of all rules and restrictions for UAV flight, which vary from country to country.

9.6.2 Steps in carrying out a remote sensing study

Step 1: Definition of goals, end users, and classification system

Remote-sensing analyses are often used to create maps that may be used by a wide range of users (e.g., scientists, managers, local communities). Since freshwater ecosystem studies intersect several disciplines such as plant ecology, limnology, conservation, and water resource management, end users may not share a common vocabulary or conceptualisation of the classes to be mapped, and may not understand the constraints of what can be achieved using remote sensing. Discussions with potential users at the outset of the project will help clarify these issues; in some cases, multiple maps and legends can be produced from the same analysis to suit user needs. Because wetland vegetation and hydrologic state are often seasonally dynamic, the temporal sampling period must be clearly defined e.g., does a water body boundary represent conditions on a single imaging date, or the maximum, minimum, or mean extent based on multiple dates?

Step 2: Choice of sensor or sensor combination

Choice of sensor depends on several factors including spatial extent of the study region; spatial and temporal resolution required for the application; cost of imagery (if not freely available); and prevalence of cloud cover during critical seasons. Very high-resolution imagery provides excellent detail, but other than for small sites requires mosaicking multiple scenes. Studies of change in vegetation extent or condition benefit from using sensors with long historic records, such as the Landsat imagery available at no cost from the USGS (earthexplorer.usgs.gov). For studies requiring imagery during cloudy seasons or detection of water beneath forest canopies, SAR data or low-altitude aerial imagery will be needed. In general, multi-temporal analysis using images from multiple seasons or years will achieve better results than analyses using a single date, and combinations of multiple sensor types (optical, radar, lidar) can also improve accuracy. In addition to satellite or aerial imagery, ancillary data sets such as digital elevation models (DEMs) can be incorporated into the image-processing stream.

Step 3: Identification of validation strategy

Although accuracy assessment is the final phase in a remote-sensing study, it is important to identify at the outset what data set will be used to validate the remote sensing product. For small and accessible study areas, it is possible to use field measurements for validation (or ground-truthing). Where that is not possible, a validation data set with finer spatial resolution than that of the imagery being analysed should be used. For example, QuickBird or UAV data could be used to validate a map based on Landsat imagery. As in field survey design (Section 9.4.1), a random stratified sample is often employed. Validation samples must not be used in developing algorithms used for image classification or modelling of biophysical variables.

Step 4: Pre-processing of imagery and choice of classification or analytic approach

Pre-processing steps include mosaicking of scenes; verifying image geocoding and accurate co-registration of multi-date imagery; performance of atmospheric correction; conversion of raw data to units of radiance or surface reflectance; and cloud masking. Analysis can be carried out using commercial image processing or GIS software, or with open-source packages such as QGIS (qgis.org), GRASS (grass.osgeo.org), or the rgdal package for R (rgdal.sourceforge.net). However, earth observation data sets are increasingly being made available as products that minimise the need for pre-processing. For vegetation studies, linear transformations of optical image bands are often used in order to emphasise vegetation properties. Two of the most common are the Normalised Difference Vegetation Index (NDVI; Pettorelli et al. 2005) and the Kauth-Thomas transformation, often known as the Tasseled Cap (Baig et al. 2014),

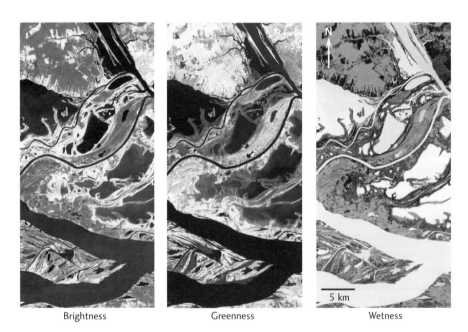

| Brightness | Greenness | Wetness |

Figure 9.5 Kauth-Thomas (Tasseled Cap) transformation of Landsat 5 Thematic Mapper scene (subset): lower Amazon floodplain at confluence with Trombetas River (30 November 2008; low water stage). Amazon River traverses lower portion of the scene; uplands including city of Oriximiná are at the top of the scene. Brightness band highlights soil, sediments, roads and buildings, or senescent herbaceous vegetation at margins of floodplain lakes and channels, in upland pastures, and in urban areas. Greenness band highlights forest and green herbaceous vegetation. Wetness band highlights open water in rivers and lakes.

which transforms input bands to a scale of Brightness, Greenness, and Wetness (Figure 9.5). Image classification can be carried out on individual pixels or on groups of pixels derived from image segmentation (object-based analysis; Blaschke et al. 2014). Lu and Weng (2007) and Adam et al. (2014) provide useful overviews of the wide variety of classification algorithms used for earth observation data.

Step 5: Accuracy assessment

Thematic accuracy assessment of classified images is typically reported in the form of a 'confusion matrix' that shows mapped versus actual class labels; from this, various accuracy measures can be derived (e.g., producer's accuracy, user's accuracy, kappa index; Foody 2002). Since errors in vegetation maps will be propagated through any estimates or modelling based on the maps, it is vital to provide a thorough account of the validation methods and results as part of the map reporting.

9.7 What questions can these methods answer?

The following case studies provide examples of a range of sampling and remote sensing approaches in contrasting wetland types and biomes. Studies such as these are extremely useful in providing guidance for new freshwater vegetation projects with similar aims and questions.

9.7.1 *Sphagnum* bog

Sphagnum bogs throughout Tasmania were studied to assess floristic gradients and environmental correlates, and identify peatlands of high conservation values (Whinam et al. 2001). Species cover abundance data (per cent) for 171 taxa were recorded in 174 plots; however, all data were converted to presence/absence data because of the high number of taxa with less than 5 per cent cover. Environmental data collected included aspect, slope, peat depth, pH, altitude, precipitation, and temperature. Ordination using hybrid multidimensional scaling revealed the strongest floristic gradients correlated with changes in altitude and moisture. Cluster analysis resulted in several new types of *Sphagnum* peatland, mostly at lower altitude, that have unusual geologies, or are geographically disjunctive. Many of these are not protected in reserves and under threat of modification, and therefore are a high priority for protection and management.

9.7.2 Restiad bog

Restiad bogs (dominated by Restionaceae) are common throughout New Zealand and in south-east Australia. Lowland bogs of different ages from the northern North Island of New Zealand were sampled to identify the major environmental determinants of vegetation pattern and successional dynamics (Clarkson et al. 2004a). Sectors within each bog were selected as representative of the vegetation patterns and 2 × 2 m plots established systematically along transects. Vegetation cover was assessed using a modified Braun-Blanquet cover scale (Mueller-Dombois and Ellenberg 1974) and height was also recorded. Plot environmental variables sampled included peat pH, N, P, K, bulk density, and decomposition (von Post scale). Classification of the vegetation data revealed groups that reflected a successional sequence derived from previous palaeo studies (i.e., from early successional sedges and shrubs, to mid-successional and late-successional groups dominated by different restiad species). The space–time relationships and species–environment correlations of ordination results indicated that as succession proceeds, peat decomposition, pH, and nutrients, particularly total P and total N, decline.

9.7.3 Fen

The relationships of vegetation to surface water chemistry and peat chemistry were examined to compare and contrast non-forested peatlands along a poor to rich fen gradient in Alberta, Canada (Vitt and Chee 1990). The vegetation was stratified visually into distinct communities, which were sampled using multiple plots (at least ten randomly placed 25 × 25 cm quadrats per stand/vegetation community). The vegetation was sampled as percentage canopy cover, with mean stand values used in data analyses. Water samples were analysed for pH, conductance, and a variety of mineral elements, including Ca, Mg, Na, K, P, Fe, and S, and peat samples were analysed for Ca, Mg, Na, K, S, Fe, and P. Multivariate analysis (classification and ordination) of the vegetation together with correlations of environmental variables revealed important differences between fen types. The extreme rich fens are dominated by brown moss species, the moderate-rich fens by different brown mosses, and the poor fens by various *Sphagnum* species. The fen types are also separated based on chemistry, with pH, conductivity, Ca, and Mg being highest in extreme rich fens and lowest in poor fens. In contrast, nutrient concentrations (N and P) generally did not differ appreciably between the types.

9.7.4 Sub-tropical river floodplain

Field surveys of aquatic macrophytes took place within the complex habitats of the riverine floodplain wetlands of the Upper Rio Parana of southern Brazil (Murphy et al. 2003). The aim was to link environmental variables (water and sediment) with morphological attributes of the vegetation to model the impacts of future altered hydrological regimes on the macrophyte assemblages. The environmental variables were used to predict species richness and biomass of aquatic vegetation for the floodplain waterbodies. A range of macrophyte sampling techniques was used to survey *in situ* macrophytes and extract whole plants (including roots) for biomass estimates, at 45 sampling points over 20 sites. Equipment used included a boat, floating quadrats, rake grapnel, and long-handled fork. Additional morphological attributes were measured in the lab. Three main macrophyte assemblages were detected and there were predictable patterns of variation in plant size and shape along gradients of water and sediment physico-chemistry.

9.7.5 Biomass estimation

The feasibility of predicting biomass of high-density papyrus-dominated wetlands from WorldView-2-derived vegetation indices was tested at three papyrus swamps in the iSimangaliso Wetland Park, a Ramsar wetland in KwaZulu-Natal province, South Africa (Mutanga et al. 2012). The three sites encompass about 7000 ha, dominated by *Cyperus papyrus* L., *Phragmites australis*, *Echinochloa pyramidalis*, and *Thelypteris interrupta*. Field-measured above-ground biomass was estimated at 82 randomly located 20 m × 20 m sites by sampling 3–5 (randomly selected) 1 m × 1 m subplots within each site. At the 2 m WorldView resolution, each site thus corresponded to about 100 pixels. Of the 82 estimates, 57 were used to train a regression model and 25 were used as an independent test data set. Vegetation indices calculated from WorldView-2 band combinations were used as predictors of biomass using the Random Forest regression algorithm. The standard NDVI index computed using the red and near-infrared bands was poorly correlated with biomass, a result that was interpreted as signal saturation at high biomass levels. However, NDVI calculated using the red-edge band was highly correlated with biomass ($R^2 = 0.76$) and yielded a root mean square error of prediction of 0.44 kg/m^2, or 13 per cent of the field-measured biomass.

9.7.6 Wetland delineation

A national system for wetland delineation in the USA is based on three diagnostic environmental criteria: wetland vegetation, hydric soils, and wetland hydrology (Environment Laboratory 1987; and subsequent US Army Corps of Engineers updates). The vegetation criterion involves the subjective selection of one or more representative plots in a typical stand or vegetation type, which is undertaken by experienced investigators following field reconnaissance and vegetation characterisation protocols (Tiner 2016). The unique combination of wetland indicator status and cover of plants in a plot is applied in an algorithm and compared with pre-determined thresholds to indicate whether a particular location is a 'wetland'. Different plot shapes and plot sizes may be used to cope with multiple strata at a site (e.g., 9-m-radius circular plot in forest stratum, 1.5-m-radius circular subplot in sapling/shrub stratum, and 1-m^2 subplots in herb stratum). With experience, this is a useful rapid method that, together with the soils and hydrology criteria, provides a regulatorily acceptable level of certainty in identifying wetlands and their boundaries.

9.7.7 Mapping aquatic vegetation

At three test sites in northern Sweden, Husson et al. (2014) evaluated the utility of aerial images taken with unmanned aircraft systems for producing species-level maps of non-submerged aquatic and riparian vegetation and for estimating species abundance. An aerial survey was flown with a miniature (1.2 m wingspan) fixed-wing UAS equipped with a digital compact camera; the flying height of 150 m resulted in an image pixel size of 5.6 m. For sites I and II, aerial images were printed at a scale of 1:800. For Sites II and III, imagery was processed into (1) high-resolution digital surface models with a relative height accuracy of 8–9 cm and (2) 30-cm orthoimages with planar accuracy of 4–5 cm. Species identification was performed by hand on paper printouts for 344 training sites and 269 validation sites. Species identification was highly accurate: 49 vegetation classes, mostly at the species level, with an overall accuracy of 95 per cent at site I; and 14 vegetation classes, mostly at the species level, with an overall accuracy of 80 per cent at site II. Classification of the digital vegetation map was performed manually using GIS software. The authors conclude that UASs generating sub-decimetre resolution orthoimages offer great potential for lake and river vegetation identification and mapping at the species level, as well as for abundance estimates.

9.7.8 Mapping riparian vegetation

By integrating airborne lidar data with 2.4 m QuickBird imagery, Arroyo et al. (2010) were able to accurately map riparian vegetation in a tropical savanna site in Queensland, Australia, where the complex vertical and multilayered structure of riparian areas poses challenges to classification using passive sensors alone. Lidar data collected along a 5 km reach was processed to derive a digital terrain model, a tree canopy model, a plant projective cover map, and a streambed map, which were combined with QuickBird data using an object-based classifier. The resulting map of five cover types (riparian vegetation, streambed, bare ground, woodlands, and rangelands) had an accuracy of 86 per cent, and map estimates of riparian zone and streambed widths were highly correlated with field measurements.

9.7.9 National and regional wetland inventories

The most extensive national wetland mapping to date based entirely on satellite remote sensing is for China, which has completed four moderate resolution (240 m) maps derived from Landsat and CBERS-2 imagery, spanning 1978–2008. A semi-automated (partly manual) classification was performed using a modified version of the Ramsar classification system (Niu et al. 2012). These maps have been used for applications including estimating rates of wetland loss (Niu et al. 2012), evaluating wetland responses to droughts (Cao et al. 2012), estimating organic carbon pools in China's wetlands (Zheng et al. 2013), and valuing ecosystem services provided by lakes and marshes (Zhang et al. 2014).

Regional wetland inventories, which may cross national boundaries, often target a narrower range of wetland types than national inventories, with classification systems tailored to regional conditions. For example, a Great Lakes coastal wetland map was produced by a consortium of US and Canadian agencies to address the lack of a consistent wetlands map for the transnational Great Lakes basin and to provide a baseline for future monitoring (Bourgeau-Chavez et al. 2015). Applying a Random Forest classifier to multi-season Landsat TM and ALOS PALSAR imagery, the coastline and a 10-km inland buffer were mapped for 23 classes, including wetland types such as aquatic beds, shrub and treed peatland, wetland shrub, forested wetland, and monospecific types, including invasive *Typha* spp. and *Phragmites australis*, with an overall accuracy of 94 per cent.

9.7.10 Monitoring

In New Zealand, a Wetland Condition Index has been developed for State of the Environment monitoring and reporting (Clarkson et al. 2004b). It comprises five ecological indicators that are compared and scored against an assumed natural or reference state; high scores reflect low modification. The indicators are based on changes in hydrology, physico-chemistry, ecosystem intactness, browsing, predation, and harvesting regimes (a measure of introduced animal and human impacts), and dominance of native plants. The methodology involves wetland classification, mapping and delineation of vegetation types, field reconnaissance and ground-truthing, establishing representative plots, and collecting data to enable informed assessment and scoring of the indicators. The Wetland Condition Index (and components) can be used to compare changes at different levels and scales (e.g., within an individual wetland, or across wetland types in a region or nationwide) to help gauge the effectiveness of management plans and policies.

9.7.11 Experimental meso- and macrocosms

Integrating field surveys with experiments using macro-, meso-, and microcosms was used to determine whether a floating *in situ* plant bed, or 'Living-Filter', could reduce or remove phytoplankton (micro-algae and cyanobacteria) before entering a potable water-treatment works at Farmoor Reservoir, Oxfordshire, UK (Castro-Castellon et al. 2016). The novel floating macrocosm contained three plant species: *Phragmites australis*, *Phalaris arundinacea*, and *Carex acutiformis* (Figure 9.2a; Figure 9.2b). Transects were used together with 1 m² sampling points to calculate the plant cover of the entire macrocosm over the monitoring period. The macrocosm was surveyed weekly at 16 sample sites (15 cm diameter × 50–70 cm length tubes) over 17 weeks for physico-chemical and biological parameters under continuous flow conditions.

Based on the field survey results, the efficiency of the mesoscale biofilter was further tested using mesocosms under continuous flow conditions and in batch experiments. The mesocosms consisted of triplicate filter plastic units with three filter media: one biofilter (*Phalaris arundinacea* roots) and two synthetic monofilament filters (plastic three-dimensional mesh), as well as control units with no filter media. A synthesised laboratory feed of *Microcystis* culture was pumped to the macrocosms over 5, 7 and 11 days (each in triplicates) and the results showed that removal efficiency was 20–25 per cent higher for the biofilters than the synthetic filters. Microscale studies further investigated biofilter allelochemical release in response to environmental stressors and *Microcystis* growth inhibition in filtered and unfiltered aqueous root exudate, using three replicates of 15 Falcon tube (50 ml) microcosms. The results confirmed that microscale and mesoscale studies identified local interactions and/or changes that would have been missed using only a macroscale experimental study or field survey. The mesocosm experiment validated the observations from the field survey.

9.7.12 Floodplain forest structure and inundation period

Ferreira-Ferreira et al. (2015) used multi-temporal ALOS PALSAR imagery to map major vegetation types and inundation patterns in *várzea* floodplain forests of the Mamirauá Sustainable Development Reserve, Brazil. The ~10,000 km² reserve, located on the Amazon floodplain at the confluence of the Japurá river, protects the habitats of several vulnerable species such as the Amazonian manatee (*Trichechus inunguis*) and the white uakari monkey (*Cacajao calvus*), while promoting sustainable use and protecting the traditional livelihoods of riverine communities. The reserve is seasonally inundated by a 10-m monomodal flood pulse that, combined with floodplain topography, is the major determinant of floodplain vegetation structure and composition. Thirteen dates of ALOS PALSAR 12.5-m imagery were segmented and classified into five land cover classes: várzea forest of high levees, várzea forest of low levees, low-lying *chavascal* forest, open water, and areas with seasonal herbaceous vegetation. Flood extent was then mapped for each date using backscatter threshold values, and inundation period maps were derived by relating flood extent to historic river stage data. Land cover accuracy was assessed using 5-m RapidEye imagery, and flood level accuracy was assessed using a network of field gauges. The study provided the first regional mapping of vegetation cover and inundation period for the reserve, and demonstrated wider ranges in inundation period than previously reported in the literature for these forest types at other sites.

References

Adam, E., Mutanga, O., Odindi, J., and Abdel-Rahman, E.M. (2014). Land-use/cover classification in a heterogeneous coastal landscape using RapidEye imagery: evaluating the performance of random forest and support vector machines classifiers. *International Journal of Remote Sensing*, 35, 3440–58.

Alahuhta, J., Ecke, F., Johnson, L.B., Sass, L., and Heino, J. (2017). A comparative analysis reveals little evidence for niche conservatism in aquatic macrophytes among four areas on two continents. *Oikos*, 126, 136–48.

Arroyo, L.A., Johansen, K., Armston, J., and Phinn, S. (2010). Integration of LiDAR and QuickBird imagery for mapping riparian biophysical parameters and land cover types in Australian tropical savannas. *Forest Ecology and Management*, 259, 598–606.

Attermeyer, K., Flury, S., Jayakumar, R., Fiener, P., Steger, K., Arya, V., Wilken, F., van Geldern, R., and Premke, K. (2016). Invasive floating macrophytes reduce greenhouse gas emissions from a small tropical lake. *Scientific Reports*, 6, 20424.

Baig, M.H.A., Zhang, L., Shuai, T., and Tong, Q. (2014). Derivation of a tasselled cap transformation based on Landsat 8 at-satellite reflectance. *Remote Sensing Letters*, 5, 423–31.

Barbour, M.G., Burk, J.H., Pitts, W.D., Gilliam, F.S., and Schwartz, F.S. (1980). Terrestrial Plant Ecology. Benjamin/Cummings Publishing, Menlo Park, CA, USA.

Barker, P. (2001). A Technical Manual for Vegetation Monitoring. Resource Management and Conservation, Department of Primary Industries, Water and Environment, Hobart, Tasmania, Australia. Available online.

Bergshoeff, J.A., Zargarpour, N., Legge, G., and Favaro, B. (2017). How to build a low-cost underwater camera housing for aquatic research. *FACETS*, 2, 150–9.

Blaschke, T., Hay, G.J., Kelly, M., Lang, S., Hofmann, P., Addink, E., Queiroz Feitosa, R., van der Meer, F., van der Werff, H., and van Coillie, F. (2014). Geographic Object-Based Image Analysis–Towards a new paradigm. *ISPRS Journal of Photogrammetry and Remote Sensing*, 87, 180–91.

Boon, M.A., Greenfield, R., and Tesfamichael, S. (2016). Unmanned Aerial Vehicle (UAV) photogrammetry produces accurate high-resolution orthophotos, point clouds and surface models for mapping wetlands. *South African Journal of Geomatics*, 5, 186–200.

Bourgeau-Chavez, L., Endres, S., Battaglia, M., Miller, M.E., Banda, E., Laubach, Z., Higman, P., Chow-Fraser, P., and Marcaccio, J. (2015). Development of a bi-national Great Lakes coastal wetland and land use map using three-season PALSAR and Landsat imagery. *Remote Sensing*, 7, 8655–82.

Bucak, T. Saraoglu, E., Levi, E.E., Tavsanoglu, U.N., Cakiroglu. A.I., Jeppesen, E., and Beklioglu, M. (2012). The influence of water level on macrophyte growth and trophic inter-actions in eutrophic Mediterranean shallow lakes: a mesocosm experiment with and without fish. *Freshwater Biology*, 57, 1631–42.

Bucas, M., Saskov, A., Siaulys, A., and Sinkeviciene, Z. (2016). Assessment of a simple hydroa-coustic system for the mapping of macrophytes in extremely shallow and turbid lagoon. *Aquatic Botany*, 134, 39–46.

Cao, C., Zhao, J., Gong, P., Ma, G., Bao, D., Tian, K., Tian, R., Niu, Z., Zhang, H., and Xu, M. (2012). Wetland changes and droughts in southwestern China. *Geomatics, Natural Hazards and Risk*, 3, 79–95.

Castro-Castellon, A.T., Chipps, M.J., Hankins, N.P., and Hughes, J.M.R. (2016). Lessons from the 'Living-Filter': An in-reservoir floating treatment wetland for phytoplankton reduction prior to a water treatment works intake. *Ecological Engineering*, 95, 839–51.

Clarkson, B.R., Schipper, L.A., and Lehmann, A. (2004a). Vegetation and peat characteristics in the development of lowland restiad peat bogs, North Island, New Zealand. *Wetlands*, 24, 133–51.

Clarkson, B.R., Sorrell, B.K., Reeves, P.N., Champion, P.D., Partridge, T.R., and Clarkson, B.D. (2004b). Handbook for Monitoring Wetland Condition. Coordinated Monitoring of New Zealand Wetlands, Ministry for the Environment, New Zealand. Available online.

Croft, M.V. and Chow-Fraser, P. (2009). Non-random sampling and its role in habitat conser-vation: a comparison of three wetland macrophyte sampling protocols. *Biodiversity and Conservation*, 18, 2283–2306.

Cronk, J.K. and Fennessy, M.S. (2001). Wetland Plants: Biology and Ecology. CRC Press, Taylor & Francis Group, Boca Raton, FL, USA.

De Souza, M.L., Pellegrini, B.G., and Ferragut, C. (2015). Periphytic algal community struc-ture in relation to seasonal variation and macrophyte richness in a shallow tropical reservoir. *Hydrobiologia*, 755, 183–96.

Dorney, J., Savage, R., Tiner, R., and Adamus P (eds) (2018). Wetland and Stream Rapid Assessment: Development, Validation and Application. Elsevier, USA.

Dörnhöfer, K. and Oppelt, N. (2016). Remote sensing for lake research and monitoring–Recent advances. *Ecological Indicators*, 64, 105–22.

Dromgoole, F.I. and Brown, J.M.A. (1976). Quantitative grab sampler for dense beds of aquatic macrophytes. *New Zealand Journal of Marine and Freshwater Research*, 10, 109–18.

Environmental Laboratory (1987). Corps of Engineers wetlands delineation manual technical report Y-87-1. US Army Engineer Waterways Experiment Station, Vicksburg, MS, USA. 100 pp. + appendices.

Ferreira-Ferreira, J., Silva, T.S.F., Streher, A.S., Affonso, A.G., de Almeida Furtado, L.F., Forsberg, B.R., Valsecchi, J., Queiroz, H.L., and de Moraes Novo, E.M.L. (2015). Combining ALOS/PALSAR derived vegetation structure and inundation patterns to characterize major vegetation types in the Mamirauá Sustainable Development Reserve, Central Amazon flood-plain, Brazil. *Wetlands Ecology and Management*, 23, 41–59.

Foody, G.M. (2002). Status of land cover classification accuracy assessment. *Remote Sensing of Environment*, 80, 185–201.

Fu, H., Zhong, J., Yuan, G., Ni, L., Xie, P., and Cao, T. (2014). Functional traits composition predict macrophytes community productivity along a water depth gradient in a freshwater lake. *Ecology and Evolution*, 4, 1516–23.

Fujiwara, A., Matsuhashi, S., Doi, H., Yamamoto, S., and Minamoto, T. (2016). Use of environmental DNA to survey the distribution of an invasive submerged plant in ponds. *Freshwater Science*, 35, 748–54.

Gallant, A.L. (2015). The challenges of remote monitoring of wetlands. *Remote Sensing*, 7, 10938–50.

Gholizadeh, M.H., Melesse, A.M., and Reddi, L. (2016). A comprehensive review on water quality parameters estimation using remote sensing techniques. *Sensors*, 16, 1298.

Goldberg, C.S., Turner, C.R., Deiner, K., Klymus, K.E., Thomsen, P.F., Murphy, M.A., Spear, S.F., McKee, A., Oyler-McCance, S.J., Cornman, R.S., Laramie, M.B., Mahon, A.R., Lance, R.F., Pilliod, D.S., Strickler, K.M., Waits, L.P., Fremier, A.K., Takahara, T., Herder, J.E., and Taberlet, P. (2016). Critical considerations for the application of environmental DNA methods to detect aquatic species. *Methods in Ecology and Evolution*, 7, 1299–1307.

Guo, M., Li, J., Sheng, C., Xu, J., and Wu, L. (2017). A review of wetland remote sensing. *Sensors*, 17, 777.

Henderson, F.M. and Lewis, A.J. (2008). Radar detection of wetland ecosystems: A review. *International Journal of Remote Sensing*, 29, 5809–35.

Henderson, P.A. and Southwood, T.R.E. (2016). Ecological Methods, 4th edition. John Wiley and Sons, Chichester.

Hess, L.L., Melack, J.M., Filoso, S., and Wang, Y. (1995). Delineation of inundated area and vegetation along the Amazon floodplain with the SIR-C synthetic aperture radar. *IEEE Transactions on Geoscience and Remote Sensing*, 33, 896–904.

Homolová, L., Malenovský, Z., Clevers, J.G., García-Santos, G., and Schaepman, M.E. (2013). Review of optical-based remote sensing for plant trait mapping. *Ecological Complexity*, 15, 1–16.

Horning, N., Robinson, J.A., Sterling, E.J., Turner, W., and Spector, S. (2010). Remote Sensing for Ecology and Conservation: A Handbook of Techniques. Techniques in Ecology and Conservation Series, Oxford University Press, Oxford, UK.

Hurford, C., Schneider, M., and Cowx, I. (eds) (2010). Conservation Monitoring in Freshwater Habitats: A Practical Guide and Case Studies. Springer, Dordrecht.

Husson, E., Hagner, O., and Ecke, F. (2014). Unmanned aircraft systems help to map aquatic vegetation. *Applied Vegetation Science*, 17, 567–77.

Interagency Freshwater Group (2015). Common Standards Monitoring Guidance for Freshwater Lakes. JNCC, Peterborough, UK.

Jeppesen, E. Sondergaard, M. Sondergaard, M., and Christoffersen, K. (eds) (1998). Structuring Role of Submerged Macrophytes in Lakes. Springer, New York.

Kalacska, M., Chmura, G., Lucanus, O., Bérubé, D., and Arroyo-Mora, J. (2017). Structure from motion will revolutionize analyses of tidal wetland landscapes. *Remote Sensing of Environment*, 199, 14–24.

Karban, R., Huntzinger, M., and Pearse, I.S. (2014). How to do Ecology: A Concise Handbook. 2nd edition, Princeton University Press, New Jersey and Oxford.

Kent, M. (2012) Vegetation Description and Data Analysis: A Practical Approach. John Wiley & Sons, New York, NY, USA.

Kershaw, K.A. (1973). Quantitative and dynamic plant ecology, 2nd edn. Edward Arnold, London.

Keruzoré, A.A., Willby, N.J., and Gilvear, D.J. (2013). The role of lateral connectivity in the maintenance of macrophyte diversity and production in large rivers. *Aquatic Conservation: Marine and Freshwater Ecosystems*, 23, 301–15.

Little, A. (2013). Sampling and analysing wetland vegetation. In J.T. Anderson and C.A. Davis (eds) Wetland Techniques: Volume 1: Foundations. Springer, Dordrecht, 273–324.

Lu, D. and Weng, Q. (2007). A survey of image classification methods and techniques for improving classification performance. *International Journal of Remote Sensing*, 28, 823–70.

Madsen, J.D. and Wersal, R.M. (2017). A review of aquatic plant management and assessment methods. *Journal of Aquatic Plant Management*, 55, 1–12.

Mahdavi, S., Salehi, B., Granger, J., Amani, M., Brisco, B., and Huang, W. (2017). Remote sensing for wetland classification: a comprehensive review. *GIScience and Remote Sensing*, DOI: 10.1080/15481603.2017.1419602

Maine Volunteer Lakes Monitoring Program (VLMP) (2015). Invasive Aquatic Plant Screening and Mapping Survey Procedures. Available online.

Malakoff, D. (2004). Measuring the significance of a scientist's touch. *Science*, 306, 801.

Malmer, N., Johansson, T., Olsrud, M., and Christensen, T.R. (2005). Vegetation, climatic changes and net carbon sequestration in a North-Scandinavian subarctic mire over 30 years. *Global Change Biology*, 11, 1895–1909.

Marcaccio, J.V., Markle, C.E., and Chow-Fraser, P. (2015). Unmanned aerial vehicles produce high-resolution, seasonally-relevant imagery for classifying wetland vegetation. *The International Archives of Photogrammetry, Remote Sensing and Spatial Information Sciences*, 40, 249.

Matthews, M.W. (2011). A current review of empirical procedures of remote sensing in inland and near-coastal transitional waters. *International Journal of Remote Sensing*, 32, 6855–99.

Moor, H., Rydin, H., Hylander, K., Nilsson, M.B., Lindborg, R. and Norberg, J. (2017). Towards a trait based ecology of wetland vegetation. *Journal of Ecology,* 105, 1623–1625.

Mueller-Dombois, D. and Ellenberg, H. (1974). Aims and Methods of Vegetation Ecology. John Wiley & Sons, New York, NY, USA.

Murphy, K.J., Dickinson, G., Thomaz, S.M., Bini, L.M., Dick, K., Greaves, K., Kennedy, M.P., Livingstone, S., McFerran, H., Milne, J.M., Oldroyd, J., and Wingfield, R.A. (2003). Aquatic plant communities and predictors of diversity in a sub-tropical river floodplain: the upper Rio Paraná, Brazil. *Aquatic Botany*, 77, 257–76.

Mutanga, O., Adam, E., and Cho, M.A. (2012). High-density biomass estimation for wetland vegetation using WorldView-2 imagery and random forest regression algorithm. *International Journal of Applied Earth Observation and Geoinformation*, 18, 399–406.

Nagasaka, M. (2004). Changes in biomass and spatial distribution of *Elodea nuttallii* (Planch.) St. John, an invasive submerged plant, in oligomesotrophic Lake Kizaki from 1999 to 2002. *Limnology*, 5, 129–39.

Natural England (2009). Guidance for sampling rare aquatic plants. Natural England, UK. Available online.

Newton, J., Sepulveda, A., Sylvester, K., and Thum, R.A. (2016). Potential utility of environmental DNA for early detection of Eurasian watermilfoil (*Myriophyllum spicatum*). *Journal of Aquatic Plant Management*, 54, 46–9.

Niu, Z., Zhang, H., Wang, X., Yao, W., Zhou, D., Zhao, K., Zhao, H., Li, N., Huang, H., and Li, C. (2012). Mapping wetland changes in China between 1978 and 2008. *Chinese Science Bulletin*, 57, 2813–23.

Novo, E.M.L.M., Filho, W.P., and Melack, J.M. (2013). LBA-ECO LC-07 Reflectance Spectra and Water Quality of Amazon Basin Floodplain Lakes. ORNL DAAC, Oak Ridge, Tennessee, USA. https://doi.org/10.3334/ORNLDAAC/1144

Odum, E.P. (1984). The mesocosm. *BioScience*, 34, 558–62.

Olsen, S., Chan, F., Li, W., Zhao, S., Søndergaard, M., and Jeppesen, E. (2015). Strong impact of nitrogen loading on submerged macrophytes and algae: a long-term mesocosm experiment in a shallow Chinese lake. *Freshwater Biology*, 60, 1525–36.

Palmer, M. and Ruhi, A. (2018). Measuring Earth's rivers – satellite images enable global tally of freshwater ecosystems and resources. *Science*, 361, 546–547.

Petersen, J.E., Cornwell, J.C., and Kemp, W.M. (1999). Implicit scaling in the design of experimental aquatic ecosystems. *Oikos*, 84, 3–18.

Petr, T. (2000). Interactions between fish and aquatic macrophytes in inland waters—A review. FAO Fisheries Technical Paper, no. 396. Food and Agricultural Organisation, Rome. Available online.

Pettorelli, N., Vik, J.O., Mysterud, A., Gaillard, J.-M., Tucker, C.J., and Stenseth, N.C. (2005). Using the satellite-derived NDVI to assess ecological responses to environmental change. *Trends in Ecology and Evolution*, 20, 503–10.

Radomski, P. and Holbrook, B.V. (2015). A comparison of two hydroacoustic methods for estimating submerged macrophyte distribution and abundance: A cautionary tale. *Journal of
Aquatic Plant Management*, 58, 151–9.

Rochefort, L., Isselin-Nondedeu, F., Boudrea, S., and Poulin, M. (2013). Comparing survey
methods for monitoring vegetation change through time in a restored peatland. *Wetlands
Ecology and Management*, 21, 71–85.

Schneider, B., Cunha, E.R., Marchese, M., and Thomaz, S.M. (2015). Explanatory variables
associated with diversity and composition of aquatic macrophytes in a large subtropical river
floodplain. *Aquatic Botany*, 121, 67–75.

Schultz, R. and Dibble, E. (2012). Effects of invasive macrophytes on freshwater fish and macroinvertebrate communities: the role of invasive plant traits. *Hydrobiologia*, 684, 1–14.

Semlitsch, R.D. and Boone, M.D. (2010) Aquatic Mesocosms. In C.K. Dodd (ed.) Amphibian
Ecology and Conservation: A Handbook of Techniques. Oxford University Press, 87–104.

Silva, T.S.F., Costa, M.P.F., Melack, J.M., and Novo, E.M.L.M. (2008). Remote sensing of aquatic
vegetation: theory and applications. *Environmental Monitoring and Assessment*, 140, 131–45.

Sliger, W.A., Henson, J.W., and Shadden, R.C. (1990). A quantitative sampler for biomass estimates of aquatic macrophytes. *Journal of Aquatic Plant Management*, 28, 100–2.

Sommer, U. (2012). Experimental systems in aquatic ecology. eLS, Citable Reviews in the Life
Sciences. Wiley.

Spears, B.M., Gunn, I.D.M., Carvalho, L., Winfield, I.J., Dudley, B., Murphy, K., and May, L.
(2009). An evaluation of methods for sampling macrophyte maximum colonisation depth in
Loch Leven, Scotland. *Aquatic Botany*, 91, 75–81.

Srivastava, D.S., Kolasa, J., Bengtsson, J., Gonzalez, A., Lawler, S.P., Miller, T.E., Munguia, P.,
Romanuk, T., Schneider, D.C., and Trzcinski, M.K. (2004). Are natural microcosms useful
model systems for ecology? *Trends in Ecology and Evolution*, 19, 379–84.

Sutherland, W.J. (2006a). Planning a research programme. In W.J. Sutherland (ed.)
Ecological Census Techniques: A Handbook. 2nd edition, Cambridge University Press,
Cambridge, UK.

Sutherland, W.J. (ed.) (2006b). Ecological Census Techniques: A Handbook. 2nd edition,
Cambridge University Press, Cambridge, UK.

Taberlet, P., Bonin, A., Zinger, L., and Coissac, E. (2018). Environmental DNA for biodiversity
Research and Monitoring. Oxford University Press, Oxford, UK.

Tiner, R.W. (2016). Wetland Indicators: A Guide to Wetland Formation, Identification,
Delineation, Classification, and Mapping. 2nd edition, CRC Press, Boca Raton, FL, USA.

Tiner, R.W., Lang, M.W., and Klemas, V.V. (eds) (2015). Remote Sensing of Wetlands:
Applications and Advances. CRC Press, Taylor and Francis Group, Florida, USA.

Titus, J.E. (1993). Submersed macrophyte vegetation and distribution within lakes: line transect sampling. *Lake and Reservoir Management*, 7, 155–64.

Thomaz, S.M. and Cunha, E.R. (2010). The role of macrophytes in habitat structuring in
aquatic ecosystems: methods of measurement, causes and consequences on animal assemblages' composition and biodiversity. *Acta Limnoloogica Brasiliensis*, 22, 218–36.

Tsyganskaya, V., Martinis, S., Marzahn, P., and Ludwig, R. (2018). SAR-based detection of
flooded vegetation–a review of characteristics and approaches. *International Journal of Remote
Sensing*, 39, 2255–93.

US Environment Protection Agency (2002). Methods for Evaluating Wetland Condition:
Using Vegetation to Assess Environmental Conditions in Wetlands. Office of Water, US
Environmental Protection Agency, Washington, DC.

Vauhkonen, J., Maltamo, M., McRoberts, R.E., and Naesset, E. (2014). Introduction to forestry applications of airborne laser scanning. In: Maltamo, M., Naesset, E., and Vauhkonen,
J. (eds) Forestry Applications of Airborne Laser Scanning. Springer, Dordrecht, Netherlands.

Vitt, D.H. and Chee, W.-L. (1990). The relationships of vegetation to surface water chemistry and peat chemistry in fens of Alberta, Canada. *Vegetatio*, 89, 87–106.

Warfe, D.M. and Barmuta, L.A. (2006). Habitat structural complexity mediates foodweb dynamics in a freshwater macrophyte community. *Oecologia*, 150, 141–54.

Webb, E.B., Smith, L.M., Vrtiska, M.P., and Lagrange, T.G. (2010). Effects of local and landscape variables on wetland bird habitat use during migration through the Rainwater Basin. *Journal of Wildlife Management*, 74, 109–19.

WFD-UKTAG (2014). UKTAG River Assessment Method—Macrophytes and Phytobenthos-Macrophytes (River LEAFPACS2). Water Framework Directive, United Kingdom Technical Advisory Group, Stirling, Scotland. Available online.

Whinam, J., Barmuta, L.A., and Chilcott, N. (2001). Floristic description and environmental relationships on Tasmanian *Sphagnum* communities and their conservation management. *Australian Journal of Botany*, 49, 673–85.

Zhang, Y., Zhou, D., Niu, Z., and Xu, F. (2014). Valuation of lake and marsh wetlands ecosystem services in China. *Chinese Geographical Science*, 24, 269–78.

Zheng, Y., Niu, Z., Gong, P., Dai, Y., and Shangguan, W. (2013). Preliminary estimation of the organic carbon pool in China's wetlands. *Chinese Science Bulletin*, 58, 662–70.

10

Freshwater Vertebrates

An overview of survey design and key methodological considerations

Stephen E.W. Green, Rosie D. Salazar, Gillian Gilbert,
Andrew S. Buxton, Danielle L. Gilroy, Thierry Oberdorff,
and Lauren A. Harrington

Corresponding author: stephen.green@cornwall.ac.uk

10.1 Introduction

10.1.1 Diversity of freshwater vertebrates

The variety of freshwater vertebrate life is vast (Figure 10.1, Tisseuil et al. 2013) and the ways in which freshwater systems are used by vertebrates are diverse. Some species/taxa will spend their entire life in water (e.g., fish, fully aquatic mammals such as river dolphin, and some species of amphibian) while others use water for critical life stages (e.g., breeding and larval development in amphibians). Of all vertebrate groups, fish and amphibians are most reliant on water. However, other taxa use water for a multitude of purposes, such as den sites, foraging, predator avoidance, dispersal, and territory delineation. Understanding the ecology of target species, and their reliance on specific aspects of freshwater systems, both spatially and temporally, is important for designing and implementing suitable surveys for data collection on freshwater vertebrates. However, where a variety of biological, physico-chemical, and ecological data are needed to answer the aims of the study a range of survey techniques may be necessary. It is worth noting that there are existing volumes in the 'Techniques in Ecology and Conservation Series' that are focused on vertebrates and include freshwater species. Examples are for birds (Sutherland et al. 2004); reptiles (Dodd 2016); amphibians (Dodd 2009); carnivores (Boitani and Powell 2012); and invasive species (Clout and Williams 2009).

10.1.2 Why survey freshwater vertebrates?

Vertebrate species are often considered indicators of environmental quality or of population trends in sympatric species (e.g., river dolphins, Solow and Kendall 2015). This is largely due to their positioning within the trophic system, being reliant on a

Green, S. E. W., Salazar, R. D., Gilbert, G., Buxton, A. S., Gilroy, D. L., Oberdorff, T., and Harrington, L. A., *Freshwater vertebrates: An overview of survey design and key methodological considerations*. In: *Freshwater Ecology and Conservation: Approaches and Techniques*. Edited by Jocelyne M. R. Hughes: Oxford University Press (2019). © Oxford University Press 2019. DOI: 10.1093/oso/9780198766384.003.0010

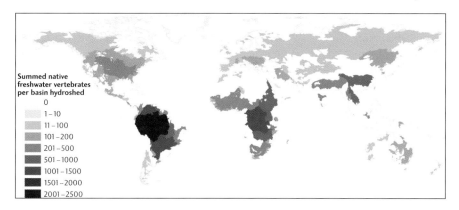

Figure 10.1 Global distribution of freshwater vertebrates-species richness by watershed (adapted from Tisseuil et al. 2013).

range of invertebrate and vertebrate species, which themselves may be sensitive to perturbations to local conditions. The highly mobile nature of many medium- to large-sized vertebrates (e.g., fish, birds, crocodilians, and dolphins) may also allow populations to respond rapidly to environmental change by simply leaving an area if it becomes degraded. For this reason, the conservation status and value of important freshwater habitats are often linked to the status of vertebrates. For example, 'wetlands of international importance' designated as Ramsar Convention sites are identified by nine criteria, four of which are specifically related to waterbirds and fish (see Ramsar website). In addition to acting as important indicators of freshwater ecosystem health, the concentration of vertebrates around water bodies provides opportunities to survey semi-terrestrial species when they are more conspicuous. For example, many amphibians are difficult to survey in the terrestrial environment, often being nocturnal and using burrows and dense vegetation or tall trees. In contrast, when in a breeding pond, detection probability may be dramatically increased due to temporary high population densities as well as conspicuous behaviours, such as courtship displays, vocalisation, and mating. Additionally, the production of eggs allows for other techniques such as egg searching to be employed. The water body in which breeding occurs provides a focal point for survey effort, increasing survey efficiency and reducing cost and time. Survey effort, however, must also be timed carefully to coincide with the target breeding season.

10.1.3 Threats facing freshwater vertebrates

Threats facing freshwater vertebrates are numerous and diverse but include reduction and degradation of freshwater habitats (e.g., drainage of wetlands and floodplains for development, agricultural run-off and pollution, loss of farmland ponds, and channelisation of rivers reducing the coverage and diversity of riparian habitats, and construction and operation of hydroelectric dams). Introduction of invasive species and disease into freshwater systems are also increasingly important threats to freshwater vertebrates. In the UK, invasive non-native species, such as the American bullfrog (*Lithobates catesbeianus*) and alpine newt (*Ichthyosaura alpestris*) are susceptible to carrying chytrid fungus which can infect native populations (Garner et al. 2006; Cunningham and Minting 2008; Miaud et al. 2016). Unsustainable harvesting practices such as fishing or hunting for pelts and skins (e.g., otters, mink, crocodilians) can cause declines in the target species as well as other non-target vertebrates. For example, although dam construction and boat collisions are implicated in the loss of the Yangtze river dolphin, its extinction is most likely attributed to unsustainable by-catch in local fisheries (Turvey et al. 2007).

Pollution and eutrophication of freshwater systems affects vertebrate communities by reducing available oxygen in the water bodies. Localised low oxygen levels in rivers can become barriers to migration of fish species. Contaminated water may also reduce food availability by affecting plants and invertebrate prey and may affect vertebrates more readily due to bioaccumulation of toxins in vertebrate species that are likely to be higher up the food chain.

10.1.4 Aim and structure of this chapter

The diversity of freshwater vertebrate life histories, and their relative dependencies on freshwater systems, is reflected by an equally diverse range of survey techniques. A detailed review of all possible techniques across all vertebrate groups is beyond the scope of this chapter. Instead, we review a selection of common methods and themes in freshwater vertebrate ecology, directing the reader to useful literature and resources. We also present several case studies to illustrate a range of survey strategies.

The chapter has been structured by survey method, as the principles of survey design will often be similar regardless of the vertebrate taxon being studied. The reader is encouraged to use Figure 10.2 to help guide survey method selection based on the aims and objectives of the study. Within each method section, specific examples are provided for different vertebrate taxa (fish, amphibians, reptiles, mammals, and birds) to illustrate some of the wide range of options available to researchers.

10.2 Setting clear aims and objectives

Before embarking on a survey, it is essential that the aims and objectives of the study are fully understood. Superficially, this may seem obvious, but selecting an appropriate study design and methodology is only possible by being clear about the specific research question and, therefore, the type of data needed. A wealth of literature exists on planning and setting objectives for field studies (e.g., Fisher 2016) and Figure 10.2 gives a brief overview of the decision processes that are needed.

It is also important to complete a rough cost–benefit analysis, weighing up the added value to the project of using any particular method or equipment. For example, is it really needed to complete a detailed population estimate, requiring resources to mark/identify individual animals, or will monitoring changes in relative abundance provide adequate information to answer the question? Monitoring populations in the long-term, as opposed to rapid snapshot assessments, often requires a considerable investment in time and resources, even when opting for the simplest 'low-tech' methodology. It is important, therefore, to consider medium to long-term sustainability of a selected

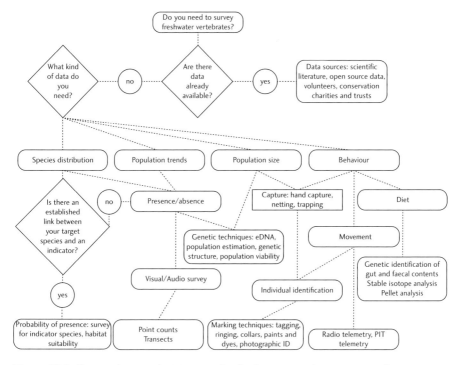

Figure 10.2 Survey planning decision tree for freshwater vertebrate surveys, illustrating a general but not exhaustive thought process when deciding on appropriate survey methods.

method, taking into account limitations on funding as well as the future availability of expertise to run surveys or analyse data. It is also important to consider whether an initial investment in equipment, for example satellite transmitters, could save time and money in the long-term by reducing the human investment in data collection.

It should be acknowledged that any survey, regardless of the method selected, has the potential to cause disturbance or harm to the individuals and populations studied. As vertebrates, ethical considerations in survey work become more important due to the increased propensity for vertebrates to experience pain, distress and lasting effects. Careful consideration should be given to whether the survey is essential, whether it could be replaced by existing data, or whether the survey methods could be refined to reduce the number of animals affected and minimise the impacts. A thorough ethical review should always form part of the survey planning stage of any project focussing on vertebrates. Also, because many freshwater vertebrates are endangered (e.g., amphibians are widely described as currently undergoing an extinction crisis with 41 per cent of amphibians worldwide threatened (IUCN website; Baillie et al. 2004)) due consideration should be given to ensuring the most appropriate methods are used to ensure that survival and recruitment are not significantly affected and that the habitat surveyed is not unduly damaged or destroyed.

At the planning stage, it is critical to establish the aims of the survey so that the most appropriate techniques can be employed (see Figure 10.2). Is it necessary to capture animals or will visual survey suffice? Are there data already available that can answer your questions? Before embarking on any survey, it is also important to first determine when is the most appropriate time to carry out field work as some species are highly seasonal in their activity patterns or use of a study area (e.g., amphibians use breeding ponds for a few weeks each year; migrating birds use wetland sites to over-winter; salmon return to their natal river to spawn; case study in Section 10.3.1).

10.3 Direct visual and audio encounter surveys

Vertebrates that are generally large, visible, or conspicuous may be well suited to sim-ple visual encounter surveys. Many birds and amphibians also vocalise and so can be monitored using acoustic techniques. Freshwater sites that are open to the public will have improved access, viewpoints and hides set up for visual encounters (particularly for birds). Existing records from these sites may yield valuable data that can augment or even replace the need for dedicated surveys. Surveyors should, however, normally consider methods that will allow survey effort to be measured such as using transects of known length, timed point counts, and distance sampling, to improve data quality. Bird point count surveys have long used a combination of both visual and audio detec-tion (e.g., Gregory et al. 2004) for monitoring bird populations.

Encounter surveys are likely to be low impact and can generate information on the presence/absence of key species as well as the diversity or richness of species inhabiting an area. When survey effort is also recorded, estimates of relative abundance and/or density between areas may be achieved. But caution is advised if attempting to com-pare relative abundance of different species within or between areas due to possible differences in detectability. Even attempting to compare relative abundance estimates of conspecifics between different locations should be approached with caution as dif-ferences in habitat complexity or behaviour of the target species may result in differences in the probability and rate of detection. Despite these considerations, rela-tive abundance estimates from visual or audio encounter surveys remain a valuable approach for surveying conspicuous freshwater vertebrates. The case study in Section 10.3.1 demonstrates the effective use of visual encounter surveys for estimating relative abundance and density of caiman and river dolphins in the Peruvian Amazon as part of a long-term volunteer research programme.

In cases where individual animals are recognisable without marking (e.g., distinct calls or natural markings), it may also be possible to estimate population size. Photographic identification methods are commonly used for this purpose and are briefly discussed later in this chapter, but a novel example of how unique calls have been used to estimate the population size of great bittern (*Botaurus stellaris*) in the UK is described in a case study in Section 10.3.2.

Playback is often used in the study of birds, where the researcher plays a recording of the call of a target species as an audio stimulus. 'Playback' is an effective method for increasing audio detection of species that are less likely to be detected simply by chance. Recorded bird calls or songs have been used since the 1950s, as detailed by Marion et al. (1981). A recent study by Freeman and Montgomery (2017) shows that playback experiments provide a robust measure of species recognition and defining species geographical ranges and limits.

Where the species of interest is highly cryptic or lives at low density, indirect detection methods, as described later in this chapter, or trapping (either physical trapping or camera trapping) may be necessary to increase detection.

10.3.1 Amazonian river surveys in Peru

This case study outlines how a voluntourism model (in this case involving Earthwatch, Operation Wallacea, and Operation Earth) has been used to monitor vertebrate populations in the Pacaya Samiria National Reserve, Peru, and to monitor the impacts of climate change on this Amazonian system. For a full account of the 20-year survey programme, see Bodmer (2012), Bodmer et al. (2012, 2014, 2018), and Chandler et al. (2017). The Pacaya-Samiria National Reserve covers an area of more than 2 million hectares in the Department of Loreto, Peru and is one of the largest protected areas in the country. This flooded forest ecosystem is driven by large seasonal fluctuations in high and low water levels. Monitoring temporal and spatial changes in a vast and dynamic system, therefore, requires a large-scale survey strategy and methodology.

The broad aim of this research is to understand how changes in climate are impacting the ecology, behaviour, and populations of aquatic and terrestrial species, as well as working with the local Cocama Indian communities and the reserve authority to determine the impacts of changes on the sustainability of fishing and bush meat hunting (Bodmer 2012; Bodmer et al. 2018). The focus of the research is on several key terrestrial, aquatic, and semi-aquatic vertebrates as indicator species of ecosystem health. Key taxa include river dolphins, giant river otters, caiman, river turtles, fish, macaws, wading birds, game birds, primates, large carnivores, ungulates, and rodents. For example, data on catch per unit effort of key fish species enable fish stocks to be monitored effectively. Data on river dolphin, giant river otter, caiman, and wading birds not only provide direct information on the abundance and spatial distribution patterns of these important target species but also they can be simultaneously used as indicators of the health of fish stocks, as highly mobile predators can respond rapidly to local declines in prey abundance by simply moving to other areas.

The seasonal nature of the flooded forest ecosystem highlights the importance of considering the timing of the survey. During the high-water season fish disperse into the flooded forest, lowering the density and reducing capture rates per unit effort in the river channel. Likewise, aquatic predators follow their prey into the flooded forests during the high-water season, and therefore encounter rates are greatly reduced. For this reason, surveys of aquatic vertebrates are most efficient during the low-water season when fish and associated predators are at high density within the rivers and lakes. Conversely, the efficiency of surveys to detect terrestrial species are generally optimised during the high-water season when terrestrial vertebrates are confined to small areas of raised land, although the densities of arboreal species tend to be less affected by these seasonal changes in water level (Bodmer et al. 2012). Annual comparisons must ensure that they are comparing same-season data e.g., high-water with high-water, or low-water with low-water, to avoid biased detection rates. Similar considerations need to be taken into account when interpreting data from flood or drought years when water levels may be highly variable.

Reporting the full breadth of survey methods used and results from the research programme is beyond the scope of this section, but we briefly summarise the methods and results from monitoring two aquatic vertebrate populations, river dolphin and caiman. The methods for both are similar as they each rely on the use of small auxiliary boats, a driver, and a small team of volunteer observers led by at least one experienced researcher and/or local guide. Caiman surveys were conducted at night, whereas dolphin surveys were conducted during the day in line with activity patterns and optimal periods of detection for these two groups.

Caiman surveys

Three species of caiman coexist within the reserve; the spectacled caiman (*Caiman crocodylus*) (Figure 10.3), black caiman (*C. niger*), and smooth-fronted caiman (*Paleosuchus trigonatus*). Both the spectacled and black caiman were intensively harvested for their skins during the 1950s–1970s (Bodmer et al. 2012) prior to the regulation of the trade in crocodilian skins by the Convention on International Trade in Endangered Species of wild flora and fauna (CITES) (Thorbjarnarson 1999). Black caiman appeared to be impacted more heavily by hunting due to their slower growth rate and, thus, delayed sexual maturity compared with the spectacled caiman (Thorbjarnarson 1999). The larger size of black caiman also increased the value of their skins in the market. Smooth-fronted caiman were the least desirable target due to their smaller size and lower value of their skins (Bodmer et al. 2012).

A long-term population-monitoring programme was initiated to assess population status of caiman within the reserve. Key objectives included: (1) to compare relative abundance of each species between survey years and between different areas of the

Figure 10.3 Spectacled caiman (*Caiman crocodylus*) during diurnal observation. Reflectance from the eyes, which are raised above the waterline, can be detected during nocturnal boat surveys. Photo courtesy of Emily Cook, Operation Wallacea.

river system, (2) to collect detailed spatial distribution and microhabitat ecological data, and (3) to collect data on diet to further assess niche overlap and competition.

Aquatic shoreline transects were completed at night using small boats, fitted with a 15–25 horsepower engine, travelling upstream or downstream on the main river and in nearby channels or lakes. Distance travelled was recorded using a GPS. Caimans were located by their eye reflections using a 12-volt spotlight and approached to a distance where the engine was silenced and the boat paddled closer. During approach, individuals were identified to species level and size noted (estimated using the distance between snout and eye as a proxy for overall size). Capture location coordinates were recorded as well as general habitat characteristics. When possible, noosing was used to capture caimans for more detailed data collection. Individuals captured in this way were brought on board, secured using standard crocodilian capture and handling techniques (Bodmer et al. 2012). Initial species identification was confirmed and sex determined. Total body length was measured from the tip of the snout to the tip of the tail, while head length was measured from the tip of the snout to the posterior edge of the orbital. Mass of the caiman was recorded in kilograms using a spring scale. Diet was also compared between caiman species by flushing stomach contents (Fitzgerald 1989; Luiselli and Amori 2016 for a review of methods).

Each year, abundance of each species was calculated using the formula N/L, where N = the number of individuals and L = the distance travelled in kilometres. This simple method allows relative abundance (the number of individuals observed per kilometre travelled) to be recorded and compared, in order to monitor long-term changes in relative abundance of each species over time. It is important to note that because no system is employed to individually mark animals, it is not possible to estimate population size or take into account differences in detectability and survival between species. Marking methods could be employed, but the additional cost of tags, the additional expertise, and time needed to process animals, as well as animal welfare issues, all need to be considered and balanced within the study design. The strength of this monitoring programme is the relatively simple methods and repeatability using a combination of experienced field staff and novice volunteers. Figure 10.4 displays annual fluctuations in relative abundance for black and spectacled caiman between 2006 and 2013.

River dolphin surveys

Diurnal river transects were completed by boat to survey pink river dolphins (*Inia geoffrensis*) and grey river dolphins (*Sotalia fluviatilis*) (Figure 10.5). Surveys were completed by drifting downstream (approximately 5 km) of the start point, with the boat engine turned off to minimise disturbance. A GPS (Global Positioning System) was used to determine the length of each aquatic census and start and end times recorded. Information collected included: species, group size, group composition, behaviour (travelling, fishing, playing), time, and any additional observations. Care was taken not to double count any dolphin sightings (Bodmer et al. 2012). Data were analysed using fixed width transect density estimates:

$$D = \frac{N}{2AL}$$

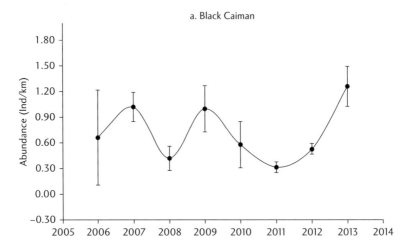

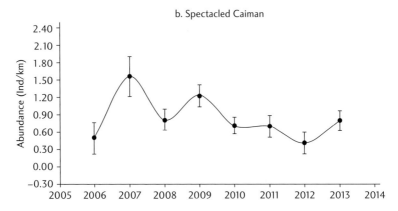

Figure 10.4 Overall abundance of black caiman (*Caiman niger*) and spectacled caiman (*Caiman crocodylus*) in the Samiria River between 2006 and 2013. Reproduced from Bodmer et al. (2014).

Where D = density, N = number of individuals, A = river width, L = distance travelled, 2 = number of margins sampled. This simple survey design allows for relative population density to be compared between years and between different areas of the reserve for each of the dolphin species (Figure 10.6).

10.3.2 Acoustic surveys of bittern booms

Vocal individuality in the advertising songs of birds has been readily demonstrated (e.g., Gonzalez-Garcia et al. 2017; Yee et al. 2016) as within species it can be advantageous to be recognised (e.g., by your neighbours in adjacent territories) (Kirschel et al. 2011). Conservationists have suggested using this as a tool to more accurately count or re-identify birds; but with few examples of success (Hoodless et al. 2008). During the

Figure 10.5 Pink river dolphin (*Inia geoffrensis*) photographed during a river survey in the Pacaya Samiria National Reserve, Peru. Photo courtesy of Fabian Mulhberger, Operation Wallacea.

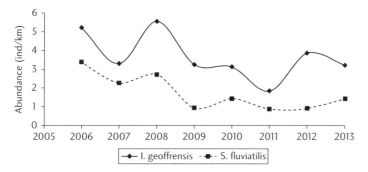

Figure 10.6 Overall abundance of pink river dolphins (*Inia geoffrensis*) and grey river dolphins (*Sotalia fluviatilis*) in the Samiria River between 2006 and 2013. Reproduced from Bodmer et al. (2014).

late 1980s the breeding great bittern (*Botaurus stellaris*) population in the UK was known to be small and mostly restricted to larger wetter reedbed (*Phragmites australis*) sites. There was no available method to accurately survey these secretive birds other than listening for the unusual booming song of the males. However, it was difficult to accurately attribute these songs to the numbers of males on each site. Between 1990 and 2002, individuality found in the distinctive boom vocalisations of male great bitterns was used to assist with the annual census of the species across the UK (McGregor and Byle 1992; Gilbert et al. 1994; Gilbert et al. 1998). Annual monitoring was based on intensive mapping using triangulation of booming locations (Poulin and Lefebvre 2003), with sound recordings and comparisons of boom spectrograms used to discriminate between individuals only in difficult counting situations (Gilbert et al. 1994;

Gilbert et al. 1998). This was a welcome additional technique, used in a standardised way and providing a reliable baseline estimate for the national bittern population each year; the lowest number of booming male bitterns counted in this way in the UK was 11 in 1997. With emergency habitat restoration work being undertaken on many wetland sites at the time, in an effort to prevent the species becoming extinct in the UK again, accurate monitoring provided a crucial foundation. The monitoring technique was used until the booming bittern population recovered, at which point the population became (gratifyingly) too large for this time-consuming technique to be useful. In 2016 there were more than 160 booming male bitterns in the UK (Wotton 2017), which is testament to the hard work and dedication of all those determined to restore and create wet reedbed habitat suitable for bitterns across the UK (White et al. 2006; Brown et al. 2012).

10.4 Indirect detection surveys

10.4.1 Sign surveys

Sign surveys can be useful for many vertebrate species. For example, presence of water voles (*Arvicola terrestris*) can be determined by searching for runs, feeding sign, footprints, and latrines (Strachan and Moorhouse 2006). Similarly, presence of European beaver (*Castor fiber*) can be detected by surveying for feeding signs, and presence of otter (*Lutra lutra*) and mink (*Mustela vison*) can be established by searching for footprints and focusing the survey effort by using equipment such as track tunnels on rafts (Reynolds et al. 2004; Harrington et al. 2008), otter holts, hair (e.g., using hair tubes/ hair snares; Kuhn 2010), spraint and scat (e.g., Bonesi and Macdonald 2004; Harrington et al. 2008). A standard methodology has been developed for otter survey (Lenton et al. 1980) by using a systematic survey of a 600-m stretch of river bank to produce results comparable across studies. For details of ecological monitoring protocols, including sign survey, for European beaver see Campbell et al. (2010). Care should be taken when determining species presence from scat/spraint as they can be easily misidentified even by experts; where possible positive identification of scat should be obtained through DNA analysis (Harrington et al. 2010). It is also not possible to use scat density to estimate population size of otters (Kruuk and Conroy 1987) and mink as their marking behaviour may be affected by the presence of predators or competitors and habitat type.

Camera traps can be used to increase detection rates of semi-aquatic mammals and have been favourably compared with scat surveys (Day et al. 2016). Placement of camera traps should be carefully considered to achieve optimal results. For example, otter detection rates by camera traps are improved by placement at latrine sites (Green et al. 2015; Wagnon and Serfass 2016), and, therefore, knowledge of target species' behaviour can increase success rates of this survey technique.

Amphibian signs in the breeding season may include eggs, larvae, or nest sites; fish eggs and spawning sites are also potentially visible depending on species and location; and bird signs may include nest sites, feathers, and regurgitated food pellets. Biological samples (e.g., faecal samples, regurgitates, hair, feathers) collected as part of sign surveys can be valuable in providing genetic data and using microscopy and stable isotope analyses can provide data on the species' diet. Environmental DNA (eDNA) is also fast becoming a standard tool for ecologists to easily and rapidly detect (genetically encountered) species which are difficult or labour intensive to detect in the wild. A series of water samples taken from a water body can quickly be tested for the presence of target species or even entire communities.

Selection of an appropriate encounter method, from direct visual or acoustic detection of the target species through to proxies such as spore, scat, and spraint, nests or even detection of a positive eDNA sample, will be determined by knowledge of the biology of the target species as well as size, complexity and accessibility of the study site and availability of resources. It must also be influenced by the question being asked. Do you need to simply confirm the presence of a target species, or do you need to achieve an estimate of relative abundance, density, or population census? For example, eDNA methods can be highly sensitive at detecting target species, but cannot, currently, give reliable estimates of abundance, density or population size (e.g., Hunter et al. 2015; Buxton et al. 2017a).

10.4.2 Freshwater species detection using environmental DNA

In relation to aquatic environments, environmental DNA often referred to as eDNA is DNA that has become separated from an organism and suspended within the water column (Jane et al. 2015) and allows for a rapid and cost-effective species distribution assessment tool. The suspended DNA, which may be intracellular or extracellular, is collected as part of a bulk sample of water, concentrated and its species of origin identified under laboratory conditions. Within the laboratory a specific species can be targeted usually using quantitative real-time PCR (Thomsen et al. 2012) or community analysis can be undertaken using metabarcoding to generate a list of species (Valentini et al. 2016). Although detection of DNA from environmental samples has been utilised with single-celled organisms, or from ancient permafrost and sediments for some time (Thomsen and Willerslev 2015), contemporary detection of macroorganisms was first demonstrated in 2005 (Martellini et al. 2005) and so eDNA is a young and rapidly advancing field. For a comprehensive background of eDNA for biodiversity research and monitoring, and separate chapter on freshwater eDNA, see Taberlet et al. (2018).

Environmental DNA has been shown to have many practical applications, and is often used in the detection of rare or endangered species at low densities, or early detection of invasive species (Thomsen and Willerslev 2015). eDNA has been in use since 2011 in the detection of the spread of invasive species of carp through the Great Lakes catchments of North America (Jerde et al. 2011). eDNA is also widely used in the UK for distribution assessments of the protected great crested newt (*Triturus cristatus*) and has been informing land-use planning applications since 2014 (Biggs et al. 2014; 2015). As with most methods of detection the efficiency varies with the activity of the target species, due to the method only being recently adopted, and influences on detection are still being identified and protocols refined (Buxton et al. 2017a; 2017b). An intensive study undertaken by Buxton et al. (2017a) identified that the

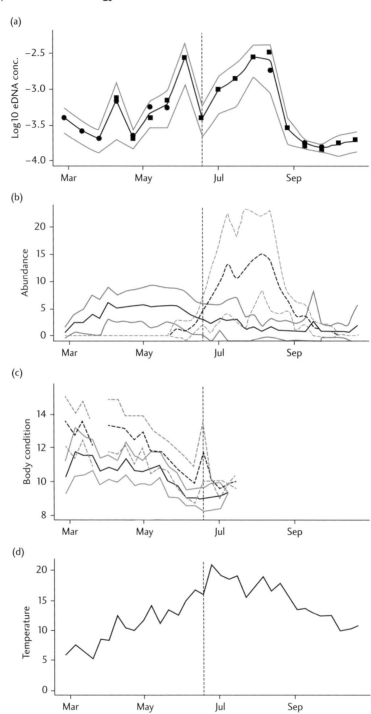

amount of target DNA of the great crested newt varied seasonally with changes in adult body condition, associated with breeding, having a greater influence on the eDNA concentration than the number of adults present (although still important) during the breeding season (Figure 10.7). High concentrations of target DNA were also identified outside the breeding season and associated with the abundance of larvae but not adults. Sharp decreases in concentration could also be associated with both the emergence of adults and the metamorphosis and emergence of larvae on entry into a terrestrial phase. Work such as this is important in refining the survey protocol and timing. The amount of DNA and, therefore, the chance of detecting the target species decreases rapidly once the species is no longer present. eDNA is only capable of detecting presence of the target species and cannot provide an estimate of population size; the timing of surveys, however, can be used to identify either the presence of the species or breeding success.

10.5 Capture methods

Depending on the research question, it may be necessary to capture freshwater animals to mark or take measurements or biological samples. Marking will likely be necessary where repeated identification of individuals is required; for example, for studies of survivorship, movement, behaviour, growth, and age. Recapture of marked individuals is also normally necessary for population estimates. It should be noted that any trapping method can be biased, for example, towards age, size, one sex or the other. Care should be taken, therefore, when interpreting data on sex ratios and other variables from trapping methods.

10.5.1 Amphibians

In addition to simple visual detection within freshwater bodies, dip netting and bottle-trapping are two commonly employed capture methods for UK amphibians, especially newts. Netting should generally be completed after initial visual inspection of the pond has been completed, as the process is likely to disturb sediment and reduce visibility within the study pond. A dip net with mesh size of 2–4 mm attached to a

←——————

Figure 10.7 Seasonal variations in eDNA concentration, in relation to adult and larval great crested newt (*Triturus cristatus*) population size, adult body condition, and temperature. (a) Log10 (x + 0.0001) of the mean eDNA concentration (ngµ L⁻¹), per pond (black line, solid circles collected using glass-microfibre filters, solid squares collected using precipitation in ethanol) with 95% confidence intervals (grey) across the eight ponds. (b) Mean estimated population size per pond (black line; adults—solid line, larvae—broken line) with 95% confidence intervals (grey line). (c) Mean body condition (males—solid line, females—dashed line) using the scaled mass index of adults caught each week throughout the survey season with 95% confidence intervals (grey line). (d) Mean weekly temperatures in degrees Celsius through the study period. The vertical dotted line represents the end of the breeding season and the start of the post-breeding season. Reproduced with permission from Buxton et al. (2017a).

sturdy frame and pole can be used to sweep through aquatic vegetation and agitate the bottom of the pond. Performing 2 m sweeps and recording the number of sample locations around the water body or recording the overall netting time allows for survey effort to be standardised (Griffiths and Brady 1996; Gent and Gibson 2003; Skelly and Richardson 2009).

Funnel-trapping, also commonly known as 'bottle-trapping' (because of the ability to make inexpensive but effective traps from 2 L plastic drinks bottles), is a standard method for capturing aquatic amphibians, especially newts (Griffiths 1985). Traps can be set up along the shoreline of the water body to be surveyed, and placed in the water with a small air pocket at the top of the trap to allow any captured amphibians to surface and breath before traps are checked and emptied. For a detailed outline of bottle-trap creation and use see Gent and Gibson (2003). Timing of trapping is critical as traps must be placed when amphibians are in the breeding ponds but must not be set if night-time temperatures are likely to exceed limits which are safe for the species, whether high or low.

10.5.2 Snakes

Many species of snakes are intimately linked with freshwater habitats and, thus, surveys of these areas can be productive for snake encounters. Capture methods are generally simple, with snakes being captured by hand, as part of visual encounter surveys (VES), with the aid of snake hooks and or tongues for venomous species (Dodd 2016). The capture and handling of venomous snakes should only be attempted by experienced herpetologists with suitable equipment and safety precautions. It should be noted that active capture methods for snakes can be highly sensitive to inter-observer differences in detection ability, environmental conditions and habitat type (Dorcas and Willson 2009). The use of artificial refugia, such as coverboards, placed in suitable terrestrial habitats adjacent to freshwater bodies can be a useful way of detecting species, such as *Natrix natrix*, while limiting observer bias (Dorcas and Willson 2009). Passive capture methods such as modified funnel traps have also been successfully used in the study of freshwater snakes (Casazza and Wylie 2000; Halstead et al. 2013), although capture and escape rates can be variable (Willson et al. 2005) and caution is advised to prevent harm to trapped animals (Wittenberg and Gifford 2008).

10.5.3 Freshwater turtles

Freshwater turtles can be captured using a range of traps, including Legler and inversion traps, hoop and fyke nets, and trammel nets (Vogt 2016). Turtles usually feed within 5–10 m along the shoreline in water 1–2 m deep, therefore traps should be set parallel to the shoreline to intercept foraging turtles. Traps must be tied in several places to keep at least part of the trap out of the water to allow turtles to lift their heads out of the water to breath (Vogt 2016). A variety of different habitats should be sampled to capture a representative sample of the resident turtle population. Traps can be baited with known food items, although it should be noted that during the breeding season turtles may not feed. For this reason, Vogt (1979) instead successfully baited traps with live painted turtles (*Chrysemys picta*) to attract other painted turtles during the breading season. This demonstrates the importance of selecting a suitable bait in relation to the behaviour and ecology of the target species.

10.5.4 Crocodilians

Crocodilians are large and dangerous animals and their capture should only be attempted by experienced surveyors. The methods by which their capture can be achieved safely for both the researcher and the animal are beyond the scope of this chapter but see Brien and Manolis (2016) for an overview. Never attempt to capture crocodilians without full prior consultation with an expert in this field. Crocodilian surveys will often take place at night from small boats, relying on reflective eye-shine to locate individual animals, although diurnal surveys can also be productive for species which bask in exposed areas along riverbanks.

10.5.5 Fish

There are a number of different methods for catching freshwater fish using nets. Before selecting the appropriate net type it is important to consider how to minimise the by-catch of non-target species, and the selection of a suitable mesh size. Net type is also influenced by type of freshwater body (lentic or lotic) and the size of the water body. Hand-thrown nets are useful in small ponds but ineffective in larger water bodies that are obstructed by vegetation (e.g., wetlands, reed beds). The behaviour of the target species will determine whether or not passive techniques using entanglement (e.g., gill and trammel nets) or entrapment (e.g., funnel traps) can be employed. See Hubert et al. (2012) for a review of passive capture techniques.

Electrofishing is a technique that can be used to assess presence/absence and relative abundance of fish species in a water body (Deacon et al. 2017). Use of a keep net can also allow for exhaustive fishing of a stretch of river to assess population density. For a population estimate, repeat surveys employing capture-mark-recapture techniques are used successfully in many studies. In the UK, the Environment Agency has standards in place to ensure safe operation of electrofishing equipment requiring a licence to perform electrofishing surveys. The standards used help to ensure that survey data are comparable between sites and surveyors. Electrofishing can be dangerous for both the practitioners and species surveyed. Care should be taken to ensure all equipment is operated safely and in teams to mitigate risks associated with electric shock, tripping, and drowning. Electrofishing equipment should deploy appropriate current type and strength based on the species to be surveyed and the habitat in which surveys are to be undertaken (Beaumont 2016).

10.5.6 Mammals

Mammals may be captured using traps (baited) including small mammal traps for species such as water shrew and water vole in Europe, and water mice in Australia (Kaluza et al. 2016). Larger live capture traps can be used for mammals such as platypus (Serena and Grant 2017), or mink and otter, including cage traps deployed on floating rafts which increase trapping success rate for mink (Reynolds et al. 2004). Deploying live traps for mink can be time consuming and requires a large number of traps (one every 300 m).

Efficiency can be significantly increased by first conducting a sign survey (ideally using track pads on floating rafts) and only deploying traps in areas where footprints have been previously found (Yamaguchi et al. 2002). Where mammals may spend some time in traps it is important to ensure measures are taken to reduce stress (e.g., by providing food, water, and bedding). It is also possible to fit larger live traps with a device that signals when the trap is occupied, which can be useful in reducing time spent in the trap. Some bats are commonly associated with freshwater systems, such as Daubenton's bat (*Myotis daubentonii*) and can be captured using mist nets and harp traps. Many live trapping techniques will require a survey and/or handling licence, depending on the country where fieldwork is being undertaken. Welfare issues associated with trapping river otters are discussed by Serfass et al. (2017).

10.5.7 Birds

The two dominant methods for capturing birds in freshwater systems are the use of mist nets or cannon nets (Sutherland et al. 2004). Mist netting is the most common survey method used for physically catching and processing individual birds. The nets are made of a fine filament that can appear invisible to birds if the nets are well placed along open rides in between vegetation and in good weather (not in wind or rain when water droplets can stick to the fine filament and make it easy to see). The nets are erected in a line and structured as a series of shelves so that when a bird flies into the net, it falls and is caught by the shelf which securely holds the bird without it getting hurt. Extracting the bird may be potentially harmful, which is why only experienced and licensed bird ringers can do this task. Mist nets catch a bigger range of species than other methods such as cannon netting or whoosh netting, and is principally used for smaller bird groups like passerines but ringers can obtain licenses to ring a whole array of birds from different orders. Mist nets are suitable to deploy near to water and monitoring data can be used to study migratory species or for population studies (e.g., Muñoz-Adalia et al. 2015).

Cannon nets can be used in the daytime to capture large numbers of wading birds simultaneously in intertidal areas, lakes, and river banks. The size of net and numbers used are worked out on a case-by-case basis depending on the target species and nature of the study site. The net is laid out in a line on the ground where the catch is planned and metal weights are connected to ropes which are attached to the net and then placed into buried cannons. When the charge is triggered the metal weights are ejected from the cannon at very high speed, carrying the net outwards and over the catch area to trap the target birds. It is possible to catch some water birds by hand (e.g., the annual Swan Upping on the River Thames in the UK, where mute swans (*Cygnus olor*) are counted, weighed, and measured and their health monitored by teams in rowing boats in association with the University of Oxford). There are also a number of different designs of funnel trap available, typically used for catching ducks using bait and decoy birds. Welfare issues associated with these different trapping techniques are reviewed by O'Brien et al. (2016).

These capture techniques allow information to be collected that extends beyond population size and trend estimation. Valuable information can be collected on bird demographics and fitness (survival and breeding success) in addition to bird movements, and some particularly outstanding migration projects have revealed the journeys of some long-distance migrants. The record is held by a wader, the eastern bar-tailed godwit (*Limosa lapponica baueri*), where one individual carried out a flight journey of 11,500 km across the Pacific Ocean from Alaska to New Zealand, with previous data showing a journey by the same bird across the Yellow Sea to China (10,000 km) and a further 5,000 km to Alaska (Gill et al. 2009).

10.6 Individual recognition and marking techniques

The key advantage of unique identification of individual animals within a population is the ability to track capture histories of individuals over time together with associated data, such as size, health, breeding condition, movement, and habitat use. For some animals, individual identification can be achieved through pre-existing unique morphological characteristics such as skin or fur patterns. However, for many species a marking system will need to be implemented to provide animals with a unique identifier. The generation of capture history and associated data over time not only allows population size estimates to be carried out through computer software such as 'Program MARK' (White and Burnham 1999) or 'multimark' R package (McClintock 2015), but also provides the opportunity to estimate survival and detectability within populations.

10.6.1 Photo identification

Advances in video and photographic methods, as well as ever more powerful software programs for analysing images, allows relatively quick and easy identification of individual animals from distinctive patterns (e.g., fur, skin, scale colouration, or unique identifiers such as scarring patterns). Although this method may not be appropriate for all species, it should be considered as a viable alternative to other marking techniques where possible. For small populations of species with distinctive markings, it may be possible to manually identify individuals from photographs to build up capture histories. However, as population size increases, the time needed to check photographic records manually quickly becomes prohibitive and investment in custom software for this purpose becomes necessary. Licences for software can be expensive, but this cost may be offset by not needing to purchase large numbers of marker tags, such as passive integrated transponder (PIT) tags, and moreover, the reduced invasiveness of the method on the target species may make this an attractive option. Freely available photo-matching software is available (e.g., HotSpotter (Crall et al. 2013)) but any software should, ideally, be tested in advance for its ability to accurately match images of the target species (e.g., Matthé et al. 2017). See Sacchi et al. (2016) for a review of digital identification methods in reptiles.

10.6.2 Marking method considerations

There are ethical issues relating to all forms of marking but some are more significant than others. Any marking method used should be fully justified and assessed as part of a robust ethical review. The least invasive option, appropriate for the question and study system, should be selected and the marking method must last for at least the duration of the study period. The following criteria for ideal marking methods have been suggested (Ferner 2007; Ferner and Plummer 2016). The method should:

- not affect survivorship or behaviour of the organism
- allow the animal to be as free from stress and pain as possible
- identify the animal as a particular individual or member of a cohort
- last indefinitely, or at least through the duration of the study
- be easily read and/or observable by all informed individuals
- be adapted to organisms of different sizes
- be easy to use in both laboratory or field
- use easily obtained materials at minimal cost
- be tested to meet these criteria before wide use
- prevent marking application tools being reused without first being thoroughly disinfected and cleaned

10.6.3 Marking fish

Fish can be marked using a wide variety of methods and extensive resources are available on this topic (see Thorsteinsson (2002) for a comprehensive review of tagging methods in fish). Methods can be broadly grouped into either external or internal marking/tagging. Fish can be marked externally via fin clipping (Gresswell et al. 1997) or by the attachment of external marker tags, such as ribbons, threads, wires, plates, disks, dangling tags, and straps, which may carry an individual code, a batch code, and/or visible instructions for reporting (Thorsteinsson 2002).

Internal tagging methods are also now commonly used and include coded wire tags (CWT), visual implant elastomer (VIE) tags, visual implant alpha (VIA) tags, and passive integrated transponder (PIT) tags. Internal tags are inserted or injected into the fish (body cavity, muscle, or cartilage) and carried internally. Which method is most appropriate will be influenced by the objectives, scope, and budget of the research but also by the body size and life stage of the target species.

CWTs are small pieces of magnetised stainless steel (size 0.5–2.0 mm × 0.25 mm) which may have a binary code engraved in the surface or laser etched Arabic numbers, either for individual or batch identification (Thorsteinsson 2002). CWTs are normally injected into the snout of a fish and are often combined with an outer mark, to aid recovery (Thorsteinsson 2002). They can be used for tagging large numbers of fish, but special detection equipment is needed. CWTs may, due to their small size, be applied to a large range of fish sizes, but routine use requires tags to be removed via dissection in order to read the identification code, thus limiting use to studies where sacrifice is intended and/or acceptable at the recapture stage.

VIEs are two-part chemical compounds, injected as a liquid, that mix to form a pliable, solid, subdermal tag that is generally visible in sunlight or clearly visible when exposed to a black light (Hale and Gray 1998). Different colour combinations can be injected to create a unique coding system. VIEs are small, flexible, and bio-compatible, making them useful for a wide variety of species and for animals smaller than most other tags can accommodate. VIA tags are similar in their application to VIEs but are small, fluorescent tags with an alphanumeric code, visible through the skin, designed to identify individual animals. Tags are implanted internally but remain externally visible for easy identification. More information about CWT, VIE, and VIA tags, as well as other tagging and detection technologies for fish can be obtained through commercial suppliers (e.g., Northwest Marine Technology website).

Passive integrated transponder (PIT) tags are small (9–12 mm) glass-encapsulated microchips which can be implanted into the body cavity (intracoelomic). Pit tags provide a means for unique identification through passive detection with a handheld or fixed position PIT tag reader. PIT tags are a common and attractive option for marking fish and have been successfully used for several study objectives, such as estimating population size, growth, survival, habitat use, predation, and sampling efficiency. Musselman et al. (2017) review the use of PIT tags in warm water fishes, showing that tag retention was generally high and mortality low, thus meeting the criteria for an appropriate marking method. Negative examples of tagging are apparent in some cases and it is suggested, therefore, that methodological testing is prudent in field studies (Musselman et al. 2017), advice that we feel should apply to the use of PIT tags in any taxonomic group.

Their small size makes them suitable for studying small-bodied fish (e.g., Pennock et al. 2016), but body size remains a limiting factor (Musselman et al. 2017). Pennock et al. (2016) tested factors influencing survival and PIT tag retention in Southern Redbelly Dace (*Chrosomus erythrogaster*) and concluded that dace and other cyprinids greater than 60 mm can be tagged reliably with 9-mm tags and individuals greater than 50 mm can be tagged with 8-mm tags.

Hanson and Barron (2017) monitored tag retention and mortality rates in lamprey ammocoetes (total length range 29–83 mm) over 137 days for CWT, VIE, and PIT tags and compared them with untagged controls. No mortality was measured in either the VIE or CWT groups but the PIT tag group experienced 62.5 per cent mortality. Although larger fish within the PIT tag group displayed higher survival, no fish under 70 mm survived. These results highlight the point that PIT tags are not suitable for some smaller size classes of fish and that VIE and CWT methods may be more appropriate. See Box 10.1 for the use of PIT tags in fish and amphibian research.

10.6.4 Marking amphibians and reptiles

Marking techniques for amphibians and reptiles are also diverse and include toe clipping, tattooing, branding, painting, scale and scute clipping, shell notching (freshwater turtles), tagging and banding, VIE and VIA tags, passive integrated transponder (PIT) tags, and photo identification. For a detailed review of amphibian and reptile marking methods, and ethical considerations of some marking techniques, see Ferner (2007), Ferner (2010), Ferner and Plummer (2016), and Brien and Manolis (2016), and for photo identification methods see Sacchi et al. (2016).

10.6.5 Marking mammals

Mammal marking techniques include using PIT tags, other tags (e.g., numbered ear tags), collaring (including use of radio-tracking collars with individual identification), fur clipping and ear notching. Marking by clipping or notching should be done using

---

BOX 10.1 PIT telemetry

Use of Passive Integrated Transponder (PIT) tags is a relatively inexpensive way to uniquely mark a very large number of individuals. When used in spatial studies, this is also important to avoid problems of spatial autocorrelation common in radio-tracking studies (e.g., Harris et al. 1990). PIT tags also have the advantage that they do not require a battery and so are long lasting and do not need to be waterproofed. PIT tags can be very small (5–11 mm in length) and weigh less than 1 g, making them suitable for use in small vertebrates (e.g., common toad *Bufo bufo* (Salazar et al. 2016)). When compared with dye marking the deleterious effects of PIT tags on the subjects (*Rana temporaria*) were found to be negligible (Brown 1997) and no effect of PIT tags was shown on breeding or survival of the golden bell frog (*Litoria aurea*) (Pyke 2005). Salazar et al. (2016) used PIT telemetry to assess habitat use by common toads by fitting 1040 adult toads with PIT tags (Trovan ID-100 tags subdermally injected into the skin on their back) and then subsequently releasing and searching for them using a PIT antenna (Dorset ID LID650 antenna and decoder) which was able to read through water, earth, and vegetation to up to approximately 30 cm depth (depending on tag orientation). By systematic searching and taking a GPS location of recaptured toads it was possible to collect presence absence data which could then be used to generate a Resource Selection Function (RSF) model for common toad habitat use.

PIT telemetry has also been used by Cucherousset et al. (2005) to track small fish in shallow streams and by Gibbons and Andrews (2004) in a study of Pyrenean brook salamanders in streams.

PIT tags can also provide spatial information from static antennas when they are arranged in an array at known locations; for example, to count and identify migrating salmon (*Salmo salar*) as they move through fish passes (Pinder et al. 2007).

---

an established coding system (based on placement of the mark) to indicate survey session or animal identity as appropriate. For short-term studies it may also be appropriate to use an animal marking spray. Collaring or radio-tagging of mammals such as otter, mink and beaver can be combined with the use of data loggers (e.g., to measure temperature or depth) or accelerometers to study animal behaviour such as diving behaviour or activity patterns (Hays et al. 2007; Graf et al. 2018). Radio transmitters can be attached to freshwater mammals for spatial studies of home range and to monitor movement behaviour (e.g., American mink; Macpherson and Bright 2010; Harrington and Macdonald 2008). GPS transmitters may be too large to be deployed on some freshwater mammals, but have been successfully used to study otters, although accuracy of the signal can be affected by submersion in water and by dense riparian vegetation (Quaglietta et al. 2012); and for tracking Amazon river dolphins for the first time in 2017 (WWF website).

10.6.6 Marking birds

Birds are most commonly marked by ringing, which should be done under licence and in accordance with national and international (for migratory species e.g., geese) ringing schemes. Numbered rings are the most commonly used method for marking; for example, the British Trust for Ornithology (BTO) use aluminium split rings with unique codes that identify each individual bird and data are stored within a large national database. When individuals are re-captured, new data can be added and unique capture histories are obtained. As long as the methods are standardised, bird populations can be reliably monitored using these capture methods if the capture effort is kept constant over time and is repeatable. For example the Constant Effort Sites scheme (CES) run by the BTO (e.g., Peach et al. 1996) has been adopted by many European countries; and north America have a similar programme, the Monitoring Avian Productivity and Survival scheme (see MAPS programme on the Institute for Bird Populations website). For the CES scheme, the capture method involves putting the same number and sizes of nets in the same locations during the same time periods each year for a number of years. All birds caught have morphometric measurements taken, are identified to species level, aged and sexed if possible, and any observations noted such as breeding condition and moulting patterns. This information is essential for making conservation and management decisions, as well as understanding species ecology.

10.7 Genetic studies of freshwater vertebrates

10.7.1 Applications

Genetic diversity is recognised as a vital component of long-term species persistence and survival, and genetic research techniques should be considered in conservation initiatives (Hedrick 2001). Molecular methods can be used to assess population viability, inbreeding risk, and to identify species hybrids. For freshwater vertebrates, habitats are often spatially discrete (e.g., breeding ponds) allowing for assessment of landscape permeability. For example, in a study by Salazar (2014) genetic samples of common toad (*Bufo bufo*) were collected from eight breeding ponds across west Oxfordshire, UK. Using microsatellite analyses from seven markers, the genetic distance (dissimilarity) between each pond population was calculated and compared with Euclidean and cost distances to determine whether isolation by distance or isolation by barrier is most effective in separating common toad breeding populations. At the landscape scale (between 2–20 km between ponds) there was no effect of isolation by distance detected, but there was a positive and significant relationship observed between increasing cost distance and genetic distance of ponds, implying that barriers of unsuitable habitat were more effective at isolating populations at this scale.

Assessment of population structure in lampreys (*Lampetra fluviatilis* and *L. planeri*) by Bracken et al. (2015) showed contrasting patterns for mitochondrial and microsatellite markers allowing the tracing of founder events (using mtDNA) and more recent genetic differentiation of populations due to limited dispersal and anthropogenic barriers (using microsatellite analyses). Microsatellite analyses were used by Jehle et al. (2005) to infer rates of recent migration of great crested newts (*Triturus cristatus*) to ponds.

Genetic techniques can be used to assess paternity or relatedness of individuals and to support and explain behavioural observations. Individual identification is also possible allowing for non-invasive capture-mark-recapture studies using biological samples (Lampa et al. 2015). Collecting biological samples (e.g., faeces) is a non-invasive method for establishing species presence and can also be used to determine sex (O'Neil et al. 2013). eDNA is rapidly becoming more frequently used to determine the presence or absence of species in a water body, as described in Section 10.4.2. Dietary analysis of gut contents and faecal samples can be biased towards prey items that are less easily digested (e.g., with hard body parts such as bones or carapaces). In contrast, using molecular techniques it is possible to identify prey remains that would otherwise not have been identified by microscopy (Symondson 2002). Molecular methods can be useful in identifying and characterising disease, such as the new strain of ranavirus affecting the critically endangered Chinese giant salamander (*Andrias davidianus*) (Zhou et al. 2013).

The nature of freshwater environments means they can act as both barriers to gene flow (e.g., isolated lakes and ponds) or dispersal corridors (in the case of rivers and streams). It is beyond the scope of this chapter to review all genetic methods available to researchers wishing to study freshwater vertebrate populations. However, we would strongly encourage the collection of non-invasive DNA samples as part of any research or monitoring programme. Such samples can be hugely valuable for answering a myriad of questions relating to community composition, population size, and population structure. Additionally, cases of wildlife disease can be monitored through genetic methods and can easily be incorporated into standard long-term monitoring protocols; for example, in identifying amphibian chytrid fungus (*Batrachochytrium dendrobatidis* and *B. salamandrivorans*).

10.7.2 Monitoring chytrid prevalence in Cusuco National Park, Honduras

Operation Wallacea has been running a long-term biodiversity monitoring programme in the cloud forest of Cusuco National Park (CNP), Honduras, since 2004. CNP has been identified as one of the world's top-50 irreplaceable protected areas for conservation of threatened amphibians, birds, and mammals (Le Saout et al. 2013), an accolade likely resulting from the intensive survey effort by Operation Wallacea within this relatively small national park. Herpetofauna diversity is particularly high, with an incredible 100 species of reptiles and amphibians having been recorded. Fifteen of these species are endemic to Honduras, nine of which are endemic to CNP. Annual surveys take place between June and August and utilise teams of volunteers led by experienced field biologists. A network of research camps, transects, and study plots have been established throughout the buffer and core zones of the park, spanning an elevational range of approximately 500–2000 m above sea level. At each survey camp an aquatic 'stream' transect of approximately 200 m (length variable between camps) has been established for the long-term monitoring of stream-breeding amphibians. Stream transects at each camp are surveyed at night at least four times per field season to determine average detection rates. Survey effort is quantified by time (marking start and end time for each survey), the number of participants, and the distance travelled.

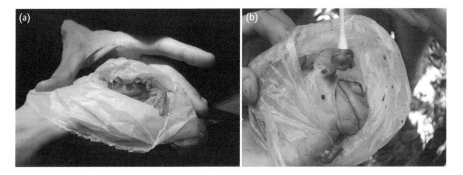

Figure 10.8 (a) *Plectrohyla exquisita being handled using a disposable plastic bag to prevent spreading disease between individuals and* (b) *P. exquisita* being swabbed for Batrachochytrium dendrobatidis (B.d).

Detection rates per unit effort can, thus, be determined for each species to monitor population trends over time (Green et al. 2012).

Within this sampling strategy, four species of tree frog (*Plectrohyla exquisita*, *P. dasypus*, *Duellmanohyla soralia*, and *Ptychohyla hypomykter*) are routinely swabbed and assessed for amphibian chytrid fungus (*Batrachochytrium dendrobatidis*), hereafter referred to as *Bd*, which is causing catastrophic amphibian population declines throughout Mesoamerica, and is a serious threat to the amphibians of CNP (Kolby et al. 2010). To date, 12 amphibian species have tested positive for *Bd* within CNP, of which eight are listed as endangered or critically endangered by the IUCN Red List of Threatened Species (Gilroy et al. 2015). Frogs are swabbed using non-lethal protocols established by Hyatt et al. (2007). Ventral surfaces of the legs and feet are each swabbed five times, applying moderate friction (Figure 10.8) and swab buds are broken off and stored in 2 mL microcentrifuge tubes containing 1 mL of 70 per cent ethanol. Samples are returned to the pop-up genetics lab at the temporary survey camp and analysed by polymerase chain reaction (PCR) and gel electrophoresis to detect the presence/absence of *Bd*. Despite the relatively basic lab setup and protocol available in the field, rapid assessment of large numbers of individuals can be performed giving an immediate result and feedback to survey teams. Summary results from 2015 show overall prevalence to be just over 10 per cent. Duplicate samples are exported to the UK for more rigorous analysis (Clake 2015).

10.8 Dietary studies of freshwater vertebrates

The diet of freshwater vertebrates can be assessed through direct observation. Where this is not possible, prey and feeding remains may be identified (e.g., using microscopy, in faecal samples, regurgitated pellets and gut contents from stomach flushing). The visual identification of prey remains may be biased towards species that have harder, less digestible parts (e.g., bone, fur, carapace, woody plant matter). The method is still useful as long as this potential source of bias is considered and has the advantage that prey remains may even be identifiable to species level using this technique (Grant and Harrington 2015). Where a complete picture of diet composition is required it is worth considering collection of biological samples for DNA barcoding or stable isotope analysis. At Lake Naivasha in Kenya, two studies used these techniques to study allochthonous inputs to the lake via hippopotamus guts (stable isotope analysis; Grey and Harper 2007) and the impacts of the diets of size-differentiated introduced carp on the lake vegetation (gut content; Britton et al. 2007).

10.9 Biosecurity

Biosecurity should be considered early in the survey planning process to reduce the likelihood of spreading wildlife disease (e.g., *Batrachochytrium dendrobatidis*, *B. salamandrivorans*, and ranavirus in amphibians), and transporting invasive non-native species, on clothing boots and survey equipment. Rigorous protocols have been established to eliminate these risks (e.g., see GB Non-native Species Secretariat website) and should be adhered to by everyone working within freshwater habitats. These include cleaning boots and equipment with Vercon or equivalent disinfectant and changing clothes between survey site visits. Multiple site visits per day may also be discouraged or even prohibited in some locations to minimise risk. We do not provide a comprehensive list of biosecurity measures here, as protocols must be adapted to the location and specific risks associated with the fieldwork. We do, however, stress the high level of importance and priority that should be given to appropriate biosecurity measures whenever working in freshwater habitats.

10.10 Concluding remarks

Vertebrates can be key indicators of freshwater ecosystem health. Careful design and implementation of long-term monitoring programmes can provide valuable information about the effectiveness of current management strategies of watersheds and wetland areas, as well as the population status and ecology of target species of conservation interest or concern.

The continued and healthy functioning of freshwater ecosystems is of critical importance to both the aquatic and terrestrial life which depend on them, including our own species. The variety of vertebrate life and the diverse ways in which they utilise and depend on freshwater systems requires an equally diverse and varied approach to survey design and implementation. This chapter highlights some key considerations but is in no means a comprehensive review of all possible survey methods and techniques available to the researcher wishing to study freshwater vertebrates. We encourage readers to use this information as a platform for exploring the wealth of literature available on this subject, only some of which could be covered here.

References

Baillie, J.E.M. Hilton-Taylor, C. and Stuart, S.N. (eds) (2004). 2004 IUCN Red List of Threatened Species. A Global Species Assessment. IUCN, Gland, Switzerland and Cambridge, UK.

Beaumont, W.R.C. (2016). Electricity in Fish Research and Management: Theory and Practice. 2nd edition, John Wiley and Sons, Hoboken.

Biggs, J., Ewald, N., Valentini, A., et al. (2014). Analytical and methodological development for improved surveillance of the great crested newt. Defra Project WC1067, Oxford.

Biggs, J., Ewald, N., Valentini, A., et al. (2015). Using eDNA to develop a national citizen science-based monitoring programme for the great crested newt (*Triturus cristatus*). *Biological Conservation*, 183, 19–28.

Bodmer, R. (2012). Earthwatch Institute Amazon Riverboat Exploration 2012 Field Report. Available online from Earthwatch Institute website.

Bodmer, R., Fang, T., Puertas, P., Antunez, M., Chota, K., and Bodmer, W. (2014). Impacts of Climate Change on Wildlife Conservation in the Samiria River Basin of the Pacaya-Samiria National Reserve, Peru. Available online from Operation Wallacea website.

Bodmer, R., Mayor, P., Antunez, M., et al. (2018). Major shifts in amazon wildlife populations from recent climatic intensification. *Conservation Biology*, 00, 1–12, doi: 10.1111/cobi.12993.

Bodmer, R., Puertas, P., Antunez, M., Fang, T., and Gil, G. (2012). Impacts of Climate Change on Wildlife Conservation in the Samiria River Basin of the Pacaya-Samiria National Reserve, Peru. Available online from Operation Wallacea website.

Boitani, L. and Powell, R.A. (eds) (2012). Carnivore Ecology and Conservation: A Handbook of Techniques, Techniques in Ecology and Conservation Series. Oxford University Press, Oxford.

Bonesi, L. and Macdonald, D.W. (2004). Evaluation of sign surveys as a way to estimate the relative abundance of American mink (Mustela vison). *Journal of Zoology*, 262, 65–72.

Bracken, F.S.A., Hoelzel, A.R., Hume, J.B., and Lucas, M.C. (2015). Contrasting population genetic structure among freshwater-resident and anadromous lampreys: the role of demographic history, differential dispersal and anthropogenic barriers to movement. *Molecular Ecology*, 24, 1188–1204.

Brien, M. and Manolis, C. (2016). Crocodilians. In C.K. Dodd (ed.) Reptile Ecology and Conservation: A Handbook of Techniques. Oxford University Press, Oxford.

Britton, J.R., Boar, R.R., Grey, J., Foster, J., Lugonzo, J., and Harper, D.M. (2007). From introduction to fishery dominance: the initial impacts of the invasive carp *Cyprinus carpio* in Lake Naivasha, Kenya, 1999 to 2006. *Journal of Fish Biology*, 71, 239–57.

Brown, L.J. (1997). An evaluation of some marking and trapping techniques currently used in the study of anuran population dynamics. *Journal of Herpetology*, 31, 410–19.

Brown, A., Gilbert, G., and Wotton, S. (2012). Bitterns and Bittern conservation in the UK. *British Birds*, 105, 58–87.

Buxton, A.S., Groombridge, J.J., Zakaria, N.B., and Griffiths, R.A. (2017a). Seasonal variation in environmental DNA in relation to population size and environmental factors. *Scientific Reports*, 7, 46294.

Buxton, A.S., Groombridge, J.J., and Griffiths, R.A. (2017b). Is the detection of aquatic environmental DNA influenced by substrate type? *PLoS ONE*, 12, 0183371.

Campbell, R.D., Feber, R., Macdonald, D.W., Gaywood, M.J., and Batty, D. (2010). The Scottish Beaver Trial: Ecological monitoring of the European beaver Castor fiber and other riparian mammals—Initial methodological protocols 2009. Scottish Natural Heritage Commissioned Report No. 383.

Casazza, M.L. and Wylie, G.D. (2000). A funnel trap modification for surface collection of aquatic amphibians and reptiles. *Herpetological Review*, 31, 91–2.

Chandler, M., See, L., Andrianandrasana, H., et al. (2017). Community- and citizen-based approaches to tropical forest biodiversity monitoring. In: GOFC-GOLD and GEO BON (eds) Sourcebook of Methods and Procedures for Monitoring Essential Biodiversity Variables in Tropical Forests with Remote Sensing. Report version UNCBD COP-13, GOFC-GOLD Land Cover Project Office, Wageningen University, The Netherlands.

Clake, D. (2015). Variation in drivers of prevalence and infection intensity of the amphibian chytrid fungus *Batrachochytrium dendrobatidis* in endangered Honduran tree frogs, MSc Thesis. University of Kent, Kent, UK.

Clout, M.N. and Williams, P.A. (eds) (2009) Invasive Species Management: A Handbook of Principles and Techniques, Techniques in Ecology and Conservation Series. Oxford University Press, Oxford.

Crall, J.P., Stewart, C.V., Berger-Wolf, T.Y., Rubenstein, D.I., and Sundaresan, S.R. (2013). HotSpotter—Patterned species instance recognition. In: Workshop on Applications of Computer Vision (WACV) 2013, Institute of Electrical and Electronic Engineers (IEEE), 230–7.

Cucherousset, J., Roussel, J.-M., Keeler, R., Cunjak, R.A., and Stump, R. (2005). The Use of Two New Portable 12-mm PIT Tag Detectors to Track Small Fish in Shallow Streams. *North American Journal of Fisheries Management*, 25, 270–4.

Cunningham, A.A. and Minting, P. (2008). National survey of Batrachochytrium dendroba-tidis infection in UK amphibians: Institute of Zoology, Zoological Society of London, Available on ARG UK website.

Day, C.C., Westover, M.D., Hall, L.K., Larsen, R.T., and McMillan, B.R. (2016). Comparing direct and indirect methods to estimate detection rates and site use of a cryptic semi-aquatic carnivore. *Ecological Indicators*, 66, 230–4.

Deacon, A.E., Mahabir, R., Inderlall, D., Ramnarine, I.W., and Magurran, A.E. (2017). Evaluating detectability of freshwater fish assemblages in tropical streams: Is hand-seining sufficient? *Environmental Biology of Fish*, 100, 839–49.

Dodd, C.K. (ed.) (2009). Amphibian Ecology and Conservation: A Handbook of Techniques, Techniques in Ecology and Conservation Series. Oxford University Press, Oxford.

Dodd, C.K. (ed.) (2016). Reptile Ecology and Conservation: A Handbook of Techniques, Techniques in Ecology and Conservation Series. Oxford University Press, Oxford.

Dorcas, M.E. and Willson, J.D. (2009). Innovative methods for studies of snake ecology and conservation. In: S.J. Mullin and R.A. Seigel (eds) Snakes: Ecology and Conservation, 5–37. Cornell University Press, New York.

Ferner, J.W. (2007). A Review of Marking and Individual Recognition Techniques for Amphibians and Reptiles A Review of Marking and Individual Recognition Techniques for Amphibians and Reptiles, Issue 35 of Herpetological circular, Society for the Study of Amphibians and Reptiles.

Ferner, J.W. (2010). Measuring and marking post-metamorphic amphibians. In C.K. Dodd Jr (ed.) Amphibian Ecology and Conservation: A Handbook of Techniques, 124–41. Oxford University Press, Oxford.

Ferner, J.W. and Plummer, M.V. (2016). Marking and measuring reptiles. In C.K. Dodd Jr (ed.) Reptile Ecology and Conservation: A Handbook of Techniques, 45–58. Oxford University Press, New York.

Fisher, R.N. (2016). Planning and setting objectives in filed studies. In C.K. Dodd Jr (ed.) Reptile Ecology and Conservation: A Handbook of Techniques, 16–31. Oxford University Press, New York.

Fitzgerald, L.A. (1989). An evaluation of stomach flushing techniques for crocodilians. *Journal of Herpetology*, 23, 170–2.

Freeman, B.G. and Montgomery, G.A. (2017). Using song playback experiments to measure species recognition between geographically isolated populations: A comparison with acoustic trait analyses. *The Auk*, 134, 857–70.

Garner, T.W.J., Perkins, M.W., Govindarajulu, P., et al. (2006). The emerging amphibian pathogen Batrachochytrium dendrobatidis globally infects introduced populations of the North American bullfrog, Rana catesbeiana. *Biology letters*, 2, 455–9.

Gent, T. and Gibson, S. (2003). Herpetofauna Workers Manual, JNCC. Available on JNCC website.

Gibbons, J.W. and Andrews, K.M. (2004). PIT tagging: Simple technology at its best. *Bioscience*, 54, 447–54.

Gilbert, G., Gibbons, D.W., and Evans, J.E. (1998). Bird monitoring methods. Sandy: RSPB.

Gilbert, G., McGregor, P.K., and Tyler, G. (1994). Vocal individuality as a census tool: practical considerations illustrated by a study of two rare species. *Journal of Field Ornithology*, 65, 335–48.

Gill, R.E. Jr, Tibbitts, T.L., Douglas, D.C., et al. (2009). Extreme endurance flights by land-birds crossing the Pacific Ocean: ecological corridor rather than barrier? *Proceedings of the Royal Society B*, 276, 447–57.

Gilroy, D., Jones, S., Vulinec, K., et al. (2015). Operation Wallacea Cusuco National Park, Honduras 2015: End of Season Report. Operation Wallacea Available on the Opwall.com website.

Gonzalez-Garcia, F., Roberto, S.-L.J., and Francisco, O.J. (2017). Individual variation in the booming calls of captive Horned Guans (Oreophasis derbianus): an endangered Neotropical mountain bird. *Bioacoustics*, 26, 185–98.

Graf, P.M., Wilson, R.P., Sanchez, L.C., Hackländer, K., and Rosell, F. (2018). Diving behavior in a free-living, semi-aquatic herbivore, the Eurasian beaver *Castor fiber. Ecology and Evolution*, 8, 997–1008.

Grant, K.R. and Harrington, L.A. (2015). Fish selection by riverine Eurasian otters in lowland England. *Mammal Research*, 60, 217–31. doi: 10.1007/s13364-015-0223-3.

Green, M.L., Monick, K., Manjerovic, M.B., Novakofski, J., and Mateus-Pinilla, N. (2015). Communication stations: cameras reveal river otter (Lontra canadensis) behavior and activity patterns at latrines. *Journal of Ethology*, 33, 225–34.

Green, S., Slater, K., Burdekin, O., and Long, P. (2012). Cusuco National Park, Honduras 2012 status report, Available on Operation Wallacea website.

Gregory, R.D., Gibbons, D.W., and Donald, P.F. (2004). Bird census and survey techniques. In W.J. Sutherland, I. Newton, and R.E. Green (eds) Bird Ecology and Conservation: A Handbook of Techniques, 17–56. Oxford University Press, Oxford.

Gresswell, R.E., Liss, W.J., Lomnicky, G.A., Deimling, E.A., Hoffman, R.L., and Tyler, T. (1997). Using Mark-Recapture Methods to Estimate Fish Abundance in Small Mountain Lakes, *Northwest Science*, 71, 39–44.

Grey, J. and Harper, D.M. (2007). Using stable isotope analyses to identify allochthonous inputs to Lake Naivasha mediated via the hippopotamus gut. *Isotopes in Environmental and Health Studies*, 38, 245–50.

Griffiths, R.A. (1985). A simple funnel trap for studying newt populations and an evaluation of trap behaviour in smooth and palmate newts, Triturus vulgaris and T. helveticus. *Herpetological Journal*, 1, 5–10.

Griffiths, R.A. and Brady, L.D. (1996). Evaluation of a standard method for surveying common frogs (Rana temporaria) and newts (Triturus cristatus, T. helveticus, and T. vulgaris). JNCC Report No 259, Peterborough. Available on JNCC website.

Hale, R.S. and Gray, J.H. (1998). Retention and Detection of Coded Wire Tags and Elastomer Tags in Trout. *North American Journal of Fisheries Management*, 18, 197–201. doi: 10.1577/1548-8675(1998)0182.0.CO;2.

Halstead, B.J., Wylie, G.D., and Casazza, M.L. (2013). Efficacy of trap modifications for increasing capture rates of aquatic snakes in floating aquatic funnel traps. *Herpetological Conservation and Biology*, 8, 65–74.

Hanson, K.C. and Barron, J.M. (2017). Evaluation of the Effects of Marking Pacific Lamprey Ammocoetes with Visual Implant Elastomer, Coded Wire Tags, and Passive

Integrated Transponders. *Transactions of the American Fisheries Society*, 146, 626–33. doi: 10.1080/00028487.2017.1290681.

Harrington, L.A., Harrington, A.L., Hughes, J., Stirling, D., and Macdonald, D.W. (2010). The accuracy of scat identification in distribution surveys: American mink in the northern highlands of Scotland. *European Journal of Wildlife Research*, 56, 377–84. DOI 10.1007/s10344-009-0328-6.

Harrington, L.A., Harrington, A.L., and Macdonald, D.W. (2008). Estimating the relative abundance of American mink Mustela vison on lowland rivers: evaluation and comparison of two techniques. *European Journal of Wildlife Research*, 54, 79–87.

Harrington, L.A. and Macdonald, D.W. (2008). Spatial and temporal relationships between invasive American mink and native European polecats in the southern United Kingdom. *Journal of Mammalogy*, 89, 991–1000.

Harris, S., Cresswell, W.J., Forde, P.G., Trewhella, W.J., Woollard, T., and Wray, S. (1990). Home-range analysis using radio-tracking data—a review of problems and techniques particularly as applied to the study of mammals. *Mammal Review*, 20, 97–123.

Hays, G.C., Forman, D.W., Harrington, L.A., Harrington, A.L., Macdonald, D.W., and Righton, D. (2007). Recording the free-living behaviour of small-bodied, shallow-diving animals with data loggers. *Journal of Animal Ecology*, 76, 183–90.

Hedrick, P.W. (2001). Conservation genetics: where are we now? *Trends in Ecology & Evolution*, 16, 629–36.

Hoodless, A.N., Inglis, J.G., Doucet, J.-P., and Aebischer, N.J. (2008). Vocal individuality in the roding calls of Woodcock *Scolopax rusticola* and their use to validate a survey method. *Ibis*, 150, 80–9.

Hubert, W.A., Pope, K.L., and Dettmers, J.M. (2012). Passive capture techniques. In A.V. Zale, D.L. Parrish, and T.M. Sutton (eds) Fisheries techniques, 3rd edition, 223–65. American Fisheries Society, Bethesda, Maryland.

Hunter, M.E., Oyler-McCance, S.J., Dorazio, R.M., et al. (2015). Environmental DNA (eDNA) sampling improves occurrence and detection estimates of invasive Burmese pythons A. R. Mahon. *PLoS ONE*, 10, 1–17.

Hyatt, A.D., Boyle, D.G., Olsen, V., et al. (2007). Diagnostic assays and sampling protocols for the detection of Batrachochytrium dendrobatidis. *Diseases of Aquatic Organisms*, 73, 175–92.

Jane, S.F., Wilcox, T.M., McKelvey, K.S., et al. (2015). Distance, flow and PCR inhibition: eDNA dynamics in two headwater streams. *Molecular Ecology Resources*, 15, 216–27.

Jehle, R., Wilson, G.A., Arntzen, J.W., and Burke, T. (2005). Contemporary gene flow and the spatio-temporal genetic structure of subdivided newt populations (Triturus cristatus, T. marmoratus). *Journal of Evolutionary Biology*, 18, 619–28.

Jerde, C.L., Mahon, A.R., Chadderton, W.L., and Lodge, D.M. (2011). 'Sight-unseen' detection of rare aquatic species using environmental DNA. *Conservation Letters*, 4, 150–7.

Kaluza, J., Donald, R.L., Gynther, I.C., Leung, L.K.-P., and Allen, B.L. (2016). The distribution and density of water mice (*Xeromys myoides*) in the Maroochy River of southeast Queensland, Australia. *PLoS ONE* 11, e0146133.

Kirschel, A.N.G., Cody, M.L., Harlow, Z.T., Promponas, V.J., Vallejo, E.E., and Taylor, C.E. (2011). Territorial dynamics of Mexican Ant-thrushes *Formicarius moniliger* revealed by individual recognition of their songs. *IBIS*, 153, 255–68.

Kolby, J., Padgett-Flohr, G., and Field, R. (2010). Amphibian chytrid fungus Batrachochytrium dendrobatidis in Cusuco National Park, Honduras. *Diseases of Aquatic Organisms*, 92, 245–51.

Kruuk, H. and Conroy, J.W.H. (1987). Surveying otter *Lutra lutra* populations: a discussion of problems with spraints. *Biological Conservation*, 41, 179–83.

Kuhn, R.A. (2010). Note on Hair-Sampling Devices for Eurasian Otters. *IUCN Otter Spec. Group Bull.*, 27, 98–104.

Lampa, S., Mihoub, J.B., Gruber, B., Klenke, R., and Henle K. (2015). Non-Invasive Genetic Mark-Recapture as a Means to Study Population Sizes and Marking Behaviour of the Elusive Eurasian Otter (*Lutra lutra*). *PLoS ONE*, 10, e0125684.

Lenton, E.J., Chanin, P.R.F., and Jefferies, D.J. (1980). Otter survey of England 1977–1979. Nature Conservancy Council, London: 1–75.

Luiselli, L. and Amori, G. (2016). Diet. In C.K.J. Dodd (ed.) Reptile Ecology and Conservation: A Handbook of Techniques, 97–109. Oxford University Press, New York.

Macpherson J.L. and Bright, P.W. (2010). Movements of radio-tracked American mink (*Neovison vison*) in extensive wetland in the UK, and the implications for threatened prey species such as the water vole (*Arvicola amphibius*). *European Journal of Wildlife Research*, 56, 855–9. doi:10.1007/s10344-010-0383-z.

Marion, W.R., O'Meara, T.E., and Maehr, D.S. (1981). Use of playback recordings in sampling elusive or secretive birds. *Studies in Avian Biology*, 6, 81–5.

Martellini, A., Payment, P., and Villemur, R. (2005). Use of eukaryotic mitochondrial DNA to differentiate human, bovine, porcine and ovine sources in fecally contaminated surface water. *Water Research*, 39, 541–8.

Matthé, M., Sannolo, M., Winiarski, K., et al. (2017). Comparison of photo-matching algorithms commonly used for photographic capture-recapture studies. *Ecology and Evolution*, 7, 5861–72.

McClintock, B.T. (2015). Multimark: an R package for analysis of capture–recapture data consisting of multiple 'noninvasive' marks. *Ecology and Evolution*, 5, 4920–31.

McGregor, P.K. and Byle, P.A. (1992). Individually distinctive Bittern booms. *Bioacoustics*, 4, 93–109.

Miaud, C., Dejean, T., Savard, K., et al. (2016). Invasive North American bullfrogs transmit lethal fungus *Batrachochytrium dendrobatidis* infections to native amphibian host species. *Biological Invasions*, 18, 2299–308.

Muñoz-Adalia, E.J., Jubete, F., Zumalacárregui, C., and Baglione, V. (2015). Hydrological Management in a Restored Wetland Affects Stopover Ecology of Aquatic Warbler: The Case of La Nava Wetland, Northern Spain. *Ornithological Science*, 15, 89–98.

Musselman, W.C., Worthington, T.A., Mouser, J., et al. (2017). Passive Integrated Transponder Tags: Review of Studies on Warmwater Fishes with Notes on Additional Species, *Journal of Fish and Wildlife Management*, 8, 353–64. doi: 10.3996/122016-JFWM-091.

O'Brien, M.F., Lee, R., Cromie, R., and Brown, M.J. (2016). Assessment of the rates of injury and mortality in waterfowl captured with five methods of capture and techniques for minimizing risks. *Journal of Wildlife Diseases*, 52, 86–95.

O'Neill, D., Turner, P.D., O'Meara, D.B., Chadwick, E.A., Coffey, L., and O'Reilly, C. (2013). Development of novel real-time TaqMan® PCR assays for the species and sex identification of otter (*Lutra lutra*) and their application to noninvasive genetic monitoring. *Molecular Ecology Resources*, 13, 877–83. doi:10.1111/1755-0998.12141.

Peach, W.J., Buckland, S.T., and Baillie, S.R. (1996). The use of constant effort mist-netting to measure between-year changes in the abundance and productivity of common passerines. *Bird Study*, 43, 142–56.

Pennock, C.A., Frenette, B.D., Waters, M.J., and Gido, K.B. (2016). Survival of and Tag Retention in Southern Redbelly Dace Injected with Two Sizes of PIT Tags. *North American Journal of Fisheries Management*, 36, 1386–94. doi: 10.1080/02755947.2016.1227403.

Pinder, A.C., Riley, W.D., Ibbotson, A.T., and Beaumont, W.R.C. (2007). Evidence for an autumn downstream migration and the subsequent estuarine residence of 0+ year juvenile Atlantic salmon *Salmo salar* L., in England. *Journal of Fish Biology*, 71, 260–4. doi:10.1111/j.1095–8649.2007.01470.x.

Poulin, B. and Lefebvre, G. (2003). Variation in booming among great bitterns *Botaurus stellaris* in the Camargue, France. *Ardea*, 91, 177–81.

Pyke, G. (2005). The Use of PIT tags in capture-recapture studies of frogs: A field evaluation. *Herpetological Review*, 36, 281–5.

Quaglietta L., Martins B.H., de Jongh, A., Mira, A., and Boitani, L. (2012). A Low-Cost GPS GSM/GPRS Telemetry System: Performance in Stationary Field Tests and Preliminary Data on Wild Otters (*Lutra lutra*). *PLoS ONE*, 7, e29235.

Reynolds, J.C., Short, M.J., and Leigh, R.J. (2004). Development of population control strategies for mink *Mustela vison*, using floating rafts as monitors and trap sites. *Biological Conservation*, 120, 533–43.

Sacchi, R., Scali, S., Mangiacotti, M., Sannolo, M., Alberto, M., and Zuff, L. (2016). Digital identification and analysis. In C.K. Dodd Jr (ed.) Reptile Ecology and Conservation: A Handbook of Techniques, 59–72. Oxford University Press, New York.

Salazar, R.D. (2014). The distribution and dispersion of herpetofauna in UK lowland farmland, with focus on the common toad Bufo bufo. DPhil thesis, University of Oxford.

Salazar, R.D., Montgomery, R.A., Thresher, S.E., and Macdonald, D.W. (2016). Mapping the Relative Probability of Common Toad Occurrence in Terrestrial Lowland Farm Habitat in the United Kingdom. *PLoS ONE*, 11, 1–14.

Le Saout, S., Hoffmann, M., Shi, Y., et al. (2013). Conservation. Protected areas and effective biodiversity conservation. *Science*, 342, 803–5. DOI: 10.1126/science.1239268.

Serena, M. and Grant, T.R. (2017). Effect of flow on platypus (*Ornithorhynchus anatinus*) reproduction and related population processes in the upper Shoalhaven River. *Australian Journal of Zoology*, 65, 130–9.

Serfass, T.L., Wright, L., Pearce, K., and Duplaix, N. (2017). Animal welfare issues pertaining to the trapping of otters for research, conservation, and fur. In A. Butterworth (ed.) Marine Mammal Welfare. Animal Welfare Series, volume 17, Springer, Cham.

Skelly, D.K. and Richardson, J.L. (2009). Larval sampling. In C.K. Dodd (ed.) Amphibian Ecology and Conservation: A Handbook of Techniques, 55–70. Oxford University Press, Oxford.

Solow, A.R. and Kendall, B.E. (2015). River dolphins can act as population trend indicators in degraded freshwater systems. *Ecology*, 96, 2027–8.

Strachan, R. and Moorhouse, T. (2006). Water vole conservation handbook. Wildlife Conservation Research Unit, University of Oxford.

Sutherland, W.J., Newton, I., and Green, R. (eds) (2004). Bird Ecology and Conservation: A Handbook of Techniques, Techniques in Ecology and Conservation Series. Oxford University Press, Oxford.

Symondson, W.O.C. (2002). Molecular identification of prey in predator diets. *Molecular Ecology*, 11, 627–41.

Taberlet, P., Bonin, A., Zinger, L., and Coissac, E. (2018). Environmental DNA for Biodiversity Research and Monitoring. Oxford University Press, Oxford.

Thomsen, P.F., Kielgast, J., Iversen, L.L., et al. (2012). Monitoring endangered freshwater biodiversity using environmental DNA. *Molecular Ecology*, 21, 2565–73.

Thomsen, P.F. and Willerslev, E. (2015). Environmental DNA—An emerging tool in conservation for monitoring past and present biodiversity. *Biological Conservation*, 183, 4–18.

Thorbjarnarson, J. (1999). Crocodile tears and skins: International trade, economic constraints, and limits to the sustainable use of crocodilians. *Conservation Biology*, 13, 465–70.

Thorsteinsson, V. (2002). Tagging Methods for Stock Assessment and Research in Fisheries. Report of Concerted Action FAIR CT.96.1394 (CATAG). Reykjavik. Marine Research Institute Technical Report (79), 179.

Tisseuil, C., Cornu, J.F., Beauchard, O., et al. (2013). Global diversity patterns and cross-taxa convergence in freshwater systems. *Journal of Animal Ecology*, 82, 365–76.

Turvey S.T., Pitman, R.L., Taylor, B.L., et al. (2007). First human-caused extinction of a cetacean species? *Biology Letters*, 3, 537–40.

Valentini, A., Taberlet, P., Miaud, C., et al. (2016). Next-generation monitoring of aquatic biodiversity using environmental DNA metabarcoding. *Molecular Ecology*, 25, 929–42.

Vogt, R.C. (1979). Spring aggregating behaviour of painted turtles, *Chrysemys picta* (Reptilia, Testudines). *Journal of Herpetology*, 13, 363–5.

Vogt, R.C. (2016). Freshwater Turtles. In C.K. Dodd (ed.) Reptile Ecology and Conservation: A Handbook of Techniques, 168–80. Oxford University Press, Oxford.

Wagnon, C.J. and Serfass, T.L. (2016). Camera traps at northern river otter latrines enhance carnivore detectability along riparian areas in eastern North America. *Global Ecology and Conservation*, 8, 138–43.

White, G.C. and Burnham, K.P. (1999). Program MARK: Survival estimation from populations of marked animals. *Bird Study*, 46, 120–39.

White, G., Purps, J., and Alsbury, S. (2006) The bittern in Europe: a guide to species and habitat management. The Royal Society for the Protection of Birds, Sandy.

Willson, J.D., Winne, C.T., and Fedewaa, L.A. (2005). Unveiling Escape and Capture Rates of Aquatic Snakes and Salamanders (*Siren* spp. and *Amphiuma means*) in Commercial Funnel Traps. *Journal of Freshwater Ecology*, 20, 397–403.

Wittenberg, R.D. and Gifford, M.E. (2008). Funnel traps may be inappropriate for many studies of semi-aquatic snakes. *Journal of Freshwater Ecology*, 23, 213–18.

Wotton, S. (2017). Summary of the 2016 Bittern breeding season. RSPB, Sandy.

Yamaguchi, N., Strachan, R., and Macdonald, D.W. (2002). Practical considerations for the field study of American mink *Mustela vison* in lowland England. *Mammal Study*, 27, 127–33.

Yee, S.A., Puan, C.L., Chang, P.K., and Azhar, B. (2016). Vocal individuality of Sunda Scops-owl (*Otus lempiji*) in peninsular Malaysia. *Journal of Raptor Research*, 50, 379–90.

Zhou, Z.Y., Geng, Y., Liu, X.X., et al. (2013). Characterization of a ranavirus isolated from the Chinese giant salamander (Andrias *davidianus*, Blanchard, 1871) in China. *Aquaculture*, 384–387, 66–73. DOI:10.1016/j.aquaculture.2012.12.018.

11

Aquatic Macroinvertebrates

Richard Marchant and Catherine M. Yule

Corresponding author: rmarch@museum.vic.gov.au

11.1 Introduction

This chapter will deal with the sampling of populations and communities of freshwater macroinvertebrates; that is, invertebrates which are generally >200 μm in length, such as aquatic insect larvae (of the orders Odonata, Ephemeroptera, Plecoptera, Trichoptera, Diptera, Coleoptera, Hemiptera, and several minor orders), crustaceans (including Amphipoda, Isopoda, and Decapoda), molluscs (Bivalvia and Gastropoda), worms (Oligochaeta), and several other phyla. These groups are widespread and often common or abundant both in running and still waters, the two major subdivisions of freshwater ecosystems (Batzer and Boix 2016). Sampling equipment can range from the simple and versatile to the complicated and specialised. The assumption here is that this equipment is being used to conduct ecological and conservation studies and not simply as a means to acquire specimens.

Many accounts of sampling freshwater macroinvertebrates concentrate on statistical analysis of the spatial distribution of species, the calculation of confidence limits for estimated mean densities and the number of samples to take to achieve a given degree of precision (see FBA 2014; Hayslip 2007). These considerations are vital for successful sampling, but the techniques have been thoroughly explained by Elliott (1977) and others, and will not be repeated here. However, several important problems encountered when sampling invertebrates are not statistical in nature, but must be addressed if accurate data are to be obtained. These can best be summarised by two ideas: efficiency and representativeness. No amount of statistical analysis can resurrect numerical data if these two ideas have been ignored.

11.2 General considerations

The major aim of any sampling technique is to catch specimens of invertebrates and, if sampling quantitatively, to catch all specimens in a given area of habitat (e.g., a square metre). This demands that the efficiency of extraction of specimens in the field from their habitat by the sampling apparatus must be close to 100 per cent or at least must be known. For many commonly used sampling devices in freshwater studies little is known of their efficiency of operation in the field. Once samples have been

Marchant, R. and Yule, C. M., *Aquatic macroinvertebrates*. In: *Freshwater Ecology and Conservation: Approaches and Techniques*. Edited by Jocelyne M. R. Hughes: Oxford University Press (2019). © Oxford University Press 2019. DOI: 10.1093/oso/9780198766384.003.0011 obtained it is also necessary to separate with accuracy the specimens of interest from the organic detritus, sediment and other material that is inevitably collected. This is commonly done in the laboratory and is usually under better control than can be achieved with sampling apparatus in the field. But seldom is any account taken of the efficiency that can be achieved in the laboratory. In many studies the accuracy of neither of these tasks is discussed and the assumption is that perfect efficiency has prevailed.

The representativeness of sampling has been considered more frequently than the efficiency of field and laboratory techniques. This is most commonly achieved by stratification of the sampling effort. In other words samples should be taken in all the sub-habitats where the species or communities of interest are known to occur; and the number of samples per sub-habitat should be proportional to the per cent of the total habitat represented by each sub-habitat. Elliott (1977) has thoroughly covered the statistical procedures required for analysis of stratified samples. In an unfamiliar ecosystem or habitat the spatial distribution of the taxa among sub-habitats may well be unknown and can only be discovered by preliminary sampling. Stratification thus ensures that samples are taken in those portions of the habitat where the bulk of the population or community occurs. There is little point expending effort on sub-habitats (particularly if these are difficult to sample) that contain only small percentages of the total population. For instance, in a population study of a lotic amphipod (*Gammarus pseudolimnaeus*) Marchant and Hynes (1981) showed that 95 per cent of the population occurred within 3 m of the bank. They assumed density in the middle 6 m of the 12-m-wide river was zero. This assumption resulted in a very slight underestimate of the true mean density, but as mean density had 95 per cent confidence limits of 40–60 per cent any underestimate was unimportant compared with the level of spatial variation.

A final important question to ask at the beginning of any sampling programme is: are there any sections of the population or community that are not vulnerable to the chosen sampling technique? For example, most lotic sampling devices sample the surface of the riverbed to a maximum depth of 10 cm. If organisms are distributed to greater depths, then their densities can be seriously underestimated. Several species of Chironomidae, a family which is ubiquitous in rivers, can be found deeper than 10 cm either seasonally or permanently (Williams and Hynes 1974; Batzer and Boix 2016). Knowledge of the general natural history of the target species or community is essential before decisions about the representativeness of the proposed sampling methods can be made. For a detailed review of insect sampling in general, but including freshwater macroinvertebrates, see Samways et al. (2010) in the Techniques in Ecology and Conservation series; and for wetland invertebrates, see Anderson et al. (2013).

11.3 Field sampling techniques

Before choosing field sampling apparatus the aim of the study must be well defined. If the aim is to examine the diversity and composition of communities, then such information can often be collected with simple, qualitative, methods that provide data on the presence or absence of taxa and perhaps on the relative abundance of the different taxa. On the other hand, if the aim is to conduct a quantitative study (e.g., to investigate population dynamics or to make production estimates), then data on the density of specimens (numbers of individuals per unit area or perhaps volume) will be necessary and such information usually requires the use of more sophisticated apparatus. Density estimates can, of course, be extrapolated to provide a measure of total population size. No mention will be made here of the use of emergence traps, drift nets or artificial substrates (e.g., tiles, litter bags) as a means of sampling macroinvertebrates (see Merritt et al. 1996 for information on these techniques). These techniques do not actively extract specimens from the habitat in which they live, but rely on behaviour to transport specimens to a trap. Also nothing will be said about specialised techniques developed for estimating population sizes of surface dwelling insects such as some of the Hemiptera.

It is also imperative to define the scope of the study. Quantitative studies on the life history and population dynamics of a single species will require regular sampling, generally at least monthly in temperate regions where univoltine or bivoltine life cycles predominate. However, in the summer months it may be necessary to increase the frequency if short-lived (1–3 month) generations are present. In tropical freshwaters where water temperatures are typically high all year (>25°C), and life cycles are rapid, sampling may need to be much more frequent. Studies on diversity and community composition (either qualitative or quantitative) generally sample numerous sites over a broad spatial scale; and to provide representative data, sites may need to be sampled in each season (in the tropics these are usually wet and dry, but some wet tropical climates are aseasonal). It is easy to appreciate that numerous samples can be accumulated in a short period during any of these sorts of studies. Thus scope must be well defined otherwise the effort to take and process samples becomes overwhelming.

11.3.1 Qualitative methods

Qualitative methods have been widely applied to benthic communities in lakes, ponds, swamps, and rivers. The most common form of qualitative sampling is to use a hand net in much the same way as terrestrial sweep sampling of bushes and grasses for insects. Hand nets have been commonly used for surveys over large regions, such as the UK (Wright 2000) and Australia (Davies 2000), to provide data on macroinvertebrate composition for assessing the biological quality of running waters. The large spatial scale of these studies emphasises the versatility and ease of use of hand nets.

In still waters hand nets are swept through marginal vegetation of ponds or along the shore of lakes, agitating bottom sediment and any plants present. In rivers the hand net (which usually has a flat bottom and a mesh size no smaller than 250 μm) is held facing the current (Figure 11.1a). The riverbed is disturbed in front of the net, usually by the feet or hands of the operator, so that bottom material is swept into the net by the current. These operations can be timed (e.g., the 3-minute sampling time used in stream surveys in the UK (Wright 2000)); or the net can be drawn over a fixed distance (Dedieu et al. 2015). In this way such sampling can be standardised and the numerical data derived for each taxon can be used as a relative measure of abundance. Samples, once taken, are retained in suitable containers (stout plastic bags can

Figure 11.1 (a) A hand net being used to take a benthic sample. The stony bottom is being disturbed by hand rather than by foot. (b) A diver operating an air-lift sampler in a riverine pool dominated by cobbles (depth 3 m). The collecting net can be seen at the top left attached to the end of the riser tube. (c) A viewing tube being used to count Trichopteran larvae on the surface of stream rocks. The water was generally clear and shallow allowing monthly counts to be made. Photos taken by R. Marchant.

accommodate a wide variety of sample sizes and shapes), preserved (usually in 70 per cent ethanol rather than formalin), and specimens extracted and identified in the laboratory (see Section 11.4). These procedures are common to most sampling techniques and detailed protocols are available (e.g., Hauer and Resh (2006); Drake et al. (2007)).

In recent decades picking of live specimens from hand net samples has also been used in large-scale monitoring operations (Chessman 1995) to shorten the length of the overall study. The primary aim is to remove at least a few specimens of all taxa present. Experienced operators have been shown to be able to carry out this technique with remarkable consistency (Metzeling et al. 2003). Its considerable advantage is that laboratory processing and identification is much faster because specimens do not have to be separated from detritus. Live picking requires: white trays for spreading out and examining the catch; pipettes and forceps for removing specimens; and vials of various sizes for retaining and preserving the catch.

Hand nets can obviously only be used in shallow water and many community surveys are restricted to the margins of lakes or to the wadeable parts of rivers. Deeper water requires different sampling apparatus. Various forms of dredge or grab have been used to obtain qualitative samples from lakes or deep pools in rivers. The efficient operation of these devices depends very much on the nature of the lake or river bottom and they mostly can only be used on soft sediments (mud or sand). Elliott and Drake (1981a, 1981b) provide a very thorough examination of the biases and limitations of these devices. Rough, stony bottoms in deep water generally defeat such equipment.

11.3.2 Quantitative methods

A much wider range of sampling equipment has been developed and used for quantitative sampling. It is not possible to itemise the very many devices which have been described in the literature. Fortunately an excellent bibliography is available (Elliott et al. 1993) in which the various samplers have been categorised broadly as: nets and quadrat samplers (e.g., the well-known Surber and Hess samplers); trawls and dredges; grabs; corers; suction and air-lift samplers; and electroshockers. Illustrations of the lighter equipment are given by Merritt et al. (1996), Hauer and Resh (2006), and Brinkhurst (2002). Heavy equipment, particularly large grabs and dredges, is essentially identical to that used by marine workers and this has been comprehensively illustrated by Eleftheriou and Holme (1984) and see Chapters 6 and 9 in this volume.

Electroshocking is most commonly used for catching fish, but can stun larger invertebrates such as crayfish and shrimp (see Box 11.1). These taxa can move quickly when approached and are likely to elude other samplers. Smaller macroinvertebrates such as the larvae of aquatic insects generally move more slowly and can be taken with a range of samplers from the categories described above. Surber samplers consist of a quadrat with a downstream net attached, while a Hess sampler is an open-ended cylinder which admits water through mesh on the upstream side and has a net attached on the downstream side. Both devices are used in running water only. The enclosed area (often 0.04–0.1 m²) is disturbed by hand and the light organic material and invertebrate fauna are washed by the current into the downstream net. Open-ended boxes with a solid wall, usually delineating a circular or square patch of habitat, are used in shallow still water. The enclosed patch is swept with a small net or the contents are sucked out and passed through a net (e.g., the Boulton suck sampler, Boulton 1985). These quadrat type samplers work on a range of bottom types in shallow water. In rocky rivers coring is also possible but this is done using frozen cores (Pugsley and Hynes 1985) or specialised mechanical devices (Williams and Hynes 1974). The aim in such studies is often the delineation of the vertical distribution of invertebrates within the stream bed, as well as to obtain estimates of mean density.

If the bottom consists of soft sediments such as mud or fine sand, then mechanical grabs or corers can be used. These are typically used in the deep waters of lakes where sediments are likely to be soft and fine; Brinkhurst 2002 gives a critical account of these devices. Grabs extract fixed areas of surface sediments, which must then be sieved to separate the invertebrate fauna. Corers retain cylindrical lengths of bottom sediments enabling the vertical distribution of invertebrates within the sediment as well as their spatial distribution within the water body to be studied. Trawls and dredges have been most commonly used in marine habitats, but have also been used in lakes to provide qualitative samples. If they are drawn over a fixed area they may provide a quantitative sample, but have been less used in this manner.

Coarse sediments in deep water either in lakes or rivers are perhaps the most difficult from which to take quantitative samples. Various types of suction or air-lift samplers can be used in this situation. Suction samplers use small water pumps to suck water and specimens from a given area. This material is then trapped in a closed container from which excess water escapes through a net or fine mesh. In fresh water habitats air-lift samplers have been used most commonly to sample deep coarse sediments (Drake and Elliott 1983). Compressed air is expelled from nozzles on the bottom end of a 1–2 m tube (known as a riser tube), stirring up the bottom sediment and creating a mixture of sediment, water, and air. This mixture is less dense than the surrounding water and flows up the riser tube into a net. Large air-lifts can be operated from a boat (Drake and Elliott 1983), such as by the UK Environment Agency for routine macroinvertebrate sampling along rivers, while smaller, lighter versions have been used by divers (Figure 11.1b) (Barton and Hynes 1978; Marchant and Grant 2015).

Sedentary bottom taxa such as bivalves may be caught in any of the devices mentioned so far. However, large bivalves may be too large and too sparsely distributed to be caught efficiently and hand collection along transects by divers has been used (Strayer and Malcom 2006) to quantify populations. Also in clear shallow water it may be possible to observe and count specimens visually using a viewing tube or glass-bottomed bucket (Figure 11.1c). If it is known that the taxa of interest are readily visible or occur only on the surface of sediments, then quantitative samples can be taken by placing a quadrat on the sediment and counting the numbers of specimens within.

Physically complex habitats such as submerged branches at the edge of water bodies or large woody debris in deeper water have also been quantitatively sampled (Benke et al. 1984). The submerged wood is either removed from the water and the invertebrate fauna collected by brushing or scraping a fixed area of the wooden surface into a net; or can be sampled *in situ* using a snag bag, which envelops a given area of the submerged structure (Growns et al. 1999). An additional consideration is that the amount of wood available for colonisation must itself be quantified before invertebrate densities can be related to a unit area of the river or lake. Similarly, many studies have examined the macroinvertebrate communities inhabiting leaf litter because leaf litter fuels many aquatic food webs (e.g., Yule et al. 2009). Hand nets are used to collect leaves from litter that is either fully or partly submerged (e.g., a 20 × 20-cm area); when leaf litter is sparse then leaves are collected from various spots within a few metres.

Material collected by any of these techniques often requires some preliminary processing in the field such as the removal of large stones, other detritus or silt. This can be achieved by washing the lighter organic material into sieves leaving behind the inorganic residue or in the case of silt by sieving the whole sample. The size of specimens retained by these techniques is obviously dependent on the mesh size of the nets or sieves: 250–500 μm mesh is commonly used. Finer mesh (100 μm) may be required if the smallest instars of certain crustaceans or aquatic insects need to be retained. (The life spans of the smallest instars are generally very brief and it may not be worth attempting to catch these quantitatively.) Coarser mesh (500 μm to 1 mm) is often used when community composition is the target because it retains few of the smallest instars. Small instars of many aquatic insects are often difficult or impossible to identify with confidence and in studies of community composition are commonly ignored.

11.3.3 Removal sampling

For quantitative samplers it is obviously important to know the efficiency with which specimens can be extracted from their habitat. Unfortunately, efficiency or probability of capture has seldom been considered during field sampling, but can be estimated by taking repeated catches (each with the same degree of effort) from a given sampling area. This procedure is known as removal sampling, whose underlying theory is well understood (Seber 1973; Henderson and Southwood 2016). The technique assumes that decreases in population size at the sampling point only occur by capture, that the probability of capture (p) is constant from catch to catch and that the population being sampled is closed (i.e., no immigration or emigration). Under these assumptions, the number in each successive catch is linearly but inversely related to the sum of the previous catches and the slope of the line is an estimate of p. Once p is known then the percentage of the population captured in a given number of catches can be estimated and the total numbers in the sampling area calculated.

Removal sampling in freshwater habitats has most frequently been used to estimate fish populations in streams (where the technique is often referred to as depletion sampling). Carle and Maughan (1980) were the first to apply this technique to benthic invertebrates. They sampled river benthos with a sampling device similar to a Hess sampler in order to estimate p for a range of lotic macroinvertebrate families. They showed that p varied considerably. Taxa living within the sediments (e.g., certain Oligochaeta, Tipulidae, Chironomidae) generally had the lowest probabilities of capture (0.19–0.47), while those on the surface of the sediment (e.g., Heptageniidae (Epehemeroptera), Hydropsychidae (Trichoptera), Physidae (Gastropoda)) had higher probabilities (0.71–0.90). Marchant and Hehir (1999a) applied the same technique of removal sampling to stream benthos, but in their case sampled with a hand net. They demonstrated that there was no significant non-linearity in the relation between the successive catches and the sum of the previous catches and could make robust estimates of p for a range of Trichoptera, Caenidae and Leptophlebiidae (Ephemeroptera), Gripopterygidae (Plecoptera), and Gammaridae. For all these groups p was high (0.66–0.81). Marchant and Hehir (1999a) concluded that it was possible to take quantitative samples with a hand net of all but the most mobile taxa. This was novel because it had been generally assumed, as noted above, that hand nets were useful for taking qualitative samples, but were not suitable for making quantitative estimates of population density.

Despite the clear utility of removal sampling, it has seldom been used for estimating population sizes of freshwater invertebrates. However, Elliott (2008) estimated the size of the medicinal leech population (*Hirudo medicinalis*, a rare freshwater invertebrate) in a small tarn (0.25 ha) by repeatedly netting the whole shoreline. Each removal operation or catch took 1 hour, with 30 minutes between catches. A total of five catches on each sampling occasion removed essentially all the leeches from the tarn. The area of

Box 11.1 Estimating crayfish abundance using removal sampling: by Dan Chadwick and Eleri Pritchard, PhD students at University College London; and Paul Bradley, of PBA Applied Ecology Ltd

Removal sampling was used to estimate the population density (and size distribution) of Signal crayfish (*Pacifastacus leniusculus*) in a stream in the Yorkshire Dales, UK. The aim was to compare the accuracy of conventional survey techniques for estimating crayfish population size (traps, hand searches) with removal sampling.

A 15 m² headwater stream (2 m wide, 7.5 m length, 10 cm average depth) section was de-watered (also called draw-down method), after which large boulders and cobbles and all accessible crayfish were removed. The survey section was allowed to re-wet twice, allowing for three crayfish removals or 'sweeps'. A strong depletion in the population was observed: 1,341, 227 and 88 crayfish were caught in the first, second, and third sweeps respectively. The number in each catch (y) was linearly related to the sum of the previous catches (x): y= 1337.5–0.81x; r^2 = 0.99; and thus the assumptions of the technique were met. The probability of capture (*p*) was 0.81 and the total population was estimated as 1,651 crayfish or 110 individuals m^{-2}. Traps (modified 5 mm mesh traps) caught only 188 individuals in 40 trap nights; while hand searching (four searches, each with 250 stones turned) caught 1,048. Juvenile crayfish (carapace length, CL < 12 mm) formed 72 per cent of the total catch, but only 20 per cent of individuals caught by hand and 0 per cent of those in traps. Thus most juvenile crayfish were not caught by hand or by traps.

This study highlights the ineffectiveness of trapping as a technique for quantifying and controlling Signal crayfish populations.

the tarn was much larger than that commonly associated with quantitative samplers in freshwater. The larger sampling area may be advantageous when dealing with taxa that can move fairly quickly such as crayfish. See Box 11.1 for an example of removal sampling applied to a population of invasive crayfish in the UK.

According to Seber (1973) *p* should be >0.5 if the numbers in the total population are to be estimated without significant negative bias. Generally, removal techniques will only provide accurate estimates of total density if a large percentage of the population is captured. Thus, the population densities of some macroinvertebrate taxa are unlikely to be estimated well if their capture efficiencies are too low. The advantage of removal sampling is that it demonstrates readily which taxa are not well sampled. Anyone using quantitative samplers needs to be convinced that their device does indeed catch their target taxa with a high probability of capture. The efficiencies of Surber or Hess samplers, for instance, have seldom been assessed despite the fact that they are typically considered to take quantitative samples. It is assumed that all the specimens trapped within these samplers are completely removed into the capture net. Often, operators state that sampling was continued until no more individuals were observed within the sampling quadrat or box. However, this is a subjective assessment, especially with very small taxa, and removal sampling would indicate with some certainty whether a sufficient amount of effort has been expended. More importantly, by adopting a standard amount of effort that catches a known percentage of individuals an operator might avoid such laborious attempts to trap all observable specimens.

Removal sampling cannot be undertaken with corers or grabs because they extract an entire portion of sediment from the benthic habitat, thus precluding any repetition. However, comparisons can be made between these sampling devices to discover how they vary in their ability to catch different taxa. Elliott and Drake (1981a) and Brinkhurst (2002) provide valuable discussions of these matters.

11.4 Laboratory processing of samples

Processing of benthic samples is a tedious and lengthy task. When undertaking regular sampling of invertebrate communities or populations, probably 90 per cent of the total effort is in the laboratory. Most specimens of freshwater macroinvertebrates are small and thus samples must be searched under low magnification in order to remove individuals from the organic and inorganic detritus that is always present. There are two ways to speed up this process: removal of inorganic material by flotation of specimens; examination of a known fraction of the whole sample, commonly known as subsampling. As many studies require repeated sampling of the same population or community through time, it is essential to synchronise effort in the lab with that in the field. Otherwise large backlogs of samples accumulate which become an ever more difficult hurdle to overcome in completing the study.

11.4.1 Flotation

If samples contain large amounts of mineral or other inorganic particles, then flotation can be a remarkably effective way of removing this material. The sample is mixed with a relatively dense liquid (i.e., one that has a specific gravity higher than fresh water, such as a saturated solution of calcium chloride or a concentrated solution of glucose). Organic material, including invertebrates, has a specific gravity close to that of water and thus will float in such solutions and can be poured off leaving the inorganic residue behind. (Further details are given by Hynes 1970 and Hellawell 1978.) If this process is repeated several times the organic fraction can be rapidly separated from the inorganic. Sorting then becomes less time consuming because the specimens are more concentrated and not dispersed through large amounts of inorganic detritus. It is important to remember that taxa enclosed in mineral cases (e.g., caddis flies) or in shells (e.g., molluscs) may not float readily and the residue will need to be searched for them. If there are large amounts of organic detritus in samples, then flotation will not separate specimens from such detritus because both will float off together. Flotation before extraction of specimens is a widely used preliminary treatment for samples from community studies (Bird and Hynes 1981; Marchant et al. 1985) or those of populations of single taxa (Marchant and Hehir 1999b).

11.4.2 Subsampling

If large numbers of individuals are expected in individual samples (say >200–300) then subsampling is always likely to save considerable time. Various devices have

Figure 11.2 (a) A box subsampler, originally described by Marchant (1989). The lid which makes the box watertight is not shown. In this version an inner box sits within a larger outer box (37.5 × 37.5 × 15.0 cm). The bottom of the inner box is divided into 100 cells and is screened with 150 μm mesh. Once the subsample has been taken the inner box is removed from the outer box and the remaining sample washed out. In some versions of this device only one box is used. (b) A Perspex sorting tray for use under a dissecting microscope. The model shown (external dimensions: 77 × 77 × 15 mm) can hold just under 50 ml. Photos taken by R. Marchant.

been designed for subsampling benthic samples, some quite complex to build or use (Waters 1969), others simpler such as a box subsampler (Marchant 1989). The latter is a watertight box, the bottom of which is divided into 100 cells, and within which a sample can be dispersed by shaking (Figure 11.2a). A subset of cells is selected (10 per cent is often convenient) and the contents removed by suction. Marchant (1989) stated that the whole operation from introducing the sample into the subsampler to flushing out the remaining 90 per cent took an experienced operator about 10 minutes. Provided specimens are at least randomly distributed among the cells then the precision of the estimate of total numbers depends only on the number of animals counted in the subsample. Thus a count of 100 has a standard error (SE) of 10 per cent; a count of 400 has a SE of 5 per cent.

Subsampling is most likely to be used for samples in which all the taxa are of interest, such as in studies of community composition. However, in such cases it is not straightforward to set a suitable level of precision if taxa with very different levels of abundance are encountered. There would be little point in counting 100 of each of the rarer taxa as this could require examination of the whole sample and thus defeat the purpose of subsampling. Differences in community composition are likely to be most reliably shown by the abundant and the moderately abundant taxa and the rare taxa will add little to such patterns. Therefore, taking a subsample of fixed size is unlikely to obscure major trends.

If the study concerns a single species or a group of related species then subsampling is unlikely to be used, unless thousands of the target group are encountered. In such a case a subsample should contain at least 100 individuals, if not 400, to ensure a high precision of the estimate of total numbers in a single sample.

11.4.3 Extraction of specimens

Specimens are most efficiently extracted under a microscope by using a sorting tray (Figure 11.2b), which holds the sample in a narrow channel about the width of the field of view under low magnification. The channel zigzags across the tray and by following the channel the observer can be confident of scanning the whole sample. The difficulty with samples placed in petri dishes or similar containers is that there is no obvious beginning or end to the sample. The other benefit of the sorting tray is that the sample can be scanned twice in opposite directions. The number of individuals extracted in the second scan divided by the number in the first scan is an estimate of the probability of missing a specimen (q) (Marchant and Hehir 2000; Henderson and Southwood 2016). Thus q^2 is the probability of missing a specimen in two scans and $1-q^2$ represents the efficiency of picking (i.e., the percentage of specimens extracted). This information is not commonly collected. Many seem to believe that there is no possibility of missing specimens provided sufficient effort is used. However, small cryptic species or small instars are easily obscured by detritus and overlooked. The two-count method allows for this loss to be estimated. If the aim of the study is an analysis of population dynamics, then underestimates of the density of early stages or instars of a generation or cohort could result in inaccurate mortality estimates.

Staining of specimens may also speed up their extraction. Rose Bengal is commonly used (Hellawell 1978) and is preferentially taken up by organic material, turning specimens a deep pink, a colour which stands out well against unstained sediment.

11.5 Some analytical considerations of data from invertebrate samples

Quantitative data are the only basis from which a deeper understanding of community or population dynamics can arise. This has long been recognised. Hynes (1970) stated that 'it is essential for a full understanding of the ecology of any environment to have some measure of the abundance...of the species'. However, he also acknowledged the uncertain basis of much numerical data on lotic invertebrates. He indicated that common sampling techniques can be biased to a greater or lesser extent against certain taxa and that species commonly showed a large amount of variation in their densities in a single apparently uniform habitat such as a riffle. The latter is undoubtedly a consequence of a common phenomenon: a clumped spatial distribution.

Since Hynes' critique a large number of quantitative studies of individual species or groups of related species have been carried out. Despite taxa showing clumped distributions, stable trends in density with time have often been shown. For instance, those populations with well-defined cohorts or generations show a continual decrease in density, as expected, if the population is closed and there is no recruitment. In some studies, densities have been portrayed on a logarithmic rather than an arithmetic scale (e.g., Elliott 1981; Marchant and Hehir 1999b). Confidence limits on a log scale, given as a percentage of the mean, are much reduced compared with those on an arithmetic scale; and the data are more normally distributed. These are beneficial characteristics before applying statistical tests. Indeed, a log scale is probably the most natural way to analyse population changes, because it emphasises relative changes in population density (e.g., a per capita rate of mortality or natality) (Henderson and Southwood 2016). An arithmetic scale, on the other hand, shows only absolute changes in population size, which are rarely the focus of analysis.

Numerical data on species abundances derived from community studies are also commonly converted to logarithms (or a similar scale such as double square root) before being subjected to multivariate analysis (Clarke and Warwick 2001). This is commonly done to lessen the influence of the few abundant species, a common situation, and increase the influence of the much more numerous moderately abundant species. Again, this illustrates the importance of considering the numerical scale on which the data will be analysed. The imposition of an arithmetic scale on community data can seriously distort the ecological picture and can lead to conclusions about community similarity being based on the abundance of one or a few very abundant taxa. In such a situation one could not claim that conclusions were derived from the community as a whole.

Qualitative data have traditionally been considered as less amenable to rigorous analysis. However, in the last three decades the composition of stream benthic communities over wide spatial scales has been studied using qualitative samples in the UK, North America, and Australia (e.g., Wright 2000). One of the results of these studies has been the creation of predictive or reference condition models (Bailey et al. 2004) that quantify the probability of encountering a specific taxon given knowledge about various environmentally stable predictor variables (i.e., those unaffected by human induced disturbance), often those relating to physical and chemical characteristics of a site within a catchment. Such models enable users such as management agencies to predict or anticipate biological conditions and then directly sample a site using standardised (e.g., hand net) techniques to determine whether the site conforms with expectation. This is a major advance in that it uses prediction to draw conclusions from field data. It was not originally envisioned that simple qualitative samples could be analysed in so sophisticated a manner. An alternative approach is to develop numerical indices or metrics that relate the tolerance of macroinvertebrate taxa to specific disturbances (e.g., silt, flow, toxins, pollutants) (e.g., Dedieu et al. 2015). These metrics are combined into a multimetric measure, which is then applied to data on community composition from qualitative samples to determine what environmental disturbances are evident.

References

Anderson, J.T., Zilli, F.L., Montalto, L., Marchese, M.R., McKinney, M., and Park, Y.-L. (2013). Sampling and processing aquatic and terrestrial invertebrates in wetlands. In: J.T. Anderson and C.A. Davis (eds) Wetland Techniques, Volume 2, Invertebrates. Springer, Dortrecht.

Bailey, R.C., Norris R.N., and Reynoldson T.B. (2004). Bioassessment of Freshwater Ecosystems: using the reference condition approach. Kluwer Academic, Boston.

Barton, D. and Hynes, H.B.N. (1978). Wave-zone macrobenthos of the exposed Canadian shores of the St Lawrence great lakes. *Journal of Great Lakes Research*, 4, 27–45.

Batzer, D. and Boix, D. (eds) (2016). Invertebrates in Freshwater Wetlands: An International Perspective on their Ecology. Springer International, Switzerland.

Benke, A.C., Van Arsdall, T.C., Gillespie, D.M., and Parrish, F.K. (1984). Invertebrate productivity in a subtropical blackwater river: the importance of habitat and life history. *Ecological Monographs*, 54, 25–36.

Bird, G.A. and Hynes, H.B.N. (1981). Movement of immature aquatic insects in a lotic habitat. *Hydrobiologia*, 77, 103–12.

Boulton, A.J. (1985). A sampling device that quantitatively collects benthos in flowing or standing waters. *Hydrobiologia*, 127, 31–9.

Brinkhurst, R.O. (2002). The Benthos of Lakes. The Blackburn Press, New Jersey.

Carle, F.L. and Maughan, O.E. (1980). Accurate and efficient estimation of benthic populations: a comparison between removal estimation and conventional sampling techniques. *Hydrobiologia*, 71, 181–7.

Chessman, B.C. (1995). Rapid assessment of rivers using macroinvertebrates: a procedure based on habitat specific sampling, family level identification and a biotic index. *Australian Journal of Ecology*, 20, 122–9.

Clarke, K.R. and Warwick, R.M. (2001). Change in Marine Communities: an approach to statistical analysis and interpretation, 2nd edition. PRIMER-E, Plymouth.

Davies, P.E. (2000). Development of a national river bioassessment system (AUSRIVAS) in Australia. In: J.F. Wright, D.W. Sutcliffe, and M.T. Furse (eds) Assessing the Biological Quality of Fresh Waters: RIVPACS and other techniques. Freshwater Biological Association, Ambleside.

Dedieu, N., Clavier, S., Vigouroux, R., Cerdan, P., and Céréghino, R. (2015). A multimetric macroinvertebrate index for the implementation of the European water framework directive in French Guiana, East Amazonia. *River Research and Applications*, 32, 505–15.

Drake, C.M. and Elliott, J.M. (1983). A new quantitative air-lift sampler for collecting macroinvertebrates on stony bottoms in deep rivers. *Freshwater Biology*, 13, 545–59.

Drake, C.M., Lott, D.A., Alexander, K.N.A., and Webb, J. (2007). Surveying Terrestrial and Freshwater Invertebrates for Conservation Evaluation. Natural England, Sheffield.

Eleftheriou, A. and Holme, N.A. (1984). Macrofauna techniques. In: N.A. Holme and A.D. McIntyre (eds) Methods for the Study of Marine Benthos, 2nd edition. Blackwell Scientific Publications, Oxford.

Elliott, J.M. (1977). Some methods for the statistical analysis of samples of benthic invertebrate. 2nd edition, Scientific Publication No. 25. Freshwater Biological Association, Ambleside, UK.

Elliott, J.M. (1981). A quantitative study of the life cycle of the net-spinning caddis *Philopotamus montanus* (Trichoptera: Philopotamidae) in a Lake District stream. *Journal of Animal Ecology*, 50, 867–83.

Elliott, J.M. (2008). Population size, weight distribution and food in a persistent population of the rare medicinal leech, *Hirudo medicinalis*. *Freshwater Biology*, 53, 1502–12.

Elliott, J.M. and Drake, C.M. (1981a). A comparative study of seven grabs used for sampling benthic macroinvertebrates in rivers. *Freshwater Biology*, 11, 99–120.

Elliott, J.M. and Drake, C.M. (1981b). A comparative study of four dredges used for sampling benthic macroinvertebrates in rivers. *Freshwater Biology*, 11, 245–61.

Elliott, J.M., Tullett, P.A., and Elliott, J.A. (1993). A New Bibliography of Samplers for Freshwater Benthic Invertebrates. Occasional Publication No. 30. Freshwater Biological Association, Ambleside.

Freshwater Biological Association (FBA) (2014). Review of Techniques for Sampling Benthic Macro-invertebrates in Deep Rivers. Science Report, Environment Agency.

Growns, J.E., King, A.J., and Betts, F.M. (1999). The Snag Bag: a new method for sampling macroinvertebrate communities on large woody debris. *Hydrobiologia*, 405, 67–77.

Hauer, F.R. and Resh, V.H. (2006). Macroinvertebrates. In: F.R. Hauer and G.A. Lamberti (eds) Methods in Stream Ecology, 2nd edition. Academic Press, Amsterdam.

Hayslip, G. (ed.) (2007). Methods for the collection and analysis of benthic macroinvertebrate assemblages in wadeable streams of the Pacific Northwest. Pacific Northwest Aquatic Monitoring Partnership, Cook, Washington.

Hellawell, J.M. (1978). Biological Surveillance of Rivers. Water Research Centre, Stevenage.

Henderson, P.A. and Southwood, T.R.E. (2016). Ecological Methods. 4th Edition, John Wiley and Sons Ltd., Chichester.

Hynes, H.B.N. (1970). The Ecology of Running Waters. Liverpool University Press, Liverpool.

Marchant, R. (1989). A subsampler for samples of benthic invertebrates. *Bulletin of the Australian Society for Limnology*, 12, 49–52.

Marchant, R. and Grant, T.R. (2015). The productivity of the macroinvertebrate prey of the platypus in the upper Shoalhaven River, New South Wales. *Marine and Freshwater Research*, 66, 1128–37.

Marchant, R. and Hehir, G. (1999a). A method for quantifying hand-net samples of stream invertebrates. *Marine and Freshwater Research*, 50, 179–82.

Marchant, R. and Hehir, G. (1999b). Growth, production and mortality of two species of *Agapetus* (Trichoptera: Glossosomatidae) in the Acheron River, south-east Australia. *Freshwater Biology*, 42, 655–71.

Marchant, R. and Hehir, G. (2000). How efficient is extraction of stream insect larvae from quantitative benthic samples? *Marine and Freshwater Research*, 51, 825–6.

Marchant, R. and Hynes, H.B.N. (1981). The distribution and production of *Gammarus pseudolimnaeus* (Crustacea: Amphipoda) along a reach of the Credit River, Ontario. *Freshwater Biology*, 11, 169–82.

Marchant, R., Metzeling, L., Graesser, A., and Suter P. (1985). The organization of macroinvertebrate communities in the major tributaries of the La Trobe River, Victoria, Australia. *Freshwater Biology*, 15, 315–31.

Merritt, R.W., Resh, V.H., and Cummins, K.W. (1996). Design of aquatic insect studies: collecting, sampling and rearing procedures. In: R.W. Merritt and K.W. Cummins (eds) An Introduction to the Aquatic Insects of North America, 3rd edition. Kendall/Hunt, Dubuque.

Metzeling, L., Chessman, B., Hardwick, R., and Wong, V. (2003). Rapid assessment of rivers using macroinvertebrates: the role of experience, and comparisons with quantitative methods. *Hydrobiologia*, 510, 39–52.

Pugsley, C.W. and Hynes, H.B.N. (1985). A modified freeze-core technique to quantify the depth distribution of fauna in stony streambeds. *Canadian Journal of Fisheries and Aquatic Sciences*, 40, 637–43.

Samways, M.J., McGeoch, M.A., and New, T.R. (2010). Insect Conservation: A Handbook of Approaches and Methods. Techniques in Ecology and Conservation Series, Oxford University Press, Oxford.

Seber, G.A.F. (1973). The Estimation of Animal Abundance and Related Parameters. Griffin, London.

Strayer, D. and Malcom, H. (2006). Long-term demography of a zebra mussel (*Dreissena polymorpha*) population. *Freshwater Biology*, 51, 117–30.

Waters, T.F. (1969). Subsampler for dividing large samples of stream invertebrate drift. *Limnology and Oceanography*, 14, 813–15.

Williams, D.D. and Hynes, H.B.N. (1974). The occurrence of benthos deep in the substratum of a stream. *Freshwater Biology*, 4, 233–56.

Wright, J.F. (2000). An introduction to RIVPACS. In: J.F. Wright, D.W. Sutcliffe, and M.T. Furse (eds) Assessing the Biological Quality of Fresh Waters: RIVPACS and other techniques. Freshwater Biological Association, Ambleside.

Yule, C.M., Leong, M.Y., Liew, K.C., Ratnarajah, L., Schmidt, K., Wong, H.M., Pearson, R.G., and Boyero, L. (2009). Shredders in Malaysia: Abundance and richness are higher in cool upland tropical streams. *Journal of the North American Benthological Society*, 28, 404–15.

Part III

Ecosystem Dynamics, Conservation, and Management

Freshwater Populations, Interactions, and Networks

David M. Harper and Nic Pacini
Corresponding author: dmh@le.ac.uk

12.1 Introduction

It is impossible, in the face of the multiple anthropogenic pressures experienced by freshwater bodies, to consider species in isolation from their environment, due to the strong interdependence of each (Gessner and Tlili 2016; Atkinson et al. 2017; Moss 2018). That is not to say that simple recording of species presence or absence is not valuable, because a single record of a species in a particular place, at a particular time, has a value directly proportional to the number of other records of the same species. Long-term records of the first occurrence each year of a species characteristic in the northern hemisphere, have proved invaluable in quantifying the effects of climate change (Thackeray et al. 2016). This study analysed 10,003 aquatic and terrestrial phenological data sets against climate records; the longest were for 50 years. Some were single-species records, such as the day of the first record each year in the English Lake District, 1965 to 1995, of the appearance of adult alderfly (*Sialis lutaria*). Most monitoring records, however, additionally involve some estimate of population size. Many of the aquatic records used by Thackeray et al. are derived from the UK's Freshwater Biological Association monitoring of the English Lakes' phytoplankton; one example of metrics which they used is the first day of the year on which 25 per cent of cumulative seasonal abundance of 13 phytoplankton classes was reached, collected from 1961 to 2005.

Large data sets produced by biological monitoring, which enable the temporal or spatial changes of species' distribution and/or population size to be appreciated, are well established in the UK (Pocock et al. 2015; Chapter 14 in this volume) and developing elsewhere (e.g., Conrad and Hilchey 2011 using volunteers). For the UK, data collected by surveyors is coordinated by a collaboration between government-funded Biological Records Centres and the charitable National Biodiversity Network (see BRC and NBN websites) encompassing voluntary and statutory organisations concerned with biodiversity conservation. The International Union for Conservation of Nature's BioFresh website brings together many international databases for freshwater

Harper, D. M. and Pacini, N., *Freshwater populations, interactions, and networks*. In: *Freshwater Ecology and Conservation: Approaches and Techniques*. Edited by Jocelyne M. R. Hughes: Oxford University Press (2019).
© Oxford University Press 2019. DOI: 10.1093/oso/9780198766384.003.0012

species and environmental variables which can be used to investigate freshwater populations, interactions, and networks.

This chapter discusses the important parameters that should be considered by anybody—professional scientist or voluntary surveyor—in studying the way freshwater ecosystems work, from species to communities and ecosystems to landscapes. The detailed methods needed will depend upon the species chosen and the habitat, especially whether lotic or lentic, and the resources available; for advice about standard methods the reader is referred to excellent practical textbooks, such as Hauer and Lamberti (2017) and Lamberti and Hauer (2017) for lotic waters, Schwoerbel (1987) for lentic waters, Wetzel (2001) for both lotic and lentic, Silk and Ciruna (2005) for biodiversity conservation and Part 2 of this volume. More recent publications, conference proceedings or technical reports, address methods needed in legal environmental quality assessments, such as the EU Water Framework Directive (Furse et al. 2006; Solimini et al. 2006); or investigations where technological methods are deployed (Taylor et al. 2017; Trebitz et al. 2017). These methodological texts are most appropriate to study after having developed an understanding of freshwater biology from a broader textbook such as by Moss (2018), Dodds and Whiles (2010), or volumes 2 to 4 by Hutchinson (1967; 1975; 1993).

12.2 Species and populations

12.2.1 Population sizes, ranges, distributions, and dynamics

The density and abundance of individuals composing a given population can be a very valuable parameter for assessing the conservation status of a species. Populations represent the biological level of choice for the detection and prevention of local adverse anthropogenic effects (USEPA 1998). A standard 'rule of thumb' in ecological risk assessment practice establishes that a 20 per cent reduction in the number of individuals of a given population, and/or a 20 per cent change in selected biological parameters relative to population survival and/or fitness, is significant at the screening level to recommend more detailed investigations. Internationally agreed criteria for the definition of threatened and endangered species (terrestrial and freshwater) were established by the International Union for Conservation of Nature (IUCN) in 1994 and updated in 2001 (IUCN 2012). Levels of threat to species' survival were established as a function of the proportion of observed population size reduction; respectively ≥ 50, ≥ 70, ≥ 90 for vulnerable, endangered and critically endangered. Many freshwater conservation efforts have been directed towards the monitoring and protection of single species; for example, the original criterion for the declaration of a wetland site of international importance under the Ramsar Convention, adopted in 1971, was that it should hold 1 per cent of the global population of a named water bird species at any time of year. Nevertheless, a drastic change in the distribution and range of a population is an important subsidiary criterion for officially establishing a species' extinction threat by the IUCN. More recent conservation effort, however, has realised that preserving a species' habitat can be even more effective than focussing attention on the number of individuals. This has led to the widely used and promoted 'ecosystem approach' defined in the Convention on Biological Diversity in 1992 (see 'ecosystems approach' on CBD website). There are now many criteria for Ramsar Convention site designation, including several ecosystem value characteristics (see Ramsar website), recognising the indirect role that sustainable human uses have in wetland conservation.

Of particular note, in the context of establishing species distributions, richness, status, and threats for a particular site or region, are Rapid Assessment (RA) techniques. These have been developed most famously since 1990 by Conservation International's RA Programme, where groups of scientists and surveyors will explore an area for 2–4 weeks recording all species that can be found using standard methods and techniques. These RAP surveys cover different habitats but many include wetlands, rivers, and lakes (e.g., Pantanal RAP). See Conservation International's RAP website, and Larsen (2016) for a review of the standard techniques including freshwater RAPs (or AquaRAP).

Migratory wetland bird species provide a good example of the importance of understanding population sizes and ranges. They are usually trans-national, so that their conservation requires collaboration, in both research and management (Baldwin 2011). For example, the International Water Bird Census run by Wetlands International (see wetlands.org website) is the longest running (51 years) 'citizen science' project in the world. It is composed of four separate regional schemes, each covering one of the major north to south migratory flyways, with the Africa to Eurasia flyway further strengthened by the international agreement on the conservation of African-Eurasian Migratory Waterbirds (AEWA), signed under the United Nations Environment Programme (UNEP) Convention on Migratory Species (CMS). Fifty years of census data provides a unique source of information about the status and dynamics of bird populations, leading to improved strategies for conservation of all habitats in a species' range (e.g., Martin et al. 2007).

The lesser flamingo (*Phoenicopterus minor*), while not a migratory bird, but a nomadic bird of the Old World tropics, provides an example of the value of long-term monitoring of population numbers, as well as the use of technological methods to support population censuses. *P. minor* is one of three deep-keeled flamingo species, the only vertebrates in the world which subsist on cyanobacteria and algae (Torres et al. 2014). It is a meta-population of about 2.6 million birds, with four distinct populations, three in Africa and one in Gujarat, India. The largest population is about 1.5 million birds across four to five countries of East Africa (Childress et al. 2008) where it is considered a 'landscape species' (Plumptre et al. 2007). It has been annually counted at its accessible lake locations in Kenya, its major country, since 1990 (birds at Tanzanian and Ethiopian lakes are counted rarely). A second population, in southern Africa, of about 0.5 million birds, is counted less intensively. Changes in the numbers of southern African birds in the late twentieth century suggested an overall decline, followed by a sudden increase through movement into the region from outside (Simmons 2000), which was attributed to movement south from East Africa. The Kenyan census does not cover large, transient pans in the north (e.g., Lake Logipi just south of Lake

Turkana) and south (e.g., Lake Natron lagoon on the Tanzanian border). Several times in the 25 years of census, bird numbers have fluctuated hugely from a few thousand to more than 1.5 million. It has been suggested that such numbers represent a population decline, particularly since the species has a single breeding site in East Africa, in the middle of the large Lake Natron in Tanzania, which is not accessible to observers; and has been linked to unexpected population mortalities that occurred (or were first recorded) at different lakes over the last five decades (Harper et al. 2003). The deaths were initially linked to heavy metal poisoning (Kairu 1996) and pesticides (Motelin et al. 2000), but after these were discredited, to cyanobacterial toxin poisoning (Lugomela et al. 2006). The most likely explanation is that all these and other stressors may weaken birds' immune systems, particularly after a long flight from a lake declining in food quality (Harper et al. 2016), making them susceptible to infectious diseases (Sileo et al. 1979; Kock et al. 1999; Oaks et al. 2006). The main feeding lakes periodically experience crashes of *Arthrospira fusiformis* populations, the primary food of *P. minor*. When this happens, they have to move to less accessible shallow lakes and lagoons to feed. Remote sensing analysis of the food quality in the lakes suggests that this happened in early 2012, when Lake Logipi had about 100 km^2 of food at a time when the main lakes, Nakuru (about 50km^2), and Bogoria (about 40km^2) had very little and the annual census showed only 9,130 birds (Tebbs et al. 2015).

There are, therefore, problems in understanding the dynamics of *P. minor*'s population size and distribution, made worse by a lack of knowledge of their potential wetland habitats. Satellite tracking of seven birds through three years showed one of those birds spending almost six months at a large temporary wetland in the middle of Tanzania where the species had never before been recorded (Childress et al. 2004; 2007). Coordinated counts of the lakes where *P. minor* exists are rare in Tanzania and almost non-existent in Ethiopia, so although birds satellite-tagged in Kenya were never recorded in Ethiopia, which may have a quasi-isolated population, there are still great difficulties in coming to a reasonable estimate of total numbers in East Africa. Genetic studies of mitochondrial DNA (Zaccara et al. 2008; 2011; Paracharya et al. 2015) show that for each generation, a few individuals of the populations of East and southern Africa exchange with each other and with that in India. Limited circumstantial evidence suggests something similar occurs with the West African population, as evidenced by a bird ringed as a chick at Lake Magadi Kenya, on 30 October 1962 was recovered on 28 September 1997 near Laayoune, Western Sahara. This not only revealed how old wild birds can live to, but it also showed their capability for moving—in this case, 3,850 miles. Very little is otherwise known about the West African population, believed to be around 20,000 (Moreno-opo et al. 2013). Estimating numbers is a problem for a species with many fixed-habitat locations, if they are inaccessible. Censusing can be done by digital analysis of aerial photographs (Morales-Roldan et al. 2010) but not yet from satellite images for flamingos, although that may happen in the near future (Fretwell et al. 2017).

The detailed example above, of a single species with a trans-national distribution, illustrates the challenges of monitoring a species' population size and distribution pattern and the extent to which modern technology (genetics, remote sensing, and satellite tagging) can add to the simple, but essential, technology of visual census. It is a species whose population dynamics are a complete mystery because fundamental population growth data—births, deaths, immigrants, and emigrants—are unknown and for the foreseeable future, unknowable. Many migratory wetland species are also poorly understood, because of the difficulties of quantifying those four parameters, and the dependence of evidence-based conservation management on having access to similar annual census data. Isolated populations can be more easily studied however, than landscape scale meta-populations. Studies of fish population dynamics (Pitcher and Hart 1982) and migration (see 'Research' on the Atlantic Salmon Federation website) have provided us with tools that can be adapted to most freshwater animals (e.g., invasive crayfish) (Scalici and Gherardi 2007). Zooplankton populations have been studied intensively for many decades, both in their natural environment and laboratory or field mesocosms, in order to understand environmental controls of population dynamics (e.g., Hazelwood and Parker 1961; Angino et al. 1973; Armitage et al. 1973; Chapters 7 and 8) as well as indicators of change (e.g., Dupuis and Hann 2009). In situations where freshwater species are cryptic, occupancy modelling and logistic regression models can be used to improve limited knowledge on distributions of species (e.g., for freshwater shrimps in lowland Costa Rican streams (Snyder et al. 2016); freshwater mussels in the Tar River Basin, North Carolina, USA (Pandolfo et al. 2016); and for amphibians (Bailey and Nichols 2009)).

Important conservation issues arise when populations become fragmented and small. In this context is the discussion by Olden et al. (2011) on managed relocation of freshwater species—a final option for a species threatened by a changing climate or environmental conditions; and research demonstrating the successful reintroduction of the Australian trout cod to the Ovens River in south-eastern Australia (Lyon et al. 2012). Lakes are semi-isolated ecosystems, connected to other lakes only by their river systems in or out; or by aquatic species that are airborne as adults (e.g., dragonflies); or species carried by water birds. Many riverine fish species (and to an unknown extent, invertebrates) have become isolated by man-made barriers, such as weirs or dams (Clavero et al. 2004). Isolation, often combined with hydrological alteration, may threaten species survival beyond the extinction of a single local population (Dodds et al. 2004). Genetic tools have become important in establishing the diversity of isolated populations, both as a contribution to knowledge of the overall gene pool of a single species (e.g., native trout in the US, Pritchard et al. 2007) and guide to management (e.g., whether to isolate from non-native invasive species, Fausch et al. 2009). Assessing connectivity between stream systems, a difficult subject to interpret (Demars and Harper 2005) and manage, is considerably aided by using techniques in population genetics (Hughes et al. 2009).

12.2.2 Measuring life-cycle fluctuations and life-history patterns

Most population studies are designed to achieve accurate estimates of numerical changes over time or space, such as the examples given above. Other purposes, however, such as understanding the reasons for population declines, may require more detailed information about population parameters (e.g., births, deaths, immigration, and emigration), and the environment. Moreover, sub-lethal effects because of lack of food or presence of contaminants such as pesticides, may initially become apparent through recruitment of juveniles into a population before its numbers decline through death or emigration. Synthetic oestrogens, from oral contraceptives, transferred into freshwaters through treated sewage effluent, are now known to cause sexual changes in fish and low level population effects in invertebrates, apparent through their life history and demographic structure (Sousa et al. 2013). A good understanding of a species' life history pattern and reproductive capacity can also assist in predicting its recovery response after chemical damage (e.g., Sherratt et al. 1999).

The most important methods and techniques used in population conservation biology are applicable to all aquatic ecosystems. Thus, population ecology methodological texts such as by Maurer (2009) should be consulted to plan a detailed study; and MacKenzie et al. (2017) review the latest methods in occupancy modelling; for both amphibians and freshwater reptiles respectively, Dodd (2010; 2016) includes a number of chapters on the practicalities of conducting population surveys. Many classical studies cited by textbooks have used species that are well suited to examination of their populations, such as those that leave annual evidence of their growth, such as fish (otoliths, scales) and wetland trees (rings). A text of particular relevance to freshwater conservation is by Morris and Doak (2003), which introduces the application of population viability analysis (PVA), as a means of estimating the probability of survival of a population and predicting future trends under different strengths of mortality factors, thus enabling conservation actions to be more targeted.

12.2.3 Population-environment associations, functional feeding groups, and stressors in aquatic habitats

Population dynamics can best be understood against the 'habitat templet' of a species' life history (Southwood 1977). This has been done extensively for lotic invertebrates since the concept was first described, providing a much better understanding of their resistance and resilience to natural extremes (e.g., Townsend et al. 1997). The identification and cataloguing of species traits evolved to fit their templet (river invertebrates, Tachet et al. 2010; pond invertebrates, Céréghino et al. 2012; zooplankton, Hébert et al. 2016) has considerably improved quantification of the impact of environmental factors upon populations. An example is the impact upon invertebrates of sediment deposition on riverbeds (Descloux et al. 2014). Identifying species traits also improves our ability to predict species most at risk from known future changes and developments (e.g., impacts from water resource developments on fish) (Rolls and Sternberg 2015), and Cernansky (2017) argues that richness of species traits rather than species richness (number of species) should be used to assess ecosystem health.

A sub-set of the use of species traits includes adaptations to feeding strategies. Stream ecologists grouped freshwater organisms under groups of species utilising similar strategies, or guilds, over 50 years ago as a means of building up our understanding of the role that macroinvertebrates and fish play in energy transfers in freshwater environments (Covich et al. 1999). Application of the complete set of traits to all the species in a community can provide a more predictive approach to understanding community dynamics and resilience (Verberk et al. 2013). These ideas on using species traits to describe functional feeding groups have acted as building blocks for subsequent freshwater ecological concepts, such as the River Continuum Concept (Vannote et al. 1980), the Serial Discontinuity Concept (e.g., Ellis and Jones 2016), and the Flood Pulse Concept (e.g., Oliveira et al. 2006). The definition of traits is applicable across different river systems; it has been demonstrated that the same functional adaptations dominate in invertebrate assemblages inhabiting the same river biotope, even though they belong to different species in rivers across England and Wales (Demars et al. 2012).

Many stressors now impact natural freshwater ecosystems, almost none remain unaffected. For example, lowland rivers in the UK are affected by phosphorus enrichment from diffuse as well as point sources (Demars and Harper 2005; Demars et al 2005). Specific impacts on individual populations or overall ecosystem effects are largely unknown, since most river systems have not been carefully monitored in enough detail or over enough time. Moreover, in field monitoring it is very hard to separate the effects of one stressor from another. The impacts of many individual stressors have been studied by 'enrichment' experiments. A 16-year experimental study by Slavik et al. (2004) adding phosphorus to an Alaskan stream, showed that bryophytes replaced epiphytic algae as the dominant primary producers after eight years and that this change impacted, firstly, nitrogen metabolism via ammonia uptake, and secondly, invertebrates with some decreasing in abundance as the space for tube-building on bare rock decreased, while others increased as the availability of deposited fine particulate organic matter increased.

Experiments have been widely designed to detect single stressors, by keeping both control and impact identical in all other respects. The most robust experiments have been done using parallel stream systems, or artificial stream flume mesocosms along the bank. Such experiments have tested the effects of contaminants such as heavy metals (Brent and Berberich 2014), pesticides (Pestana et al. 2009), or antibiotics (Quinlan et al. 2011), upon target species. Standard procedures for establishing the toxicity of new compounds are based on the definition of threshold concentrations; for example, the concentration that kills ≥50 per cent of an experimental population. Ecotoxicologists rely on invertebrate species that are easy to culture in laboratory tanks such as *Daphnia pulex* (lentic species) or *Gammarus pulex* (lotic species) (Brock and Wijngaarden 2012).

The synergistic effects caused by multiple stressors in nature have been addressed by applying multivariate analyses to datasets including a wide a variety of environmental factors, but this has been very rarely carried out at population level or with long-term data sets (e.g., Bordalo et al. 2011); a European Union Framework research programme (Navarro-Ortega et al. 2015) seeks to unravel the effects of multiple stressors on ecosystems where water demand will create future water scarcity. Multiple stressors need to be understood at the community and ecosystem level if their management or prevention is to lead to restoration or amelioration of the ecosystem (e.g., Lake Victoria, Hecky et al. 2010). The combined impact of two or more stressors, even at 'moderate' individual concentrations, can be dramatic; it has been recently shown (Everall et al. 2017a) that populations of a mayfly (Ephemeroptera; *Serratella ignita*), in UK rivers, suffered severe mortality in egg hatching from suspended solids (SS) and soluble orthophosphate (ortho-P) in combination, greater than the effects of either alone—45 per cent in mesocosm treatments where SS was 25 mg l^{-1} and ortho-P 300 µg l^{-1} (6 per cent in controls); these are concentrations common in rural rivers in UK. In some freshwaters, a sequential effect of multiple stressors occurs, rather than synergistic interactions amongst multiple stressors, as demonstrated by Kelly et al. (2017) for communities of phytoplankton at Lake Simcoe. In this study, multivariate analyses were used to identify interannual changes in phytoplankton community structure, and biological stressors accounted for the most variation, followed by water quality and temperature stressors.

The impact of climate change induced stressors on freshwater organisms has been investigated via an elegant study by Weiss et al. (2018) using a combination of monitoring data and lab experiments. They researched whether the long-term effects of increased anthropogenic CO_2 caused an increase in pCO_2 in freshwaters. They used long-term data over 35 years for four freshwater reservoirs and found an increase in pCO_2 associated with a decrease in pH, demonstrating an accumulation of CO_2 in freshwater bodies. The effects of acidification/increase in pCO_2 on freshwater organisms was then investigated using the keystone zooplankton species, *Daphnia*, as a model organism. Laboratory experiments demonstrated that increased levels of pCO_2 affected the sensory abilities of *Daphnia* to react to specific chemical cues and form adequate inducible defences to predators. It was an increase in pCO_2 and not a decrease in pH that impaired predator perception, thus demonstrating that increased pCO_2 could significantly affect chemical communication between freshwater species and impact all trophic levels.

12.3 Communities and ecosystems

12.3.1 Species richness, diversity, rarity, and eDNA

Much of what has been written in the previous section has also referred to communities, rather than strictly to populations, because of the difficulty of studying a single population in isolation from other species. Even mesocosms, which seek to simplify environmental influences, inevitably replicate the communities of lakes or streams, unless they are maintained in a laboratory where mortality and reproduction can be controlled. Freshwater conservation frequently focuses on biodiversity as the 'number of species', because it is necessary to give simple messages to the public and politicians (Pacini et al. 2013). Biodiversity means, in its simplest form, species richness, either as the number identified or as a higher number estimated from the sampling effort (Huang et al. 2011). If the species richness from a surveyed site is used to predict the value in other sites, it often leads to limited results; but it acquires greater value when used on a large scale (e.g., Dehling et al. 2010), or with a long-term data set (see Section 12.1). Most systems that have been developed to quantify an indicator value from species presence at a location, have combined this with either a subjective or a semi-quantitative estimate of abundance. Water quality estimates in rivers worldwide have used this approach. The original 'Trent Biotic Index' (TBI) in the UK (Woodiwiss 1964; 1980) used the number of species found in a taxonomic group matched against a numerical value or score; this was higher for groups sensitive to pollution (oxygen depletion) such as the insect order Plectoptera, and highest if more species were found in that group. A trained biologist could thus obtain a single score for a site within about an hour, by recognising (not scientifically naming) the number of species present within the highest scoring taxon (Table 12.1). Two decades after its first development, as pollution monitoring became nationwide in the UK and also widespread within the EU and other industrialised countries, this index was considerably modified to enable lowland rivers to be included, which even in a clean natural state do not support Plectoptera. A system was developed which allocated a score between 1 and 10 to each family present, with 10-scoring families considered to be good-quality indicators as Plecoptera but also including lowland lentic families from orders such as Odonata and Coleoptera. This BMWP score system (Biological Monitoring Working Party) was the basis for the development of many other systems around the world (e.g., Gutiérrez-Fonseca and Lorion 2014). In the UK, the system has been developed further as a predictive tool—RIVPACS (River Invertebrate Prediction And Classification System; Clarke et al. 2003)—and to provide a measure of the impact of other stressors, such as low flows—LIFE (Lotic invertebrate Index for Flow Evaluation; Dunbar et al. 2010). It has been used further to develop a conservation index (Chadd and Extence 2004) and refined with use and re-examination of data as the number of samples built up (Martin 2004; Table 12.2). Other indices, also based upon the principles of the TBI, are used; the most common is the EPT (E = Ephemeroptera, P = Plecoptera, T = Tricoptera), particularly in the USA, which estimates water quality through the relative abundance of these orders of insects, calculated as the percentage of these orders to total taxa in a sample.

Abundance measures—or estimates—are rarely used in aquatic invertebrate monitoring or even in research, because of the difficulties of collecting quantitative or semi-quantitative samples in all habitats of lotic waters (Everall 2017b; see Chapter 11 in this volume). A very simple survey of a site may contain a list of species and subjective abundance data using the 'DAFOR' scale (Dominant, Abundant, Frequent, Occasional, Rare). Freshwater plants, however, are extensively measured using semi-quantitative scales and surveys of plants in aquatic environments may use quadrats of varying size as the sampling unit (see Chapter 9 in this volume). The widespread system of plant sociology (Whittaker 1962; Ellenberg 1988) identifies plant species and their relative abundance using a semi-quantitative scale; this has been used in conservation for many decades as plants strongly and predictably indicate a site's hydrogeochemistry. The Domin scale, named after its inventor, is widely used to describe the cover-abundance of a plant species from 1 (simple presence) through to a score of 10 (91–100 per cent cover) (Table 12.3). This has formed the basis of the UK's National Vegetation Classification, which for over 40 years has classified 295 plant communities in the British Isles, including aquatic communities, and enabled conservation to rapidly assess a site's value and its change over time (e.g., Fojt and Harding 1995; Rodwell 1997; Rodwell 1998; Rodwell 2006). It has enabled trans-national comparison of plant

Table 12.1 *The Trent Biotic Index for river pollution monitoring. Modified from Nuffield Foundation (2008).*

Indicator* taxon**		Total number of groups present									
		0–1	2–5	6–10	11–15	16–20	21–25	26–30	31–35	36–40	41–45
		Trent Biotic Index									
Plecoptera nymph present	More than one species	–	7	8	9	10	11	12	13	14	15
	One species only	–	6	7	8	9	10	11	12	13	14
Ephemeroptera nymphs present	More than one species	–	6	7	8	9	10	11	12	13	14
	One species only	–	5	6	7	8	9	10	11	12	13
Trichoptera larvae present	More than one species	–	5	6	7	8	9	10	11	12	13
	One species only	4	4	5	6	7	8	9	10	11	12
Gammarus present	All above species absent	3	4	5	6	7	8	9	10	11	12
Asellus present	All above species absent	2	3	4	5	6	7	8	9	10	11
Tubificid and/or red chironomid larvae present	All above species absent	1	2	3	4	5	6	7	8	9	10
All the above types absent	Some species that do not require oxygen may be present	0	1	2	–	–	–	–	–	–	–

Table 12.2 *The revised BMWP Scoring system for river pollution monitoring. From Walley and Hawkes (1997).*

Common Name	Family	Original BMWP Score	Revised BMWP Score	Habitat Specific Scores		
				Riffles	Riffle/Pools	Pools
Flatworms	Planariidae	5	4.2	4.5	4.1	3.7
	Dendrocoelidae	5	3.1	2.3	4.1	3.1
Snails	Neritidae	6	7.5	6.7	8.1	9.3
	Viviparidae	6	6.3	2.1	4.7	7.1
	Valvatidae	3	2.8	2.5	2.5	3.2
	Hydrobiidae	3	3.9	4.1	3.9	3.7
	Lymnaeidae	3	3.0	3.2	3.1	2.8
	Physidae	3	1.8	0.9	1.5	2.8
	Planorbidae	3	2.9	2.6	2.9	3.1
Limpets and Mussels	Ancylidae	6	5.6	5.5	5.5	6.2
	Unionidae	6	5.2	4.7	4.8	5.5
	Sphaeriidae	3	3.6	3.7	3.7	3.4
Worms	Oligochaeta	1	3.5	3.9	3.2	2.5
Leeches	Piscicolidae	4	5.0	4.5	5.4	5.2
	Glossiphoniidae	3	3.1	3.0	3.3	2.9
	Hirudididae	3	0.0	0.3	-0.3	
	Erpobdellidae	3	2.8	2.8	2.8	2.6
Crustaceans	Asellidae	3	2.1	1.5	2.4	2.7
	Corophiidae	6	6.1	5.4	5.1	6.5
	Gammaridae	6	4.5	4.7	4.3	4.3
	Astacidae	8	9.0	8.8	9.0	11.2
Mayflies	Siphlonuridae	10	11.0	11.0		
	Baetidae	4	5.3	5.5	4.8	5.1
	Heptageniidae	10	9.8	9.7	10.7	13.0
	Leptophlebiidae	10	8.9	8.7	8.9	9.9
	Ephemerellidae	10	7.7	7.6	8.1	9.3
	Potamanthidae	10	7.6	7.6		
	Ephemeridae	10	9.3	9.0	9.2	11.0
	Caenidae	7	7.1	7.2	7.3	6.4
Stoneflies	Taeniopterygidae	10	10.8	10.7	12.1	
	Nemouridae	7	9.1	9.2	8.5	8.8
	Leuctridae	10	9.9	9.8	10.4	11.2
	Capniidae	10	10.0	10.1		
	Perlodidae	10	10.7	10.8	10.7	10.9
	Perlidae	10	12.5	12.5	12.2	
	Chloroperlidae	10	12.4	12.5	12.1	
Damselflies	Platycnemidae	6	5.1	3.6	5.4	5.7
	Coenagriidae	6	3.5	2.6	3.3	3.8
	Lestidae	8	5.4			5.4
	Calopterygidae	8	6.4	6.0	6.1	7.6

(Continued)

Table 12.2 *Continued*

Common Name	Family	Original BMWP Score	Revised BMWP Score	Habitat Specific Scores		
				Riffles	Riffle/Pools	Pools
Dragonflies	Gomphidae	8				
	Cordulegasteridae	8	8.6	9.5	6.5	7.6
	Aeshnidae	8	6.1	7.0	6.9	5.7
	Corduliidae	8				
	Libellulidae	8	5.0			5.0
Bugs	Mesoveliidae *	5	4.7	4.9	4.0	5.1
	Hydrometridae	5	5.3	5.0	6.2	4.9
	Gerridae	5	4.7	4.5	5.0	4.7
	Nepidae	5	4.3	4.1	4.2	4.5
	Naucoridae	5	4.3			4.3
	Aphelocheiridae	10	8.9	8.4	9.5	11.7
	Notonectidae	5	3.8	1.8	3.4	1.4
	Pleidae	5	3.9			3.9
	Corixidae	5	3.7	3.6	3.5	3.9
Beetles	Haliplidae	5	4.0	3.7	4.2	4.3
	Hygrobiidae	5	2.6	5.6	−0.8	2.6
	Dytiscidae	5	4.8	5.2	4.3	4.2
	Gyrinidae	5	7.8	8.1	7.4	6.8
	Hydrophilidae	5	5.1	5.5	4.5	3.9
	Clambidae	5				
	Scirtidae	5	6.5	6.9	6.2	5.8
	Dryopidae	5	6.5	6.5		
	Elmidae	5	6.4	6.5	6.1	6.5
	Chrysomelidae *	5	4.2	4.9	1.1	4.1
	Curculionidae *	5	4.0	4.7	3.1	2.9
Alderflies	Sialidae	4	4.5	4.7	4.7	4.3
Caddisflies	Rhyacophilidae	7	8.3	8.2	8.6	9.6
	Philopotamidae	8	10.6	10.7	9.8	
	Polycentropidae	7	8.6	8.6	8.4	8.7
	Psychomyiidae	8	6.9	6.4	7.4	8.0
	Hydropsychidae	5	6.6	6.6	6.5	7.2
	Hydroptilidae	6	6.7	6.7	6.8	6.5
	Phryganeidae	10	7.0	6.6	5.4	8.0
	Limnephilidae	7	6.9	7.1	6.5	6.6
	Molannidae	10	8.9	7.8	8.1	10.0
	Beraeidae	10	9.0	8.3	7.8	10.0
	Odontoceridae	10	10.9	10.8	11.4	11.7
	Leptoceridae	10	7.8	7.8	7.7	8.1
	Goeridae	10	9.9	9.8	9.6	12.4
	Lepidostomatidae	10	10.4	10.3	10.7	11.6
	Brachycentridae	10	9.4	9.3	9.7	11.0
	Sericostomatidae	10	9.2	9.1	9.3	10.3
True flies	Tipulidae	5	5.5	5.6	5.0	5.1
	Chironomidae	2	3.7	4.1	3.4	2.8
	Simuliidae	5	5.8	5.9	5.1	5.5

Table 12.3 *The Domin scale, which is used to estimate percentage cover-abundance of plants in a quadrat sample.*

Cover	Domin score
91–100%	10
76–90%	9
51–75%	8
34–50%	7
26–33%	6
11–25%	5
4–10%	4
<4% (many individuals)	3
<4% (several individuals)	2
<4% (few individuals)	1

community values, such as those of floodplain forests within Europe (Douda et al. 2016) in addition to accurate assessment of the value of the UK's natural wetland capital, and assessment of future changes with climate change (Fry 2015).

True measures of diversity are quantitative; a numerical index based upon the number of different biological units and the number of individuals of each unit in a sample. The usual unit of diversity is the species, but higher taxonomic units (genera, families) and smaller units, most often alleles (genetic diversity), are also used. Indices have also been used to measure physical diversity in studies of river conservation and restoration (Kemp et al. 2000) There are many indices of diversity (Magurran and Gill 2014), but the commonest are those invented by C.E. Shannon and N. Weiner (known as the Shannon-Weiner Index) and E.H. Simpson, both in 1949. Choosing exactly which one to use depends upon the aims of the study, but there are many publications to provide guidance, which test different indices against each other (e.g., Morris et al. 2014). Diversity indices hide the taxonomic information that goes in to calculating them, so it is important that any study comparing diversity of different wetland locations also includes taxonomic information.

The success of a study that involves diversity measurements is highly dependent on the effort put into collecting accurate data from the sample units, because rare species represented by just 1–2 individuals can make a large difference to some indices. A diversity index may also be one part of a study that is seeking to establish the total biodiversity of a location (species richness), in order to make a species list. Recent technical developments in genetics have dramatically reduced the cost of genome sequencing, and have made investigations of environmental DNA, or eDNA, feasible. This is DNA shed from individuals in faeces or slime, which can be detected by analysing nuclear or mitochondrial DNA in quantitative Polymerase Chain Reaction (qPCR), for 1–2 weeks after their presence; thus very rare or mobile species can be detected by traces left behind. The methodology is a development from DNA barcoding, which was invented over a decade ago as a means of identifying species more simply and reliably than the traditional visual morphometric methods and can also include many rare species missed by conventional surveys (Janzen et al. 2009). It is not simple to use, because it depends upon the life span of the DNA in the environment, which varies by species and by location conditions; it depends upon a comprehensive reference database, and it depends on effective statistical analyses of the genetic data produced (Yoccoz 2012; Taberlet et al. 2018). eDNA metabarcoding does, however, seem likely to become a valuable tool in trying to monitor the serious decline in freshwater biodiversity (Valentini et al. 2016).

12.3.2 Food chains and webs

Environmental DNA is one of the new tools that can greatly enhance the manual, microscopic methods of stomach analysis to understand a species' diet and thus build up information about food chains in a community. Before this, immunological methods had been used, such as antigen–antibody interactions, but these are very labour intensive. DNA metabarcoding can use stomach, gut or faecal contents to produce a complete list of prey items, including rarer species which might be missed by other means of gut analyses. In this way, closely related species with overlapping prey items could be separated out, as was done by Corse et al. (2010) for three freshwater European cyprinid fish species with overlapping niches.

Another important method is stable isotope analysis (Hobson 1999). This measures the ratio of 15N to 14N (known as $\delta 15N$ (‰)) and the ratio of 13C to 12C (known as ($\delta 13C$(‰)) in plant and animal proteins (e.g., small samples of muscle tissue, because those in consumers reflect the proteins of species they have fed upon in a predictable fashion). Typically, there is a step-by-step increase in $\delta 15N$ ‰ at each trophic level by between 2.5‰ and 5.0‰ with a smaller increase in $\delta 13C$ of around 1‰. Although $\delta 13C$ values show less enrichment along a food chain, they also often vary between photosynthetic sources (e.g., aquatic phytoplankton vs. terrestrial plants). In this way, a food web can be built up on a 2-dimensional biplot showing the values for $\delta 13C$‰ against $\delta 15N$ ‰ (Figure 12.1). Differences can also be shown between the basal resources, which are either plants or detritus, from different sources; thus, food chains built on allochthonous detritus inputs differ from autochthonous algal inputs (Lau et al. 2009). Most stable isotope studies enable descriptive links to be built in a food web, without quantification, but where comparisons are needed, over time or spatial gradients, it is possible to use circular statistics to test for significant differences in the areas of the biplot occupied by the error bars from the replicates analysed for a species. Schmidt et al. (2007) demonstrated this method for changes in arthropod communities across a spatial gradient of salinity in a salt marsh; and for changes in the fish community of Lake Tahoe across a temporal gradient.

A combination of the two methods produces added value in studies, because they are able to offer different perspectives. For example, a study of a pool in the lower Murray-Darling river in Australia by Hardy et al. (2010) was able to separate the sources of allochthonous detritus using stable isotope methods and then identify the diet of small fish species, using genetic analysis of small aquatic and terrestrial arthropods, rotifers, algae and other plant material. Unexpected isotope ratios indicated that carbon generated by bacterial methanotrophs in sediments was made available to the food web through benthic invertebrate feeding. The study gave the authors a new

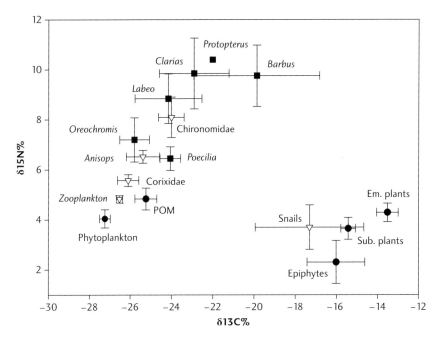

Figure 12.1 A stable isotope bi-plot of δ15N against δ13C from Lake Baringo, Kenya, showing the step-wise increase in isotopic values (carbon being more conservatively transferred than nitrogen) with trophic transfer up the food chain. The two main basal resources, phytoplankton and epiphytes associated with macrophytes, were isotopically distinct. The majority of biomass within the food web is derived from planktonic resources. Only *Barbus* appears to assimilate resources from both the planktonic and epiphytic chains (the latter via snails; as demonstrated by the variance in δ13C). Modified after Britton et al. (2009).

perspective on the ecosystem changes of the river and its low-flow pools, as a consequence of loss of environmental flows caused by a severe drought.

12.3.3 Energy flows—quantifying food webs in lakes, streams, wetlands

Understanding the carbon—and thus the energy—flow through a freshwater community and ecosystem is achieved by quantification of its food webs. This field—Ecological Energetics –was important in the 1960s, when many countries established teams under the International Biological Programme to measure ecosystem energetics as the 'biological basis of human welfare'. The intensity of many of these studies has never been matched again and many publications, especially the manuals, are still relevant (e.g., Ricker 1968; Vollenveider 1974; Downing and Rigler 1984). The theory of energy flow in freshwater systems dates back to the 1940s with the seminal work of Raymond Lindeman on Cedar Bog lake, Minnesota (Lindeman 1942); followed by Odum (1957) on Silver Springs, Florida; Teal (1957) on a cold spring in Massachusetts; and Fisher and Likens (1973) on Bear Brook in New Hampshire. Many studies since then, all over the world, have built up our understanding of the practical differences in energy flow through different systems, all underpinned by the same principles of thermodynamics (e.g., Atkinson et al. 2017). Important issues in modern conservation ecology are the partitioning of energy between different biotic components. For example, eutrophication of freshwaters (Harper 1992) moves most autochthonous production away from aquatic macrophytes into phytoplankton, with consequent changes in the consumer pathways because small zooplankton grazers are favoured over littoral macroinvertebrate decomposers. Kuiper et al. (2015) use a combination of food-web and ecosystem modelling to demonstrate that aquatic food-web instability precedes critical transitions in temperate shallow lakes. In ecosystems important to humans, such as fisheries, the energy transfer to the upper levels may then go into species of low economic value, as happened in Lake Erie in the 1950s–60s. Alien invasive species also often change the magnitude and pathways of energy flow by changing rates of decomposition (Macneil et al. 2011; Kuglerová et al. 2017) or through the timing of the introductions and number of species (Ellis et al. 2011).

12.4 Landscape freshwater ecology

All freshwater ecosystems are a component of the landscape, linked together by the downhill or below-ground passage of water, and the movement of species between different ecosystems. Wetland, lotic, and lentic conservation efforts all need to take a landscape or catchment approach, in recognition that the pressures on our freshwater ecosystems can never be fully solved by small piecemeal actions. The rapid loss of freshwater biodiversity, which is occurring at a greater rate than for any other ecosystem type, can only be halted by acting at the catchment scale (Pacini et al. 2013) despite the political difficulties this may bring. These ideas are reinforced by the concept of Integrated Freshwater Ecology and Biodiversity Conservation (IFEBC) that advocates for a holistic approach to understanding freshwater biodiversity functioning and management, using a methodological combination of molecular and ecological tools (Geist 2011).

Freshwater 'patches' are recognisable in the landscape at a variety of scales. Demars et al. (2012) have shown that 'biotopes' are discrete physical units in river channels, defined by dominant substrate or vegetation type; while their integrity is demonstrated by the macroinvertebrate assemblages present (Harper et al. 1992; Kemp et al. 2000), they are visually recognisable as landscape units in the size order of one to tens m². Similar units, on a larger scale, have also been shown in more lentic environments, such as the mosaic of channels and ponds on natural floodplains that are cut off from the main river channel other than during flood events, and therefore have a temporal as well as a spatial gradient. Kobayashi et al. (2015) showed that there was biological uniqueness of three semi-arid floodplain water body types, measured by their plankton structure (density, richness) and function (respiration, photosynthesis): they demonstrated a need for environmental flows in managed river systems to consider the whole floodplain, not only the river channel by inundating diverse, hydro-geomorphically distinct habitats; with environmental flows achievable only at the catchment scale.

Individual ecosystem types form patches, often across gradients governed by water. The Okavango inland delta, Botswana, is a huge landscape of interlinked ecosystems totally dependent upon the annual flood event, which originates in the mountains of Angola in two headwater rivers, the Cubango and the Cuito. These flow south and east and inundate the delta March–June each year; annually depositing 9.4 km^3 of water over 40,000 km^2 (about 10 per cent of the catchment) into Kalahari desert sands. This event maintains a mosaic of permanent and temporary wetland ecosystems with terrestrial ecosystems in the ecotone between wetland and desert. The delta is under increasingly heavy threats from upstream water abstraction and agricultural encroachment from the delta edges (King and Chonguiça 2016). A recent study of the terrestrial large mammal, zebra (*Equus burchelli*), based upon 14 GPS-collared mares tracked for over 4,000 fixes over just under 200 days' observations, identified home range and habitat use across the delta in the dry season (Bartlam-Brooks et al. 2013). Five habitats (each on a scale of hundreds of m^2) were identified—(a) floodplain grasslands, (b) savannah shrub grasslands, (c) *Acacia* woodland, (d) Riparian woodland, and (e) Mopane woodland; zebra movements and distribution compared with characteristics that showed habitat frequency and shape (see paper for the methods used). The study found home range halved in size in the central delta compared with the peripheral area and the implications for conservation are clear—high landscape quality is provided by the annual flooding, and both water loss and encroachment will decrease habitat patches, increase resource competition, thus decreasing population size.

Considering a landscape by its ecosystem patches, their interactions with each other and with species, misses one very important point, which is an appropriate way to conclude this chapter. That is on the processes of energy flow across the whole landscape (and higher scales too; biome and biosphere). Photosynthesis and biological production in ecosystems—the accumulation of the sun's energy into biomass of plants, microbes and animals—is a small fraction of the total solar radiation that enters the earth's atmosphere. This direct effect of energy, with the smaller fraction that is used in heating the biomass, together only account for <10 per cent of incoming energy. The effects of the remaining, indirect, solar radiation are far greater in both magnitude and importance and their transformations only really understood at the larger scales. These are the directions in which solar energy is dissipated, either largely by 'latent heat' of evapotranspiration (effectively trapping the energy in water vapour molecules, the dominant process in natural landscapes), or largely by 'sensible heat' directly heating the atmosphere, rocks, soil, and non-biological components (dominant in anthropogenic landscapes; Huryna et al. 2014; Pokorný et al. 2016).

One example, from a wetland-dominated semi-natural landscape in southern Bohemia, shows this effect (Huryna et al. 2014). The authors measured energy fluxes (see their paper for methods)—net radiation in, latent heat flux and sensible heat flux—on clear summer days from four types of land use: floodplain meadow, natural pasture, arable field, and an urban artificial surface, all with similar incoming radiation. They found that the meadow transformed 30 per cent more energy by evapotranspiration than the pasture or the arable and 70 per cent more than the urban surface, trapping energy of 600 MW km^{-2}, which, over several square kilometres, is equivalent to a medium-sized power station. This energy is thus not available to increase ambient temperature. When temperature drops below the dew point, such as at night or inside vegetation, water vapour condenses, releasing the latent heat and also causing a decrease of air pressure, such that air from surrounding areas is sucked in. Thus, wetland evapotranspiration can be very important in cooling anthropogenic landscapes at increasing risk of desiccation due to climate change.

On a larger scale, Makarieva and Gorshkov (2007; 2010) have shown that natural vegetation and forests act as a 'biotic pump', sustaining the water cycle. Forests achieve this by creating a temperature inversion within their stands, with higher temperatures in the crown than lower down, and this cooler, heavier air remains with the stand. Pokorný et al. (2018) have shown that wetlands behave in the same way as forests, producing a temperature inversion within their vegetation. Natural vegetation types with a vertical structure—wetlands and forests—thus cycle water continuously and play vital roles in both energy dissipation and biogeochemical cycling in a landscape.

In areas without evapotranspiration as the major process, sensible heat causes landscapes and their components to dry out in summer (Ripl 2003; Ripl and Eiseltová 2010; Eiseltová et al. 2012). The Czech Republic lost about a million ha of wetlands by conversion to agriculture between 1950 and 1989 under the Communist Regime. This means that 1×10^6 MW of sensible heat is now released on sunny days that was not previously, increasing soil temperature and transporting water vapour upwards. Soil heating makes it prone to decomposition of organic material, leading to loss of nutrients and particle erosion. Currently, nutrient and sediment pollution are the greatest threat to freshwaters in almost every country of the world (Ripl 2003). Wetland conservation and restoration is vitally important to the future of this planet due to the dependence of human populations on freshwater and wetland services. Understanding how freshwater ecosystems, communities, and aquatic populations work and interact at the site and landscape scales will greatly benefit their management and conservation.

References

Angino, E., Armitage, K., and Saxena, B. (1973). Population dynamics of pond zooplankton II *Daphnia ambigua* Scourfield. *Hydrobiologia*, 42, 491–507.

Armitage, K. Saxena, B., and Angino, E. (1973). Population dynamics of pond zooplankton, I. *Diaptomus pallidus* Herrick. *Hydrobiologia*, 42, 295–333.

Atkinson, C.L., Capps, K.A., Rugenski, A.T., and Vanni, M.J. (2017). Consumer-driven nutrient dynamics in freshwater ecosystems: from individuals to ecosystems. *Biological Reviews*, 92, 2003–2023.

Bailey, L.L. and Nichols, J.D. (2009) Capture-mark-recapture, removal sampling and occupancy models. In Dodd, C.K. (ed.) Amphibian Ecology and Conservation, Techniques in Ecology and Conservation Series. Oxford University Press, Oxford, UK.

Baldwin, E.A., (2011). Twenty-five years under the Convention on Migratory Species: migration conservation lessons from Europe. *Environmental Law*, 41, 535–571.

Bartlam-Brooks, H.L.A., Bonyongo, M.C., and Harris, S. (2013). How landscape scale changes affect ecological processes in conservation areas: external factors influence land use by zebra (Equus burchelli) in the Okavango Delta. *Ecology and Evolution*, 3, 2795–805.

Bordalo, M., Ferreira, S., Cardoso, P., Leston, S., and Pardal, M. (2011). Resilience of an isopod population (Cyathura carinata) to multiple stress factors in a temperate estuarine system. *Hydrobiologia*, 671, 13–25.

Brent, R. and Berberich, D. (2014). Use of artificial stream mesocosms to investigate mercury uptake in the South River, Virginia, USA. *Archives of Environmental Contamination and Toxicology*, 66, 201–12.

Britton, J.R. Jackson, M.C., Muchiri, M. Tarras-Wahlberg, H., Harper, D.M., and Grey, J. (2009). Status, ecology and conservation of an endemic fish, *Oreochromis niloticus baringoensis*, in Lake Baringo, Kenya. *Aquatic Conservation: Marine and Freshwater Ecosystems*, 19, 487–96.

Brock, T. and Wijngaarden, R. (2012). Acute toxicity tests with *Daphnia magna*, *Americamysis bahia*, *Chironomus riparius* and *Gammarus pulex* and implications of new EU requirements for the aquatic effect assessment of insecticides. *Environmental Science and Pollution Research*, 19, 3610–18.

Céréghino, R., Oertli, B., Bazzanti, M., Coccia, C., Compin, A., Biggs, J., Bressi, N., Grillas, P., Hull, A., Kalettka, T., and Scher, O. (2012). Biological traits of European pond macroinvertebrates. *Hydrobiologia*, 689, 51–61.

Cernansky, R. (2017). Biodiversity moves beyond counting species. *Nature*, 546, 22–4.

Chadd, R. and Extence, C. (2004). The conservation of freshwater macroinvertebrate populations: a community-based classification scheme. *Aquatic Conservation Marine and Freshwater Ecosystems,* 14, 597–624.

Childress, B., Harper, D.M., Hughes, B., Van Den Bossche, W. Berthold, P., and Querner, U. (2004). Satellite tracking Lesser Flamingo movements in the Rift Valley, East Africa: pilot study report. *Ostrich*, 75, 57–65.

Childress, B., Hughes, B., Harper, D.M., and Van Den Bossche, W. (2007). East African flyway and key site network of the Lesser Flamingo (*Phoenicopterus minor*) documented through satellite tracking. *Ostrich*, 78, 463–8.

Childress, B., Nagy, S., and Hughes, B. (2008). *International single species action plan for the conservation of the Lesser Flamingo (Phoeniconaias minor).* AEWA Technical Series, Bonn, Germany.

Clarke, R.T., Wright, J.F., and Furse, M.T (2003). RIVPACS models for predicting the expected macroinvertebrate fauna and assessing the ecological quality of rivers. *Ecological Modelling*, 160, 219–33.

Clavero, M., Blanco-Garrido, F., and Prenda, J. (2004). Fish fauna in Iberian Mediterranean river basins: biodiversity, introduced species and damming impacts. *Aquatic Conservation: Marine and Freshwater Ecosystems*, 14, 575–85.

Conrad, C.C. and Hilchey, K.G. (2011). A review of citizen science and community-based environmental monitoring: issues and opportunities. *Environmental Monitoring and Assessment*, 176, 273–91.

Corse, E., Costedoat, C., Chappaz, R., Pech, N., Martin, J-F., and Gilles, A. (2010). A PCR-based method for diet analysis in freshwater organisms using 18S rDNA barcoding on faeces. *Molecular Ecology Resources*, 10, 96–108.

Covich, A.P., Palmer, M.A., and Crowl, T.A. (1999). The role of benthic invertebrate species in freshwater ecosystems: Zoobenthic species influence energy flows and nutrient cycling. *BioScience*, 49, 119–27.

Dehling, D.M., Hof, C., Brändle, M., and Brandl, R. (2010). Habitat availability does not explain the species richness patterns of European lentic and lotic freshwater animals. *Journal of Biogeography*, 37, 1919–26.

Demars, B.O.L. and Harper, D.M. (2005). Distribution of aquatic vascular plants in lowland rivers: separating the effects of local environmental conditions, longitudinal connectivity and river basin isolation. *Freshwater Biology*, 50, 418–37.

Demars, B.O.L., Harper, D.M., Pitt, J.-A., and Slaughter, R. (2005). The impact of climate variability and phosphorus control measures on river phosphorus retention associated with point source pollution. *Hydrology and Earth System Sciences*, 9, 1–24.

Demars, B.O.L., Kemp, J.L., Friberg, N., Usseglio-Polatera, P., and Harper, D.M. (2012). Linking biotopes to invertebrates in rivers: Biological traits, taxonomic composition and diversity. *Ecological Indicators*, 23, 301–11.

Descloux, S., Datry, T., and Usseglio-Polatera, P. (2014). Trait-based structure of invertebrates along a gradient of sediment colmation: Benthos versus hyporheos responses. *Science of the Total Environment*, 466–7, 265–76.

Dodd, C.K. (ed.) (2010). Amphibian Ecology and Conservation: A Handbook of Techniques. Techniques in Ecology and Conservation Series, Oxford University Press, UK.

Dodd, C.K. (ed.) (2016). Reptile Ecology and Conservation: A Handbook of Techniques. Techniques in Ecology and Conservation Series, Oxford University Press, UK.

Dodds, W.K., Gido, K., Whiles, M.R., Fritz, K.M., and Matthews, W.J. (2004). Life on the edge: the ecology of Great Plains prairie streams. *BioScience*, 54, 205–17.

Dodds, W.K. and Whiles, M.R. (2010). Freshwater Ecology: Concepts and Environmental Applications of Limnology. Elsevier, USA.

Douda, J., Boublík, K., Slezák, M., Biurrun, I., Nociar, J., Havrdová, A., Doudová, J., Aćić, S., Brisse, H., Brunet, J., Chytrý, M., Claessens, H., Csiky, J., Didukh, Y., Dimopoulos, P., Dullinger, S., Fitzpatrick, Ú., Guisan, A., Horchler, P. J., Hrivnák, R., Jandt, U., Kącki, Z., Kevey, B., Landucci, F., Lecomte, H., Lenoir, J., Paal, J., Paternoster, D., Pauli, H., Pielech, R., Rodwell, J. S., Roelandt, B., Svenning, J.C., Šibík, J., Silc, U., Škvorc, Ž., Tsiripidis, I., Tzonev, R. T., Wohlgemuth, T., and Zimmermann, N.E. (2016). Vegetation classification and biogeography of European floodplain forests and alder cars. *Applied Vegetation Science*, 19, 147–63.

Downing, J.A. and Rigler F.H. (eds) (1984). A Manual on Methods for the Assessment of Secondary Production in Fresh Waters, IBP Handbook No.17, 2nd edition. Blackwell Publishing, Oxford.

Dunbar, M.J., Pedersen, M.L., Cadman, D., Extence, C., Waddingham, J., Chadd, R., and Larsen, S.E. (2010). River discharge and local-scale physical habitat influence macroinvertebrate LIFE scores. *Freshwater Biology*, 55, 226–41.

Dupuis, A.P. and Hann, B.J. (2009). Climate change, diapause termination and zooplankton population dynamics: an experimental and modelling approach. *Freshwater Biology*, 54, 221–35.

Eiseltová, M., Pokorný, J., Hesslerová, P., and Ripl, W. (2012). Evapotranspiration—A Driving Force in Landscape Sustainability. In: Irmak, A. (ed.) Evapotranspiration—Remote Sensing and Modelling. InTech, Croatia.

Ellenberg, H. (1988). Vegetation Ecology of Central Europe. Cambridge University Press, Cambridge, UK.

Ellis, B.K., Stanford, J.A., Goodman, D., Stafford, C.P., Gustafson, D.L., Beauchamp, D.A., Chess, D.W., Craft, J.A., Deleray, M.A., and Hansen, B.S. (2011). Long-term effects of a trophic cascade in a large lake ecosystem. *PNAS*, 108, 1070–5.

Ellis, L.E. and Jones, N.E. (2016). A test of the Serial Discontinuity Concept: Longitudinal trends of benthic invertebrates in regulated and natural rivers of northern Canada. *River Research and Applications*, 32, 462–72.

Everall, N.C., Johnson M.F., Wood P., and Mattingley, L. (2017a). Sensitivity of the early life stages of a mayfly to fine sediment and orthophosphate levels. *Environmental Pollution*, 10, 131.

Everall, N.C., Johnson, M.A., Wood, P., Farmer A., Wilby, R.L., and Measham, N. (2017b). Comparability of macroinvertebrate biomonitoring indices of river health derived from semi-quantitative and quantitative methodologies. *Ecological Indicators*, 78, 437–48.

Fausch, K.D., Rieman, B.E., Dunham, J.B., Young M.K., and Peterson D.P. (2009). Invasion versus Isolation: Trade-offs in Managing Native Salmonids with Barriers to Upstream Movement. *Conservation Biology*, 23, 859–70.

Fisher, S.G. and Likens, G.E. (1973). Energy Flow in Bear Brook, New Hampshire: An Integrative Approach to Stream Ecosystem Metabolism. *Ecological Monographs*, 43, 421–43.

Fojt, W. and Harding, M. (1995). Thirty Years of Change in the Vegetation Communities of Three Valley Mires in Suffolk, England. *Journal of Applied Ecology*, 32, 561–77.

Fretwell, P.T., Scofield, P., and Phillips, R.A. (2017). Using super-high-resolution satellite imagery to census threatened albatrosses. *Ibis*, 159, 481–90.

Fry, M. (2015). Wetland tool for climate change. Available on the Centre for Ecology and Hydrology (CEH) website.

Furse, M.T., Hering, M., Brabec, K., Buffagni, A., Sandin, L., and Verdonschot, P.F.M. (eds.) (2006). The Ecological Status of European Rivers: Evaluation and Intercalibration of Assessment Methods. *Developments in Hydrobiology*, 188, whole issue.

Geist, J. (2011). Integrative freshwater ecology and biodiversity conservation. *Ecological Indicators*, 11, 1507–16.

Gessner, M.O. and Tlili, A. (2016). Fostering integration of freshwater ecology with Ecotoxicology. *Freshwater Biology*, 61, 1991–2001.

Gutiérrez-Fonseca, P.E. and Lorion, C.M. (2014). Application of the BMWP-Costa Rica biotic index in aquatic biomonitoring: sensitivity to collection method and sampling intensity. *Revista de Biología Tropical*, 62, 275–89.

Hardy C.M., Krull E.S., Hartley D.M., and Oliver, R.L. (2010). Carbon source accounting for fish using combined DNA and stable isotope analyses in a regulated lowland river weir pool. *Molecular Ecology*, 19, 197–212.

Harper, D.M. (1992). Eutrophication of Freshwaters; Principles, Problems and Restoration. Chapman and Hall, London.

Harper, D.M. Childress, R.B., Harper, M.M., Boar, R.R., Hickley, P., Mills, S.C., Otieno, N., Drane, A., Vareschi, E., Nasirwa, O., Mwatha, W.E., Darlington, J.P.E.C., and Escuté-Gasulla, X. (2003). Aquatic biodiversity and soda lakes: Lake Bogoria National Reserve, Kenya. *Hydrobiologia*, 500, 259–76.

Harper, D.M. Smith, C.D., and Barham, P.J. (1992). Habitats as the building blocks for river conservation assessment. In: Boon, P.J. Calow, P., and Petts, G.E. (eds) River Conservation and Management. John Wiley, Chichester.

Harper, D.M. Tebbs, E. Bell, O., and Robinson V.J. (2016). Conservation and Management of East Africa's Soda Lakes. In: M. Schagerl (ed.) Soda Lakes of East Africa. Springer International, Switzerland.

Hauer, F.R. and Lamberti, G.A. (eds.) (2017). Methods in Stream Ecology, Volume 1, Ecosystem Structure. 3rd edition, Academic Press, Elsevier, Burlington, USA.

Hazelwood, D.H. and Parker, R.A. (1961). Population dynamics of some freshwater zooplankton. *Ecology*, 42, 266–74.

Hébert, M.P., Beisner, B.E., and Maranger, R. (2016). A compilation of quantitative functional traits for marine and freshwater crustacean zooplankton. *Ecology*, 97, 1081.

Hecky, R.E., Mugidde, R., Ramlal, P.S., Talbot, M.R., and Kling, G.W. (2010). Multiple stressors cause rapid ecosystem change in Lake Victoria. *Freshwater Biology*, 55, 19–42.

Hobson, K.A. (1999). Tracing origins and migration of wildlife using stable isotopes: a review. *Oecologia*, 120, 314–26.

Huang, J., Cao. Y., and Cummings, K.S. (2011). Assessing sampling adequacy of mussel diversity surveys in wadeable Illinois streams. *Journal of the North American Benthological Society*, 30, 923–34.

Hughes, J.M., Schmidt, D.J., and Finn, D.S. (2009). Genes in streams: Using DNA to understand the movement of freshwater fauna and their riverine habitat. *BioScience*, 59, 573–83.

Huryna, H., Brom, J., and Pokorny, J. (2014). The importance of wetlands in the energy balance of a landscape. *Wetlands Ecology and Management*, 22, 363–81.

Hutchinson, G.E. (1967). A Treatise on Limnology, Volume 2: Introduction to Lake Biology and the Limnoplankton. John Wiley and Sons, New York.

Hutchinson, G.E. (1975). A Treatise on Limnology, Volume 3: Limnological Botany. John Wiley and Sons, New York.

Hutchinson, G.E. (1993). A Treatise on Limnology, Volume 4: The Zoobenthos. John Wiley and Sons, New York.

IUCN (2012). Guidelines for Application of IUCN Red List Criteria at Regional and National Levels, Version 4. International Union for Conservation of Nature, Gland, Switzerland and Cambridge, UK.

Janzen, D.H., Hallwachs, W., Blandin, P., et al. (2009). Integration of DNA barcoding into an ongoing inventory of complex tropical biodiversity. *Molecular Ecology Resources*, 9, 1–26.

Kairu, J.K. (1996). Heavy metal residues in birds of Lake Nakuru, Kenya. *African. Journal of Ecology*, 34, 397–400.

Kelly, N.E., Young, J.D., Winter, J.G., Palmer, M.E., Stainsby, E.A., and Molot, L.A. (2017). Sequential rather than interactive effects of multiple stressors as drivers of phytoplankton community change in a large lake. *Freshwater Biology*, 62, 1288–302.

Kemp J.L., Harper D.M., and Crosa, G.A. (2000). The habitat-scale ecohydraulics of rivers. *Ecological Engineering*, 16, 17–29.

King, J. and Chonguiça, E. (2016). Integrated management of the Cubango-Okavango River Basin. *Ecohydrology and Hydrobiology*, 16, 263–71.

Kobayashi, T., Ralph, T.J., Ryder, D.S., Hunter, S.J., Shiel, R.J., and Segers, H. (2015). Spatial dissimilarities in plankton structure and function during flood pulses in a semi-arid flood-plain wetland system. *Hydrobiologia*, 747, 19–31.

Kock, N.D., Kock, R.A., Wambua, J., Kamau, G.J., and Mohan, K. (1999). *Mycobacterium avium*-related epizootic in free-ranging lesser flamingos in Kenya. *Journal of Wildlife Diseases*, 35, 297–300.

Kuglerová, L., García, L., Pardo, I., Mottiar, L., and Richardson, J.S. (2017). Does leaf litter from invasive plants contribute the same support of a stream ecosystem function as native vegetation? *Ecosphere*, 8, 1–18.

Kuiper, J.J., van Altena, C., de Ruiter, P.C., van Gerven, L.P.A., Janse, J.H., and Mooji, W.M. (2015). Food-web stability signals critical transitions in temperate shallow lakes. *Nature Communications*, 6, 7727.

Lamberti, G.A. and Hauer, F.R. (eds) (2017). Methods in Stream Ecology, Volume 2, Ecosystem Function. 3rd edition, Academic Press, Elsevier, Burlington, USA.

Larsen, T.H. (ed.) (2016). Core Standardized Methods for Rapid Biological Field Assessment. Conservation International, Arlington, VA.

Lau, D.C.P., Leung, K.M.Y., and Dudgeon, D. (2009). What does stable isotope analysis reveal about trophic relationships and the relative importance of allochthonous and autochthonous resources in tropical streams? A synthetic study from Hong Kong. *Freshwater Biology*, 54, 127–41.

Lindeman, R.L. (1942). The trophic-dynamic aspect of ecology. *Ecology*, 23, 399–418.

Lugomela, C., Pratap, H.B., and Mgaya, Y.D. (2006). Cyanobacteria blooms—A possible cause of mass mortality of Lesser Flamingos in Lake Manyara and Lake Big Momela, Tanzania. *Harmful Algae*, 5, 534–41.

Lyon, J.P., Todd, C., Nicol, S.J., MacDonald, A., Stoessel, D., Ingram, B.A., Barker, R. J., and Bradshaw, C.J.A. (2012). Reintroduction success of threatened Australian trout cod (*Maccullochella macquariensis)* based on growth and reproduction. *Marine and Freshwater Research*, 63, 598–605.

MacKenzie, D., Nichols J., Royle, J., Pollock, K., Bailey, L., and Hines J. (2017). Occupancy Estimation and Modeling: Inferring Patterns and Dynamics of Species Occurrence, 2nd edition. Elsevier.

Macneil, C., Dick, J.T.A., Platvoet, D., and Briffa, M. (2011). Direct and indirect effects of species displacements: an invading freshwater amphipod can disrupt leaf-litter processing and shredder efficiency. *Journal of the North American Benthological Society,* 30, 38–48.

Magurran, A.E. and McGill, B.J. (2014). Biological Diversity: Frontiers in Measurement and Assessment. Oxford University Press, Oxford.

Makarieva, A.M. and Gorshkov, V.G. (2007). Biotic pump of atmospheric moisture as driver of the hydrological cycle on land. *Hydrology and Earth System Sciences*, 11, 1013–33.

Makarieva, A.M. and Gorshkov, V.G. (2010). The biotic pump: condensation, atmospheric dynamics and climate. *International Journal of Water,* 5, 365–85.

Martin, R. (2004). Origin of the Biological Monitoring Working Party System. Centre for Intelligent Environmental systems, University of Staffordshire.

Martin, T.G., Chadès, I., Arcese, P., Marra, P.P., Possingham, H.P., Norris, D., and Ryan Jones, P. (2007). Optimal Conservation of Migratory Species (Migratory Species Conservation). *PLoS ONE,* 2, e751.

Maurer, B.A. (2009). Geographical Population Analysis: Tools for the Analysis of Biodiversity. John Wiley and Sons, New York.

Morales-Roldan, H.L., Mwinami, T., and Harper, D.M. (2010). Is ground census of flamingo lakes worthwhile? *Flamingo: Bulletin of the Flamingo Specialist Group*, 18, 53–6.

Moreno-opo, R., Ould Sidaty, Z.E., Baldò, J.M., Garcìa, F., Ould Sehla Daf, D., and Gonzàlez, L.M. (2013). A breeding colony of the Near Threatened Lesser Flamingo *Phoeniconaias minor* in western Africa: a conservation story of threats and land management. *Bird Conservation International*, 23, 426–36.

Morris, E.K., Caruso, T., Buscot, F., Fischer, M., Hancock, C., Maier, T.J., Meiners, T., Müller, C., Obermaier, E., Prati, D., Socher, S.A., Sonnemann, I., Wäschke, N., Wubet, T., Wurst, S., and Rillig, M.C. (2014). Choosing and using diversity indices: insights for ecological applications from the German Biodiversity Exploratories. *Ecology and Evolution*, 4, 3514–24.

Morris, W.F., and Doak, D.F. (2003). Quantitative Conservation Biology: Theory and Practice of Population Viability Analysis. Sinauer Associates, Sunderland, Massachusetts, USA.

Moss, B.R. (2018). Ecology of Freshwaters—Earth's Bloodstream. Wiley-Blackwell. In press.

Motelin, G., Thampy, R., and Doros, D. (2000). An eco toxicological study of the potential roles of metals, pesticides and algal toxins on the 1993/5 lesser flamingo mass die-offs in Lake Bogoria and Nakuru, Kenya; and the health status of the same species of birds in the Rift Valley Lakes during the 1990s. In *Proceedings of the East African Environmental Forum*, Nairobi, 11–12 May 2000.

Navarro-Ortega, A., Acuña, V., Belli, A., Burek P., Cassiani G., Choukr-Allah, R., Dolédec, S., Elosegi, A., Ferrari, F., Ginebreda, A., Grathwohl, P., Jones C., Ker Rault, P., Kok, K., Koundouri, P., Ludwig, R.P., Merz, R., Milacic, R., Muñoz, I., Nikulin, G., Paniconi, C., Paunović, M., Petrovic, M., Sabater, L., Sabater, S., Skoulikidis, N.T., Slob, A., Teutsch, G., Voulvoulis, N., and Barceló, D. (2015). Managing the effects of multiple stressors on aquatic ecosystems under water scarcity—the GLOBAQUA project. *Science of the Total Environment,* 503, 3–9.

Nuffield Foundation (2008). Estimating environmental damage in freshwater. Worksheet published by The Nuffield Foundation. Available from the nuffieldfoundation.org website.

Oaks, J.L., Walsh, T., Bradway, D., Davis, M., and Harper, D.M. (2006). Septic arthritis and disseminated infections caused by *Mycobacterium avium* in Lesser Flamingos, Lake Bogoria, Kenya. *Flamingo: Bulletin of the Flamingo Specialist Group*, 14, 30–2.

Odum, H.T. (1957). Trophic structure and productivity of Silver Springs, Florida. *Ecological. Monographs*, 27, 55–112.

Olden, J.D., Kennard, M.J., Lawler, J.J., and Poff, N.L. (2011). Challenges and opportunities in implementing managed relocation for conservation of freshwater species. *Conservation Biology*, 25, 40–7.

Oliveira, A., Soares, M., Martinelli, L., and Moreira, M. (2006). Carbon sources of fish in an Amazonian floodplain lake. *Aquatic Sciences*, 68, 229–38.

Pacini, N., Harper, D.M., Henderson, P., and Le Quesne, T. (2013). Lost in muddy waters: freshwater biodiversity. In: D.W. Macdonald and K.J. Willis (eds) Key Topics in Conservation Biology 2. John Wiley and Sons Ltd., London.

Pandolfo, T.J., Kwak, T.J., Cope, W.G., Heise, R.J., Nichols, R.B., and Pacifici, K. (2016). Species traits and catchment-scale habitat factors influence the occurrence of freshwater mussel populations and assemblages. *Freshwater Biology*, 61, 1671–84.

Paracharya, B.M., Rank, D.N., Harper, D.M. Crosa, G., Zaccara, S., Patel N., and Joshi, C.G. (2015). Long-distance dispersal capability of Lesser Flamingo Phoeniconaias minor between India and Africa: genetic inferences for future conservation plans. *Ostrich*, 86, 221–9.

Pestana, J.L.T., Alexander, A.C., Culp, J.M., Baird, D.J., Cessna, A.J., and Soares, A.M.V.M. (2009). Structural and functional responses of benthic invertebrates to imidacloprid in outdoor stream mesocosms. *Environmental Pollution*, 157, 2328–34.

Pitcher, T.J. and Hart, P.J.B. (1982). Fisheries Ecology. Kluwer Academic Publishers, The Netherlands.

Plumptre, A.J., Kujirakwinja, D., Treves, A., Owiunji, I., and Rainer, H. (2007). Transboundary conservation in the greater Virunga landscape: its importance for landscape species. *Biological Conservation*, 134, 279–87.

Pocock, M.J.O, Roy, H.E., Preston, C.D., and Roy, D.B. (2015). The Biological Records Centre: a pioneer of citizen science. *Biological Journal of the Linnaean Society*, 115, 475–93.

Pokorný, J., Hesslerová, P., Huryna, H., and Harper, D.M. (2016). Indirect and direct thermodynamic effects of wetland ecosystems on climate. In: J. Vymazal (ed.) Natural and Constructed Wetlands. Springer, The Netherlands.

Pokorný, J. Huryna, H. Hesslerová, P., and Harper, D.M. (2018). Wetlands regulate climate via evapotranspiration. In: J. Verhoeven (ed.) Proceedings of the INTECOL Conference, China. Springer International, Switzerland.

Pritchard, V.L., Jones K., and Cowley, D.E. (2007). Genetic Diversity within Fragmented Cutthroat Trout Populations. *Transactions of the American Fisheries Society*, 136, 606–23.

Quinlan, E.L., Nietch, C.T., Blocksom, K., Lazorchak, J.M., Batt, A.L., Griffiths, R., and Klemm, D.J. (2011). Temporal dynamics of periphyton exposed to tetracycline in stream mesocosms. *Environmental Science and Technology*, 45, 10684–90.

Ricker, W.E. (ed.) (1968). Methods for Assessment of Fish Production in Fresh Waters, IBP Handbook No. 3. Blackwell, Oxford.

Ripl, W. (2003). Water: the bloodstream of the biosphere. *Philosophical Transactions of the Royal Society London B*, 358, 1921–34.

Ripl, W. and Eiseltová, M. (2010). Criteria for sustainable restoration of the landscape. In: M. Eiseltová (ed.) Restoration of Lakes, Streams, Floodplains, and Bogs in Europe. Springer, The Netherlands.

Rodwell, J.S. (ed.) (1997). British Plant Communities: Volume 4. Aquatic Communities, Swamps and Tall-Herb Fens. Cambridge University Press, Cambridge, UK.

Rodwell, J.S. (ed.) (1998). British Plant Communities: Volume 2. Mires and Heaths. Cambridge University Press, Cambridge, UK.

Rodwell, J.S. (2006). National Vegetation Classification: Users' Handbook. Joint Nature Conservation Committee, Peterborough, UK. Available online from the jncc.defra.gov.uk website.

Rolls, R. and Sternberg, D. (2015). Can Species Traits Predict the Susceptibility of Riverine Fish to Water Resource Development? An Australian Case Study. *Environmental Management*, 55, 1315–26.

Scalici, M. and Gherardi, F. (2007). Structure and dynamics of an invasive population of the red swamp crayfish (Procambarus clarkii) in a Mediterranean wetland. *Hydrobiologia*, 583, 309–19.

Schmidt, S.N., Olden, J.D., Solomon, C.T., and Vander Zanden, J. (2007). Quantitative Approaches to the analysis of stable isotope food web data. *Ecology*, 88, 2793–2802.

Schwoerbel, J. (1987). Handbook of Limnology. Ellis Horwood, Chichester.

Silk, N. and Ciruna, K. (eds) (2005). Practitioners Guide to Freshwater Biodiversity Conservation. Island Press.

Sherratt, T.N., Roberts, G., Williams, P., Whitfield, M., Biggs, J., Shillabeer, N., and Maund, S.J. (1999). A life-history approach to predicting the recovery of aquatic invertebrate populations after exposure to xenobiotic chemicals. *Environmental Toxicology and Chemistry*, 18, 2512–18.

Sileo, L., Grootenhuis, J.G., Tuite, C.H., and Hopcraft, J.B.D. (1979). Mycobacteriosis in the lesser flamingos of Lake Nakuru, Kenya. *Journal of Wildlife Diseases*, 15, 387–9.

Simmons, R.E. (2000) Declines and Movements of Lesser Flamingos in Africa. *Waterbirds: The International Journal of Waterbird Biology*, 23, 40–6.

Slavik, K., Peterson, B.J., Deegan, L.A., Bowden, W.B., Hershey, A.E., and Hobbie, J.E. (2004). Long-term responses of the Kuparuk river ecosystem to phosphorus fertilization. *Ecology*, 85, 939–954.

Snyder, M.N., Freeman, M.C., Purucker, S.T., and Pringle, C.M. (2016). Using occupancy modeling and logistic regression to assess the distribution of shrimp species in lowland streams, Costa Rica: does regional groundwater create favorable habitat? *Freshwater Science*, 35, 80–90.

Solimini, A.G., Cardoso, A.C., and Heiskanen, A.-S. (2006). Indicators and methods for the ecological status assessment under the Water Framework Directive: linkages between chemical and biological quality of surface waters. European Commission; Joint Research Centre, Institute for Environment and Sustainability. EUR—Scientific and Technical Research Series.

Southwood, T.R.E. (1977). Habitat, the templet for ecological strategies? *Journal of Animal Ecology*, 46, 337–65.

Souza, M.S., Hallgren, P., Balseiro, E., and Hansson, L-A. (2013). Low concentrations, potential ecological consequences: synthetic estrogens alter life-history and demographic structures of aquatic invertebrates. *Environmental Pollution*, 178, 237–43.

Taberlet, P., Bonin, A., Zinger, L., and Coissac, E. (2018). Environmental DNA for Biodiversity Research and Monitoring. Oxford University Press, Oxford.

Tachet, H., Richoux, P., Bournaud, M., and Usseglio-Polatera, P. (2010). Invertébrés d'eau douce: systématique, biologie, écologie. CNRS editions, Paris.

Taylor, M.D., Babcock, R.C., Simpfendorfer, C.A., and Crook, D.A. (2017). Where technology meets ecology: acoustic telemetry in contemporary Australian aquatic research and management. *Marine and Freshwater Research*, 68, 1397–402.

Teal, J.M. (1957). Community metabolism in a temperate cold spring. *Ecological Monographs*, 27, 282–302.

Tebbs, E.J., Remedios, J.J., Avery, S.T., Rowland, C.S., and Harper, D.M. (2015). Regional assessment of lake ecological states using Landsat: A classification scheme for alkaline–saline, flamingo lakes in the East African Rift Valley. *International Journal of Applied Earth Observation and Geoinformation*, 40, 100–8.

Torres, C.R., Ogawa, L.M., Gillingham, M.A.F., Brittney, F., and van Tuinen, M. (2014). A multi-locus inference of the evolutionary diversification of extant flamingos (Phoenicopteridae). *BMC Evolutionary Biology*, 14, 36.

Townsend, C.R., Doledec, S,. and Scarsbrook, M.R. (1997). Species traits in relation to temporal and spatial heterogeneity in streams: a test of habitat templet theory. *Freshwater Biology*, 37, 367–87.

Trebitz, A.S. et al. (2017). Early detection monitoring for aquatic non-indigenous species: Optimizing surveillance, incorporating advanced technologies, and identifying research needs. *Journal of Environmental Management*, 202, 299–310.

USEPA (1998) Guidelines for Ecological Risk Assessment. EPA/630/R-95/002F. Risk Assessment Forum. Washington DC, USA.

Valentini, A., Taberlet, P., Miaud, C., Civade, R., Herder, J., Thomsen, P.F., Bellemain, E., Besnard, A., Coissac, E., Boyer, F., Gaboriaud, C., Jean, P., Poulet, N., Roset, N., Copp, G.H., Geniez, P., Pont, D., Argillier, C., Baudoin, J-M., Peroux, T., Crivelli, A.J., Olivier, A., Acqueberge, M., Le Brun, M., Møller, P.R., Willerslev, E., and Dejean, T. (2016). Next-generation monitoring of aquatic biodiversity using environmental DNA metabarcoding. *Molecular Ecology*, 25, 929–42.

Vannote, R.R., Minshall, G.W., Cummins, K.W., Sedell, J.R., and Cushing, C.E. (1980). The river continuum concept. *Canadian Journal of Fisheries and Aquatic Sciences*, 37, 130–7.

Verberk, W.C.E.P., van Noordwijk, C.G.E., and Hildrew, A.G. (2013). Delivering on a promise: integrating species traits to transform descriptive community ecology into a predictive science. *Freshwater Science*, 32, 531–47.

Vollenveider, R.A. (1974). A manual on methods for the assessment of primary production in fresh waters. International Biological Programme Handbook 12, Blackwells Sci., Oxford.

Walley, W.J. and Hawkes, H.A. (1997). A computer-based development of the Biological Monitoring Working Party score system incorporating abundance rating, biotope type and indicator value. *Water Research*, 31, 201–10.

Weiss, L.C., Potter, L., Steiger, A., Kruppert, S., Frost, U., and Tollrian, R. (2018). Rising pCO$_2$ in freshwater ecosystems has the potential to negatively affect predator-induced defenses in *Daphnia*. *Current Biology*, 28, 327–32.

Wetzel, R.G. (2001). Limnology: Lake and River Ecosystems, 3rd edition. Academic Press.

Whittaker, R.H. (1962). Classification of Natural Communities. *Botanical Review,* 28, 1–239.

Woodiwiss, F. (1964). The biological system of stream classification used by the Trent River Board. *Chemistry and Industry,* 14, 443–7.

Woodiwiss, F.S. (1980). Biological Monitoring of Surface Water Quality, Summary Report. Environment and Consumer Protection Service, Commission of the European Communities.

Yoccoz, N.G. (2012). The future of environmental DNA in ecology. *Molecular Ecology*, 21, 2031–8.

Zaccara, S., Crosa, G., Childress, B., McCulloch, G., and Harper, D.M. (2008). Lesser Flamingo *Phoenicopterus minor* populations in eastern and southern Africa are not genetically isolated. *Ostrich*, 79, 165–70.

Zaccara, S., Crosa, G., Vanetti, I., Binelli, G., Childress, B., McCulloch, G., and Harper, D.M. (2011). Lesser Flamingo *(Phoeniconaias minor)* as a nomadic species in African shallow alkaline lakes and pans: genetic structure and future perspectives. *Ostrich*, 82, 95–100.

13

Changes Over Time

Peter A. Gell, Marie-Elodie Perga, and C. Max Finlayson

Corresponding author: p.gell@federation.edu.au

13.1 Introduction

Freshwater ecosystems vary continuously over a range of time frames. These range from diurnal variations that are driven by factors that are tied with changes in light and temperature, as well as tidal variations driven by lunar cycles, through to forces associated with the orbital cycles of the earth which vary over millennial time scales. The temporal context draws one to see trends in the state of a system as either a change, or part of an ongoing pattern of variation (Semeniuk 2013). Understanding this context is critical as it accounts for factors that will allow a system to recover and so enable us to distinguish cyclical from directional change.

Aquatic systems have responded to even longer cycles of change such as the rifting and coalescing of land masses through forces associated with continental drift. While knowledge of these deep-in-time changes enable us to explain patterns of evolution and phylogeny, they concern us less than more proximal factors that have driven the patterns we observe today and the challenges we face in managing the impacts of people. That said, efforts to manage freshwater systems today, without understanding of the ecological history that a waterway and waterbody has experienced, is likely to lead to misinterpretation of the drivers of change, the prospects of recovery and the likely future path (Finlayson 2013; Gell et al. 2016). The present condition of a lake, river, or wetland is a function of all the events of its past. A challenge lies in establishing which of these events remain influential and how the combination of past and present factors will shape the nature of the system into the future (Finlayson et al. 2013). This is an important issue for the implementation of the Ramsar Convention on Wetlands that provides the framework for wetland conservation and wise use globally (Finlayson et al. 2011, 2017b). On account of the relevance of the legacy of the past this understanding needs to draw from the approaches of many disciplines (e.g., geology, ecology, chemistry, etc.). This chapter aims to present some of the methodological approaches needed to quantify, understand, and attribute the ultimate controls of natural and human causes driving ecological variability in freshwater systems over time.

Gell, P. A., Perga, M. –E., and Finlayson, C. M., *Changes over time*. In: *Freshwater Ecology and Conservation: Approaches and Techniques*. Edited by Jocelyne M. R. Hughes: Oxford University Press (2019).
© Oxford University Press 2019. DOI: 10.1093/oso/9780198766384.003.0013

13.2 Change in freshwater systems

Given the wet phase of aquatic environments the organisms they support tend to be less easily detected than their terrestrial counterparts and their environments are also relatively foreign to people. So, understanding change and variability in aquatic ecosystems both temporally and spatially often poses a greater challenge than the monitoring of land-based biological systems of which we are mostly a part.

13.2.1 Cyclical change and variability

Interpretations from contemporary monitoring will be drawn to attributing any shift in state as a change in the system. This view is based on an assumption that the state is stable at the commencement of the monitoring programme and so any variation is viewed as a departure from this perceived baseline. However, the system may be in one of many possible states at a single point in time including:

i. a phase of response or recovery from a past perturbation;
ii. in the peak, trough, rise or fall phases of a cyclical pattern of variation; or
iii. undergoing directional change through evolutionary processes or the establishment of new feedback mechanisms.

The lack of consideration of these different timescales in the understanding of the evolution of aquatic systems can lead to misleading conclusions and prevent reliable prediction of future changes. For instance, the increasing trend in organic carbon concentrations in surface waters across parts of Europe and North America during the past decades has been essentially assimilated to one disturbance due to climate change. More recent palaeostudies on Swedish lakes, however, revealed that this was preceded by a landscape-wide, long-term decrease in DOC beginning as early as AD 1450–1600 (Meyer-Jacob et al. 2015). Thereby, instead of a climate change disturbance, this increasing DOC trend could in part be the result of watershed recovery to past human disturbances.

Freshwater systems are subject to variations in the hydroclimate. Orbital forcing of climate drives substantial, global scale variations in moisture availability. During glacial phases large volumes of water are locked up as ice and sea levels are in the order of 130 m lower than present. However, temperatures are lower and so are evaporation rates. This reduced energy can dampen the intensity of tropical climatic systems and lead to adjustment in global circulation patterns and ocean currents. Conversely warmer interglacial periods see the contraction of ice caps and glacier fields, the rise in sea levels, and the increase in heat in the tropics which may intensify circulation cells. These combine to drive changes in effective moisture and run-off which vary regionally on account of location and topography. While early human populations experienced the substantial rise in sea level and the inundation of freshwater systems by the sea, this remains part of a cyclical change with a precedent that pre-dates the emergence of humans from Africa. Cycles of short return times also exist explaining much of the variation through the Holocene. These include the Medieval Climate Anomaly (800–1200 AD; Lamb 1965; Bradley et al. 2003) and the Little Ice Age (1400–1850 AD;

see in Bradley and Jones 1993). The latter represents a cool baseline which emphasises the effect of recent anthropogenic warming (Bradley and Jones 1993). The latter represents a directional shift (at least until emissions reductions take hold) which has the capacity to subdue the effect of orbital forcing to the extent that Earth may avoid the next glacial phase in 53,000 years (Stager 2011).

Freshwater systems are subject to hydroclimatic variations that are both cyclical at higher frequencies, and stochastic. Climate cycles such as those driven by phenomena that recur over multi-decadal (Interdecadal Pacific Oscillation) and shorter (*El Niño* Southern Oscillation; North Atlantic Oscillation) time frames can bring extended periods of wet or dry conditions. These influence lake levels, river–floodplain connection, and flood and drought frequencies. For instance, positive phases of the Atlantic Multidecadal Oscillation manifested in the USA through a warmer and drier climate, that caused the Midwest droughts in the 1930s and 1950s (Enfield et al. 2001). The Mississippi River outflow can vary by 10 per cent between positive and negative phases, and the inflow to Lake Okeechobee, Florida by up to 40 per cent. Further, these extreme states can increase fluvial or aeolian sediment transport, nutrient release through erosion or fire and the incidence of stratification. Independent of these major droughts or floods events can occur which may induce changes in channel course, impound valleys, or cause aggradation (Figure 13.1). The almost 13 years of drought in south-eastern Australia at the start of the twenty-first century associated with a positive IPO (Gergis et al. 2011) had a major impact on the rivers and wetlands in the Murray-Darling Basin with low water levels exposing acid sulphate sediments (Hall et al. 2006) and resulting in prolonged dry conditions with major ecological

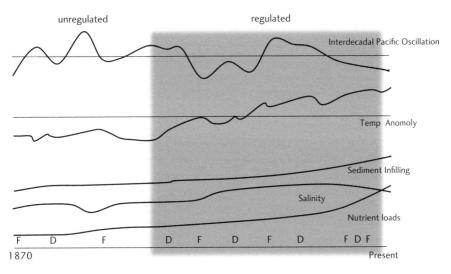

Figure 13.1 Theoretical patterns of change in the drivers of the state of a wetland over time (e.g., a floodplain lake in the southern Murray River Basin). These drivers can act to subdue others or generate synergistic effects amplifying the range of variation (Gell 2017b).

consequences prompting large declines in populations of many species (Kingsford et al. 2011). The so-called 'boom and bust' ecological conditions that characterise these landscapes (Bunn et al. 2006) were exhibited when the drought was broken by massive flooding in 2010.

On the other hand of the time spectrum, aquatic ecosystems undergo small (seasonal) to very small (diurnal) scale variability in their physical, chemical and biological components. Cyclicity of solar irradiance is the ultimate driver of such variability, the sun being the major source of heat for fresh waters. Water temperature varies between day and night or between season, with cascading effects on the physics of the waterway or waterbody. Diurnal and seasonal variations in solar radiation and temperature regulate river discharges through their effects on evapotranspiration, infiltration, and snowmelt. The amplitude itself of this diurnal cyclicity in discharges depends on season, the dominant climate over the watershed and the location of the river. Snowmelt-dominated rivers have their highest flows and largest diurnal fluctuations during the spring melt season. Evapotranspiration/infiltration-dominated rivers have seasonal maximum flows during the winter rainy season but exhibit strongest diurnal cycles during the summer months when discharge is lowest (Erup 1982). Temporal variability in river discharge can also manifest in terms of cyclicity in sediment transport and riverbed erosion (Retelle and Child 1996). In lakes, heat losses and wind forcing at night can trigger vertical convection (Anderson 1968), the deepening of the surface layer while, during the day, the diurnal thermocline re-establishes because of solar heating, and the daytime decay of turbulence within the subsurface layer between the diurnal thermocline and the seasonal thermocline (MacIntyre et al. 2002). These conditions have a marked impact on the productivity of the water column in warm polymictic lakes, such as Lake Moondarra in inland northern Australia where thermal stratification can occur during the day but breaks down in the night in the cooler months, while in the warmer months intense rainstorms prevent the establishment of a persistently stratified water column (Finlayson et al. 1980). Light cyclicity triggers variability in photosynthetic activities and subsequent gas concentrations in waters, such as their redistribution and transport within the waterbodies through currents and convections (Eugster et al. 2003). As also shown in Lake Moondarra the stratification also affects the distribution of nutrients and is an important factor in determining the trophic status and productivity throughout the water column (Farrell et al. 1979; Finlayson et al. 1984). In higher latitudes seasonal variations in lake ice cover conspire to produce marked annual band pairs (winter/summer) in sediments (varves) that enable sequences to be accurately aged (Lamoureux 2001).

13.2.2 Accounting for variability to understand change

Changes at human-relevant timescales (decadal to sub-decadal time scales) superimpose on all these different time levels of ecosystem physical, chemical and biological variability. Because palaeo-records can be analysed at different time resolutions, from annual time resolution for varved records up to millennial time scales, they give access to many of these embedded cycles of ecological variability for the waterbody of interest. When reconstructed over the last centuries, the ecological history can help targeting a reference state of the waterbody, as for pre-eutrophication phosphorous concentrations (Bennion et al. 2011). Palaeo-records also comply with the concept of natural variability that accounts for the range of ecological states a system can undertake under natural external forcings and thereby assorts the reference state with a confidence interval (i.e., limits of acceptable changes under natural forcings). Besides, provided with an adequate time scale of the changes, palaeo-records provide a quantification of the speed at which the ecosystem is being altered and this can discriminate, for instance, between processes of natural versus cultural eutrophication (Battarbee and Bennion 2012). Last, dynamics of changes in the ecological state of a waterbody can be quantified in terms of variance and autocorrelation, two estimates for ecosystem resilience (Dakos et al. 2015). Altogether, palaeo-records document, at human-relevant timescales, how far the ecosystem has departed from its natural reference range, how fast changes occurred, and whether its resistance and resilience to ongoing cumulative disturbances have been compromised (Bunting et al. 2016). Further, as they are effective at reconstructing changes to regulating services through time, they can be used to critically assess the sustainability of the intensification of the use of provisioning services and so potentially warn of future system collapse (Dearing et al. 2012).

Palaeoecological approaches have been used to demonstrate the causes of change in wetlands with examples outlined in Gell and Finlayson (2016) covering the acidification and eutrophication of lakes, salinisation of wetlands, and pollution from local and global sources of mercury. As sediments contain a diverse array of biological and chemical materials they can often be used to reconstruct the ecological condition of wetlands and lakes over time. Traditionally, fossil diatoms (Battarbee et al. 2001), pollens (Bennett and Willis 2001), and ostracods (Holmes 2001) have been used, and these techniques and their applications are well reviewed. Increasingly, plant macrofossils (Birks 2001), algal pigments (Leavitt and Hodgson 2001), testate amoebae (Beyens and Meisterfeld 2001), cladocerans (Korhola and Rautio 2001), and chironomids (Walker 2001) are also being applied. While these palaeoecological approaches can rarely reconstruct the nature of, for example, waterbird or fish populations over time, they can provide information on long-term changes in water quality, sediment load, plant and algal communities, and invertebrates, particularly ostracods, cladocerans, and chironomids (Gell et al. 2016), from which deductions can be made as to the nature of the wetland ecosystem over time. While inferences can be drawn as to the cause of changing bird populations from palaeolimnological evidence for changing prey availability (Brookes et al., 2012) the recent emergence of the DNA extraction and analysis from sediment records has the capacity to reveal more about not only the nature of catchment change (Giguet-Covex et al. 2014) but also estimating the colonisation timing of fish taxa (Olajos et al. 2017) and establishing whether certain species are native or exotic (Stager et al. 2015). Recent applications of C and N stable isotope techniques on organic sedimentary components (bulk sediment, fossil remains, or compound specific) can complement the time-picture by documenting the role of anthropogenic nutrient inputs on observed changes in biodiversity (atmospheric nitrogen deposition, Wolfe et al. 2001) or whether biodiversity changes came along with further modifications of food webs and lacustrine carbon cycle (Perga et al. 2010; Frossard et al. 2015; or Van Hardenbroek et al. 2014).

Palaeoecological records such as these can extend the knowledge of change in the ecological character of lakes and wetlands back centuries and even millennia, although with decreasing resolution for the latter. There are now many palaeoecological records taken from sediment sequences for many Ramsar Sites and other wetlands. In several instances these studies have revealed that the ecological character at the time of listing as a Ramsar Site have misrepresented the condition and variability that has occurred. Century-long sediment records from the Peace-Athabasca Delta site, for example, show lakes changing in response to engineering works which preceded listing, as well as a gradual trend in reduced flooding under a drying hydroclimate (Wolfe et al. 2008). Identification of such changes can create a tension when considering the 'natural ecological character', 'limits of acceptable change', and 'degraded state' of any particular wetland, and can cause confusion when determining how to manage for change in ecological character. This misrepresentation can result in perverse or strongly contested management decisions about the steps needed to manage a particular wetland, as well illustrated for the lakes and wetlands at the mouth of the Murray River in South Australia where a human-determined condition is being purposefully maintained (Gell 2017a) with implications for managing water flows throughout the upstream river system (Gell and Reid 2016).

13.2.3 Understanding directional change

Within a subset of time waterways undergo directional change, even if that change is merely part of variation nested within a cycle with long return intervals. So, the formation of a cut-off meander or billabong arises from the isolation of a section of river as a consequence of the ongoing migration of the river within the bounds of its floodplain. Once isolated, the lake will undergo directional change while the morphology of the river will continue to vary in response to hydroclimatic forces. Kashouh (2012), for example, dated palaeochannels in the Missouri River floodplain and revealed the continuous evolution of meander loops which ultimately transitioned from a meandering system to braided ~1600 years BP. Over the longer time frame the directional change occurring within the cut-off is also part of a geomorphic cycle.

The newly abandoned meander cut-off will inherit its form from the fluvial forces that shaped its parent river. So, it will conceivably be deep with steep sides. It will continue to receive clastic and organic material directly from the floodplain and that transported by the river will enter the cut-off when connectivity is re-established during flood events. It will also host its own biota which will also contribute to the mix of materials which will settle in the benthos under the still water conditions of the lake. Depending on the light environment of the lake the open water sections may be dominated by pelagic forms such as plankton and fish and submerged aquatic plants may be confined to the littoral margins where they are able to gain light but remain rooted to the substrate. As the productivity of aquatic plants is often greater than pelagic systems, the seston accumulates slowly in the bottom of the lake at first. In systems isolated from disturbance from industrialised people sediments may accumulate at less than 1 mm/yr and so the morphology of the lake may remain functionally similar for centuries or even millennia.

Ultimately, however, infilling will bring the benthos to within the photic zone whereupon submerged plants may begin to colonise the floor of the lake accelerating net accumulation in the sediments through organic production and sediment trapping. Large floods may act to reverse this process through scour events but net accumulation sets the wetland on a path towards terrestrialisation. The shallowing of the lake means that it holds less volume when inundated and so is more likely to dry when isolated. So, it evolves from a deep, plankton-dominated system to one with rich plant beds with littoral fauna and ultimately to a shallow plant-dominated, intermittent or seasonal semi-terrestrial wetland, perhaps after several thousand years after it was first abandoned by the river.

13.3 Long-term ecological change

Lakes are formed through a range of geomorphic processes ranging from glaciations, channel meandering, deflation, volcanic and aeolian activity, and sea level change, among others. As noted above, clastic, chemical, and biological material then begins to accumulate in the benthos more or less continuously, unless flooding or drying events lead to the removal of some layers. While the rate of accumulation may vary over time this steady build-up of material represents an archive of the state of the lake through time. At any point in time these sediments can be accessed by sediment coring (see Skilbeck et al. 2017 for a range of coring techniques) which extracts a column of the lake sediments (Figure 13.2). Each sediment depth increment represents a point in time and a range of radiometric and other dating techniques can be applied to establish the chronology for the entire sequence.

Traditionally, such sediment sequences were analysed for sediment particle size and geochemistry and fossil pollen were used to reconstruct the aquatic, local, and regional vegetation through time. Increasingly, fossil diatom algal remains have been extracted as bioindicators of past water quality (e.g., pH, salinity). Contemporary, 'multi-proxy' studies of lake sediment sequences may employ any of a wide range of preserved indicators (Table 13.1) that are the preserved evidence of the physical, chemical or biological conditions of the lake through time.

As climate changes continuously so do lake ecosystems and so long sediment sequences can reveal evidence for climate cycles in the past. Long sequences in tectonic basins, and crater lakes, may archive records that extend over one million years and provide evidence for orbital forcing on climate. The fossil remains of temperature dependent biota can be combined with temperature responsive isotopes to gauge thermal variability, which indicators of lake depth and salinity can together be used to derive moisture budget variations. Multi-indicator analyses of a 569-m core from Lake Ohrid in Albania for example reveal broad hydroclimate patterns in the Mediterranean and the response of the lake ecosystem and its catchment over 630,000 years (Wagner et al. 2017). Large lake fields allow for the reconstruction of multiple climate records from the same region allowing for the partitioning of the response of each individual lake

Figure 13.2 A short sediment core extracted manually from a wetland using a d-section or Russian corer (Jowsey 1966) (photo taken by P. Gell).

from the regional pattern. Syntheses of multiple sites across large areas can generate evidence for the response of lakes and rivers to global-scale climate changes as in the review of intertropical Africa by Barker et al. (2004) as part of the PEP III (Pole-Equator-Pole) project of IGBP-PAGES. Also, long sequences also record evidence for hydroseral evolution in lakes which also needs to be taken into account when reconstructing hydroclimate.

Lake sequences have been used extensively to document the impact of humans on lakes and their catchments. Such studies benefit from an understanding of the ethnohistory of the landscape and the technological evolution of the inhabitants (Dubois et al. 2018). European records attest to the influence of people on lake eutrophication over millennia (Bradshaw et al. 2006) while rapid shifts in aquatic systems through the construction of impoundments are evident from the early nineteenth century (e.g., Reeves et al. 2016). As the settlement of the planet was time transgressive, so the patterns of the first detectable impact of people vary through time, not least because new immigrants arrived with a different suite of tools depending upon when a land was first colonised (Mills et al. 2017; Dubois et al. 2018).

The arrival of people can be independently detected by the first arrival of pollen of domesticated plants (e.g., *Olea*) or weeds (*Plantago*) or the plants deliberately transported and planted in new lands (e.g., *Eucalyptus*, *Pinus*). Corroborative evidence may come in the form of changed sedimentation rates or sediment markers such as mineral magnetics, increase in nitrogen associated with animal wastes, or abrupt changes in proxies of water chemistry, including pH, nutrients, and salinity. Where documentary evidence for first human arrival is available then radiometric techniques can be used to identify particular ages; however, accuracy diminishes with depth.

Table 13.1 *A range of proxies used in palaeolimnology that can be used to reconstruct change in freshwater systems through time.*

	Indicator	Measure	Inference
Physical	Particle size	Clay, silt, sand	Fluvial energy
	Sediment accumulation rate	210Pb, 137Cs, luminescence, 14C	Erosion inputs
	Mineral magnetics	Magnetic susceptibility	Sediment source
	Laminations, varves		Hypoxia
Chemical	Organic content	Loss-on-ignition	Productivity
	C & N isotopes of biological remains	$\delta13C; \delta15N$	C & N biogeochemistry
	O isotopes of carbonaceous or siliceous shells	$\delta16O, \delta18O$	Temperature
	Elements	Sr, Ti, Ca, Si	Sediment source and detrital origin
Biological	Pollen; plant macrofossils	Species; presence of exotic taxa	Vegetation, dating
	Charcoal	Count	Fire history
	Diatoms	Species assemblage, indices; transfer functions	Salinity, nutrients, pH, turbidity, water source, habitat structure
	Ostracods	Species assemblage; shell chemistry	Salinity, anoxia
	Cladocerans	Species assemblage	Habitat, food-web structure
	Chironomids	Species assemblage	Anoxia
	Testate amoebae	Species assemblage	Marsh hydrology
	aDNA	Presence/absence or relative change in abundance of a given taxon	Extinctions; invasions; population dynamics and diversity

Significant human impacts have been detected from palaeolimnological approaches, including the impact and cause of lake acidification from acid rain (Battarbee and Charles 1986), the widespread increase in hypoxia in the nineteenth century through release of urban wastes into receiving waters (Jenny et al. 2016), the salinisation of floodplain systems through irrigated agriculture (Gell et al. 2007), the impact of heavy metals on lake ecology (Sayer et al. 2006), and the existence of regime shifts in eutrophic waters (Wang et al. 2012). Consistently, these records attest to substantial change to lake ecosystems well before detection by contemporary monitoring methods. Importantly, they also document and quantify the efficiency of the recovery after management actions (Perga et al. 2015), such as nutrient mitigation (French peri-alpine lakes) and liming of acidified systems (Battarbee et al. 2014).

13.4 Detecting ecological response to disturbance

13.4.1 Limnological response over time

Lakes and rivers respond to forces in their catchment over time. These can be discreet events such as the accidental release of pollution such as oil but are more often the input of materials or biota over an extended period. These may be direct inputs from an identifiable source and, like individual events, the effects can be readily determined with a before-after-control-impact (BACI) design assessment, provided there was sufficient foresight to monitor the impacted, and control, systems before impact, as done in Kakadu National Park, Australia (Humphrey et al. 1999). More challenging and pervasive are the chronic effects of diffuse inputs to aquatic systems which can arise through multiple release points, or unidentifiable inlet points as in sediment input through catchment erosion. As these are difficult to identify both in space and time, it is often challenging to attribute a detected ecological change to specific disturbances, and in separating the effect of events or new regimes of disturbance given ongoing cyclical and directional change and variability.

Monitoring programmes are increasingly a basis for structured research and there has been a recent focus in these in drawing from citizen science. The compounding of natural variability and the increasing regime of disturbance through into the Anthropocene make detecting limnological change and attributing that change to a cause increasingly challenging (Capon et al. 2015; Finlayson et al. 2017a). Necessarily most monitoring programmes are short, hampered by the short duration of training and grant programmes and the ephemeral nature of government commitments to invest in this work. Despite this there are a vast number of studies which document the response of aquatic systems to disturbance reported in the literature and this volume. In particular where there has been a long-term commitment to monitoring new insights into variability and change arise. Important examples include the Hubbard Brook (Likens 2013) which has continued for in excess of 50 years and the lake observatories in the UK which commenced in the 1930s and continue to the present (Maberly and Elliot 2011).

Where long-term monitoring programmes have been implemented, particularly if coupled with palaeolimnological records, it is possible to characterise variability and so better identify the response of the system to particular events. This is well shown for a number of wetlands including those listed as Ramsar sites, such as the Everglades in the USA and the Gippsland Lakes in south-eastern Australia (Gell et al. 2016). However, despite some notable exceptions, long-term monitoring programmes are rare as limnological research is a relatively young field and cost and time demands often leave well planned programmes abandoned. Further, palaeolimnological records may not be of sufficient resolution to identify the first point of ecological change and their chronology may not be sufficiently accurate to attribute a detected change to a known disturbance. Given the antiquity of human impact on aquatic ecosystems (Dubois et al. 2018) the identification of cause and effect often remain unclear or presumed, and generally not supported by hypothesis-based assessment and monitoring.

13.4.2 Ecosystem recovery

The term recovery implies that a system will return to a previous condition after being in a degraded or disrupted one. This condition can be assessed through different indicators, in different terms, depending upon the questions being asked. Recovery or distance to full recovery is quantified by examining changes in specific, targeted community structures or fundamental ecological processes in response to human activities; or even by monetising extractable ecosystem services. Major human driven disturbances on lakes and rivers include eutrophication, acidification and hydromorphological degradation that stem primarily from human population growth and increases in urbanisation (changes in hydrology and of nutrients and other substances), industrialisation (air pollution/acidification and flows of substances), land use (agricultural intensification affecting flows of water, landscape morphology, and flux of substances), and water-use changes (e.g., drinking water, recreation) (Verdonschot et al. 2013). Although anthropogenic pressures are usually assumed to be rather similar between lakes and rivers, eutrophication and acidification have received the most attention in lake studies, while hydromorphological changes were more topical for rivers. Long-term studies of recovery in rivers and lakes based on regular monitoring surveys are scarce yet differ on the considered system and also indicator. Recovery after riparian buffer instalment may take at least 30–40 years (Jowett et al. 2009). In lakes, time for recovery from eutrophication varies from 10 to 20 years for macroinvertebrates, 2 to 40 years for macrophytes, and 2 to 10 years for fish (Jeppesen et al. 2005). Natural recovery from acidification takes much longer compared with recovery after liming, and like eutrophication, biological recovery is taxon-specific and often decades are needed to achieve pre-disturbed conditions (Verdonschot et al. 2013). A comparison of concepts and models used in both degradation and restoration studies clearly revealed that processes following restoration do not mirror those during degradation (i.e., the trajectories of degradation and recovery differ). Because they extend the instrumental time span, palaeoecological approaches can ascertain whether these discrepancies result from true hysteresis (Wang et al. 2012; Randsalu-Wendrup et al. 2016) or is due to a new combination of confounding factors such as climate change (Perga et al. 2015; Bunting et al. 2016) but they almost inevitably result in a different overall status for a water body as before the disturbance regime, referred to as the recovery debt (Moreno-Mateos et al. 2017).

13.4.3 Regime shifts

Aquatic ecosystems can be expected to respond smoothly, almost linearly, to external disturbances, or instead, through non-linear, catastrophic shifts. Linear responses are, for instance, observed when the mean phosphorus concentration of lake water decreases as inputs are reduced. This has been the case for large lakes with limited release of phosphorus from the sediment (as for Lake Geneva, for instance) and in this case recovery is expected to occur with limited delays. Yet, ecosystems are likely to experience abrupt and drastic changes in their structure and functions beyond a certain threshold of an environmental disturbance. These catastrophic changes are usually referred as to critical transitions (Capon et al. 2015), and include 'regime shifts' (i.e., large, non-linear, persistent changes in the structure and function of a system) (Scheffer and Carpenter 2003). Such regime shifts are expected to become more frequent as the human influence on the planet increases. Since regime shifts imply non-reciprocal trajectories as the disturbing factor is reversed, efficient recovery requires that the environmental disturbance is decreased even below the initial values at which the threshold was passed. These hysteretic responses thereby generate critical consequences for the cost and success of ecosystem restoration and management.

Shallow lakes have long been model systems for testing regime shift theories; they exist in only the macrophyte-dominated, clear-water state at low nutrient concentrations, in a turbid, phytoplankton-dominated state at high nutrient concentrations, or in either state at the intermediate nutrient range (Scheffer and van Nes 2007). Specific feedback mechanisms, mainly related to macrophyte interaction with water clarity and sediment, operate to maintain the ecosystem in this particular stable state (i.e., its current attraction basin). If the system drifts away from the attraction basin due to an external forcing, feedbacks become less efficient, resilience (defined as the capacity of the system to recover its function after a perturbation) decreases, and the system becomes directed toward the route of a regime shift as it reaches a threshold. Mathematically, these feedback mechanisms generate increasing autocorrelation and variance for state variables until the system reaches a threshold (Andersen et al. 2009). These dynamics of a 'critical slowing down' or 'flickering' provide an early warning signal of one form of catastrophic transition, which is the fold bifurcation (Dakos et al. 2015). Therefore, an important assumption for alternative stable states is that the endogenous feedback processes counteract exogenous forces and thus explain hysteretic responses to restoration measures. In larger and deeper lakes, stabilising mechanisms are generally assumed to be limited. As the relative macrophyte coverage decreases with lake size and depth, linear rather than discontinuous vegetation responses to changes in turbidity are favoured (Scheffer and van Nes 2007). Indeed, some modelling approaches have reported no remarkable effect of submerged macrophytes on water clarity for lakes deeper than 10 m (Genkai-Kato and Carpenter 2005), although further studies have indicated the possibilities of alternative stable states in lakes as deep as 20–30 m (Wang et al. 2012).

13.5 Recent anthropogenic change

While humans have impacted freshwater systems for centuries and even millennia, anthropogenic change continues at an increasing rate under great acceleration. Human stressors have increased from the 1950s, most exponentially: human populations, use of fertilisers, construction of impoundments, diversion of freshwaters, use of groundwater, have all increased through this time. This period also coincides with an expansion of research on the nature and state of freshwater systems so, while much of the impact of people is largely only available through palaeolimnological approaches, much of the recent anthropogenic change has been detected through monitoring and structured research, and summarised in various environment or biodiversity assessments, such as the Millennium Ecosystem Assessment (Finlayson and D'Cruz 2005; MEA 2005) and Global Environment Outlook (Armenteras and Finlayson 2012). These have identified the major drivers of recent change, such as changes in agricultural land practices, including the increased use of agro-chemicals, and the diversion of water for irrigation, and in some areas, increasingly for urban use, and the spread and establishment of invasive alien species.

Gordon et al. (2010) reviewed the trade-offs that occurred in freshwater systems, including with multiple ecosystem services, which have been generated by agriculture-induced changes to the water quality and quantity. They highlighted three main strategies by which agricultural water management could be used to deal with such trade-offs: (a) improving water management practices on agricultural lands, (b) better linkage with management of downstream aquatic ecosystems, and (c) paying more attention to how water can be managed to create multifunctional agro-ecosystems. This could only be done if the ecological landscape processes that often shape the ecological character of aquatic ecosystems were better understood, and the values of ecosystem services other than food production also recognised. The need to manage at landscape scales in order to sustain aquatic ecosystems and their species, especially those that move across landscapes, both locally and inter-continentally, has been seen as critical but not universally implemented (Finlayson et al. 2017b). In contrast many aquatic ecosystems have been managed as if they were isolated from the surrounding landscapes and land use practices, including many freshwater protected areas that almost by definition are connected to the surrounding landscape or support species that migrate across the landscape. This reality caused Falkenmark et al. (2007) to recommend the following steps to address the widespread social and environmental inequities and failures in governance and policy as well as on-ground management that are the fundamental drivers of anthropogenic change:

- Rehabilitate degraded ecosystems and, where possible, restore lost ecosystems;
- Develop institutional and economic measures to prevent further loss and to encourage further changes in the way we do business;
- Increase transparency in decision-making about agriculture-related water management and increase the exchange of knowledge about the consequences of these decisions.

13.6 Understanding change for managing freshwater systems

Understanding incremental change in freshwater systems requires the investment of long-term monitoring with regular assessment of the nature of the system relative to long-term baselines. Where this is unavailable a perceived change in an aquatic system may be considered deleterious when, in fact, it may represent one of many possible states in a variable system. Further, where a system has been degraded for some time, this condition may be considered natural and management may pursue measures to retain the system in a degraded state. Clearly, the management of an aquatic system would benefit from the provision of evidence of ecosystem condition over the long term to fully describe the range of historical condition. This would inform management of whether to act, which drivers of change to address and to identify targets for management actions.

13.6.1 Understanding change in Ramsar wetlands

In 1971 the United Nations established the Convention of the Protection of the world's most significant wetlands as an international mechanism to promote the conservation and wise use of wetlands. Deemed the Ramsar Convention today over 2,200 sites are proclaimed as internationally important by 169 nations, arguably representing around 13 per cent of the global area of wetlands (Milton and Finlayson 2019). In the nomination of a wetland for international importance signatory nations must make a case that it satisfies at least one of nine listing criteria, which mostly relate to their role in providing habitat for waterbird and fish populations, particularly those of rare or vulnerable taxa.

Declaration of a wetland under the Convention requires the signatory nation to identify its boundaries and to ascribe its 'natural ecological character'. Evidence for long-term change in fresh waterways dictates that many systems were impacted by human activities well before the establishment of Ramsar. So, in some instances the perceived natural character of these wetlands does not represent that which existed for extended periods in the past (Finlayson et al. 2016; Gell et al. 2016). As an example, the Lower Lakes at the mouth of the Murray Darling Basin in Australia was an estuary influenced by tidal inflows, but at the time of listing in 1985 it was described as a mostly freshwater system (Gell 2017a).

Contracting Parties to The Ramsar Convention on Wetlands have accepted a number of commitments to ensure the maintenance of the ecological character of all wetlands (Finlayson et al. 2011, 2017a; Pittock et al. 2010). However, despite extensive guidance that has been provided by the Convention (Finlayson et al. 2005; Davidson 2016) there remain gaps in the advice that is available to support wetland managers in measuring and assessing the extent of change in character. Fundamental to the approaches adopted by the Convention is a baseline description of the ecological character of a wetland. A key issue in this approach is the choice of a baseline or reference point against which change can be compared, including accounting for the natural variability and successional processes that shape the ecological character. Although the need to address the variability within wetlands when describing its character was recognised some years ago (Finlayson 1996), there has been little subsequent attention given to the issues (Finlayson et al. 2016; Gell et al. 2016). As a consequence those elements of the Convention that specifically address changes in the ecological character of wetlands have not been sufficiently developed and it seems that they are still largely addressed as if they were a static or non-evolving state (Finlayson et al. 2017).

The process of describing the ecological character of a wetland and assessing change in such character is strongly circumscribed by the condition of the wetland at the time it was designated by the Contracting Party, as well as the extent of ecological knowledge available, usually collected through traditional ecological surveys and surveillance. Various techniques, notably palaeoecology, have the potential to contribute to filling

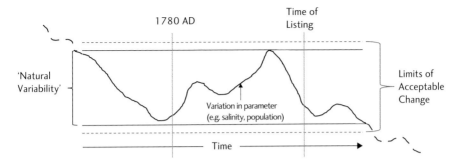

Figure 13.3 Theoretical approach to developing Limits of Acceptable Change within Ramsar wetlands (redrawn from Phillips 2006). The dashed lines are added to illustrate that longer cycles of change may challenge even this record of natural variability. More realistically, there is little continuous monitoring data that extends far beyond the time of listing to realistically assess variation in any parameter and so palaeoecological records are essential in defining limits (Gell 2017b).

this knowledge gap, and to improve understanding of the trajectories of past and recent change, so providing the context for fluctuations and trajectories of likely future change.

Further to the challenges of identifying natural ecological character, contracting parties have been encouraged to identify the 'limits of acceptable change' in the nature of their listed wetlands. While this attempts to characterise the nature of variability in condition (Figure 13.3) it is also limited by the availability of evidence of sufficient duration to capture the drivers of change that may continue to remain at play in future change. So again, palaeoecological evidence can extend the instrumental record and strengthen the limits the understanding of historical variability. This line of evidence may enable nations to gain a better understanding of the management options and the 'limits of acceptable change' and can identify whether causes of change are outside the jurisdiction of the party, perhaps relieving them for responsibility for reparation. Alternatively, in the absence of a longer-term view, nations may be excused for all impacts prior to listing thereby hampering the charter's goals of restoring the world's most significant wetlands. Records of past change, together with future climate scenarios, was used by Newall et al. (2016) to explore the suitability of the natural ecological character of the Riverland site in South Australia, and review the prospects of limits of acceptable change given projected catchment drying.

13.6.2 The novel ecosystem

In recent years, the term 'novel ecosystems' has been increasingly used in the literature when describing ecosystems that have been significantly altered or completely transformed in terms of their historical ecological structure, composition, and function, and yet remain self-sustaining. These ecosystems are characterised by new species assemblages, distributions, and abundances that change or modify the traditional or historical structure and functions of the ecosystem. Intensely cultivated, engineered, or otherwise disturbed ecosystems that have been abandoned are prime candidates for the creation of novel ecosystems. In their seminal paper, Hobbs et al. (2006) have framed the discussion on novel ecosystems and its implications for the theory and practice of ecosystem restoration, and specifically how do we conserve, manage, and/or restore these ecosystems.

A widely used definition of ecological or ecosystem restoration is provided by the Society for Ecological Restoration International & Policy Working Group (2004): 'Ecological restoration is the process of assisting the recovery of an ecosystem that has been degraded, damaged or destroyed.' While this definition emphasises the need to re-establish the historical trajectory of an impaired ecosystem with respect to its structure and function, it is generally recognised that many degraded ecosystems can no longer feasibly be restored to any particular pre-disturbance or historical condition. Indeed, within certain circles, restoration projects/programmes are often considered effective when managing or manipulating biotic and/or abiotic variables successfully reinstates the provisioning and regulatory services desired by society and assists in the recovery of a mature, resilient ecosystem, whether historical, hybrid, or novel.

There are a combination of biotic and abiotic drivers—environmental (e.g., climate change, fire, and drought), ecological (e.g., invasive species, salinisation, nutrient imbalances), and human (e.g., urbanisation, land degradation and fragmentation), and their feedback loops—that are contributing to the increasingly rapid transformation and establishment of ecosystems that have never before existed.

Abiotic changes, such as climate change, nutrient imbalances, urbanisation, pollution, and land-use changes, rarely occur without impacting the living components of an ecosystem. Biotic drivers, such as species decline and extinction, coupled with increases in the prevalence of non-natives may initially result in the emergence of a hybrid ecosystem. However, as new species assemblages begin to dominate and functional traits begin to change, an ecosystem may continue its slide along the continuum and achieve novel status. This could be the case in the wetlands in the Carlos Anwandter Sanctuary in Chile where a wetland formed after an earthquake led to the establishment of the invasive submerged macrophyte *Egeria densa* which in turn led initially to the establishment of a thriving ecosystem with large population of black-necked swans, later collapsing with the swans departing. The macrophtye species concerned is further seen as an ecological engineer that decreases water turbulence, decreasing resuspension of sediments, and increasing sedimentation (Yarrow et al. 2009; Marin et al. 2017). Other examples of modified ecosystems occurring as a consequence of invasion by alien plant species include the widespread changes caused by the floating weed water hyacinth (*Eichhornia crassipes*) and the shrub *Mimosa pigra* in tropical and sub-tropical wetlands (Finlayson 2018).

This brings us to a further conceptual step: how to identify biotic and abiotic thresholds, whether the ecosystem has indeed crossed these thresholds, and how this informs our management options. This is an issue that points towards reassessing the concept of stationarity that has underpinned the implementation and operational outcomes from instruments such as the Ramsar convention (e.g., see Pittock et al. 2010; Finlayson et al. 2017a). In some hybrid ecosystems, interventions to restore important structures and functions can be helpful as part of an overall management strategy. However, in novel ecosystems, where thresholds have been crossed and a mature and resilient ecosystem persists, conservation and management focusing on the delivery of goods and services are generally the only options available.

Invasive or non-native species are now well entrenched and often provide essential habitat or resource functions in many ecosystems. Finlayson (2009) documented the effect of alien species, including floating weeds such as *Salvinia molesta* and *Eichornia crassipes* and exotic fauna such as the Nile Perch (*Lates nilotica*) and Asian Water Buffalo (*Bubalis bubalis*), on the ecological functions in freshwater systems. Some would argue that these non-natives can and should have an important role in providing essential ecosystem goods and services in the future. For example African *Tilapia* spp. have now become widely established in the world's tropical waters and, while they have impacted many native fish species, have become a staple food source. Therefore, conservation practices to prevent or reduce changes in ecological character could be used to maintain the health of an ecosystem in its hybrid state.

Distinguishing between hybrid and novel ecosystems will probably be a greater challenge for scientists than determining whether ecosystems are outside historical ranges of variability. One outstanding question is whether hybrid ecosystems represent a stable state or whether they are an unstable transition between Anthropocene and historical baselines. We hypothesise that hybrid river ecosystems are likely to be among the least disturbed rivers in many areas, including Cooper Creek and the Ovens and Paroo Rivers in Australia; the Fraser, South Fork Salmon, Buffalo, Upper Pitt, and Usumacinta Rivers in North America; and the unregulated tributaries in the Okavango and Congo Basin in Africa. These types of river ecosystems may be candidates for intensive restoration and conservation activities, but in many cases, anthropogenic baselines may still represent the most appropriate point of reference.

13.7 Final remarks

Reviews of the state of the world's freshwater systems identify them to be under considerable pressure to change suffering from many stressors. Increasingly, they are subject to anthropogenic warming, which, while birthing new glacial lakes from melt water, is an additional stressor likely to exacerbate the more direct impacts of catchment and waterway disturbance. The long-term record attests to considerable cyclical change as well as ongoing evolution of freshwater systems, and so the biodiversity of these systems has demonstrated considerable resilience, and resistance, over time. However, the additional and accelerating influence of humanity will inevitably impact gravely on freshwater biodiversity and limit the ecosystem services that these systems have provided people for millennia. Restoration of these systems (see Chapter 18) will be faced with increasing challenges and the best solutions may be a focus on the ecosystem function of, and ecosystem services provided by, what have now evolved into hybrid or novel ecosystems.

References

Anderson, D.V. (1968). Nocturnal heat loss of a lake and seasonal variation in its vertical thermal structure. *Hydrological Sciences Journal*, 13(3), 33–40.

Andersen, T., Carstensen, J., Hernandez-Garcia, E., and Duarte, C.M. (2009). Ecological thresholds and regime shifts: approaches to identification. *Trends in Ecology and Evolution*, 24, 49–57.

Armenteras, D., and Finlayson, C.M. (2012). Biodiversity. In UNEP. *Keeping Track of Our Changing Environment: From Rio to Rio+20 (1992–2012)*. Division of Early Warning and Assessment (DEWA), United Nations Environment Programme (UNEP), Nairobi.

Barker, P.A., Talbot, M.R., Street-Perrott, F.A., Marret, F., Scourse, J., and Odada, E.O. (2004). Late Quaternary climatic variability in Intertropical Africa. in Battarbee, R.W., Gasse, F., and Stickley, C.E. (eds) Developments in Paleoenvironmental Research vol 6., Past Climate Variability through Europe and Africa. Springer, Dordrecht, The Netherlands.

Battarbee, R.W., and Bennion, H. (2012). Using palaeolimnological and limnological data to reconstruct the recent history of European lake ecosystems: introduction. *Freshwater Biology*, 57, 1979–85.

Battarbee, R.W., Carvalho, L., Jones, V.J., Flower, R.J., Cameron, N.G., Bennion, H., and Juggins, S. (2001). Diatoms. In: Smol, J.P., Birks, H.J.B., and Last, W.M. (eds) Tracking Environmental Change using Lake Sediments, vol 3: terrestrial, algal and siliceous indicators. Kluwer Academic, Dordrecht, The Netherlands.

Battarbee, R.W., and Charles, D.F. (1986). Diatom-based pH reconstruction studies of acid lakes in Europe and North America—A synthesis. *Water Air and Soil Pollution*, 30, 347–54.

Battarbee, R.W., Simpson, G.L., Shilland, E.M., Flower, R.J., Kreiser, A., Yang, H., and Clarke, G. (2014). Recovery of UK lakes from acidification: An assessment using combined palaeoecological and contemporary diatom assemblage data. *Ecological Indicators*, 37, 365–80.

Bennett, K.D., and Willis, K.J. (2001). Pollen. In: Smol, J.P., Birks, H.J.B., and Last, W.M. (eds) Tracking Environmental Change using Lake Sediments, vol 3: terrestrial, algal and siliceous indicators. Kluwer Academic, Dordrecht, The Netherlands.

Bennion, H., Battarbee, R.W., Sayer, C.D., Simpson, G.L., and Davidson, T.A. (2011). Defining reference conditions and restoration targets for lake ecosystems using palaeolimnology: a synthesis. *Journal of Paleolimnology*, 45, 533–44.

Beyens, L., and Meisterfeld, R. (2001). Protozoa: testate amoebae. In: Smol, J.P., Birks, H.J.B., and Last, W.M. (eds) Tracking Environmental Change using Lake Sediments, vol 3: terrestrial, algal and siliceous indicators. Kluwer Academic, Dordrecht, The Netherlands.

Birks, H.H. (2001). Plant macrofossils. In: Smol, J.P., Birks, H.J.B., and Last, W.M. (eds) Tracking Environmental Change using Lake Sediments, vol 3: terrestrial, algal and siliceous indicators, Kluwer Academic, Dordrecht, The Netherlands.

Bradley, R.S., and Jones, P.D. (1993). 'Little Ice Age' summer temperature variations: their nature and relevance to recent global warming trends. *The Holocene*, 3, 367–76.

Bradley, R.S., Hughes, M.K., and Diaz, H.F., (2003). Climate in Medieval Time. *Science*, 302, 404–5.

Bradshaw, E.G., Nielsen, A.B., and Anderson, NJ (2006). Using diatoms to assess the impacts of prehistoric, pre-industrial and modern land-use on Danish lakes. *Regional Environmental Change*, 6, 17–24.

Brooks, S.J., Jones, V.J., Telford, R.J., Appleby, P.G., Watson, E., McGowan, S., and Benn, S. 2012. Population trends in the Slavonian grebe *Podiceps auritus* (L.) and Chironomidae (Diptera) at a Scottish loch. *Journal of Paleolimnology*, 47, 631–44.

Bunn, S.E., Thoms, M.C., Hamilton, S.K., and Capon, S.J. (2006). Flow variability in dryland rivers: boom, bust and the bits in between. *River Research and Applications*, 22, 179–86.

Bunting L., Leavitt, P.R., Simpson, G.L., Wissel, B., Laird, K.R., Cumming, B.F., St. Arnand, A., and Engstrom, D. (2016). Increased variability and sudden ecosystem state change in Lake Winnipeg, Canada, caused by 20th century agriculture. *Limnology and Oceanography*, 61, 2090–2107.

Capon, S.J., Lynch, J.J., Bond, N., Chessman, B.C., Davis, J., Davison, N., Finlayson, C.M., Gell, P.A., Hohnberg, D., Humphrey, C., Kingsford, R.T., Nielsen, D., Thomson, J.R., Ward, K., and MacNally, R. (2015). Regime shifts, thresholds and multiple stable states in freshwater ecosystems; a critical appraisal of the evidence. *Science of the Total Environment*, 534, 122–30.

Dakos, V., Carpenter, S., van Nes, E.H., and Scheffer, M. (2015). Resilience indicators: prospects and limitations for early warnings of regime shifts. *Philosophical Transactions of the Royal Society B-Biological Sciences*, 370, 20130263.

Davidson, N.C. (2016). Editorial: Understanding change in the ecological character of internationally important wetlands. *Marine and Freshwater Research*, 67, 685–6.

Dearing, J.A., Yang, X., Dong, X., Zhang, E., Chen, X., Langdon, P.G., Zhang, K., Zhang, W., and Dawson, T.P. (2012). Extending the timescale and range of ecosystem services through paleoenvironmental analyses, exemplified in the lower Yangtze basin. *Proceedings of the National Academy of Sciences*, 109, E1111–1120.

Dubois, N., Saulnier-Talbot, E., Mills, K., Gell, P., Battarbee, R., Bennion, H., Chawchai, S., Dong, X., Francus, P., Flower, R., Gomes, D.F., Gregory-Eaves, I., Humane, S., Kattel, G., Jenny, J.-P., Langdon, P., Massaferro, J., McGowan, S., Mikomägi, A., Ngoc, N.T.M., Ratnayake, A.S., Reid, M., Rose, N., Saros, J., Schillereff, D., Tolotti, M., and Valero-Garcés, B. (2018). First human impacts and responses of aquatic systems: A review of palaeolimnological records from around the world. *The Anthropocene Review*, 5, 28–68.

Enfield, D.B., Mestas-Nuñez, A.M., and Trimble, P.J. (2001). The Atlantic Multidecadal Oscillation and its relation to rainfall and river flows in the continental U.S. *Geophysical Research Letters*, 28, 2077–80.

Erup, J. (1982). Diurnal fluctuations of stage and discharge in the Danish River Suså. *Hydrology Research*, 13, 293–8.

Eugster, W., Kling, G., Jonas, T., McFadden, J.P., Wüest, A., MacIntyre, S., and Chapin, F.S. III (2003). CO2 exchange between air and water in an Arctic Alaskan and mid-latitude Swiss lake: Importance of convective mixing, *Journal of Geophysical Research*, 108, 4362.

Falkenmark, M., Finlayson, C.M., and Gordon, L. (coordinating lead authors) (2007). Agriculture, water, and ecosystems: avoiding the costs of going too far. In: Molden, D. (ed.) Water for food, water for life: a comprehensive assessment of water management in agriculture. Earthscan, London, UK.

Farrell, T.P., Finlayson, C.M., and Griffiths, D.J. (1979). Studies of the hydrobiology of a tropical lake in north-western Queensland: I. Seasonal changes in chemical characteristics. *Australian Journal of Marine and Freshwater Research*, 30, 579–595.

Finlayson, C.M. (1996). The Montreux Record. A mechanism for supporting the wise use of wetlands. In: *Proceedings of the 6th Meeting of the Conference of the Contracting Parties of the Convention on Wetlands. Technical Sessions: Reports and Presentations*. Vol 10/12 B, 32–8. Ramsar Convention Bureau, Gland, Switzerland.

Finlayson, C.M. (2009). Biotic pressures and their effect on wetland functioning. In: Maltby, E., and Barker, T. (eds) The Wetlands Handbook. Wiley-Blackwells, Oxford, UK.

Finlayson, C.M. (2013). Climate change and the wise use of wetlands—information from Australian wetlands. *Hydrobiologia*, 708, 145–52.

Finlayson, C.M. (2018). Alien plants and wetland biotic dysfunction. In Finlayson, C.M., Milton, G.R., Prentice, R.C., and Davidson, N.C. (eds) The Wetland Book II: Distribution, Description and Conservation. Springer Publishers, Dordrecht, 383–9. DOI 10.1007/978-94-007-4001-3_48.

Finlayson, C.M., Bellio, M.G., and Lowry, J.B. (2005). A conceptual basis for the wise use of wetlands in northern Australia—linking information needs, integrated analyses, drivers of change and human well-being. *Marine and Freshwater Research*, 56, 269–77.

Finlayson, C.M., Capon, S.J., Rissik, D., Pittock, J., Fisk, G., Davidson, N.C., Bodmin, K.A., Papas, P., Robertson, H.A., Schallenberg, M., Saintilan, N., Edyvane, K., and Bino, G.

(2017a). Policy considerations for managing wetlands under a changing climate. *Marine and Freshwater Research* 68, 1803–1815.

Finlayson, C.M., Clarke, S.J., Davidson, N.C., and Gell, P. (2016). Role of palaeoecology in describing the ecological character of wetlands. *Marine and Freshwater Research, 67,* 687–94.

Finlayson, C.M., Davidson, N., Pritchard, D., Milton, R., and MacKay, H. (2011). The Ramsar Convention and ecosystem-based approaches for the wise use and sustainable development of wetlands. *Journal of International Wildlife Law and Policy*, 14, 176–98.

Finlayson, C.M., Davis, J.A., Gell, P.A., Kingsford, R.T., and Parton, K.A. (2013). The status of wetlands and the predicted effects of global climate change: the situation in Australia. *Aquatic Sciences*, 75, 73–93.

Finlayson, C.M., Davidson, NC, Gell, P.A., Kumar, R., and McInnes, R.J. (2017b). Managing freshwater protected areas in the global landscape. In: Finlayson, C.M., Arthington, A.H., and Pittock, J. (eds) Freshwater Ecosystems in Protected Areas: Conservation and Management. Taylor and Francis, Oxford, UK.

Finlayson, C.M., and D'Cruz, R. (2005). Inland water systems. In: Hassan, R., Scholes, R., and Ash, N. (eds) Ecosystems and Human Well-being: Current State and Trends: Findings of the Condition and Trends Working Group. Island Press, Washington, DC.

Finlayson, C.M., Farrell, T.P. and Griffiths, D.J. (1980). Studies of the hydrobiology of a tropical lake in north-western Queensland: II. Seasonal changes in thermal and dissolved oxygen characteristics. *Australian Journal of Marine and Freshwater Research,* 31, 589–96.

Finlayson, C.M., Farrell, T.P., and Griffiths, D.J. (1984). Studies of the hydrobiology of a tropical lake in the north-western Queensland: III. Growth, chemical composition and potential for harvesting of the aquatic vegetation. *Australian Journal of Marine and Freshwater Research*, 35, 525–536a.

Gell, P.A. (2017a). Paleolimnological history of the Coorong: identifying the natural ecological character of a Ramsar wetland in crisis. In: Weckström, K., Saunders, K.M., Gell, P.A., and Skilbeck, C.G. (eds) Applications of Paleoenvironmental Techniques in Estuarine Studies Developments in Paleoenvironmental Research, volume 20. Springer, Dordrecht.

Gell, P.A. (2017b). Understanding change and variability to inform the Ramsar Convention on the Conservation of Wetlands of International Importance. *PAGES Magazine*, 25, 86–7.

Gell, P., Baldwin, D., Little, F., Tibby, J., and Hancock, G. (2007). The impact of regulation and salinisation on floodplain lakes: the lower River Murray, Australia. *Hydrobiologia*, 591, 135–46.

Gell, P., and Reid, M. (2016). Muddied waters: the case for mitigating sediment and nutrient flux to optimise restoration response. *Frontiers in Ecology and Evolution*, 4, 16.

Gell, P. and Finlayson, C.M. (eds) (2016). Understanding change in the ecological character of internationally important wetlands. *Marine and Freshwater Research*, 67, 683–879.

Gell, P.A., Finlayson, C.M., and Davidson, N.C. (2016). Understanding change in the ecological character of Ramsar wetlands: perspectives from a deeper time—synthesis. *Marine and Freshwater Research,* 67, 869–79.

Genkai-Kato, M., and Carpenter, S.R. (2005). Eutrophication due to phosphorus recycling in relation to lake morphometry, temperature, and macrophytes. *Ecology*, 86, 210–19.

Gergis, J., Gallant, A.J.E., Braganza, K., Karoly, D.J., Allen, K., Cullen, L., D'Arrigo, R., Goodwin, I., Grierson, P., and McGregor, A. (2011). On the long-term context of the 1997–2009 'Big Dry' in South-Eastern Australia: insights from a 206-year multi-proxy rainfall reconstruction. *Climatic Change*, 111, 923–44.

Gordon, L., Finlayson, C.M., and Falkenmark, M. (2010). Managing water in agriculture to deal with trade-offs and find synergies among food production and other ecosystem services. *Agricultural Water Management*, 97, 512–19.

Hall, K.C., Baldwin, D.S., Rees, G.N., and Richardson, A.J. (2006). Distribution of inland wetlands with sulphidic sediments in the Murray-Darling Basin, Australia. *Science of the Total Environment*, 370, 235–44.

Hobbs, R.J., Arico, S., Aronson, J., Baron, J.S., Bridgewater, P., Cramer, V.A., Epstein, P.R., Ewel, J.J., Klink, C.A., Lugo, A.E., Norton, D., Ojima, D., Richardson, D.M., Sanderson, E.W., Valladares, F., Vilà, M., Zamora, R., and Zobel, M. (2006). Novel ecosystems: theoretical and management aspects of the new ecological world order. *Global Ecology and Biogeography*, 15, 1–7.

Holmes, J. (2001). Ostracoda. In: Smol, J.P., Birks, H.J.B., and Last, W.M. (eds) Tracking Environmental Change using Lake Sediments. Vol 4, Zoological Indicators. Kluwer Academic Publishers, Dordrecht.

Humphrey, C.L., Thurtell, L., Pidgeon, R.W.J., van Dam, R.A., and Finlayson, C.M. (1999). A model for assessing the health of Kakadu's streams. *Australian Biologist*, 12, 33–42.

Jenny, J-P., Francus, P., Normadeau, A., Lapointe, F., Perga, M-E., Ojala, A., Schemmelmann, A., and Zolitschka, B. (2016). Global spread of hypoxia in freshwater ecosystems during the last three centuries caused by rising local human pressure. *Global Change Biology*, 22(4), 1481–9. doi: 10.1111/gcb.13193.

Jeppesen, E., Sondergaard, M., Jensen, J.P., Havens, K.E., Anneville, O., Carvalho, L., Coveney, M.F., Deneke, R., Dokulil, M.T., Foy, B., Gerdeaux, D., Hampton, S.E., Hilt, S., Kangur, K., Kohler, J., Lammens, E., Lauridsen, T.L., Manca, M., Miracle, M.R., Moss, B., Noges, P., Persson, G., Phillips, G., Portielje, R., Schelske, C.L., Straile, D., Tatrai, I., Willen, E., and Winder, M. (2005). Lake responses to reduced nutrient loading—an analysis of contemporary long-term data from 35 case studies. *Freshwater Biology*, 50, 1747–71.

Jowett, I.G., Richardson, J., and Boubée, J.A.T. (2009). Effects of riparian manipulation on stream communities in small streams: Two case studies. *New Zealand Journal of Marine and Freshwater Research*, 43, 763–74.

Jowsey, P.C. (1966). An improved peat sampler. *New Phytologist*, 65, 245–8.

Kashouh, M.V. (2012). A Late Holocene Meander-braid transition of the Lower Missouri River Valley. Doctoral Dissertation, University of Texas, Arlington, USA.

Kingsford, R.T., Walker, K.F., Lester, R.E., Young, W.J., Fairweather, P.G., Sammut, J., and Geddes, M.C. (2011). A Ramsar wetland in crisis—the Coorong, Lower Lakes and Murray Mouth, Australia. *Marine and Freshwater Research*, 62, 255–65.

Korhola, A., and Rautio, M. (2001). Cladocera and other Branchiopod Crustaceans. In: Smol, J.P., Birks, H.J.B., and Last, W.M. (eds) Tracking Environmental Change using Lake Sediments. Vol 4, Zoological Indicators. Kluwer Academic Publishers, Dordrecht.

Lamb, H.H. (1965). The early medieval warm epoch and its sequel, *Palaeogeography, Palaeoclimatology, Palaeoecology*, 1, 13–37.

Lamoureux, S.F. (2001). Varve chronology techniques. In: Last, W.M. and Smol, J.P. (eds) Tracking Environmental Change using Lake Sediments. Vol 2. Physical and Geochemical Methods, 247–60. Kluwer Academic Publishers, Dordrecht.

Leavitt, P.R., and Hodgson, D.A. (2001). Sedimentary Pigments. In: Smol, J.P., Birks, H.J.B.. and Last, W.M. (eds) Tracking Environmental Change using Lake Sediments, vol 3: terrestrial, algal and siliceous indicators, 295–325. Kluwer Academic, Dordrecht, The Netherlands.

Leng, M.J., Lamb, A.L., Heaton, T.H.E., Marshall, J.D., Wolfe, B.B., Jones, M.D., Holmes, J.A., and Arrowsmith, C. (2005). Isotopes in lake sediments. In: Leng, M.J. (ed.) Isotopes in Paleoenvironmental Research. Developments in Paleoenvironmental Research, volume 10, 147–84. Springer, Dordrecht.

Likens, G.E. (2013). The Hubbard Brook Ecosystem Study: Celebrating 50 years. *Bulletin of the Ecological Society of America*, 94, 336–7.

Maberly, S.C. and Elliot, J.A. (2011). Insights from long-term studies in the Windermere catchment: external stressors, internal interactions and the structure and function of lake ecosystems. *Freshwater Biology*, 57, 233–43.

MacIntyre, S., Romero, J.R., and Kling, G.W. (2002). Spatial-temporal variability in surface layer deepening and lateral advection in an embayment of Lake Victoria, East Africa. *Limnology and Oceanography*, 47, 656–71.

Meyer-Jacob, C., Tolu, J., Bigler, C., Yang, H., and Bindler, R. (2015) Early land use and centennial scale changes in lake-water organic carbon prior to contemporary monitoring. *Proceedings of the National Academy of Sciences*, 112, 6579–84.

Millennium Ecosystem Assessment (2005). Ecosystems and human well-being: water and wetland synthesis. Washington, DC: Island Press. 68pp.

Mills, K., Schillereff, D., Saulnier-Talbot, E., Gell, P., Anderson, N., Arnaud, F., Dong, X., Jones, M., McGowan, S., Massafero, S., Moorhouse, H., Perez, L., and Ryves, D. (2017). Deciphering long-term records of natural variability and human impact as recorded in lake sediments: the palaeolimnological conundrum. *WIREs Water*, 4, e1195.

Milton, G.R., and Finlayson, C.M. (2019). Diversity of freshwater ecosystems, global distributions, adaptations. In: Hughes, J.M.R. (ed.) Freshwater Ecology and Conservation: Approaches and Techniques; Oxford University Press, Oxford, 3–21.

Moreno-Mateos D., Barbier, E.B., Jones, P.C., Jones, H.P., Aronson, J., López-López, J.A., McCrackin, M.L., Meli, P., Montoya, D., and Benayas, J.M.R. (2017). Anthropogenic ecosystem disturbance and the recovery debt. *Nature Communications,* 8, 14163.

Newall, P., Lloyd, L., Gell, P., and Walker, K. (2016). Implications of environmental trajectories on limits of acceptable change: a case study of the Riverland Ramsar Site, South Australia. *Marine and Freshwater Research*, 67, 738–47.

Olajos, F., Bokma, F., Bartels, P., Myrstener, E., Rydberg, J., Öhlund, G., Bindler, R., Wang, X-R., Zale, R., and Englund, G. (2017) Estimating species colonization dates using DNA in lake sediment, *Methods in Ecology and Evolution*, 9(3): 535–43. DOI: 10.1111/2041-210X.12890.

Perga M.-E., Frossard, V., Jenny, J-P., Alric, B., Arnaud, F., Berthon, V., Black, J.L., Domaizon, I., Giguet-Covex, C., Kirkham, A., Magny, M., Manca, M., Marchetto, A., Millet, L., Paillès, C., Pignol, C., Poulenard, J., Reyss, J.-L., Rimet, F., Sabatier, P., Savichtcheva, O., Sylvestre, F., and Verneaux, V. (2015). High-resolution paleolimnology opens new management perspectives for lakes adaptation to climate warming. *Frontiers in Ecology and Evolution*, 3. doi.org/10.3389/fevo.2015.00072.

Perga, M-E., Desmet, M., Enters, D., and Reyss, J.L. (2010). A century of bottom-up- and top-down-driven changes on a lake planktonic food web: A paleoecological and paleoisotopic study of Lake Annecy, France. *Limnology and Oceanography*, 55 (2), 803–16.

Phillips, B. (2006). Critique of the Framework for describing the ecological character of Ramsar Wetlands. Mainstream Environmental Consulting Pty Ltd, Waramanga ACT.

Pittock, J., Finlayson, C. M., Gardner, A., and McKay, C. (2010). Changing character: the Ramsar Convention on Wetlands and climate change in the Murray–Darling Basin, Australia. *Environmental and Planning Law Journal,* 27, 401–42.

Randsalu-Wendrup, L., Conley, D.J., Carstensen, J., and Fritz, S.C. (2016). Paleolimnological records of regime shifts in lakes in response to climate change and anthropogenic activities. *Journal of Paleolimnology*, 56, 1–14.

Reeves, J.M., Gell, P.A., Reichman, S.M., Trewarn, A.J., and Zawadzki, A. (2016). Industrial past, urban future: using palaeo-studies to determine the industrial legacy of the Barwon estuary, Victoria, Australia. *Marine and Freshwater Research*, 67, 837–49.

Retelle, M.J., and Child, J.K. (1996). Suspended sediment transport and deposition in a high arctic meromictic lake. *Journal of Paleolimnology*, 16, 151–67.

Sayer, C.D., Hoare, D.J., Simpson, G.L., Henderson, A.C.G., and Liptrot, E.R. (2006). TBT causes regime shift in shallow lakes. *Environmental Science and Technology*, 40, 5269–75.

Scheffer, M., and Carpenter, S.R. (2003). Catastrophic regime shifts in ecosystems: linking theory to observation. *Trends in Ecology and Evolution*, 18, 648–56.

Scheffer, M., and van Nes, E.H. (2007). Shallow lakes theory revisited: various alternative regimes driven by climate, nutrients, depth and lake size. *Hydrobiologia*, 584, 455–66.

Semeniuk, V. (2013). Predicted response of coastal wetlands to climate changes: a Western Australian model. *Hydrobiologia*, 708, 23–43.

Society for Ecological Restoration International & Policy Working Group, (2004). The SER International Primer on Ecological Restoration. Society for Ecological Restoration International, 13 pp.

Skilbeck, C.G., Trevathan-Tackett, S., Apichanangkool, P., and Macreadie, P.I. (2017). Sediment sampling is estuaries: site selection and sampling techniques. In: Weckström, K., Saunders, K.M., Gell, P.A., and Skilbeck, C.G. (eds) Applications of Paleoenvironmental Techniques in Estuarine Studies. Developments in Paleoenvironmental Research, vol. 20, 89–120. Springer, Dordrecht.

Stager, C. (2011). Deep Future: The next 100,000 years of life on Earth. Scribe Publications, Carlton North, Australia.

Stager, J.C., Sporn, L.A., Johnson, M., and Regalado, S. (2015). Of paleo-genes and perch: what if an 'alien' is actually a native? *PLoS One*, 10 (3), doi: 10.1371/journal.pone.0119071.

van Hardenbroek, M., Heiri, O., Parmentier, F.J.W., Bastviken, D., Ilyashuk, B.P., Wiklund, J.A., Hall, R.I., and Lotter, A.F. (2013). δ13C of chitinous invertebrate remains provides evidence for past variations in methane availability in a Siberian thermokarst lake. *Quaternary Science Reviews*, 66, 74–84.

Verdonschot P.F.M., Spears, B.M., Feld, C.K., Brucet, S., Keizer-Vlek, H., Borja, A., Elliot, M., Kernan, M., and Johnson, R.K. (2013). A comparative review of recovery processes in rivers, lakes, estuarine and coastal waters. *Hydrobiologia*, 704, 453–74.

Wagner, B., Wilke, T., Wagner-Cremer, F., and Middelburg, J. (2017). Integrated perspectives on biological and geological dynamics in ancient Lake Ohrid. *Biogeosciences*, 13 (special issue).

Walker, I.R. (2001). Midges: Chironmidae and related Diptera. In: Smol, J.P., Birks, H.J.B., and Last, W.M. (eds) Tracking Environmental Change using Lake Sediments. Vol 4, Zoological Indicators, 43–66. Kluwer Academic Publishers, Dordrecht.

Wang, R., Dearing, J.A., Langdon, P.G., Zhang, E., Yang, X., Dakos, V., and Scheffer, M. (2012). Flickering gives early warning signals of a critical transition to a eutrophic lake state. *Nature*, 492, 419–22.

Wolfe, A.P., Baron, J.S., and Cornett, R.J. (2001). Anthropogenic nitrogen deposition induces rapid ecological changes in alpine lakes of the Colorado Front Range (USA). *Journal of Paleolimnology*, 25, 1–7.

Wolfe, B.B., Hall, R.I., Edwards, T.W.D., Vardy, S.R., Falcone, M.D., Sjunneskog, C., Sylvestre, F., McGowan, S., Leavitt, P.R., and van Driel, P. (2008). Hydroecological responses of the Athabasca Delta, Canada, to changes in river flow and climate during the 20th century. *Ecohydrology*, 1, 131–48.

Yarrow, M., Marín, V.H., Finlayson, M., Tironi, A., Delgado, L.E., and Fischer, F. (2009). The ecology of *Egeria densa* Planchon (Liliopsida: Alismatales): a wetland ecosystem engineer? *Revista Chilena de Historia Natural*, 82, 299–313.

14

Secondary Data

Taking advantage of existing data and improving data availability for supporting freshwater ecology research and biodiversity conservation

Aaike De Wever, Astrid Schmidt-Kloiber,
Vanessa Bremerich, and Joerg Freyhof

Corresponding author: aaike.dewever@gmail.com

14.1 Introduction

Freshwater ecosystems are under pressure worldwide in parallel with a noticeable and documented decline in biodiversity (Vörösmarty et al. 2010; Dudgeon et al. 2006). Detecting changes in biodiversity requires systematic monitoring, but an integrated observation system for biodiversity, such as proposed by Scholes et al. (2012), remains to be established. Species monitoring is still the most clear-cut methodological approach for biodiversity surveillance, but analyses of community structure, traits, genetic diversity, phenology, or ecosystem services are other important components required to measure the 'biodiversity pulse' of the planet. These components have been recognised as 'Essential Biodiversity Variables' (EBVs) proposed by the Group on Earth Observation Biodiversity Observation Network (GEO BON) and aim to enable the harmonised observation of biodiversity change (Pereira et al. 2013; Kissling et al. 2015; Turak et al. 2016). In order to understand what is driving biodiversity change, large-scale data sets need to be analysed and modelled in relation to different anthropogenic stressors. Such broad approaches are also used in other fields of freshwater research (e.g., in the assessment of climate change or for defining conservation priorities).

Though the call for open-access data is becoming louder, long-term and large-scale data are still difficult to obtain. This may be due to the fact that only few water managers, policy-makers, or even scientists are aware that globally shared open access (biodiversity) data can be useful. Freshwater ecosystems are often observed and managed by local authorities and non-governmental organisations and assessment data are kept locally or nationally to serve local and national needs. Due to the lack of a larger perspective, local activists often do not even consider that their data might be useful

De Wever, A., Schmidt-Kloiber, A., Bremerich, V., and Freyhof, J., *Secondary data: Taking advantage of existing data and improving data availability for supporting freshwater ecology research and biodiversity conservation.* In: *Freshwater Ecology and Conservation: Approaches and Techniques,* Edited by Jocelyne M. R. Hughes: Oxford University Press (2019). © Oxford University Press 2019. DOI: 10.1093/oso/9780198766384.003.0014 for others. Furthermore, making these data freely available poses an extra effort on data holders, an action that is rarely paid for by funding agencies. The consequence of this 'local-serves-local' approach leads—as explained above—to the unavailability of a (global) comprehensive data source and is therefore seriously affecting freshwater biodiversity research, conservation, assessment, and management. In this chapter, we emphasise the importance of secondary data, give an overview about existing databases (e.g., taxonomy, molecular or occurrence databases), discuss problems in understanding and caveats when using such data, and discuss the need to make primary data publicly available.

14.2 Importance of secondary data

As outlined in the introduction, understanding biodiversity change and addressing questions in freshwater management and conservation requires access to biodiversity data and information from all parts of the world. Often, the research questions envisage spatial and temporal scales that typically extend beyond the scope of a single data set, campaign, or sampling event. Therefore, it is necessary to look into relevant existing freshwater data.

Original or 'primary' data are generated to address a specific research question or respond to monitoring requirements (e.g., European reporting on the ecological status of water bodies according to the Water Framework Directive, WFD). Data compiled from different sources, which have been generated and used in another context, are referred to as 'secondary' data. Section 14.4 discusses where to find secondary data and focuses on larger initiatives and data types for which large repositories exist.

When evaluating the status of freshwater ecosystems at global, continental, or sometimes even national scale, researchers have to build on existing (secondary) data. Studying long-term trends and establishing baselines that represent a good ecological status, clearly require building on available data. Collecting new data from scratch, while ignoring existing data, is often unrealistic and might represent a waste of resources. Nevertheless, data compilation for answering large-scale or multidisciplinary questions can be challenging, on the one hand due to limited availability of data (need for targeted data mobilisation and digitisation of information) and on the other hand because of 'issues with the data' (e.g., differences in sampling methodology, taxonomy, or data quality) (see Section 14.5).

Although—as outlined in Section 14.3—the call for open data is increasing, a large amount of data are still unavailable or require at least serious efforts to obtain access and permissions for use. This is why Wilkinson et al. (2016) call for using the 'FAIR Guiding Principles' and Open Data advocates are referring to the 5-star system (see 14.3.2 and 14.3.3). By adhering to these principles, at least the process of compiling data could be largely automated (especially once the data have been rendered machine readable). The issue of limited data availability remains a difficult one, and in practice, many data holders are still puzzled and reluctant about making data publicly available, as illustrated in Box 14.1. We try to address some of the questions raised by explaining key concepts and terms in the next section and by guiding researchers on what to do with their data in Figure 14.1.

Box 14.1 Lost in translation: A freshwater biologist's view on making data publicly available

I was first confronted with the question of 'making data publicly available' when one of my scientific papers came back from the reviewers. Comments included critique on the non-standard methodology and a request to make the underlying data publicly available through an online repository.

As a researcher new to this 'open data world', I was completely puzzled: on the one hand I wasn't sure what to do about the methodology. I had just used the protocols proposed by my supervisor and wasn't aware of a central repository of standard methods. On the other hand, I had no clue where to look and who to turn to for advice on where to deposit my data and how to do so. After all, my data were just organised in a couple of simple (Excel) spreadsheets.

I basically had no idea what to do. This was my first contact with the question of data access, data standards, and open data, and presented a major challenge. Finding out 'How to do it' confronted me with a wide range of questions:

• Which kinds of data do I have that should be made available?
• What is the difference between all those repositories and databases? How do I find the right repository for my data? To whom should I submit the data? Is there someone who can help me out?
• What are these 'metadata' the repositories ask for?
• How should the data be organised?
• Is it useful to publish the data in one of these 'Data Journals'?

Finding the answer to these questions in the literature was not easy and the information available seemed to be written in 'data-English'. I also did not want to ask my supervisor as he seemed to have only a vague idea of the topic. Furthermore, after searching the Internet for hours, I was frustrated about the time lost on researching this topic, which could have been used to do 'real' science. In the end I was tempted to simply submit the data as supplementary material of the paper, but I believe the number of initiatives to support scientists in the process of sharing data is increasing, which should reduce the burden on researchers.

Figure 14.1 Decision flow chart to identify how to document and publish different types of data. The numbers below are footnotes listed in the flow chart.

1. Wieczorek, J., Bloom, D., Guralnick, R., Blum, S., Döring, M., Giovanni, R., et al. (2012). Darwin Core: An evolving community-developed biodiversity data standard. *PLoS ONE*, 7(1), e29715.
2. Darwin Core Terms – A quick reference guide http://rs.tdwg.org/dwc/terms/
3. Draft data policy of the Freshwater Biodiversity Data Portal (Freshwater Information Platform) http://data.freshwaterbiodiversity.eu/datapolicy
4. Darwin Core data policy https://www.gbif.org/darwin-core
5. Integrated Publishing Toolkit (IPT) to publish and share biodiversity datasets through the GBIF network http://www.gbif.org/ipt

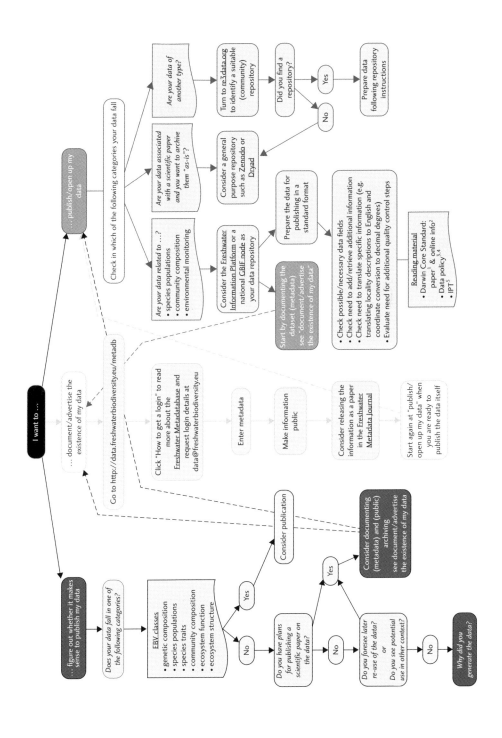

I want to ...

... figure out whether it makes sense to publish my data

Does your data fall in one of the following categories?

EBV classes
- genetic composition
- species populations
- species traits
- community composition
- ecosystem function
- ecosystem structure

Yes → Consider publication

No →

Do you have plans for publishing a scientific paper on the data?

Yes → Consider publication

No →

Do you foresee later re-use of the data?
or
Do you see potential use in other context?

Yes → Consider documenting (metadata) and (public) archiving see document/advertise the existence of my data

No →

Why did you generate the data?

... document/advertise the existence of my data

Go to http://data.freshwaterbiodiversity.eu/metadb

Click "How to get a login" to read more about the Freshwater Metadatabase and request login details at data@freshwaterbiodiversity.eu

Enter metadata

Make information public

Consider releasing the information as a paper in the Freshwater Metadata Journal

Start again at "publish/ open up my data" when you are ready to publish the data itself

... publish/open up my data

Check in which of the following categories your data fall

Are your data related to ...?
- species populations
- community composition
- environmental monitoring

Consider the Freshwater Information Platform or a national GBIF node as your data repository

Start by documenting the dataset (metadata) see "document/advertise the existence of my data"

Prepare the data for publishing in a standard format

- Check possible/necessary data fields
- Check need to add/retrieve additional information (e.g. translating locality descriptions to English and coordinate conversion to decimal degrees)
- Evaluate need for additional quality control steps

Reading material
- Darwin Core Standard: paper[1] & online info[2]
- Data policy[3,4]
- IPT[5]

Are your data associated with a scientific paper and you want to archive them "as-is"?

Consider a general purpose repository such as Zenodo or Dryad

Are your data of another type?

Turn to re3data.org to identify a suitable (community) repository

Did you find a repository?

Yes → Prepare data following repository instructions

No →

Further discussion on the reasons for data not being made public is beyond the scope of this chapter, but the reluctance of data holders to make data available is largely understandable given the relatively limited perceived advantages (see Costello 2009; Piwowar et al. 2007 and Michener 2015 for benefits), the effort involved in preparing the data, as well as the lack of funding and proper citation. This unavailability of data restricts many potentially high-impact studies, which could, in turn, increase the visibility of the data used and justify the investment in standardised routine monitoring for longer time spans. We therefore strongly advocate the adoption of systematic approaches towards both data generation and data publication and sharing. Both issues can be addressed by implementing a data management planning procedure which covers the use of standardised methodologies, documenting the data (i.e., producing metadata), use of exchange standards, publishing, and long-term preservation of data.

14.3 Open data in environmental sciences

During the past years, open access to data, which enables the reuse of data for addressing large-scale or trans-disciplinary research problems, has been loudly called for (Reichman et al. 2011). In the context of environmental science, the Aarhus Convention, for instance, establishes the right to access environmental information held by public authorities. Secondly, the Berlin Declaration on 'Open Access to Knowledge in the Sciences and Humanities' promotes sharing of scientific knowledge covering all open-access contributions, including raw data and metadata, and has been signed by a large number of institutes, governments, and funding agencies. Finally, in its Horizon 2020 funding scheme, the EU is running an 'Open Research Data Pilot' which aims to improve the access to and reuse of research data (see openaire.eu/h2020-oa-data-pilot; and 'Guidelines to the rules on open access to research data', H2020 Programme, on the European Commission website). In the following subsections we explain a number of important terms and concepts related to open data.

14.3.1 Metadata

In short, metadata is information characterising data (sets). Extensive metadata that describe all aspects of a specific data set (i.e., the who, why, what, when, and where) help to improve the understanding and discoverability of datasets and also contain information on how to acquire, use, and cite the data (Michener 2015). In addition, the availability of metadata facilitates the evaluation of the data reuse potential in a specific context—their so-called 'fitness for purpose' (Schmidt-Kloiber et al. 2012).

14.3.2 Open data, licences, and the '5-stars of Open Data'

Data are considered open if they are freely available to use and republish without any restrictions. In practice, this means that data are either dedicated to the public domain through the Creative Commons CC0 or Open Data Commons Public Domain Dedication and Licence (PDDL) or available under an attribution license such as the Creative Commons CC-BY or Open Data Commons Attribution Licence. Attribution licences enforce the users to give proper credit and citation. Data in the public domain do not have this requirement, but often request citation through community norms (i.e., it is considered good scientific practice to cite the data used).

The '5-stars of Open Data' (5stardata.info) refer to the level to which data under an open licence can easily be reused. An image of a data table from which one cannot readily use data for analysis gets one star, a data table in a comma-separated values (CSV) file gets 3 stars, and fully machine-readable Linked Open Data get 5.

14.3.3 FAIR data

Wilkinson et al. (2016) propose 'The FAIR Guiding Principles for scientific data management and stewardship' to enhance re-usability of data. FAIR stands for findable, accessible, interoperable, and reusable. These principles (force11.org/fairprinciples) have also been adopted by the European H2020 funding scheme.

14.3.4 Data repositories and (community) standards

Data are typically made publicly available through online data repositories. The re3data.org (Registry of Research Data Repositories) initiative aims to facilitate identifying suitable data repositories for specific research questions. These repositories range from general purpose ones accepting a wide variety of data types and formats to very specific community repositories adhering to strict data standards.

In the biodiversity realm, the Global Biodiversity Information Facility (GBIF) network enables the publication of species occurrence data in standard formats, the most common one being the Darwin Core standard, which is governed by the Biodiversity Information Standards (TDWG). This standard is basically a collection of terms, and has recently been extended to cover sample-based data. The extension of the standard was implemented in 2015 with the purpose of facilitating the publication of data from systematic sampling and monitoring campaigns and is commonly referred to as 'event core' (links.gbif.org/ipt-sample-data-primer). The Darwin Core 'event core' standard offers the possibility to document and refer to the sampling methodology, report absences, and include density values.

14.3.5 Data papers and the role of journal publishers

Data papers are scientific articles describing data sets, which are made publicly available. The content of such papers is generally an elaborated version of the metadata, providing, for instance, details on the construction of the dataset, its significance, and summary statistics. These papers provide a means to further advertise the existence of the data and a straightforward mechanism for referring to (details about) the data.

A growing number of journal publishers and editors are accepting data papers or have dedicated data journals (e.g., Nature Scientific Data, Biodiversity Data Journal). The Freshwater Metadata Journal was established to specifically publish metadata, even if the data are not publicly available. Given their central role in the data lifecycle, scientific journals can play an important role in encouraging data publication by requiring data to be deposited in a data archive or in a community repository following specific data standards. Sequence deposition in GenBank has become a mandatory and widely accepted practice. To ensure that primary biodiversity data gets systematically integrated in the global GBIF network, several authors (Huang and Qiao 2011; De Wever et al. 2012) have suggested the need for data submission in parallel to publishing of scientific papers.

14.3.6 Citizen science and crowdsourcing

Involving volunteers in generating data for scientific research is becoming increasingly popular as its tremendous potential for improving data availability is evident (Chandler et al. 2017). 'Citizen Science' is an umbrella term for such collaborations and covers a wide range of activities from reporting species observations and Secchi-disk measurements to deciphering herbarium species labels. Open participation to such initiatives is almost universal and the generated data are generally made publicly available. Two types of projects stand out with regards to freshwater ecology and biodiversity: (1) observation projects (from generic platforms to those focussed on specific organism groups or even individual species or areas); (2) crowdsourcing projects either involving the public in physical activities such as water sampling or in virtual activities employing the 'superior human perceptual capacities to computationally difficult image recognition and classification tasks' (Wiggins and Crowson 2011). Observation projects are the most common type (see also next section). Public participation in physical activities often involves the use of mobile phone apps to record measurements (e.g., the 'Secchi Disk' and 'Lake Observer' projects). Other initiatives focus on the collection of water samples for scientists to analyse or provide participants with sensors or analysis kits to do so themselves (e.g., Clean Water for Wildlife project in the UK (Freshwater Habitats Trust)). The Freshwater Watch project (Earthwatch) mobilises a large global community of volunteers and organises a wide variety of projects contributing to a broad range of EBVs.

14.4 Obtaining data relevant to freshwater ecology and conservation

14.4.1 Types of data

Species observed in freshwaters are typically good indicators of the health and status of these ecosystems and are therefore frequently analysed as part of ecological monitoring. Data generated through such monitoring routines, in combination with data from other ecological studies in freshwaters, can form a valuable source of information to support sustainable management and conservation of aquatic ecosystems. But there doesn't exist a large 'freshwater ecology database', and, therefore, it is often not clear which types of data exist and where to get them. As outlined above, we can distinguish between primary and secondary data in freshwater ecological research. In addition to the raw and descriptive data collected through field studies and surveys, data sets are logically complemented by processed data, including computed, analysed and simulated data.

The Essential Biodiversity Variables (EBVs) are organised into six broad EBV classes: (1) genetic composition, (2) species populations, (3) species traits, (4) community composition, (5) ecosystem structure, and (6) ecosystem function. Each of these classes may require a different approach to the type of data collection, structure, and storage, but not all EBVs are reflected in specific databases. In this section we focus on the following data types: freshwater relevant taxonomic data, molecular and genetic data, observation and distribution data, and other information sources for data retrieval.

14.4.2 Freshwater relevant taxonomic data

On a global scale, the Catalogue of Life (catalogueoflife.org) currently combines 169 databases into a gateway of the world's known species of animals, plants, fungi, and microorganisms, but this database does not offer a freshwater species filter. In 2008, the Freshwater Animal Diversity Assessment (FADA; fada.biodiversity.be) produced the first comprehensive global overview of species diversity in inland waters. Existing information was compiled for major species groups and resulted in an estimate of at least 126,000 plant and animal species considered to be freshwater species (Balian et al. 2008a, 2008b). FADA is an informal network of scientists specialised in freshwater biodiversity, and provides an information system with access to authoritative species lists and information on global distributions. The data are also integrated in the Freshwater Biodiversity Data Portal, to which FADA acts as a taxonomic backbone. The species checklists in the database are compiled from different sources, and are in some cases also available elsewhere. The fish checklist was originally generated based on FishBase, a major resource providing information on both marine and inland water fishes including taxonomy (see also details under 14.5). Not all freshwater taxa groups are currently covered by FADA, as is the case for amphibians, for which AmphibiaWeb and the Amphibian Species of the World act as major information sources.

Another source for taxonomic data—specifically focussing on Europe—is the Pan-European Species directories Infrastructure (PESI; eu-nomen.eu) that aims to deliver an integrated, annotated checklist of species occurring in Europe. The PESI web portal is formed by databases from Euro+Med PlantBase, Fauna Europaea, European Register of Marine Species, and Species Fungorum Europe. PESI includes interactions with the geographic focal point networks, a network of taxonomic experts, and global species databases. At this stage, there is no specific search filter for freshwater species, but information can be extracted by searching for specific species or species groups. Results link to GBIF, the Biodiversity Heritage Library (BHL, biodiversitylibrary.org), GenBank, and BOLD.

Regarding alien species in freshwaters, which has taken on greater significance over the past years, the North American Nonindigenous Aquatic Species database (nas.er.usgs.gov) as well as its European counterpart, the European Alien Species Information Network (EASIN; easin.jrc.ec.europa.eu), are valuable sources for taxonomic as well as ecological information.

14.4.3 Molecular and genetic data

Knowledge of the genetic composition plays an important role in freshwater research, as river catchments and lakes can be spatially separated and isolated from each other. This might limit gene flow such that populations of the same species may vary considerably in their genetic composition, which could then have major consequences for ecosystem service provision.

Most molecular data can be accessed through the International Nucleotide Sequence Database Collaboration initiative (INSDC; insdc.org), which is a long-standing, foundational initiative that operates between three major genetic databases; namely, the DNA DataBank of Japan (DDBJ), the European Molecular Biology Laboratory (EMBL), and GenBank at the National Center for Biotechnology Information (NCBI) in the USA. These three organisations exchange data on a daily basis. Genetic data of freshwater species can be extracted through these data sources. Further, partial ribosomal sequence data can be retrieved through the SILVA (arb-silva.de) or RDP (rdp.cme.msu.edu) databases, which offer a suite of applications to handle, analyse, and classify genetic data. Additionally, the Barcode of Life Data Systems (BOLD; boldsystems.org) aim at supporting the generation and application of DNA barcode data by aiding the acquisition, storage, analysis, and publication of DNA barcode records. It assembles molecular, morphological, and distributional data. The platform consists of four main modules: a data portal, a database of barcode clusters, an educational portal, and a data collection workbench. International and national Barcode of Life (BOL; e.g., iBOL) initiatives contribute sequence data to BOLD. It needs to be mentioned that these databases allow unrestricted upload of data without quality control of reference species, potentially complicating the assessment of genetic diversity.

Recent advances in high-throughput sequencing technology greatly increase the potential to assess genetic biodiversity in freshwaters through community metabarcoding or environmental DNA (eDNA) approaches. Such methods generate large amounts of raw data that are processed to eventually obtain a final community data set. Due to their size, these data sets are difficult to store in international repositories. Also, application of eDNA mostly relies on shorter DNA fragments and therefore requires different background databases. Apart from ongoing discussions on standardised workflows, there is also neither a commonly agreed reference database to identify species within metabarcoding or eDNA samples, nor a database where data stemming from such analyses are uploaded. Data can be stored in NCBI and BOLD, but are also deposited in repositories like Dryad. In terms of quality control and reliability checks an unrestricted access to these data is essential.

14.4.4 Observation and distribution data

Biodiversity research is to a large extent focussed on species distributions, populations, and community composition. The best-known platform of distribution records is the Global Biodiversity Information Facility (GBIF; gbif.org), which is an international open data infrastructure, funded by governments and supported by member countries and other associated participants. In 2001 GBIF started its efforts to collate global diversity data with the aim of providing free and open access to species occurrence data from one single online (web) gateway. Currently, GBIF offers more than 768 million occurrence records related to 1.6 million species provided by about 890 data publishers (April 2017). The data portal covers all realms, and freshwater data can be extracted via the species search tool by looking for specific species, data sets, or keywords.

There are several citizen science initiatives (e.g., Artportalen, eBird, iNaturalist, etc.) that are already contributing millions of observation records to the GBIF network and have become a major source of biodiversity data. When working on specific organism groups or areas, it is worthwhile checking for more specific initiatives as not all of them automatically release data to GBIF (see Chandler et al. 2017 for an overview).

Related to freshwater, the Freshwater Biodiversity Data Portal (data.freshwater-biodiversity.eu) provides access to information on freshwater data sets, species, and occurrence data. For species data, the portal links with the Freshwater Animal Diversity Assessment database (FADA, see Section 14.4.2), whereas for occurrence data, it provides access to freshwater data on GBIF and acts as a data publishing platform for freshwater data. The portal is connected to the Freshwater Metadatabase which assembles explanatory descriptions of existing datasets, thus making them discoverable regardless of whether the data are publicly available or not. The metadatabase provides an overview on hundreds of major data sources related to freshwater research and management, including the option to easily explore access rights of relevant datasets.

A particular type of occurrence data constitutes monitoring data, which serve as an ecosystem health indicator (to the benefit of the society at large). These data are either gathered continuously or at specific intervals and are often accompanied by so-called abiotic data (e.g., physico-chemical data, hydromorphological data, land-use data, etc.). The Water Framework Directive (WFD) monitoring data include, for instance, occurrence data of species in combination with abiotic data resulting in calculated metrics and indices to assess the quality of water bodies in member states. While the results of these assessments are intercalibrated throughout Europe (Birk et al. 2012) and the ecological status of waterbodies is stored in the Water Information System for Europe (WISE; water.europa.eu), a central storage system of raw data is missing. Central availability is hampered by the impracticality of combining data stemming from different collection methods, using different data structures and storage methods, as well as by concerns about a consistent quality of data regarding the underlying taxonomy, identification, and taxonomic resolution (Hering et al. 2010). The public accessibility of these data would allow for additional large-scale analyses and would therefore be worth aiming for.

14.4.5 Other information sources for data retrieval

Apart from the data sources mentioned above, a wide range of general data portals exist that focus on specific data types in freshwater research. Table 14.1 gives an overview of selected data sources including their geographic coverage. This list is far from being complete and focuses on initiatives with a large geographic scope. Overall, the EBVs for genetic composition, species populations, species traits, and community

Table 14.1 *Freshwater specific data sources.*

Name	Geographic coverage	URL
general data portals		
WISE—Water Information System for Europe	EU	http://water.europa.eu
BISE—Biodiversity Information System for Europe	EU	http://biodiversity.europa.eu
EEA Data Centre	EU	http://www.eea.europa.eu/themes/water/dc
JRC Science Hub	EU	https://ec.europa.eu/jrc/
FIP—Freshwater Information Platform	global, EU focus	http://www.freshwaterplatform.eu
GEOSS—Global Earth Observation System of Systems	global	http://www.geoportal.org
EU BON	EU	http://biodiversity.eubon.eu
occurrence data		
GBIF—Global Biodiversity Information Facility	global	http://www.gbif.org
Freshwater Metadatabase and Biodiversity Data Portal	global, EU focus	http://data.freshwaterbiodiversity.eu
taxonomy data		
Catalogue of Life	global	http://www.catalogueoflife.org
FADA—Freshwater Animal Diversity Assessment	global	http://fada.biodiversity.be
PESI—Pan-European Species directories Infrastructure	EU	http://www.eu-nomen.eu
trait data		
freshwaterecology.info	EU	http://www.freshwaterecology.info
EPA	US	https://www.epa.gov/risk/freshwater-biological-traits-database-traits
gene data		
BOLDSYSTEMS—Barcode of Life Data Systems	global	http://www.boldsystems.org
INSDC—International Nucleotide Sequence Database Collaboration	global	http://www.insdc.org
GenBank	global	http://www.ncbi.nlm.nih.gov/genbank
spatial data		
FEOW—Freshwater Ecoregions of the World	global	http://www.feow.org
HydroSHEDS and HydroBASINS	global	http://www.hydrosheds.org/page/hydrobasins
Freshwater Key Biodiversity Areas	EU	http://www.birdlife.org/datazone/freshwater
Global Freshwater Biodiversity Atlas	global	http://atlas.freshwaterbiodiversity.eu
Protected Planet	global	http://www.protectedplanet.net
other data		
IUCN Red Lists	global	http://www.iucnredlist.org
EASIN—European Alien Species Information Network	EU	http://easin.jrc.ec.europa.eu
NAS—Nonindigenous Aquatic Species	US	https://nas.er.usgs.gov

composition are covered by the data sources discussed in this chapter. Selected spatial data products related to the EBVs for ecosystem structure and ecosystem function can be accessed through the GEOSS portal, but we are not aware of specific portals focussing on these data types.

14.5 Understanding secondary data

In this chapter we, on the one hand, encourage scientists to use publicly available data and, on the other hand, contribute to open data publishing, thereby supporting biodiversity research. Nevertheless we would like to point out some common criticisms towards publicly available data and pitfalls with regards to their use, most of which can be overcome by improving our understanding of the mechanisms involved in generating, collating, and sharing data and information.

Publicly available data are readily accessible and generally do not require issuing requests to the people who generated them. This leaves the responsibility of understanding how the data were generated, collated, and shared with the user. While this may sound obvious, it is important to point this out given that the interpretation of data generated by others might be difficult, and misinterpretation of data cannot be entirely ruled out. This fear of misinterpretation is often cited as a reason for not sharing data (Costello 2009; Goddard et al. 2011; Enke et al. 2012; Roche et al. 2015). Goddard et al. (2011) suggest that the chances of misinterpretation can be reduced by ensuring that raw data are complete and sufficiently well documented, which is normally done through metadata documentation. Additionally to this description of data characteristics, Edwards et al. (2011) suggest including metadata processes—that is, communication about the data with the data provider—as an inevitable tool to the correct use of secondary data. With respect to the responsibility of understanding data as provided, we can draw some parallels with the information in Wikipedia, which is a recognised and important online resource, but not all topics are equally well covered or curated and information should always be cross checked.

Different data providers and compilers often apply different quality control procedures, so that the quality of aggregated data can be expected to be quite variable. GBIF indicates that it performs specific automatic checks and conversions before integrating the data, but stresses that data quality issues are most easily and reliably addressed at the source. Additionally, different approaches and update cycles for aggregating data from living data sets can potentially lead to sources being out of sync for a certain time (Mesibov 2013). In a reply to this paper, Belbin et al. (2013) highlight how the community can help to address data-quality issues and the need for combining automatic corrections with community assistance. While such community curation for biological data would be highly valuable (Goddard et al. 2011), it is also an extremely complex issue that has been under discussion for several years (see Page 2014) and solutions are unlikely to materialise in the near future. It remains, therefore, up to the users to acknowledge and address these issues.

As an example of the complexity of the relations between different initiatives and the importance to gain a better understanding of these systems, we can refer to FishBase.

FishBase is a major database initiative which incorporates available information from the literature and a variety of online sources. For taxonomic information FishBase consults, among others, the Catalogue of Fishes, which is recognised as a major taxonomic resource in the field. Given the differences in scope and update cycle, discrepancies may exist as detailed by Bailly (2010). As a major source on fish names, information from FishBase is propagated to aggregators such as the World Register of Marine Species (WoRMS), Catalogue of Life, Encyclopedia of Life, and GBIF. Similarly, the fish checklist in the FADA database was originally constructed based on a FishBase extract and has been curated and amended with faunistic information during its creation, but has not been updated since. While some of this information may be updated automatically by certain aggregators, this is not the case for others. As a user of these data, one should always check the date of last update, gain a basic understanding of the data flows, and be aware of the risk of error propagation.

Typically, different uses of data require different conditions to be met. Anderson (2012) sums up the major pitfalls when using data for ecological niche modelling. They include (1) incorrect taxonomic identification; (2) limited representation and absence of geographic coordinates for a significant percentage of records; and (3) the effects of sampling bias. Similarly, Franklin et al. (2016) discuss challenges in using publicly available data in global change plant ecology, and list types of uncertainties to be considered when using such data. These authors also warn not to take for granted the claim that the size of a data set can overcome problems in individual records (representing noise), especially if such errors are systematic.

From a practical perspective, in addition to performing basic quality control and fitness-for-use checks of different component data sets, modellers should: (a) check the up-to-dateness of data and the presence of duplicate records when aggregating data from different sources; (b) consult with taxonomists or colleagues with a good understanding of species distributions to spot any suspect records (e.g., mis-identifications) or obvious data gaps (e.g., as data mobilisation is often nationally organised, the absence of a species (group) in a specific country where all neighbouring countries show the presence of the species (group) may only represent a lack of published data for this country); (c) be aware of the effect of possible 'outlier' records on the model outcome in relation to the methods used; and (d) check for the required citation of different component data sets (e.g., GBIF generates a DOI for downloaded data that can be used as overall citation for the data, but specific data sets may require to be explicitly cited).

The need to check fitness for use is especially important when considering sample-based data, species presence/absence data, or densities. As the methodologies and measurement units are bound to vary widely among data sets (and even within), it is very important to check and account for this when selecting data for analysis.

Despite the multitude of possible pitfalls and limitations in the usage of data from different sources, the benefits of having them publicly available clearly outweigh the potential issues (e.g., 'Benefits of data sharing' in Michener 2015). Public availability of data exposes them to possible scrutiny by peers, opens the potential to reuse including integration in large-scale analyses, represents an increased resource efficiency (not requiring new investments in data generation for well-covered areas), and results in a better understanding of gaps in the data. The value of public data availability is also illustrated by the large collection of papers which reference GBIF and which can be consulted on GBIF's resources pages. We do, however, stress the importance of gaining a good understanding of where data originally come from, for what purpose they were generated, and how they were compiled, aggregated, or combined with other data before using them further.

References

Anderson, R.P. (2012). Harnessing the world's biodiversity data: promise and peril in ecological niche modeling of species distributions. *Annals of the New York Academy of Sciences*, 1260, 66–80.

Bailly, N. (2010). Why there may be discrepancies in the assessment of scientific names between the Catalog of Fishes and FishBase. Available on fishbase.org website.

Balian, E.V., Segers, H., Lévêque, C., and Martens, K. (2008a). The Freshwater Animal Diversity Assessment: an overview of the results. *Hydrobiologia*, 595, 627–37.

Balian, E.V., Segers, H., Martens, K., and Lévêque, C. (2008b). An introduction to the freshwater animal diversity assessment (FADA) project. *Hydrobiologia*, 595, 3–8.

Belbin, L., Daly, J., Hirsch, T., Hobern, D., and LaSalle, J. (2013). A specialist's audit of aggregated occurrence records: An 'aggregator's' perspective. *ZooKeys*, 305, 67–76.

Birk, S., Bonne, W., Borja, A., Brucet, S., Courrat, A., Poikane, S., Solimini, A., van de Bund, W., Zampoukas, N., and Hering, D. (2012). Three hundred ways to assess Europe's surface waters: an almost complete overview of biological methods to implement the Water Framework Directive. *Ecological Indicators*, 18, 31–41.

Chandler, M., See, L., Copas, K., Bonde, A.M.Z., López, B.C., Danielsen, F., Legind, J.K., Masinde, S., Miller-Rushing, A.J., Newman, G., Rosemartin, A., and Turak, E. (2017). Contribution of citizen science towards international biodiversity monitoring. *Biological Conservation*, 213, 280–94.

Costello, M.J. (2009). Motivating online publication of data. *Bioscience*, 59, 418–27.

De Wever, A., Schmidt-Kloiber, A., Gessner, M.O., and Tockner, K. (2012). Freshwater journals unite to boost primary biodiversity data publication. *Bioscience*, 62, 529–30.

Dudgeon, D., Arthington, A.H., Gessner, M.O., Kawabata, Z.-I., Knowler, D.J., Lévêque, C., Naiman, R.J., Prieur-Richard, A.-H., Soto, D., Stiassny, M.L.J., and Sullivan, C. (2006). Freshwater biodiversity: importance, threats, status and conservation challenges. *Biological Reviews of the Cambridge Philosophical Society*, 81, 163–82.

Edwards, P.N., Mayernik, M.S., Batcheller, A.L., Bowker, G.C., and Borgman, C.L. (2011). Science friction: Data, metadata, and collaboration. *Social Studies of Science*, 41, 667–90.

Enke, N., Thessen, A., Bach, K., Bendix, J., Seeger, B., and Gemeinholzer, B. (2012). The user's view on biodiversity data sharing—Investigating facts of acceptance and requirements to realize a sustainable use of research data. *Ecological Informatics*, 11, 25–33.

Franklin J., Serra-Diaz, J.M., Syphard, A.D., and Regan, H.M. (2016). Big data for forecasting the impacts of global change on plant communities. *Global Ecology and Biogeography*, 26, 6–17.

Goddard, A., Wilson, N., Cryer, P., and Yamashita, G. (2011). Data hosting infrastructure for primary biodiversity data. *BMC Bioinformatics*, 12(Suppl 15), S5.

Hering, D., Borja, A., Carstensen, J., Carvalho, L., Elliott, M., Feld, C.K., Heiskanen, A.-S., Johnson, R.K., Moe, J., Pont, D., Solheim, A.L., and van de Bund, W. (2010). The European Water Framework Directive at the age of 10: A critical review of the achievements with recommendations for the future. *Science of the Total Environment*, 408, 1–13.

Huang, X. and Qiao, G. (2011). Biodiversity databases should gain support from journals. *Trends in Ecology and Evolution*, 26, 377–8.

Kissling, W.D., Hardisty, A., García, E.A., Santamaria, M., De Leo, F., Pesole, G., Freyhof, J., Manset, D., Wissel, S., Konijn, J., and Los, W. (2015). Towards global interoperability for supporting biodiversity research on essential biodiversity variables (EBVs). *Biodiversity*, 16, 99–107.

Mesibov, R. (2013). A specialist's audit of aggregated occurrence records. *ZooKeys*, 293, 1–18.

Michener, W.K. (2015). Ecological data sharing. *Ecological Informatics*, 29, 33–44.

Page, R. (2014). Annotating GBIF: some thoughts. Available on iphylo.blogspot.be

Pereira, H.M., Ferrier, S., Walters, M., Geller, G.N., Jongman, R.H.G., Scholes, R.J., Bruford, M.W., Brummitt, N., Butchart, S.H.M., Cardoso A.C., Coops, N.C., Dulloo, E., Faith, D.P., Freyhof, J., Gregory, R.D., Heip, C., Hoft, R., Hurtt, G., Jetz, W., Karp, D.S., McGeoch, M.A., Obura, D., Onoda, Y., Pettorelli, N., Reyers, B., Sayre, R., Scharlemann, J.P.W., Stuart, S.N., Turak, E., Walpole, M., and Wegmann, M. (2013). Essential Biodiversity Variables. *Science*, 339, 277–8.

Piwowar, H.A., Day, R.S., and Fridsma, D.B. (2007). Sharing detailed research data is associated with increased citation rate. *PLoS ONE*, 2, e308.

Reichman, O.J., Jones, M.B., and Schildhauer, M.P. (2011). Challenges and opportunities of open data in ecology. *Science*, 331, 703–5.

Roche, D.G., Kruuk, L.E.B., Lanfear, R., and Binning, S.A. (2015). Public data archiving in ecology and evolution: How well are we doing? *PLoS Biology*, 13, e1002295–12.

Schmidt-Kloiber, A., Moe, S.J., Dudley, B., Strackbein, J., and Vogl, R. (2012). The WISER metadatabase: the key to more than 100 ecological datasets from European rivers, lakes and coastal waters. *Hydrobiologia*, 704, 29–38.

Scholes, R.J., Walters, M., Turak, E., Saarenmaa, H., Heip, C.H.R., Tuama, E O., et al. (2012). Building a global observing system for biodiversity. *Current Opinion in Environmental Sustainability*, 4, 139–46.

Turak, E., Harrison, I., Dudgeon, D., Abell, R., Bush, A., Darwall, W., Finlayson, C.M., Ferrier, S., Freyhof, J., Hermoso, V., Juffe-Bignoli, D., Linke, S., Nel, J., Patricio, H.C., Pittock, J., Raghavan, R., Revenga, C., Simaika, J.P., and De Wever, A. (2016). Essential Biodiversity Variables for measuring change in global freshwater biodiversity. *Biological Conservation*, 213, 272–9.

Vörösmarty, C.J., McIntyre, P.B., Gessner, M.O., Dudgeon, D., Prusevich, A., Green, P., Glidden, S., Bunn, S.E., Sullivan, C.A., Liermann, C.R., and Davies, P.M. (2010). Global threats to human water security and river biodiversity. *Nature*, 467, 555–61.

Wiggins, A. and Crowston, K. (2011). From conservation to crowdsourcing: A typology of citizen science. Proceedings of the Forty-fourth Hawai'i International Conference on System Sciences (HICSS-44), Koloa, Hawai'i.

Wilkinson, M.D., Dumontier, M., Aalbersberg, I.J., Appleton, G., Axton, M., Baak, A., et al. (2016). The FAIR Guiding Principles for scientific data management and stewardship. *Scientific Data*, 3, 160018–19.

15

Freshwater Ecosystem Services and Functions

C. Max Finlayson, Rudolph S. de Groot, Francine M.R. Hughes,
and Caroline A. Sullivan

Corresponding author: mfinlayson@csu.edu.au

15.1 Introduction

Freshwater ecosystems, also broadly referred to as freshwater wetlands (Milton and Finlayson 2017), as defined by the Ramsar Convention, and their services are highly valued by many people worldwide for their biodiversity and existence values, and more critically, for livelihoods and economic benefits (Millennium Ecosystem Assessment 2005; TEEB 2010). However, many of these services have been degraded and destroyed, particularly over the past century (Millennium Ecosystem Assessment 2005; Gardner et al. 2015). Lack of recognition and understanding of ecosystem services is seen as a key driver of decisions and management actions leading to continued conversion and degradation (Millennium Ecosystem Assessment 2005; de Groot et al. 2016). This has continued despite ample evidence that their Total Economic Value is often greater than that of the alternative land uses (de Groot et al. 2012).

The links between nature, including wetlands, and humans have been explored in an increasing body of literature. Westman (1977) examined the link between ecological and economic systems in a paper entitled 'How much are Nature's services worth?' while Ehrlich and Ehrlich (1981) may have been the first to use the term 'ecosystem services' with ecologists and economists further elaborating the notion that ecosystems provided a variety of services and economic benefits for many people. In the late 1990s the concept of ecosystem services and their importance for people received increasing attention with the publication of seminal papers by Costanza et al. (1997) and Daily (1997). At around the same time the concept of natural capital was developed through the interdisciplinary and expanding field of ecological economics (Costanza and Daly 1992). These concepts were examined in the Millennium Ecosystem Assessment (2005) which agreed four general categories of services; namely, provisioning services (including food, fresh water, wood, fibre, and fuel), regulating services (including the regulation of climate, floods and diseases, and water purification), supporting services (including nutrient cycling, soil formation, and primary production), and cultural services (including aesthetic, spiritual, educational, and recreational).

Finlayson, C. M., de Groot, R. S., Hughes, F. M. R., and Sullivan, C. A., *Freshwater ecosystem services and functions.* In: *Freshwater Ecology and Conservation: Approaches and Techniques.* Edited by Jocelyne M. R. Hughes: Oxford University Press (2019). © Oxford University Press 2019. DOI: 10.1093/oso/9780198766384.003.0015

Taking into account the established connections between people and wetlands in many parts of the world, Horwitz and Finlayson (2011) proposed an integrated strategy to help decision-makers embellish if not restore the benefits that can accrue from healthy wetlands. This comprised three parts:

i) making assessments of the ecosystem services provided by wetlands more routine;
ii) adopting a 'settings' approach wherein wetlands are one of the settings for human health and provide a context for health policies; and
iii) layering of a suite of health issues in wetland settings.

Finlayson and Horwitz (2015) extended these concepts and explored the metaphorical association of healthy wetlands and enhanced ecosystem services with improved outcomes for human health, and similarly, the association of unhealthy wetlands and degraded ecosystem services with poor outcomes for human health. They also explored the paradoxical situations where some direct benefits for human health could lead to the loss of other ecosystem services. This was termed a *health paradox* whereby there was a loss of regulating and supporting services from steps taken to enhance human health. A *wetland paradox* also occurs when there were poor outcomes for human health as a consequence of the maintenance or enhancement of ecosystem services.

To understand the benefits of ecosystems to people, *The Economics of Ecosystems and Biodiversity* initiative extended the understanding of the benefits of ecosystems to people (TEEB 2010; Finlayson 2018a). More recently, TEEB has evolved into the *Natural Capital Coalition* (Natural Capital Coalition website) bringing together global initiatives and organisations to harmonise approaches to the benefits we obtain from nature. The UK Ecosystem Assessment further provided an in depth analysis of freshwater ecosystems and determined that these ecosystems provided major services, but their benefits to people were inadequately identified and valued (Maltby et al. 2011). Further assessments have been undertaken (see TEEBweb website) and in 2012 the Intergovernmental science–policy Platform on Biodiversity and Ecosystem Services (IPBES) was established to provide a mechanism to strengthen the science-policy interface for biodiversity and ecosystem services (Finlayson 2018b). IPBES has presented a rationale for a valuation of Nature's Contributions to People in decision-making, as well as undertaking a number of assessments (see IPBES.net website).

In spite of the attention that ecosystem services have received there is still debate about definitions and classifications. In this respect we may do better to accept that it is very unlikely that a single definition or classification could capture the many ways in which ecosystems support human wellbeing and contribute to livelihoods (de Groot et al. 2012). With these limitations in mind an appraisal of how to measure ecosystem services is provided, building on the approaches presented by the Ramsar Convention (de Groot et al. 2006). This is accompanied by a review of open-access toolkits for evaluating ecosystem services. Issues associated with the identification and awareness of ecosystem services provided by wetlands are illustrated through two cases: (i) the Richmond river catchment in eastern Australia, and (ii) a national assessment undertaken in Colombia.

15.2 Measuring ecosystem services and functions

15.2.1 Background

The study of ecosystem services from freshwater wetlands is a multidisciplinary undertaking, requiring an understanding of both the biophysical provision of services and the socio-economic benefits that are derived from them. The multidisciplinary study of wetland ecosystem services is, however, increasingly challenging due to differences in paradigms and the language used across disciplines. Because these are dynamic ecosystems, their functionality and capacity to provide ecosystem services fluctuate both inter- and intra-annually. This functional understanding of wetlands in particular exposes a mismatch with economic approaches to ecosystem valuation. Although ecological processes that are important in delivering services are identified, these cannot easily be incorporated into economic assessments of dynamic biophysical systems.

Approaches to ecosystem service measurement and valuation have tended to be reductionist, focussing on individual services or suites of services and generally unable to accurately associate service values and biophysical function. Direct monitoring of changing ecosystem services is also difficult (Hughes et al. 2016) and most valuation tools either calculate services for a single point in time or adopt a scenario-building approach based on both known and assumed direct and indirect relationships between biophysical components and processes and ecosystem services. Measurements at a single point in time generally do not incorporate an assessment of the status of the underpinning natural capital providing these services, nor do they indicate how sustainable the ecosystem service provision is into the future. Different services also pose different measurement problems. There is a considerable literature on valuing ecosystem services that focuses on monetary valuations, but more recently there has been a stronger emphasis on valuations including both qualitative and quantitative units; for example, the integrated valuation approaches advocated by Sullivan (2002) and Gomez-Baggethun et al. (2014).

In this chapter, we clarify the steps that are necessary in any attempt to value ecosystem services from freshwater wetlands. These are outlined in more detail in the guidance produced for the Ramsar Convention on Wetlands for valuing the benefits derived from wetland ecosystem services (de Groot et al. 2006, 2016).

15.2.2 Function analysis: inventory of wetland services

Freshwater wetlands are composed of a number of physical, biological, and chemical components. Interactions among and within these enable such systems to perform certain functions that have been defined as '*the capacity of ecosystem processes and components to provide goods and services that satisfy human needs, directly or indirectly*' (see de Groot 1992; de Groot et al. 2002). The Millennium Ecosystem Assessment has further defined ecosystem services as '*the benefits people obtain from ecosystems*' (Millennium Ecosystem Assessment 2005). This more anthropocentric viewpoint includes both goods (i.e., tangible resources) and services (ecological processes). Given the interactions that occur between the components of a freshwater wetland and the benefits that people obtain from these, the Ramsar Convention responded to the Millennium Ecosystem Assessment and incorporated ecosystem services into the definitions of

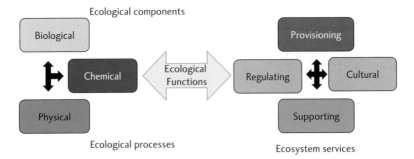

Figure 15.1 Relationships between the ecological components and processes that comprise a wetland and the ecosystem services they provide (adapted from de Groot et al. 2006).

ecological character, and wise use of wetlands that forms the basis of the international policy framework for wetlands (Finlayson et al. 2011). This change increased the importance of understanding wetland functions and how they provide benefits to people.

The first part of the function analysis should translate wetland characteristics (ecological processes and components) into a comprehensive list of services which can then be quantified in appropriate units (biophysical or otherwise) to determine their value (importance) to human society (Figure 15.1). The degree to which the ecological components and processes within a freshwater wetland provide ecosystem services depends on the functional properties of the ecosystem (e.g., biomass production, nutrient cycling, and food-chain dynamics) as well as the needs of the society.

15.2.3 Identification and selection of wetland services

Depending on the purpose of the valuation and assessment, the ecological and socio-economic setting and the interests of stakeholders must be taken into account. Different services will be relevant in different contexts. The first step is the development of a checklist of the main services of the wetland being assessed. de Groot et al. (2006) provide a list of the main services provided by different types of wetland (both inland and coastal), and their general relative magnitude, the latter having been developed by expert opinion based on the rapidly developing number of assessments that were being done. Depending on the complexity of the wetland being valued, the services should be described for each of the main components of the ecosystem (e.g., rivers, lakes, marshes, etc.). Where possible, maps of the spatial distribution of each service should be used to support this process. Selection of what are considered the key components should be done in close consultation with key stakeholders, as demonstrated in a national ecosystem service assessment in Colombia (Ricaurte et al. 2017).

15.2.4 Quantification of the capacity of wetlands to provide ecosystem services on a sustainable basis

Once the main services delivered by the wetland have been selected, the magnitude of the (actual and potential) availability of these should be determined, based on sustainable use levels. de Groot et al. (2006) provide a list of example indicators suitable for determining the sustainable use of wetland services. The capacity of ecosystems to provide services in a sustainable manner depends on the biotic and abiotic characteristics which should be quantified with appropriate indicators. For example, the capacity of wetlands to provide fish can be measured by maximum sustainable harvest levels (in terms of biomass or other unit), the capacity to store water by hydrological parameters (e.g., water volume, flow velocity, etc.), and the capacity for recreational use by aesthetic quality indicators and carrying capacity for visitor numbers.

As most functions and related ecosystem processes are inter-linked, sustainable use levels should be determined under complex system conditions taking due account of the dynamic interactions between functions, values, and processes (Limburg et al. 2002). Further references and data sources on the application of methods to assess each of the wetland services and indicators can be obtained from existing information sources, such as those mentioned in the text below.

15.2.5 Valuing wetland ecosystem services

In order to incorporate ecosystem process and services into financial accounting frameworks, economic values associated with the natural capital of freshwater systems and their services can be categorised into distinct components of the 'Total Economic Value' (Figure 15.2).

Direct Use Values are derived from the various uses made of these freshwater resources and their associated habitats. For example, water, for domestic, municipal, and irrigation uses, would derive different monetary values than those derived from environmental flows. Similarly, recreational use of water resources can generate high monetary values in some areas, but not in others. Timber used for energy or building also have high monetary values. Direct Use Values are consistently reported in monetary terms.

Indirect Use Values are derived from various ecological functions and include storm protection, nutrient retention, and erosion prevention, all of which are much more

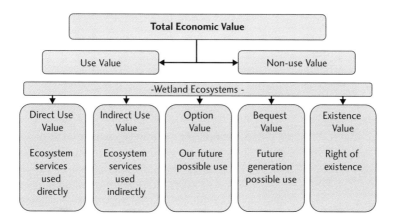

Figure 15.2 The Total Economic Value framework (adapted from Millennium Ecosystem Assessment 2003).

difficult to value in monetary terms. It is important, however, to recognise these values and incorporate them into monetary systems. Failing to do this will reduce the effectiveness of their inclusion. A multi-criteria approach is essential for any attempt in assessing all kinds of values. Using money as the only numeraire vastly underestimates the true value of the ecosystem.

Option Value extends the conceptualisation of use to cover a society's preference to maintain the possibility of future uses. The concept of option value includes preferences for preserving an environmental benefit for possible use by current (philanthropic value) or future generations (bequest value).

Non-Use Values of ecological systems reflect the societal values the actual existence of that system, unrelated to any direct, indirect, or future use (de Groot et al. 2006). This is also expressed as *Existence Value*.

To incorporate all of these values into our financial accounting processes, several techniques have been used with varying success. Most commonly, these values may be expressed as 'willingness to pay' for the ecological attribute in question, and this in turn is interpreted as the economic value of that resource. It is important to note that all methodologies designed to attribute a value to ecosystems, are inherently flawed, since at present, our scientific understanding of these systems is imperfect.

In wetlands, for example, some goods and services may be traded directly in well-functioning markets generating readily observable values. However, market failure resulting from undefined property rights or the 'public good' characteristics of some wetland functions, means that many valuable wetland services are not traded directly or even indirectly through markets. The term 'public good' describes a benefit to society at large, equally available to all (non-excludable), and where the level of consumption of the good by one beneficiary does not reduce the level of service received by another. For example, the reduction in downstream flood risk provided by wetlands in a water catchment is received by the entire population living downstream irrespective of their recognition of that service or the number of other beneficiaries. Due to these 'non-excludability' and 'non-rivalry' characteristics, markets for such services generally do not exist, so market-based instruments cannot be used to manage them. As a result, ecosystem services are frequently undervalued in decisions regarding the conversion of privately owned freshwater systems (flooded forests, riparian wetlands, etc.) to other productive uses generating marketable goods and services (e.g., agriculture). The lack of understanding of, and information on, the value of such ecosystems has thus led to their omission in both private and public decision-making, regarding land conversion.

In response to this lack of information on the (direct) market value of freshwater systems and their services, there is an expanding literature that employs indirect- and non-market valuation approaches to estimate the economic value (i.e., welfare contribution) of such freshwater services. In the case of wetlands, these studies demonstrate that the estimated value of wetland ecosystem services varies enormously across wetlands depending on both biophysical and socio-economic characteristics.

In an attempt to consolidate these studies, a total of 458 value-estimates were examined (de Groot et al. 2012), generating average annual values (2007 prices) for the (potential) Total Economic Value (US$ hectare⁻¹) of six main wetland types. These were as follows: open ocean $490; coral reefs $350,000; coastal systems (including beaches) $29,000; coastal wetlands (including mangroves) $190,000; inland wetlands $25,000; and rivers and lakes $4,300.

Sometimes, in decision-making situations, values such as these may be used in a process of *Benefit Transfer*, which serves to assign a value to an attribute, even though that attribute itself has not been measured. Much care should be taken when using these values for Benefit Transfer since the actual value of a wetland is, of course, strongly time- and context-dependent and just like the 'average person' does not exist, there is no 'average wetland'.

15.3 Approaches and toolkits for measuring ecosystem services

As discussed above, the services offered by freshwater wetlands are not always easy to identify. Any monetary valuation of them must be seen as no more than an indicative representation of their full value. The types of services provided by a wetland result from the complex interplay between climate, soils, topography, vegetation, land use, and, crucially, on the location both of the wetland within the water catchment and of beneficiaries. Hydrological processes occurring upstream may provide services to beneficiaries who might live locally, downstream, or elsewhere. In addition, the same ecological function in two different places may deliver different services depending on how such services are perceived and used by the service beneficiaries. Thus it is clear that the essential first step in measuring the ecosystem services provided by a wetland, is to work out what services are offered and to which beneficiaries.

Approaches to identifying services offered by a wetland usually fall into one of two categories:

1. Consideration of the biophysical attributes of the wetland and of the services that are likely to be provided. In many tools this approach is carried out using GIS layers in which each cell will be allocated biophysical attributes. Based on existing data, expert opinion, or on modelling outputs, levels of service will be derived from these biophysical attributes and allocated to the cells. For example, GIS layers of soil type and flooding depth might allow assessment of the likely CO_2 or methane emissions of a wetland based on expert knowledge and known values from the literature. Dangers in this approach include the current paucity of appropriate data sets for many wetlands and the site-specific nature of most wetland services.

2. Consultation with a wide range of stakeholders. This is an essential approach in many cases where the existence of services is not obvious on the basis of biophysical information alone. Most tools include some dimension of stakeholder consultation within their methods.

A combination of these two approaches is likely to be the most useful. Once services have been identified, the choice of tool has to be made depending on:

- the types of services that are to be assessed;
- how important it is to have quantitative values;
- the available budget;
- the spatial scale of the wetland; and
- the technical capacity to use a tool.

Some tools can only give a snapshot of a chosen year, which is useful, but gives little information about how sustainable the services are over the long term, or how representative the assessment is of other years. Scenario building for predictive assessments into the future is available in some tools, but is only as good as the modelling on which it is based. Some tools are suitable for use over larger areas and may be used for regional planning (e.g., InVEST; Shan et al. 2016), while others are more suitable for site-based assessments (e.g., TESSA; Peh et al. 2013). The difference in technical expertise and facilities needed to use these toolkits also varies hugely. Some deal with a range of ecosystem services while others specialise in one area only (e.g., GRACE is only for assessing cultural ecosystem services). Useful overviews of many of these tools are given by Peh et al. (2013), by the UK Tool Assessor website of the Ecosystems Knowledge Network and by the EU-funded OPERAS project. In addition, the Natural Capital Coalition is piloting a new database (called the Natural Capital Protocol Toolkit) to help navigate the numerous toolkits available for identifying and measuring ecosystem services.

Table 15.1 highlights the most commonly used tools that can be used in freshwater wetlands, which also openly accessible. Many individual companies and organisations have developed their own tools for their own use, many primarily to identify rather than measure ecosystem services. There are also a number of tools that have been developed for catchment-based assessments of water resources rather than specifically for wetland ecosystem services (e.g., WaterWorld for water resources assessment on the PolicySupport.org website, WEAP for water resources planning on the Weap21.org website, and AQUEDUCT for water risk planning on the wri.org website). Co$ting Nature is also a useful web-based tool for natural capital accounting and analysing the ecosystem services provided by natural environments on PolicySupport.org website.

The descriptions of toolkits in Table 15.1 are based largely on their own websites. Most tools require a training period before it is possible to use them effectively. The TESSA toolkit is perhaps the most easily used toolkit that engages in quantitative assessments, without prior training, although training is advocated. It is aimed at managers of sites with nature conservation interest that are often remote and with intermittent Internet access. Its methods are generally non-technical, requiring basic mathematics skills rather than GIS or modelling skills. It has been successfully used in numerous protected areas. The tools or approaches aimed at cultural ecosystem services such as GRACE and the Ramsar Convention's rapid cultural inventories also provide insights into the cultural values that people place on rivers, wetlands, and other freshwater systems.

Table 15.1 Open-access toolkits suitable for measuring or evaluating freshwater wetland ecosystem services. Descriptions based on Waage and Stewart (2008); Peh et al. (2013), the Ecosystems Knowledge Network website http://ecosystemsknowledge.net/, the OPERAS project output http://operas-project.eu/sites/default/files/resources/ms2.1 reviewofexistingprotocols.pdf, as well as directly from toolkit websites and author experience.

Tool	Description	Spatial scale of use	Technical contexts	Case studies using river or wetland examples
TESSA (Toolkit for Ecosystem Service Site-based Assessment) http://tessa.tools/	A practical suite of methods for rapid measurement and assessment of wetland services at a site scale.	Sites up to 10,000 km². Extrapolation from smaller areas is recommended for larger sites. Many case studies are based in protected areas.	An interactive pdf makes this tool usable with limited Internet access. It requires identification and assessment of an alternative state for comparison with the site of interest. Calculations of values are given as net benefits compared with the alternative state. TESSA gives methods for ecosystem service data collection *in situ* in both biophysical and/or monetary units. It includes modules for water-related services, cultivated goods, harvested wild goods, nature-based recreation, and global climate regulation, with further modules in development.	Peh et al. 2014. Blaen et al. 2015.
ARIES (ARtificial Intelligence for Ecosystem Services) http://aries. integratedmodelling. org/	A modelling approach for quantifying ESS and factors influencing their values in a defined geographical area with priorities set by users.	Local to national. A number of case studies include watershed scale assessments or regional assessments that include river and wetland areas.	ARIES is web-based and enables simple use of complex models. It allows researchers to contribute models and scientific data that simulate and integrate social-ecological systems. Training is recommended. Many outputs are map-based. ARIES has embedded data sets but users can add their own. It aims to use local, high-resolution spatial data sets whenever available to populate models that account for a broad range of ecosystem services in a variety of ecological and socio-economic settings. Eleven ecosystem services across provisioning, regulating, and cultural categories are included.	Villa et al. 2015.

(Continued)

Table 15.1 *Continued*

Tool	Description	Spatial scale of use	Technical contexts	Case studies using river or wetland examples
InVEST (Integrated Valuation of Ecosystem Services and Trade-offs) http://www.naturalcapitalproject.org/invest/	A spatially explicit modelling approach for mapping and valuing scenarios of ecosystem services and their benefits.	The spatial resolution of analyses is flexible. Users can address questions at local, regional, or global scales.	InVEST models are spatially-explicit, using maps as information sources and producing maps as outputs. InVEST returns results in either biophysical or economic terms. InVEST models can be run independently, or as script tools in ArcGIS ArcToolBox. Running InVEST requires basic to intermediate skills in GIS software. The toolset includes 18 distinct ecosystem service models designed for terrestrial, freshwater, marine, and coastal ecosystems.	Watson et al. 2016.
LUCI https://catalogue.ceh.ac.uk/documents/adead10b-a6a0-45fa-b350-acd16b23c3fe	A catchment-based spatially explicit ecosystem service model.	Field to national scale.	It is a spatially explicit ecosystem service model, which uses look-up tables, combined with topographic routing of water, sediment and nutrients. It is spatially explicit: at the resolution of the topographic data layer used: model applications to date have used a 5 m by 5 m resolution. Outputs include: Ecosystem service condition, opportunity, and trade-off maps.	Bagstad et al. 2013.
WESPUS (Wetland Ecosystem Services Protocol for the United States) https://www.novascotia.ca/nse/wetland/docs/Manual_WESPUS.pdf	A standardised method intended for rapidly assessing ecosystem services of all wetland types throughout temperate North America.	Services are assessed primarily at the scale of an individual wetland rather than across landscapes.	A WESPUS assessment requires completing a single three-part data form, taking about 1–3 hours. Responses to questions on that form are based on review of aerial imagery and observations during a single site visit; GIS is not required. Logic models automatically generate scores intended to reflect a wetland's ability to support a series of wetland functions and their values. The numeric estimates that WESPUS provides for wetland functions, values and other attributes are estimates of those attributes arrived at using standardised criteria (models).	Raudsepp-Hearne et al. 2011.

A Guidance Manual for Assessing Ecosystem Services at Natura 2000 Sites http://ww2.rspb.org.uk/Images/natura_2000_guidance_manual_tcm9-399208.pdf	A practical guide for practitioners involved in the management of Natura 2000 sites but applicable to other protected areas.	Primarily for use at site-scale but intended to support planning at regional or larger scale.	Draws on a number of existing 'toolkits' for assessing the value of ecosystem services, including Kettunen et al. (2009) and the TESSA Toolkit produced by Peh et al. (2013). The Toolkit aims to help to understand, assess, and communicate the total overall socio-economic benefits and value of a site, including qualitative, quantitative, and monetary estimates. It includes both market-based and non-market economic valuation techniques and methods designed to indicate how to approach putting biophysical or monetary values on the Natura 2000 sites.	Wetland examples are included in the manual.
GRACE (Guidance for the Rapid Assessment of Cultural Ecosystem services) https://live-fauna-flora-international.pantheonsite.io/wp-content/uploads/old-images/grace_report_final.pdf	Aimed at conservation and development NGOs working with communities, it allows assessment of cultural ecosystem services.	Primarily for use in a conservation context and aimed at community-level work rather than specified spatial scales.	GRACE is not a toolkit but a set of approaches to enable stakeholder engagement and provides a suite of seven participatory tools designed to facilitate discussion and help analyse relationships between people and nature, how nature contributes to well-being, and how benefits may be affected by changes in land use, policies, practices, and other external drivers of change. The guidance includes a mix of qualitative and quantitative data collection approaches.	Infield et al. 2015. Wetland and river examples are included in the manual.
Ramsar's Rapid Cultural Inventories for wetlands https://www.ramsar.org/sites/default/files/documents/library/guidance_-_rapid_cultural_inventories_for_wetlands.pdf	Specifically for cultural ecosystem service values of wetlands with an aim to support conservation action.	Generally for use at individual wetland sites.	Rapid Cultural Inventories are a simple and practical way to identify, document, and make available information about notable cultural values and practices associated with identified wetland areas. It is a step-wise process involving stakeholder consultation and documentation of their cultural values and of the status of these valued services.	Pritchard et al. 2016. Case studies began in 2016 and not yet available.

15.4 Awareness of ecosystem services

In keeping with the concepts outlined by the Millennium Ecosystem Assessment (2005), and expanded by others, the ecosystem services provided by wetlands and freshwaters are closely linked to their effective functioning and management. These services characterise catchments in their present state, and the way catchments are managed to conserve these services will have a major impact on their sustainability. Despite increased interest in ecosystem services, and consequent increased sophistication of methodologies, the evidence base for evaluating their importance in many wetlands is still poor with a need to refine wetland evaluation techniques. By way of illustration we present two different responses to the increasing awareness of the value of ecosystem services from wetlands.

The first is the Richmond River in eastern Australia, which highlights a previous lack of awareness about the importance of ecosystem services in a catchment where the natural resources are being heavily exploited and the ecosystems degraded. The second highlights the situation in Colombia where a national response and recognition of the importance of ecosystem services provided by wetlands followed catastrophic flooding. In neither case has prior knowledge about the ecosystem service been collected despite the importance of such services being increasingly expressed and documented in the wider literature. Gaining further traction to raise awareness of the critical importance of the ecosystem services provided by wetlands and freshwaters is seen as an essential part of the process needed to assess and measure their value, and to adequately manage these ecosystems to ensure that their benefits are realised. With such awareness it is anticipated that further evaluation could occur and quantitative analyses undertaken (e.g., using the open-access toolkits illustrated in Table 15.1).

15.4.1 Richmond catchment, Australia

The Richmond River in eastern Australia meanders eastward from the Great Dividing Range to the Pacific Ocean and formerly contained a multitude of valuable and highly diverse forest and riverine ecosystems. Over the past 150 years these have been adversely affected by land clearing, including for forestry products and for agriculture, with the loss of large amounts of the native vegetation (Dawson 2002; Eyre 1997; Logan et al 2011). Additionally, a highly complex set of drains and sluices were constructed with catastrophic outcomes for the wetlands. Cropping activities are now dominated by plantations of Macadamia nuts and sugar cane, with cattle grazing also a major activity. The fragmentation of the riparian and estuarine ecosystems has given rise to river hypoxia, leading to fish kills, and the expansion of areas affected by acid sulphate soils (Eyre et al 2006; Corfield 2000).

The impacts of these changes on the riparian and wetlands ecosystems has somewhat belatedly been recognised with major impacts on the water flows through the catchment and the water purification function provided by river catchment processes (Hossain and Eyre 2002). The rivers are also adversely affected by high levels of sediments from the orchards and fields during the large rainfall events that are common in this area (McKee et al. 2001). Although today the forests and their riparian zones are greatly degraded, the region is nevertheless highly valued and is recognised by UNESCO as a World Heritage Biodiversity Hotspot and there is a steadily growing level of tourism. However, little attention has been hitherto directed towards the loss of regulating ecosystem services and ecosystem connectivity arising from the development and exportation of the natural resources of the catchment.

Across the whole of this multifunctional catchment there are many ecosystem functions and services, and steps to measure these have only started in recent years. Given the scale of change across the catchment it is assumed but not supported by any measured evidence that current ecosystem services have been greatly reduced. Further, there is no effective way to accurately calculate and represent their monetary value within current economic frameworks (Plant and Ryan 2013). In response steps have been taken to work with local community organisations to help identify and qualitatively value the important ecosystem services and functions associated with the river system, including the remaining coastal swamps. These steps are seen as essential for raising the awareness of the extent of the ecosystem services that formerly occurred in the catchment and which in some circumstances could be restored. Given a history of ecosystem services being largely ignored and degraded by the natural resource exploitation that has occurred and still dominates economic activity there has been a low appreciation of what has been lost, and what could be gained through changes in land- and water-management practices. A key to getting better outcomes is increasing the awareness of ecosystem services. With this in mind the words of the local indigenous Wijubal people of the Bunjalung nation is worth considering: '*Garima gala nyabay. Gala nyabay garima ngali ngih.*' When translated, this means 'Look after the water ... The water looks after us' (Rous Water and Sustainable Futures Forum 2005).

15.4.2 Colombian wetland ecosystem service assessment

The interest of Colombian policymakers in the condition of the country's wetlands and their ecosystem services was raised after extreme floods covered approximately 31 per cent of the country in 2010–2011, adversely affecting some 3.2 million people (CEPAL 2012). The flooding and economic losses indicated that not only was Colombia highly vulnerable to the impacts of extreme climactic events, but that the loss and degradation of wetlands had contributed to the adverse consequences for local people. The realisation that wetlands and associated ecosystems played a vital role in the regulation of flooding contributed to a decision to initiate a national wetland project including the assessment and mapping of the ecosystem services provided by wetlands.

While wetland inventory and mapping exercises were also conducted an assessment of the ecosystem services provided by wetlands across the country was undertaken using expert knowledge and participatory processes to outline the importance of ecosystem services in 19 wetland types (Ricaurte et al. 2017). The assessment was based on a qualitative analysis in the absence of quantitative information about the wetlands and in particular the benefits they provided. The success of the assessment therefore depended on the knowledge of experts familiar with the range of wetlands across the country and the range of services they provided.

The most important wetland types for the provision of ecosystem services were floodplain forests, riparian wetlands, freshwater lakes, rivers, and mangroves. The results showed that water regulation and habitat for wetland species were the most important. Freshwater lakes were seen as particularly important, possibly due to them being important for water and food supply, flood regulation, and recreational uses. Water reservoirs, which provided water supply, fishing, and recreational services, were of intermediate importance. These outcomes show the diversity of Colombian wetlands and their importance for the delivery of cultural and regulating services, including the mitigation of future floods, and provide a sound basis for valuation and further management. They have not included the more quantitative analyses that could be performed, with reference to the toolkit presented above in Table 15.1. The wider purposes of wetland conservation and management in Colombia, and elsewhere, could be advanced by making use of such tools.

Acknowledgements

Support for this work has been provided by the Australian Research Council, under Grant Number LP130100498.

References

Bagstad, K.J., Semmens, D.J., Waage, S., and Winthrop, R. (2013). A comparative assessment of decision-support tools for ecosystem services quantification and valuation. *Ecosystem Services*, 5, 27–39.

Blaen, P.J., Li, J., Peh, K.S-H., Field, R.H., Balmford, A., MacDonald, M.A., and Bradbury, R.B. (2015). Ecosystem services provided by two mineral extraction sites restored for nature conservation in an agricultural landscape in eastern England. *PLoS ONE*, 10(4). DOI:10.1371/journal.pone.0121010.

CEPAL (20120. Valoración de daños y pérdidas. Ola invernal en Colombia, 2010–2011. Comisión Económica para América Latina y el Caribe. Cepal, Bogotá: Misión BID.

Corfield, J. (2000). The effects of acid sulphate run-off on a subtidal estuarine macrobenthic community in the Richmond River, NSW, Australia. *ICES Journal of Marine Science*, 57, 1517–23.

Costanza, R. and Daly, H.E. (1992). Natural Capital and Sustainable Development. *Conservation Biology*, 6, 37–46.

Costanza, R., d'Arge, R., de Groot, R., Farber, S., Grasso, M., Hannon, B., Limburg, K., Naeem, S., O'Neill, R.V., Paruelo, J., Raskin, R.G., Sutton, P., and van den Belt, M. (1997). The value of the world's ecosystem services and natural capital. *Nature*, 387, 253–60.

Daily, G.C. (1997). Nature's Services: Societal Dependence on Natural Ecosystems. Island Press, Washington, DC.

Dawson, K. (2002). Fish kill events and habitat losses of the Richmond River, NSW, Australia: An overview. *Journal of Coastal Research*, 36, 216–21.

de Groot, R.S. (1992). Functions of nature: evaluation of nature in environmental planning, management and decision-making. Wolters Noordhoff BV, Groningen, The Netherlands.

de Groot, D., Brander, L., and Finlayson, M. (2016). Wetland Ecosystem Services. In Finlayson, C.M., Everard, M., Irvine, K., McInnes, R.J., Middleton, B.A., van Dam, A.A., and Davidson, N.C. (eds) The Wetland Book I: Structure and Function, Management and Methods. Springer Publishers, Dordrecht.

de Groot, R.S., Wilson, M., and Boumans, R. (2002). A typology for the description, classification and valuation of ecosystem functions, goods and services. *Ecological Economics*, 41, 367–567.

de Groot, R., Stuip, M., Finlayson, M., and Davidson, N. (2006). Valuing wetlands: guidance for valuing the benefits derived from wetland ecosystem services. Ramsar Technical Report No. 3, CBD Technical Series No. 27.

de Groot, R., Brander, L., van der Ploeg, S., Bernard, F., Braat, L., Christie, M., Costanza, R., Crossman, N., Ghermandi, A., Hein, L., Hussain, S., Kumar, P., McVittie, A., Portela, R., Rodriguez, L.C., ten Brink, P., and van Beukering, P. (2012). Global estimates of the value of ecosystems and their services in monetary terms. *Ecosystem Services*, 1, 50–61.

Ehrlich, P.R. and Ehrlich, A.H. (1981). Extinction: the Causes and Consequences of the Disappearance of Species. Random House, New York.

Eyre, B. (1997). Water quality changes in an episodically flushed sub-tropical Australian estuary: A 50-year perspective. *Marine Chemistry*, 59. 177–87.

Eyre, B.D., Kerr, G., and Sullivan, L.A. (2006). Deoxygenation potential of the Richmond River Estuary floodplain, northern NSW, Australia. *River Research and Applications*, 22, 981–92.

Finlayson, C.M. (2018a). The Economics of Ecosystems and Biodiversity (TEEB). In: Finlayson, C.M., Everard, M., Irvine, K., McInnes, R.J., Middleton, B.A., van Dam, A.A. and Davidson, N.C. (eds) The Wetland Book I: Structure and Function, Management and Methods. Springer Publishers, Dordrecht, The Netherlands. pp. 335–339. DOI.org/10.1007/978-90-481-9659-3_80

Finlayson, C.M. (2018b). Intergovernmental Panel for Biodiversity and Ecosystem Services (IPBES). In: Finlayson, C.M., Everard, M., Irvine, K., McInnes, R.J., Middleton, B.A., van Dam, A.A., and Davidson, N.C. (eds) The Wetland Book I: Structure and Function, Management and Methods. Springer Publishers, Dordrecht, The Netherlands. pp. 349–353. DOI.org/10.1007/978-90-481-9659-3_82

Finlayson, C.M. and Horwitz, P. (2015). Wetlands as settings for human health—the benefits and the paradox. In: Finlayson, C.M., Horwitz, P., and Weinstein, P. (eds) Wetlands and Human Health, 1–13. Springer Publishers, Dordrecht, The Netherlands.

Finlayson, C.M., Davidson, N., Pritchard, D., Milton, G.R., and MacKay, H. (2011). The Ramsar Convention and ecosystem-based approaches to the wise use and sustainable development of wetlands. *Journal of International Wildlife Law and Policy*, 14, 176–98.

Gardner, R.C., Barchiesi, S., Beltrame, C., Finlayson, C.M., Galewski, T., Harrison, I., Paganini, M., Perennou, C., Pritchard, D.E., Rosenqvist, A., and Walpole, M. (2015). State of the World's Wetlands and their Services to People: A compilation of recent analyses. Ramsar Convention Secretariat, Ramsar Scientific and Technical Briefing Note No. 7, Gland, Switzerland.

Gómez-Baggethun, E.B., Martín-López, B., Barton, D., Braat, L., Saarikoski, H., Kelemen, E., García-Lorente, M., van den Bergh, J., Arias, P., Berry, P., Potschin, M., Keene, H., Dunford, R., Schröter-Schlaack, C., and Harrison, P. (2014). State-of-the-art report on integrated valuation of ecosystem services. EU FP7 OpenNESS Project Deliverable 4.1, European Commission FP7.

Horwitz, P. and Finlayson, C.M. (2011). Wetlands as settings: ecosystem services and health impact assessment for wetland and water resource management. *BioScience*, 61, 678–88.

Hossain, S. and Eyre, B. (2002). Suspended Sediment Exchange through the Sub-tropical Richmond River Estuary, Australia: a Balance Approach. *Estuarine, Coastal and Shelf Science*, 55, 579–86

Hughes, F.M.R., Adams, W.M., Butchart, S.H.M., Field, R.H., Peh, K.S.-H., and Warrington, S. (2016). The challenges of integrating biodiversity and ecosystem services monitoring and evaluation at a landscape-scale wetland restoration project in the UK. *Ecology and Society*, 21, 10.

Infield, M., Morse-Jones, S., and Anthem, H. (2015). Guidelines for the Rapid Assessment of Cultural Ecosystem Services (GRACE), Version 1. Fauna and Flora International, Cambridge.

Kettunen, M., Bassi, S., Gantioler, S., and ten Brink, P. (2009). Assessing Socio-economic Benefits of Natura 2000—a Toolkit for Practitioners (September 2009 edition). Output of the European Commission project Financing Natura 2000: Cost estimate and benefits of Natura 2000 (Contract No.: 070307/2007/484403/MAR/B2). Institute for European Environmental Policy (IEEP), Brussels, Belgium.

Limburg, K.E., O'Neil, R.V., Costanza, R., and Farber, S. (2002). Complex systems and valuation. *Ecological Economics*, 41, 409–20.

Logan, B., Taffs, K.H., Eyre, B.D., and Zawadski, A. (2011). Assessing changes in nutrient status in the Richmond River estuary, Australia, using paleolimnological methods. *Journal of Paleolimnology*, 46, 597–611.

Maltby, E., Ormerod, S.J., Acreman, M., Blackwell, M., Durance, I., Everard, M., Morris, J., and Spray, C. (2011). *Freshwaters—openwaters, wetlands and floodplains*. UK National Ecosystem Assessment: Technical Report, 295–360. United Nations Environment Programme—World Conservation Monitoring Centre (UNEP-WCMC), Cambridge.

McKee, L.J., Eyre, B.D., Hossain, S., and Pepperell, PR (2001). Influence of climate, geology and humans on spatial and temporal nutrient geochemistry in the subtropical Richmond River catchment, Australia. *Marine and Freshwater Research*, 52, 235–48.

Millennium Ecosystem Assessment (2003). Ecosystems and human well-being: a *framework* for assessment. Island Press, Washington, DC

Millennium Ecosystem Assessment (2005). Ecosystems and human well-being: water and wetland synthesis. Washington, DC: Island Press.

Milton, G.R. and Finlayson, C.M. (2017). Freshwater ecosystem types and extents. In: Finlayson, C.M., Arthington, A.H., and Pittock, J. (eds) Freshwater Ecosystems in Protected Areas: Conservation and Management. Taylor and Francis, Oxford, UK.

Peh, K.S-H., Balmford, A., Bradbury, R.B., Brown, C., Butchart, S.H.M., Brown, C., Hughes, F.M.R., Stattersfield, A., Thomas, D.H.L., Walpole, M., Bayliss, J., Gowing, D., Jones, J.P.G., Lewis, S.L., Mulligan, M., Pandeya, B., Stratford, C., Thompson, J.R., Turner, K., Vira, B., Willcock. S., and Birch, J. (2013). TESSA: a toolkit for rapid assessment of ecosystem services at sites of biodiversity conservation importance. *Ecosystem Services*, 5, 51–7.

Peh, K.S-H., Balmford, A., Field, R.H., Lamb, A., Birch, J.C., Bradbury, R. B., Brown, C., Butchart, S.H.M., Lester, M., Morrison, R., Sedgwick, I., Soans, C., Stattersfield, A.J., Stroh, P.A., Swetnam, R.D., Thomas, D.H.L., Walpole, M., Warrington, S., and Hughes, F.M.R. (2014). Benefits and costs of ecological restoration: Rapid assessment of changing ecosystem service values at a UK wetland. *Ecology and Evolution*, 20, 3875–86.

Plant, R. and Ryan, P. (2013). Ecosystem services as a practicable concept for natural resource management: Some lessons from Australia. *International Journal of Biodiversity Science, Ecosystems Services and Management*, 9, 44–53.

Pritchard, D., Ali, M., and Papayannis, T. (2016). Guidance: Rapid Cultural Inventories for Wetlands. Ramsar Convention website.

Raudsepp-Hearne, C., Claesson, G., and Kerr, G. (2011). Ecosystem Services Approach Pilot on Wetlands. Government of Alberta, Canada.

Ricaurte, L.F., Olaya-Rodrígueza, M.H., Cepeda-Valenciaa, J., Lara, D., Arroyave-Suárez, J., Finlayson, C.M., and Palomo, I. (2017). Future impacts of drivers of change on wetland ecosystem services in Colombia. *Global Environmental Change*, 44, 158–69.

Rous Water and Sustainable Futures Australia (2005). The Water Walk at Rocky Creek Dam. A Users Guide to the Far North Coast Water Cycle. Rous Water, Lismore.

Shan, M., Duggan, J.M., Eichelberger, B.A., McNally, B.W. Foster, J.R., Pepi, E., Conte, M.N., Daily, G.C., and Ziv, G. (2016). Valuation of ecosystem services to inform management of multiple-use landscapes. *Ecosystem Services*, 19, 6–18.

Sullivan, C.A. (2002). Using an income accounting framework to value non-timber forest products. In: Pearce, D. (ed.) Valuation Methodologies. Edward Elgar, Cheltenham.

TEEB (2010). The Economics of Ecosystems and Biodiversity: Mainstreaming the Economics of Nature: A Synthesis of the Approach, Conclusions and Recommendations of TEEB.

Villa, F., Portela, R., Onofri, L., Nunes, P.A.L.D., and Lange, G. (2015). Assessing biophysical and economic dimensions of societal value: An example for water ecosystem services in Madagascar. In: Martin-Ortega, J., Ferrier, R.C., Gordon, I.J., and Khan, S. (eds) Water ecosystem services: A global perspective, 110–118. Cambridge: University Press Cambridge, UK.

Waage, S. and Stewart, E. (2008). Ecosystem services management: A briefing on relevant public policy developments and emerging tools. Fauna and Flora International, Cambridge, UK.

Watson, K.B., Ricketts, T., Galford, G., Polasky, S., and O'Niel-Dunne, J. (2016). Quantifying flood mitigation services: The economic value of Otter Creek wetlands and floodplains to Middlebury, VT. *Ecological Economics*, 130, 16–24.

Westman, W. (1977). How much are nature's services worth? *Science*, 2, 960–4.

16

Invasive Aquatic Species

Julie A. Coetzee, Martin P. Hill, Andreas Hussner,
Ana L. Nunes, and Olaf L.F. Weyl

Corresponding author: Julie.coetzee@ru.ac.za

16.1 Introduction

Freshwater lakes, rivers, and impoundments are particularly susceptible to invasions by invasive non-native species (INNS) across a range of floral and faunal taxa, from algae, to flowering plants, and jellyfish to mammals, largely as a consequence of anthropogenic influences on these systems (Dudgeon et al. 2006; Ricciardi and MacIsaac 2011). The alteration of hydrological flows, trade, and intentional stocking have resulted in homogenous flora and fauna (Rahel 2002) with ecological and socio-economic impacts. Arguably, freshwater communities have been altered to a greater magnitude than terrestrial communities (Dudgeon et al. 2006). In this chapter, we use the legal US federal definition of 'invasive': a non-native species whose introduction does or is likely to cause economic or environmental harm or harm to human health, while non-native species, with respect to a particular ecosystem, is any species that is not native to that ecosystem.

Despite their susceptibility to invasion across the globe, and across taxa, there are three main biases in the assessment of freshwater invasions: (1) an ecological bias, where the majority of studies focus on terrestrial invasions (Jeschke et al. 2012) despite data suggesting that freshwater ecosystems harbour the highest level of biodiversity per surface area under greater threat from invasions (Sala et al. 2000; Rahel 2002); (2) a geographical bias, where most studies and reviews on freshwater invasions are focused on temperate Northern Hemisphere systems, mainly from Europe and North America, while many tropical and subtropical freshwater ecosystems have been invaded but remain largely unassessed (Thomaz et al. 2015); and (3) a taxon bias, where studies and reviews on faunal invasions, particularly fish invasions, populate the literature (e.g., Ricciardi and MacIsaac 2011), with relatively fewer general reviews on aquatic plant invasions (e.g., Thomaz et al. 2015), and even fewer on invertebrates (e.g., Cowie 2001).

It is important to understand the invasion biology of an organism if effective survey, evaluation, and management measures are to be implemented. Several authors (e.g., MacDougall and Turkington 2005; Bauer 2012) have grouped INNS into three

Coetzee, J. A., Hill, M. P., Hussner, A., Nunes, A. L., and Weyl, O. L. F., *Invasive aquatic species*. In: *Freshwater Ecology and Conservation: Approaches and Techniques*. Edited by Jocelyne M. R. Hughes: Oxford University Press (2019). © Oxford University Press 2019. DOI: 10.1093/oso/9780198766384.003.0016 categories *viz.* (1) drivers of biodiversity loss, which include species that do not need any disturbance to establish; (2) passengers, which are solely dependent on a disturbance for establishment and proliferation, and if the disturbance is removed, the invasion and associated impacts cease; and (3) back-seat drivers whereby an initial disturbance is required for an INNS to establish, but once established, even if the disturbance is removed, the invasion continues. In freshwater ecosystems, INNS fall into all three categories. In some systems, population declines are the direct result of INNS introductions, especially in isolated systems that are naïve to the effects of a broad range of invaders, such as alpine lakes, desert pools, isolated springs, and oligotrophic waters (Moyle and Light 1996). Here, and even in species-rich freshwater systems, INNS drive the destruction of indigenous fauna and flora through predation, competition, and habitat alteration; for example, the zebra mussel (*Dreissena polymorpha*) has driven local extinctions in the Great Lakes region of North America (Ricciardi et al. 1998). Habitat alteration through canalisation, and the formation of impoundments and reservoirs, has increased the connectedness of watersheds, to the benefit of invaders (Johnson et al. 2008), resulting in homogenisation of these systems (Rahel 2002). Invaders in these situations are likely to be passengers which would not have established had the system not been disturbed. Examples of such invaders are mostly fish (Marchetti et al. 2004), while aquatic plant invasions are more likely to be back-seat drivers which rely on the broad ecosystem disturbance of slow-flowing permanent waters caused by impoundments and eutrophication which facilitates establishment, and linked with enemy release, allows them to proliferate, thereby gaining a competitive advantage over indigenous aquatic plants (Coetzee and Hill 2012).

This chapter briefly reviews freshwater invasive non-native species (INNS) across the globe focusing on fish, invertebrates, floating macrophytes, and submerged macrophytes; emphasising the knowledge gaps in particular that have resulted in the biases described above; and highlighting some of the approaches needed to survey, monitor, and manage INNS. These approaches include: (a) biosecurity; (b) anticipating and discovering an INNS; (c) monitoring the spread; (d) monitoring the impacts on freshwater ecosystem structure and function; and (e) management and control. The techniques used to survey the different categories of aquatic INNS presented here are documented in Part 2 of this book (see Chapters 4–11).

16.2 Fish

16.2.1 Introduction

With some 624 introduced species worldwide, fish are among the most intentionally introduced organisms in the world (Gozlan 2008). The primary drivers for these introductions are as food species in the global aquaculture industry (51 per cent), aquarium fish (21 per cent), and for the enhancement of wild fisheries for sport or food (19 per cent); the rest are the result of accidental introductions (e.g., goby invasions in the

Laurentian Great Lakes) (Gozlan 2008). While not all introduced fish become invasive via establishing, spreading, and subsequently impacting on native ecosystems, eight species are listed among the World's 100 worst invaders (Lowe et al. 2000).

One of the most introduced species in the world is the Nile tilapia, *Oreochromis niloticus*, which has been introduced to 102 countries for aquaculture, and its global production now exceeds 2 million tons. Their broad diet, aggressive spawning behaviour, high levels of parental care, and their ability to spawn multiple broods throughout the year has facilitated their successful invasion of many tropical and subtropical regions worldwide (Canonico et al. 2005). Perhaps one of the earliest introductions for fisheries enhancement was common carp, *Cyprinus carpio*, which has been domesticated and used for extensive aquaculture for more than 2,000 years. In South Africa, for example, carp was introduced as a food fish in the late 1700s and has since spread to almost every river basin in the country (Ellender and Weyl 2014).

The global pet trade is another major pathway for introductions (Nunes et al. 2015). Although aquarium fish are not intended for release into the wild, the release of unwanted pets, escape from garden ponds during flooding and ritualistic release during religious practices, have all contributed to facilitating invasions by aquarium fish, such as goldfish, *Carassius auratus*, guppies, *Poecilia reticulata*, and loricariid catfishes, *Pterygoplichthys* spp. Despite this, relatively few species have become invasive, probably as a result of poor climate matching between native and recipient environments.

The largest pathway for invasions by alien fish is via the intentional introduction of fishes into the wild to enhance fisheries or to create opportunities for angling. Well-known angling species introductions include the largemouth bass, *Micropterus salmoides*, and rainbow trout, *Oncorhynchus mykiss*, that have been introduced to ca. 80 countries across all continents. Both species are highly invasive, having established populations in parts of most countries where they have been introduced and as a result of predatory impacts on native biota, they are listed among the world's worst INNS (Lowe et al. 2000).

16.2.2 Impacts

There are examples of alien fish impacts across multiple levels of biological organisations ranging from the genome to the ecosystem (Cucherousset and Olden 2011). Common carp, for example, are considered problematic in many countries because their bottom grubbing habits during feeding suspends sediments which increases nutrient availability and turbidity, and subsequently suppresses macrophyte growth and can result in eutrophication. Similarly, Nile tilapia can directly modify nutrient regimes by increasing nitrogen and phosphorus availability in a reservoir via excretion, promoting algae growth, and contributing to eutrophication and cyanobacteria blooms. Predation on and competition with alien fish can result in changes in aquatic invertebrate communities and the fragmentation or extirpation of native fish and amphibian populations. In South Africa, for example, such extirpations have resulted in many native fishes now only persisting in small headwater tributaries (Ellender and Weyl 2014).

The introduction of novel parasites and diseases into environments can be particularly severe when fish and novel parasite communities have not coevolved and hence have inadequate immune responses to infection (Gozlan 2008). The global spread of the oomycete, *Aphanomyces invadans*, from South East Asia, for example, causes epizootic ulcerative syndrome, a disease that causes epidermal lesions, ulceration, and death of cultured and wild fish, resulting in both biodiversity and economic losses (Oidtmann 2012).

On a genetic level, human-mediated hybridisation (the mating between individuals from two genetically distinct populations) and introgression (when the offspring are fertile and backcross to parental populations) is considered the leading cause of global biodiversity loss (Allendorf et al. 2013). In fishes, an excellent example is the devastating impact of Nile tilapia introductions on other *Oreochromis* species in Africa. In southern Africa, for example, Nile tilapia introductions have resulted in extensive hybridisation and introgression with native *Oreochromis mossambicus* in the Limpopo River system, with *Oreochromis andersonii* and *Oreochromis macrochir* in the Kafue River in Zambia, and in Lake Kariba, Zimbabwe, they have almost replaced the native *O. mortimeri* (Ellender et al. 2014).

16.2.3 Control

Once established, the control of invasive fishes from large water bodies is extremely difficult and mechanical removals have had little success. As a result, the eradication of alien fish populations requires either complete dewatering or the use of piscicides such as rotenone. The use of rotenone (Box 16.1), which has been used successfully to eradicate alien fish from invaded streams and reservoirs in the USA, Britain, Australia, New Zealand, and South Africa, is, however, restricted to site-specific applications on a small spatial scale. This situation may be changing, as a result of rapidly evolving technologies for controlling alien fish over larger spatial scales (Britton et al. 2011). Researchers in Australia, for example, are investigating the use of the cyprinid herpes virus-3 and genetically modified carp to produce male-only offspring as potential agents to control common carp.

16.3 Invertebrates

16.3.1 Introduction

A significant number of freshwater invasions has been caused by invertebrates, mainly from the Arthropoda, Mollusca, Annelida, and Platyhelminthes. In Europe, Crustacea

---

Box 16.1 Management of Smallmouth bass in the Rondegat River: A case study by Weyl et al. (2014)

Smallmouth bass *Micropterus dolomieu* is highly invasive in South Africa. Where it has invaded, this species has been consistently found to eliminate small-bodied native cyprinid fishes, which has demonstrable impacts to the structure and function of macroinvertebrate communities. As a result, smallmouth bass are listed as Category 1b INNS in the South African INNS regulations (Regulations on Alien and Invasive Species (A&IS Regulations 2014) and were made into law in terms of Section 97(1) of the National Environmental Management: Biodiversity Act (NEM:BA, Act 10 of 2004)) meaning that 'control via a management plan' is the legally mandated management strategy for invasive populations.

Box 16.1 *Continued*

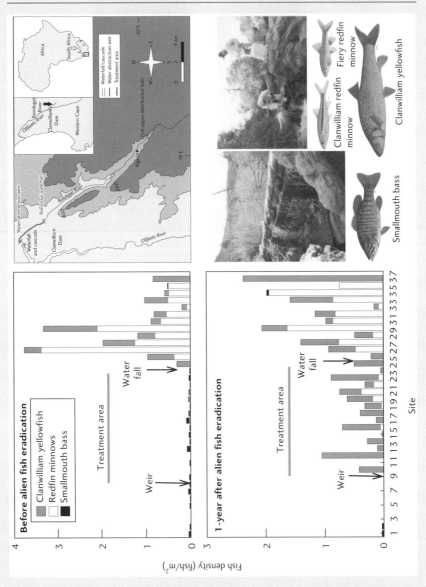

Figure 16.1 In the Rondegat River invasive smallmouth bass occurred only with large Clanwilliam yellowfish. The subsequent construction of a small 2-m high weir 4 km downstream of the waterfall, effectively isolated a portion of the smallmouth bass population in this stretch of river. In 2012, this isolated section of river was treated using the piscicide rotenone to remove smallmouth bass. Within a year following smallmouth bass removal, threatened redfin minnows began to utilise the rehabilitated section of river, and native fish abundance and diversity increased significantly (data source Weyl et al. 2014); within three years of the eradication, the native fish communities in parts of the lower Rondegat River had attained densities similar to those in pristine reaches. (Fish artwork courtesy South African Institute for Aquatic Biodiversity.)

are the most invasive group of the Arthropoda, many of which are considered as 'high-impact' species (Gherardi et al. 2009, Nunes et al. 2015). Non-native species from both the Arthropoda and Platyhelminthes have been introduced to Europe largely through aquaculture (Nunes et al. 2015). The shipping and inland canals pathways have also been responsible for a high number of Arthropoda introductions to Europe (Hulme et al. 2008), while invasive freshwater Mollusca have been introduced mostly via the pet and aquarium trade (Padilla and Williams 2004), and Annelida mostly through shipping (Nunes et al. 2015).

Invasions by freshwater invertebrates to Europe and the USA have been well documented (e.g., the invasion by Ponto-Caspian Species in Great Lakes of North America (Ricciardi and MacIsaac 2011; and crayfish invasion in Europe (Momot 1995; see Section 16.3.2)). However, there are few documented cases from other parts of the world. For example, Appleton (2003) listed ten gastropod species that had been introduced to South Africa through the aquarium trade, of which four species (*Lymnaea columella*, *Physa acuta*, *Tarebia granifera*, and *Aplexa marmorata*) have become invasive; and the New Zealand mud snail Potamopyrgus antipodarum has been reported as invasive in parts of South America (Collado 2014).

16.3.2 Impacts

Invasions of freshwater habitats may adversely affect biodiversity, alter the structure and functioning of ecosystems and cause economic losses (Gherardi et al. 2009). More specifically, when invading new habitats, introduced freshwater invertebrates can act as novel predators, competitors, habitat-changers, vectors of pathogens, or can promote hybridisation with indigenous species (Nunes et al. 2015). Famous cases of freshwater invertebrate introductions are known globally. Many of these refer to the introduction of freshwater crayfish species, a highly diverse group, whose species have been introduced to all continents except Antarctica (Lodge et al. 2012; Kouba et al. 2014). Freshwater crayfish, which are considered keystone species, often cause strong alterations at multiple trophic levels. For example, *Procambarus clarkii* (red swamp crayfish), native to north-eastern Mexico and south-central USA, has been introduced into the Iberian Peninsula, where it is a successful invader (Holdich et al. 2009). It has caused a reduction in food resources and in habitat complexity, such as refugia and spawning sites for fish and amphibians by consuming and destroying aquatic macrophytes and decreasing macroinvertebrate abundance. It is also an efficient predator of a wide array of aquatic organisms, such as insects, crustaceans, snails, and both eggs and larvae of fish and amphibians (Ilhéu et al. 2007). Due to particle resuspension, its bioturbation activity reduces light penetration and primary production, followed by losses in biodiversity. The species also consumes rice seedlings, causing negative economic impacts. *P clarkii* is a carrier of the crayfish plague fungus (*Aphanomyces astaci*), and has caused dramatic declines and even extinctions of populations of the indigenous crayfish *Austropotamobius pallipes* (Holdich et al. 2009). Another crayfish species with similar impacts is *Pacifastacus leniusculus*, the signal crayfish, present in 29 European countries. (Kouba et al. 2014).

Other cases of freshwater invertebrate species invasions, famous for their negative ecological and economic impacts, are the introduction of zebra (*Dreissena polymorpha*) and quagga (*Dreissena bugensis*) mussels, native to the Ponto-Caspian region, into the Laurentian Great Lakes in the USA. These species were introduced into the lakes through discharge of ballast water transported by transoceanic ships from Europe. Both species are considered ecosystem engineers, significantly modifying habitats where introduced, changing both ecosystem structure and functioning, causing declines of native species, and facilitation of the rapid spread of other Ponto-Caspian INNS (Ricciardi et al. 1998; Vanderploeg et al. 2002). Other than ecological impacts, these species clog intake pipes in power stations and water treatment plants, foul boat and ship hulls, deteriorate dock pilings and corrode steel and concrete (Vanderploeg et al. 2002; Benson et al. 2015).

16.3.3 Control

Although prevention is the most cost-effective way of mitigating invasion, once established, control or eradication strategies are needed to reduce their impacts (see Box 16.2). Several methods have been applied to manage or control invasive aquatic arthropod

---

Box 16.2 Monitoring the killer shrimp (*Dikerogammarus villosus*) and demon shrimp (*Dikerogammarus haemobaphes*) in English waterways. A case study by Tim Johns, Environment Agency, England, UK

The invasive Ponto-Caspian amphipod *Dikerogammarus villosus* first appeared in a reservoir in England in 2010, followed by *D. haemobaphes* in a river in 2012. The Environment Agency implemented new annual sampling methods (outlined below) to identify and track the spread of these species, focusing on high-risk sites (e.g., marinas, boating waters and fish farms) not normally covered by routine macroinvertebrate monitoring.

Baited traps

These consist of a wire mesh cage (photo below) allowing the free movement of invertebrates, with a textile mesh cover and a pebble/cobble substrate base mixed with a suitable bait. The trap is submerged, preferably on to a concrete/hard rock substrate and left for 24–48 hours, then recovered and examined for amphipods.

Multi-habitat sampling

This is a modification of a standard kick sampling method, whereby a pond net is used to collect a kick and/or a sweep sample (Chapter 11). The duration is not specified but rather the process aims to cover the range of substrates and habitats available, 'hunting' for shrimps rather than collecting a representative invertebrate community.

These methods have been successful in mapping the spread of *Dikerogammarus* spp. through waterbodies outside the river monitoring network, and provide a detection system for the possible arrival of other INNS. They are, however, limited spatially and temporally by the practicalities of monitoring all potential locations. Wider coverage of all river basins is achieved through the Environment Agency routine macroinvertebrate monitoring programme, where around 2,500 sites are sampled biannually for multiple monitoring purposes. In future, DNA and eDNA techniques are likely to be utilised to enhance coverage and improve detection of INNS.

Box 16.2 *Continued*

Figure 16.2 Four stages of baited-trap preparation.

species, used alone or in combination: mechanical (trapping, electrofishing, hand removal); physical (drainage, barriers); chemical (biocides); biological (natural predators or pathogens); and autocidal (sterile male release or sex pheromones) techniques. These methods and their successes and failures have been reviewed for freshwater crayfish species (Gherardi et al. 2011). In general, most of these methods have proved ineffective or cost-prohibitive, with mechanical and physical techniques usually being the methods of choice. For freshwater crayfish species that are widely established in the wild, eradication is expensive and virtually impossible, except for populations in small, enclosed waterbodies (Gherardi et al. 2011). Applying chemical methods, mostly using rotenone or synthetic pyrethroids, seems to be the only effective technique in eradicating crayfish populations from small water bodies. However, these products are not specific to crayfish, having marked environmental impacts on other species and on water quality. As such, it is important that whichever control techniques are employed, they are specific to the INNS in question, in order to avoid or greatly reduce impacts on non-target native species or the environment (Galil 2009).

The annual cost of controlling invasive freshwater arthropod species and compensating for their impacts has been estimated around €9.6–12.7 billion in Europe (see Invasive Alien Species page on European Commission website) and $120 billion in the USA (Pimentel et al. 2005), although these numbers are most likely an underestimate of the real cost. In the USA, estimated costs of damage caused by and control of the zebra mussel, the Asiatic clam (*Corbicula fluminea*), and the European green crab (*Carcinus maenas*) are approximately US$4.4 billion (Galil 2009).

16.4 Floating aquatic macrophytes

16.4.1 Introduction

The presence of invasive aquatic macrophytes threaten aquatic ecosystems throughout the world, costing governments vast sums of money to control. While a number of key traits are exhibited by the majority of invasive aquatic macrophytes which increase their invasiveness, such as rapid vegetative and sexual reproduction leading to fast population build-up, the ability to regenerate from fragments, high phenotypic plasticity and efficient dispersal mechanisms, their presence in a system is usually a symptom of anthropogenic spread combined with increasing urbanisation, industry, and agriculture, resulting in nutrient enrichment and ultimately eutrophication. Floating aquatic macrophytes, such as the notorious water hyacinth, *Eichhornia crassipes*, giant salvinia, *Salvinia molesta*, and water lettuce, *Pistia stratiotes*, are considered back-seat drivers of such human-mediated disturbance, invading systems largely across the tropics and subtropics, negatively impacting rural communities whose livelihoods depend on access to fresh water (Coetzee and Hill 2009).

The availability of nitrogen and phosphorus in the water column is the primary factor influencing the abundance and composition of floating aquatic plant assemblages. When readily available, growth of floating macrophytes can increase unchecked to the detriment of native aquatic flora because they absorb available nutrients through their root systems directly from the water column. For example, water hyacinth and red water fern, *Azolla filiculoides*, can double their biomass in less than two weeks under ideal nutrient conditions (Edwards and Musil 1975; Lumpkin and Plunkett 1982). Reducing agricultural, industrial and urban run-off high in nitrates, ammonium, and phosphates is key to reducing the impact of these invaders.

16.4.2 Impacts

Excessive growth of invasive floating macrophytes negatively affects freshwater systems in a variety of ways, with most invasions exhibiting similar impacts. Large floating macrophyte mats limit access to water for commercial, recreational, and subsistence purposes; reduce water quality for human and animal use; and alter habitat quality for native flora and fauna (Coetzee et al. 2014), all with negative economic consequences. In addition, the presence of floating macrophyte mats enhances habitat for vectors of disease, such as malaria, encephalitis, bilharzia, and other filariae; as well as property damage during floods as a result of plants building up against bridges, fences, walls, etc., which obstruct water flow and increase flood levels.

Ecologically, reductions in light and thus oxygen depletion beneath floating macrophyte mats alters submerged plant, plankton, invertebrate, and vertebrate communities (Coetzee et al. 2014), with knock-on trophic cascades. Further, competition from these plants alters native plant communities, often displacing wildlife forage and habitat.

16.4.3 Control

A number of management options are available for the control of invasive floating macrophytes, with varied success, often requiring integrated control to obtain acceptable control. Utilisation of the excessive biomass of floating aquatic plants infestations, particularly in poorer rural areas, is often encouraged as a management option, where local communities are perceived to benefit from their use. Unfortunately, this is rarely effective due to the effort required to remove significant amounts of high water content biomass, and may even promote their spread. Water hyacinth, for example, is nearly 95 per cent water, and to gain 1 t of dry material, 9 t of fresh material is required, decreasing the commercial viability of such harvesting operations (Julien et al. 1999).

Small infestations of floating macrophytes may be removed manually by hand, or mechanically using specialised harvesters, but this is labour intensive, requiring frequent follow-up treatments because not all plants are removed, allowing the regeneration of the infestation via vegetative reproduction.

Herbicidal control using 2,4 D, glyphosate, paraquat, and diquat, is a widely used method to control floating aquatic macrophytes, particularly in the southern USA where species such as water hyacinth, giant salvinia, and water lettuce clog waterways, impoundments, and natural lakes. This method of control costs States such as Florida, Louisiana, Texas, and Mississippi millions of dollars annually (Netherland 2014), but is limited in its success as it is temporary. New infestations regenerate from untreated plants, and seeds germinate from the hydrosoil following clearing, therefore requiring repeated applications. Where affordable, herbicidal control programmes also occur in African countries, such as South Africa, Zimbabwe, Uganda, Tanzania, and Kenya, as well as in China, Australia, and Papua New Guinea, with limited long-term success.

Large infestations of floating macrophytes can be controlled effectively through biological control, which is both economically and environmentally sustainable. Floating macrophytes are particularly susceptible to biological control with a number of successful cases throughout the world, particularly in the warmer tropics (e.g., water lettuce, salvinia, and red water fern have all been brought under complete biological control by a single agent in as little as two years, to a point where they no longer threaten aquatic ecosystems (Hill 2003)) (see Box 16.3). Biological control of water hyacinth is met with varied success, depending on water nutrient quality, cold winter temperatures, and interference from herbicide operations (Coetzee et al. 2011). In the warmer tropics, control has been achieved, most notably on Lake Victoria where the introduction of biological control agents reduced the infestation from 20,000 ha to 2,000 ha in less than five years, while control in more temperate areas requires longer time frames in the absence of herbicidal control (Hill 2003).

Box 16.3 Biological control of *Salvinia molesta*: a case study

---

Salvinia molesta (see photos below) is a sterile polyploid floating freshwater fern native to Brazil (Julien et al. 2009). It has established outside of its native range throughout the tropics, subtropics, and warm temperate areas, and has been recorded in 55 counties where it grows rapidly and is often regarded as a major aquatic weed (Hill 2003). Where the weed has become problematic, control measures include chemical, mechanical, and biological control. Biological control, using the host-specific weevil *Cyrtobagous salviniae* has proved to be the most effective (Julien et al. 2009; Coetzee et al. 2011), and is recognised throughout the world as the method of choice for *S. molesta* management (photo B). The insect was first released in Australia in 1980 and has since been released in 22 countries around the world including: Fiji, India, Kenya, Namibia, South Africa, Sri Lanka, USA, Zambia, and Zimbabwe. The impacts of *C. salviniae* on *S. molesta* are overviewed by Julien et al. (2009), with Coetzee et al. (2011) giving a case-study example for South Africa.

Salvinia molesta was first recorded in South Africa in the early 1900s, and by the 1960s was regarded as one of South Africa's worst aquatic weeds, second only to *Eichhornia crassipes*. *Cyrtobagous salviniae* was released against the weed in South

Box 16.3 *Continued*

Figure 16.3

(a) *Salvinia molesta* D.S. Mitchell (Salvinaceae). Drawn by R. Weber; first published in Stirton (1978).

(b) The biological control agent, *Cyrtobagous salviniae.*

(c) Lake Moondarra, Mt Isa, Queensland, Australia in June 1980. Lake Moondarra had a 400 ha mat of salvinia, weighing >50 000 t fresh weight. (Source: CSIRO, with permission.)

(d) The first release of *C. salviniae* was made at Lake Moondarra. The damage caused by the huge population of weevils that developed following release resulted in spectacular destruction of the mat within 15 months, reducing it to less than 1 t. (Source: CSIRO, with permission.)

(e) Farm Dam near Knysna, Western Cape Province, South Africa before introduction of *C. salvinae*, and

(f) two years later in 2009 following successful biological control.

Box 16.3 *Continued*

Africa in 1985, after undergoing host specificity trials. Annual quantitative surveys of South African freshwater systems were undertaken between 2008 and 2016, and of the 57 *S. molesta* sites visited annually, the weevil has established at all of them, and has successfully brought the weed under control to the point that no other interventions are required (photos E and F). The weed is now considered under complete biological control by *C. salviniae*. This biological control intervention was unusually quick in that most infestations were controlled within two years (see graph below). Methods for the mass-rearing and dissemination of the weevils have greatly facilitated the biological control effort.

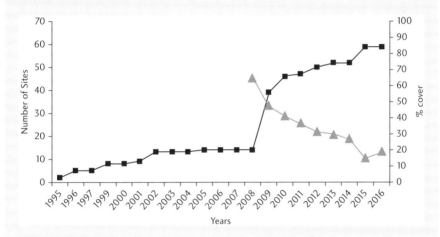

Figure 16.4 Cumulative number of *Salvinia molesta* sites recorded per year in South Africa. Records from 1995 to 2007 were historical records, while those from 2008 to 2016 were based on active annual surveys throughout the country. The second axis shows the mean per cent *S. molesta* cover of all sites surveyed throughout the country since 2008, indicating a significant decline in per cent cover of sites due to biological control by the weevil *Cyrtobagous salviniae*.

Ultimately, the long-term success of floating macrophyte control requires the integration of a variety of methods, with the most emphasis on reducing nitrate and phosphate pollution into aquatic environments (Hill 2003).

16.5 Submerged aquatic macrophytes

16.5.1 Introduction

Although the number of submerged invasive aquatic plant species is small, with <20 species being highly invasive at the larger geographical scale, including a high proportion of Hydrocharitaceae (e.g., *Elodea* spp., *Hydrilla verticillata*, *Egeria densa*, *Lagarosiphon major*, and *Vallisneria* spp.), these species are among the worst invasive aquatic weeds worldwide with significant economic and ecosystem impacts. In contrast to floating macrophyte species whose presence is known at the early stages of invasion, submerged plant invasions are often undetectable for long periods of time (Wells et al. 1997; Hussner et al. 2014), at which point eradication is impossible.

The primary introduction pathway of submerged plants into a new area is the horticultural and aquarium trade (Brunel 2009; Maki and Galatowitsch 2004); for example, *H. vertillata* into the USA and *L. major* into New Zealand. Alien submerged plants are traded either under their correct names, their synonyms, or common names (Hussner et al. 2014). They are released intentionally or unintentionally into waterbodies and subsequently spread via plant fragments, with water flow and water sport equipment identified as the major vectors (Heidbüchel et al. 2016). Most invasive submerged plants rely on vegetative modes of reproduction through stem fragmentation, which is of particular concern in systems frequented by anglers and recreational boaters. New infestations regenerate from broken stem fragments that root in the substrate, often near boat ramps, which are then continuously broken up and spread (Langeland 1996).

16.5.2 Impacts

Once introduced and established, elevated water and nutrient sediment conditions enhance the growth and spread of many submerged macrophyte invasions, in the same way that floating macrophyte invasions are driven by nutrient loading. These dense submerged plant infestations cause various ecological and socio-economic impacts, including negative effects on flora, fauna, and ecosystem services. Invasive submerged plants are able to become dominant within native plant beds or they can colonise novel, formerly unvegetated habitats (Wells et al. 1997). Mass infestations of invasive submerged plants cause a decline in native vegetation and can lead to increased sedimentation rates and increased flood risks, can block irrigation and drainage systems and can negatively impact the recreational use of the water body, which can result in reduced lakefront property values (Wilcock et al. 1999; Halstead et al. 2003). Water quality is also reduced in the presence of species such as *Cabomba caroliniana*, requiring additional purification steps for drinking water production (van Oosterhout 2009).

16.5.3 Control

As with floating macrophyte invasions, a combination of management strategies is required to mitigate the impacts of these species, because each method varies in its effectiveness and application scale. The decision of the best management option in a specific case is driven by the management goal, the species attributes, the habitat type, and also by the national legislation for management, as not all management methods are permitted in each country (e.g., the use of herbicides is prohibited in most European countries (Hussner et al. 2017)).

The easiest and cheapest way to control submerged invasive plants is the prevention of further introductions through regulation of the horticultural trade of the target species (Hussner et al. 2017). In Europe, in addition to voluntary agreements between plant producers and national authorities, legislation and regulations have been implemented both nationally and internationally to stop the trade (Hussner et al. 2017). These legislations and regulations require risk assessments of the species, which are prepared for several submerged aquatic weeds using, for example, the Aquatic Weed Risk Assessment Model (AWRAM) (Champion and Clayton 2000). In Europe, two invasive submerged plants (*Cabomba caroliniana* and *Lagarosiphon major*) are on the list of alien INNS of European Union concern, which must be controlled according to the regulation on the prevention and management of the introduction and spread of invasive alien species in Europe (Reg. EU no. 1143/2014).

Once introduced and established, manual and mechanical control can be successful in small oligotrophic systems that use hand harvesting, suction harvesting, and benthic barriers, for example, *Myriophyllum spicatum* removal in the USA. However, control is usually short term as infestations regenerate from fragments that remain behind. In most European countries, control of submerged plant invasions relies on mechanical removal, with varied success, because the use of herbicides is prohibited (except in the UK) (Hussner et al. 2017). Cutter boats is the most widespread method of control in Europe (Zehnsdorf et al. 2015), while large mechanical harvesters are frequently used in the USA (Gettys et al. 2014). None of these methods is species specific, and eradication is usually not achieved.

Herbicidal control of submerged aquatic species is practised at large scales in the USA and Canada against species such as *H. verticillata* and *M. spicatum*. Control can be achieved in certain water body types, such as canals, and irrigation channels, and even rivers, ponds, and lakes if selectivity and water-use profile are considered (Netherland 2014). The most widely used products include diquat, endothall, fluridone, and 2,4 D, which have been used for control over decades, and so their behaviour in aquatic systems is well known, allowing for improved management of submerged species. Nonetheless, overreliance on certain herbicides can lead to weed resistance; for example, *H. verticillata* has developed resistance to fluridone, the mainstay of its control in the southern USA (Netherland 2011). As with herbicidal control of floating plants, frequent reapplication is required to maintain control as regeneration from untreated plants and fragments may occur.

While biological control of floating aquatic plants has many successful examples, the biological control of submerged aquatic macrophytes has been variable. Most biological control programmes against submerged species have been implemented in the USA, against *H. verticillata* and *M. spicatum*, while agents for control of *E. densa* and *L. major* are under consideration in South Africa, and Ireland and New Zealand respectively. Fish predation, overwintering strategies of the agents, climate mismatching, and interference from herbicide operations may challenge the success of classical biological control, requiring integrated management plans under these circumstances (Hussner et al. 2017). In addition to classical biological control, control using grass carp (*Ctenopharyngodon idella*) is practised in many invaded systems in the USA (Pipalova 2006). Although complete eradication of submerged plants may be obtained under high stocking densities, grass carp are not selective and may remove all plant species from a system.

16.6 Discussion

Moyle and Light (1996) effectively attributed the patterns of freshwater invasions to three main factors, *viz.* anthropogenic habitat alteration; the frequency of introduction to freshwater systems, both intentionally and as accidental introductions; and the successful matching of invasive organisms to local environments. Thus, through human disturbance, the physical convergence of habitats has facilitated the biotic homogenisation of fauna and flora by invasive aquatic species in many aquatic ecosystems (Rahel 2002). Perhaps one of the most notorious examples of invasion-induced biotic homogenisation is the introduction of the Nile perch, *Lates niloticus*, into Lake Victoria in 1954, and the trophic cascade that followed. Predation by this species caused the extinction of approximately 200 species of haplochromine cichlids from the lake and is considered to represent the largest extinction event among vertebrates during the twentieth century (Goldschmidt et al. 1993). In addition to the extinction of the cichlids, the invasion by Nile perch resulted in nutrient enrichment of the shallow waterbody through the attraction of communities to the lake shores who could derive benefit from the Nile perch industry. This resulted in the widespread invasion of the lake by water hyacinth which had been introduced onto the system in the 1980s, which covered 20,000 ha in the mid-1990s (Albright et al. 2004), homogenising the flora of the lake, and illustrating a classical 'back-seat driver' process.

Unquestionably, the single most important mitigation measure to reduce further impacts of aquatic INNS is prevention of invasions (Tamayo and Olden 2014). This relies on an effective biosecurity approach that requires knowledge of potential invaders and invasible systems, and pathways of introduction and spread, incorporated into Early Detection and Rapid Response programmes (Hussner et al. 2017). Recent advances in development of environmental DNA (eDNA) techniques have greatly facilitated detection of freshwater INNS (Nathan et al. 2014; Klymus et al. 2017; Taberlet et al. 2018). Since only trace DNA amounts are required, it is particularly attractive to detect low abundance taxa. Thus, eDNA tools can be used for early detection of INNS when populations are small and confined, which can facilitate successful eradication outcomes. Reductions in propagule pressure of aquatic INNS; that is, the quality, quantity, and frequency of invading organisms that improve their chances of establishment success, are key to preventing the widespread economic and ecological losses associated with these invasions (Thomaz et al. 2015).

Further, reducing human disturbance to aquatic ecosystems is essential to diminish their invasibility because many aquatic INNS are passengers and back-seat drivers of disturbance, such as eutrophication and alterations in hydrology. Tamayo and Olden (2014) found that invasion vulnerability was positively correlated with urban land use surrounding lakes prone to invasion by the submerged macrophytes *E. densa*, *Potamogeton crispus*, and *M. spicatum*. Both increased propagule pressure and increased eutrophication and sedimentation are the causative factors resulting in invasibility of these lakes. Thus, in areas with high numbers of lakes and invasion-prone waterbodies, focused management efforts that prevent future introductions and limit negative impacts of established invasives should be implemented (Tamayo and Olden 2014) (see Box 16.4).

Although legislation on the prevention and management of invasive alien species does exist, often the lack of financial resources and manpower to implement them on the ground makes it challenging to enforce these regulations. Furthermore, it is important to coordinate the enforcement of actions against INNS in neighbouring countries, because otherwise a species that is being controlled or eradicated in one country might simply reinvade from an invaded neighbouring country, making all efforts hopeless. An excellent example of regional legislative collaboration is the European and Mediterranean Plant Protection Organisation (EPPO), responsible for European cooperation in plant health. Its objectives are 'to protect plants, to develop

---

Box 16.4 Control techniques in the context of freshwater INNS invasions at various stages of invasive species invasion and control (adapted from Chippendale 1991)

---

1. Prevention:
   a. Pre-border interventions: evaluation of all newly proposed introductions for possible invasiveness; e.g., pest risk analyses
   b. At-border interventions: customs control, quarantine
   c. Some problems:
      i. aquatic species invasiveness not easy to predict
      ii. many unintentional or unnoticed introductions, propagules difficult to detect
      iii. conflicts with other interest groups; e.g., horticulture and pet trade, which depend on introduced species.
   d. Suggestion: importer has to include the cost of controlling escapes in the business plan.
2. Eradication in the lag phase:
   a. Detect new invasions: coordinate surveillance for and manage records of new instances of naturalisation
   b. Provide post-border risk assessments: evaluate species in enough detail to make a decision as to whether regulation is required and, if so, in what form
   c. Plan eradication: estimate the feasibility of eradicating species and develop and implement an eradication plan if possible
   d. E.g., Black Striped Mussel in Darwin Harbour, Australia—highly invasive nature well known. Drastic control measures upon first discovery resulted in early and complete eradication.
3. Control in the already invaded phase
   a. Control most commonly implemented here, but it is the most difficult and expensive stage to control
   b. Options for INNS include manual and mechanical removal, chemical control, biological control, integrated control
   c. Cost of control increases sharply with density, and if remote, inaccessible areas are infested.
   d. Need to prevent reinfestation; i.e., follow-up treatments are vital.
   e. Other effects on the invaded system (altered nutrient cycling, extinction of native flora and fauna) may already be advanced, therefore expensive rehabilitation required.

Box 16.4 *Continued*

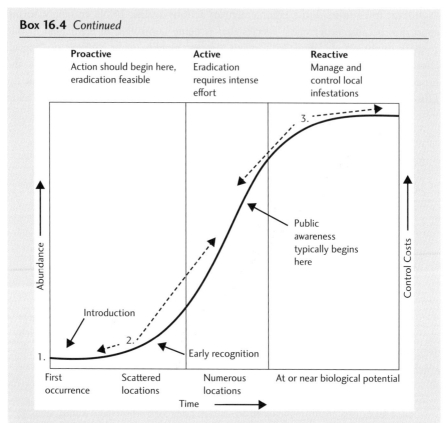

Figure 16.5 Phases of invasive species invasion and control. Numbers refer to:
1. prevention; 2. eradication; and 3. control techniques, in the context of freshwater
INNS (from Chippendale 1991).

international strategies against the introduction and spread of dangerous pests and to promote safe and effective control methods'. The EPPO, which was founded in 1951 by 15 member states, has grown to today's 51 member countries, including nearly every country in the European and Mediterranean region.

The freshwater economy globally is under threat as a burgeoning human population puts an increasing demand on the quantity and quality of water. Biotic invasion invariably reduces both the supply and quality of freshwater, directly competing with this human and ecological demand. Therefore, understanding the processes of invasion and their impacts is vital to addressing this challenge. Future research on aquatic invasions should thus aim to reduce the three biases highlighted in the introduction. More studies on *aquatic* invasions need to be conducted to address the ecological bias, conducted in the tropics and Southern Hemisphere, addressing invasions by a greater number of invertebrate and plant taxa.

References

Albright, T.P., Moorhouse, T.G., and McNabb, T.J. (2004). The rise and fall of water hyacinth in Lake Victoria and the Kagera River Basin, 1989–2001. *Journal of Aquatic Plant Management*, 42, 73–84.

Allendorf, F.W., Luikart, G.H., and Aitken, S.N. (2013). Conservation and the Genetics of Populations (2nd edition). Wiley-Blackwell, Oxford, UK.

Appleton, C. (2003). Alien and invasive fresh water Gastropoda in South Africa. African *Journal of Aquatic Science*, 28, 69–81.

Bauer, J., (2012). INNS: 'Back-seat drivers' of ecosystem change? *Biological Invasions*, 14, 1295–304.

Benson, A.J., Raikow, D., Larson, J., Fusaro, A., and Bogdanoff, A.K. (2015). *Dreissena polymorpha*. US Geological Survey Non-indigenous Aquatic Species Database, Gainesville, Florida, USA. Available online.

Britton, J.R., Gozlan, R.E., and Copp, G.H. (2011). Managing non-native fish in the environment. *Fish and Fisheries*, 12, 256–74.

Brunel, S. (2009). Pathway analysis: aquatic plants imported in 10 EPPO countries. *European and Mediterranean Plant Protection Organization Bulletin*, 39, 201–13.

Canonico, G.C., Arthington, A., McCrary, J.K., and Thieme, M.L. (2005). The effects of introduced tilapias on native biodiversity. *Aquatic Conservation: Marine and Freshwater Ecosystems*, 483, 463–83.

Champion, P.D. and Clayton, J.S. (2000). Border Control for Potential Aquatic Weeds. Stage 1 Weed Risk Model. Science for Conservation 141. Department of Conservation, Wellington (available at http://www.doc.govt.nz/upload/documents/science-and-technical/sfc141.pdf).

Chippendale, J.F. (1991). Potential returns to research on rubber vine *(Cryptostegia grandiflora)*. M.Sc. Thesis, University of Queensland, Brisbane, Australia.

Coetzee, J.A. and Hill, M.P. (2009). Aquatic Weeds. In: Clout, M.N. and Williams, P.A. (eds) Invasive Species Management- A Handbook of Principles and Techniques. Techniques in Ecology and Conservation Series, Oxford University Press, Oxford, UK.

Coetzee, J.A. and Hill, M.P. (2012). The role of eutrophication in the biological control of water hyacinth, *Eichhornia crassipes*, in South Africa. *BioControl*, 57, 247–61.

Coetzee, J.A., Hill, M.P., Byrne, M.J., and Bownes, A. (2011). A review of the biological control programmes on *Eichhornia crassipes* (C. Mart.); Solms (Pontederiaceae), *Salvinia molesta* D.S. Mitch. (Salviniaceae), *Pistia stratiotes* L. (Araceae), *Myriophyllum aquaticum* (Vell.) Verdc. (Haloragaceae) and *Azolla filiculoides* Lam. (Azollaceae) in South Africa. *African Entomology*, 19, 451–68.

Coetzee, J.A., Jones, R.W., and Hill, M.P. (2014). Water hyacinth, *Eichhornia crassipes* (Pontederiaceae), reduces benthic macroinvertebrate diversity in a protected subtropical lake in South Africa. *Biodiversity and Conservation*, 23, 1319–30.

Collado, G.A. (2014). Out of New Zealand: molecular identification of the highly invasive freshwater mollusk *Potamopyrgus antipodarum* (Gray, 1843) in South America. *Zoological Studies*, 53, 70.

Cowie, R.H. (2001). Invertebrate Invasions on Pacific Islands and the Replacement of Unique Native Faunas: A Synthesis of the Land and Freshwater Snails. Contribution number 2001-001 of Bishop Museum's Pacific Biological Survey. *Biological Invasions*, 3, 119–36.

Cucherousset, J. and Olden, J.D. (2011). Ecological impacts of non-native freshwater fishes. *Fisheries*, 36, 215–30.

Dudgeon, D., Arthington, A.H., Gessner, M.O., Kawabata, Z.-I., Knowler, D.J., Lévêque, C., Naiman, R.J., Prieur-Richard, A.H., Soto, D., Stiassny, M.L.J., and Sullivan, C.A. (2006). Freshwater biodiversity: importance, threats, status and conservation challenges. *Biological Reviews*, 81, 163–82.

Edwards, D. and Musil, C.J. (1975). *Eichhornia crassipes* in South Africa—a general review. *Journal of the Limnological Society of Southern Africa*, 1, 23–7.

Ellender, B.R. and Weyl, O.L.F. (2014). A review of current knowledge, risk and ecological impacts associated with non-native freshwater fish introductions in South Africa. *Aquatic Invasions*, 9, 117–32.

Ellender, B.R., Woodford, D.J., Weyl, O.L.F., and Cowx, I.G. (2014). Managing conflicts arising from fisheries enhancements based on non-native fishes in southern Africa. *Journal of Fish Biology*, 85, 1890–906.

Galil, B.S. (2009). Control and eradication of invasive aquatic invertebrates. In: Gherardi, F., Corti, C., and Gualtie, M. (eds) Biodiversity Conservation and Habitat Management, Volume 2, Eradication and control of INNS. Encyclopedia of Life Support Systems (EOLSS) Publications, Oxford, UK.

Gettys, L.A., Haller, W.T., and Petty, D.G. (eds) (2014). Aquatic Ecosystem. 3rd edition, Aquatic Ecosystem Restoration Foundation, Marietta, Georgia, USA.

Gherardi, F., Aquiloni, L., Diéguez-Uribeondo, X., and Tricarico, E. (2011). Managing invasive crayfish: Is there a hope? *Aquatic Sciences*, 73, 185–200.

Gherardi, F., Gollasch, S., Minchin, D., Olenin, S., and Panov, V.E. (2009). Alien invertebrates and fish in European inland waters. In: DAISIE- Handbook of Alien Species in Europe, Volume 3, Invading Nature. Springer Series in Invasion Ecology, Springer.

Goldshmidt, T., Witte, F., and Wanink, J. (1993). Cascading effects of the introduced Nile perch on the detritivorous/phytoplanktivorous species in the sublittoral areas of Lake Victoria. *Conservation Biology*, 7, 686–700.

Gozlan, R.E. (2008). Introduction of non-native freshwater fish: is it all bad? *Fish and Fisheries*, 9, 106–15.

Halstead, J.M., Michaud, J., and Hallas-Burt, S.H. (2003). Hedonic analysis of effects of anon-native invader (*Myriophyllum heterophyllum*) on New Hampshire (USA) lakefront properties. *Environmental Management*, 32, 391–8.

Heidbüchel, P., Kuntz, K., and Hussner, A. (2016). Alien aquatic plants do not have higher fragmentation rates than native species—a field study from the River Erft. *Aquatic Science*, 78, 767–77.

Hill, M.P. (2003). The impact and control of alien aquatic vegetation in South African aquatic ecosystems. *African Journal of Aquatic Science*, 28, 19–24.

Holdich, D.M., Reynolds, J.D., Souty-Grosset, C., and Sibley, P.J. (2009). A review of the ever increasing threat to European crayfish from non-indigenous crayfish species. *Knowledge and Management of Aquatic Ecosystems*, 11, 394–5.

Hulme, P.E., Bacher, S., Kenis, M., Klotz, S., Kühn, I., Minchin, D., Nentwig, W., Olenin, S., Panov, V., Pergl, J., and Pyšek, P. (2008). Grasping at the routes of biological invasions: a framework for integrating pathways into policy. *Journal of Applied Ecology*, 45, 403–14.

Hussner, A., Nehring, S., and Hilt, S. (2014). From first reports to successful control: a plea for improved management of alien aquatic plant species in Germany. *Hydrobiologia*, 737, 321–31.

Hussner, A., Stiers, I., Verhofstad, M.J.J.M., Bakker, E.S., Grutters, B.M.C., Haury, J., van Valkenburg, J.L.C.H., Brundu, G., Newman, J., Clayton, J.S., Anderson, L.W.J., and Hofstra, D. (2017). Management and control methods of invasive alien freshwater aquatic plants: a review. *Aquatic Botany*, 136, 112–37.

Ilhéu, M., Bernardo, J.M., and Fernandes, S. (2007). Predation of invasive crayfish on aquatic vertebrates: the effect of *Procambarus clarkii* on fish assemblages in Mediterranean temporary streams. In: Gherardi, F. (ed.) Biological Invaders in Inland Waters: Profiles, Distribution, and Threats. Springer, Dordrecht, The Netherlands.

Jeschke, J.M., Aparicio, L.G., Haider, S., Heger, T., Lortie, C.J., Pyšek, P., and Strayer, D.L. (2012). Support for major hypotheses in invasion biology is uneven and declining. *NeoBiota*, 14, 1–20.

Johnson, P.T.J., Olden, J.D., and Van der Zanden, M.J. (2008). Dam invaders: impoundments facilitate biological invasions into freshwaters. *Frontiers in Ecology and the Environment*, 6, 357–63.

Julien, M.H., Griffiths, M.W., and Wright, A.D. (1999). Biological control of water hyacinth. The weevils *Neochetina bruchi* and *N. eichhorniae*: biologies, host ranges, and rearing, releasing and monitoring techniques for biological control of *Eichhornia crassipes*. Australian Centre for International Agricultural Research, Monograph Number 60.

Julien, M.H., Hill, M.P., and Tipping, P.W. (2009). *Salvinia molesta* DS Mitchell (Salviniaceae). Weed Biological Control with Arthropods in the Tropics. Cambridge University Press, Cambridge, UK.

Klymus, K.E., Marshall, N.T., and Stepien, C.A. (2017). Environmental DNA (eDNA) metabarcoding assays to detect invasive invertebrate species in the Great Lakes. *PloS One*, 12, e0177643.

Kouba, A., Petrusek, A., and Kozák, P. (2014). Continental-wide distribution of crayfish species in Europe: update and maps. *Knowledge and Management of Aquatic Ecosystems*, 413, 05.

Langeland, K.A. (1996). *Hydrilla verticillata* (L.F.) Royle (Hydrocharitaceae). The perfect aquatic weed. *Castanea*, 61, 293–304.

Lodge, D.M., Deines, A., Gherardi, F., Yeo, D.C.J., Arcella, T., Baldridge, A.K., Barnes, M.A., Chadderton, W.L., Feder, J.L., Gantz, C.A., Howard, G.W., Jerde, C.L., Peters, B.W., Peters, J.A., Sargent, L.W., Turner, C.R., Wittmann, M.E., and Zeng, Y. (2012). Global introductions of crayfishes: evaluating the impact of species invasions on ecosystem services. *Annual Review of Ecology, Evolution and Systematics*, 43, 449–72.

Lowe, S., Browne, M., Boudjelas, S., and De Poorter, M, (2000). 100 of the world's worst invasive alien species: A selection from the global INNS database. INNS Specialist Group of the International Union for World Conservation (IUCN), Auckland, Australia.

Lumpkin, T.A. and Pluknett, D.L. (1982). Azolla as a green manure: Use and Management in crop production. Westview Tropical Agriculture, Series 5. Westview Press, Boulder, Colorado, USA.

MacDougall, A.S. and Turkington, R. (2005). Are INNS the drivers or passengers of change in degraded ecosystems? *Ecology*, 86, 42–55.

Maki, K. and Galatowitsch, S. (2004). Movement of invasive aquatic plants into Minnesota (USA) through horticultural trade. *Biological Conservation*, 118, 389–96.

Marchetti, M.P., Moyle, P.B., and Levine, R. (2004). Alien fishes in California watersheds: characteristics of successful and failed invaders. *Ecological Applications*, 14, 587–96.

Momot, W.T. (1995). Redefining the role of crayfish in aquatic ecosystems. *Reviews in Fisheries Science*, 3, 33–63.

Moyle, P.B. and Light, T. (1996). Biological invasions of freshwater: empirical rules and assembly theory. *Biological Conservation*, 78, 149–61.

Nathan, L.M., Simmons, M., Wegleitner, B.J., Jerde, C.L., and Mahon, A.R. (2014). Quantifying environmental DNA signals for aquatic INNS across multiple detection platforms. *Environmental Science and Technology*, 48, 12800–6.

Netherland, M.D. (2011). Comparative susceptibility of fluridone resistant and susceptible hydrilla to four ALS inhibiting herbicides under laboratory and greenhouse conditions. *Journal of Aquatic Plant Management*, 49, 100–6.

Netherland, M.D. (2014). Chemical control of aquatic weeds. In: Gettys, L., Haller, W., and Petty, D. (eds) Biological Control of Aquatic Plants. A Best Management Practices Handbook. 3rd edition, Aquatic Ecosystem Restoration Foundation, Marietta, Georgia, USA.

Nunes, A.L., Tricarico, E., Panov, V.E., Cardoso, A.C., and Katsanevakis, S. (2015). Pathways and gateways of freshwater invasions in Europe. *Aquatic Invasions*, 10, 359–70.

Oidtmann, B, (2012). Review of biological factors relevant to import risk assessment for epizootic ulcerative syndrome (*Aphanomyces invadans*). *Transboundary and Emerging Diseases*, 59, 26–39.

Padilla, D.K. and Williams, S.L. (2004). Beyond ballast water: aquarium and ornamental trades as sources of INNS in aquatic ecosystems. *Frontiers in Ecology and the Environment*, 2, 131–8.

Pimentel, D., Zuniga, R., and Morrison, D. (2005). Update on the environmental and economic costs associated with alien-INNS in the United States. *Ecological Economics*, 52, 273–88.

Pipalova, I. (2006). A review of grass carp use for aquatic weed control and its impact on water bodies. *Journal of Aquatic Plant Management*, 44, 1–12.

Rahel, F.J. (2002). Homogenization of freshwater faunas. *Annual Review of Ecology and Systematics*, 33, 291–315.

Ricciardi, A. and MacIsaac, H.J. (2011). Impacts of biological invasions on freshwater ecosystems. In: Richardson, D.M. (ed.) Fifty Years of Invasion Ecology: The Legacy of Charles Elton. Wiley-Blackwell, New Jersey, USA.

Ricciardi, A., Neves, R.J., and Rasmussen, J.B. (1998). Impending extinctions of North American freshwater mussels (Unionoida) following the zebra mussel (*Dreissena polymorpha*) invasion. *Journal of Animal Ecology*, 67, 613–19.

Sala, O.E., Chapin, F.S., Armesto, J.J., Berlow, E., Bloomfield, J., Dirzo, R., Huber–Sanwald, E., Huenneke, L.F., Jackson, R.B., Kinzig, A., Leemans, R., Lodge, D.M., Mooney, H.A., Oesterheld, M., Poff, N.L., Sykes, M.T., Walker, B.H., Walker, M., and Wall, D.H. (2000). Global biodiversity scenarios for the year 2100. *Science*, 287, 1770–4.

Taberlet, P., Bonin, A., Zinger, L., and Croissac, E. (2018) Environmental DNA for Biodiversity Research and Monitoring. Oxford University Press, Oxford.

Tamayo, M. and Olden, J.D. (2014). Forecasting the vulnerability of lakes to aquatic plant invasions. *Invasive Plant Science and Management*, 7, 32–45.

Thomaz, S.M., Mormul, R.P., and Michelan, T.S. (2015). Propagule pressure, invasibility of aquatic ecosystems by non-native macrophytes and their impacts on populations, communities and ecosystems: a review of tropical freshwater ecosystems. *Hydrobiologia*, 746, 39–59.

Van Oosterhout, E. (2009). Cabomba control manual. New South Wales Department of Primary Industries, Orange, New South Wales, Australia.

Vanderploeg, H.A., Nalepa, T.F., Jude, D.J., Mills, E.L., Holeck, K.T., Liebig, J.R., and Ojaveer, H. (2002). Dispersal and emerging ecological impacts of Ponto-Caspian species in the Laurentian Great Lakes. *Canadian Journal of Fisheries and Aquatic Sciences*, 59, 1209–28.

Wells, R.D.S., de Winton, M., and Clayton, J.S. (1997). Successive macrophyte invasions within the submerged flora of Lake Tarawera, Central North Island, New Zealand. *New Zealand Journal of Marine and Freshwater Research*, 31, 449–59.

Weyl, O.L., Finlayson, B., Impson, N.D., Woodford, D.J., and Steinkjer, J. (2014). Threatened endemic fishes in South Africa's Cape Floristic Region: a new beginning for the Rondegat River. *Fisheries*, 39, 270–9.

Wilcock, R.J., Champion, P.D., Nagel, J.W., and Croker, G.F. (1999). The influence of aquatic macrophytes on the hydraulic and physic-chemical properties of a New Zealand lowland stream. *Hydrobiologia*, 416, 203–14.

Zehnsdorf, A., Hussner, A., Eismann, F., Rönicke, H., and Melzer, A. (2015). Management options of invasive *Elodea nuttallii* and *Elodea canadensis*. *Limnologica*, 51, 110–17.

17

Freshwater Ecosystem Security and Climate Change

Jamie Pittock, C. Max Finlayson, and Simon Linke

Corresponding author: Jamie.pittock@anu.edu.au

17.1 Introduction

Climate change is recognisably already impacting on the distribution, composition, and conservation of biodiversity including major impacts on freshwater wetlands and their species (Daufresne and Boet 2007; Field et al. 2014; Parmesan and Yohe 2003). A recent example is the ongoing thinning and retreat of the Kaskawulsh Glacier in Canada that led in 2016 to a redirection of meltwater to the south-flowing Alsek River from the north-flowing Slims River, which has dried up (Headley 2017). Another is the acidification of the Gnangara Mound wetlands and Lake Mariginiup in south-west Western Australia that is attributed in large part to a marked reduction in precipitation since the mid-1970s (Appleyard and Cook 2009; Searle et al. 2010). Most biodiversity conservation institutions, however, such as national protected area systems and treaties like the Ramsar Convention on Wetlands, were established at times when the climate was assumed to be stationary. While there is analysis of how conservation mechanisms such as protected areas can help conserve terrestrial (Hannah et al. 2007; Hole et al. 2009) and marine (McLeod et al. 2008) ecosystems with climate change, there has been less work on freshwater ecosystems, which are particularly vulnerable to climate-related changes in hydrology and habitat connectivity (Field et al. 2014; Kernan et al. 2011).

Currently, freshwater biodiversity is among the most threatened and least well conserved (MEA 2005; Strayer and Dudgeon 2010; WWF 2014). Freshwater ecosystems are a focus of human habitation and at least two of their ecosystem services, capture fisheries and freshwater use, are exploited beyond sustainable levels (Falkenmark et al. 2007; MEA 2005). Uniquely, the conservation of freshwater biodiversity depends on the maintenance of particular water regimes; in other words, maintaining water flows of particular quantities, quality, and timing, which have in many cases been altered by human activities (Arthington 2012). Further, the extent or absence of connectivity between freshwater habitats constrains the distribution of some components of the freshwater biota (Pringle 2003).

Pittock, J., Finlayson, C. M., and Linke, S., *Freshwater ecosystem security and climate change*. In: *Freshwater Ecology and Conservation: Approaches and Techniques*. Edited by Jocelyne M. R. Hughes: Oxford University Press (2019).
© Oxford University Press 2019. DOI: 10.1093/oso/9780198766384.003.0017

The development of biodiversity conservation institutions for terrestrial ecosystems and then for marine ecosystems has resulted in them being least well suited to conserving freshwater ecosystems (Abell et al. 2007). The application of conservation mechanisms has lagged for freshwater ecosystems (MEA 2005; Pittock et al. 2015). International norms for the conservation of biodiversity as represented by the Convention on Biological Diversity (CBD), the Convention on International Trade in Endangered Species (CITES), the Ramsar Convention on Wetlands, and various migratory species agreements are focused on three different strategies: (1) protection of viable populations of indigenous species across their ranges; (2) protection of distinct 'ecological communities' or assemblages of species and their associated physical environment; and (3) conservation of the ecological processes and functions of ecosystems that produce services valued by people (Groves et al. 2002). The need to conserve the world's biodiversity in the face of unrelenting pressures and decline underpins all three strategies (Chopra et al. 2005).

The Millennium Ecosystem Assessment describes the primary direct drivers of degradation and loss of freshwater biodiversity as including: infrastructure development, land conversion, water withdrawal, pollution, over-harvesting and overexploitation, the introduction of invasive alien species, and global climate change (MEA 2005). Pressures on freshwater ecosystems continue to grow with the human population, increasing wealth, and global environmental change processes, and these pressures and drivers are synergistic in their impacts. To date, interventions to ameliorate these pressures have been unsuccessful in halting and reversing the rate of loss of freshwater biodiversity as evidenced by a loss of 64–71 per cent of wetlands since 1900 AD (Davidson 2014) and a 76 per cent decline in known populations of freshwater species between 1970 and 2010 (WWF 2014). However, interventions such as the Ramsar Convention site listing have enabled better management of wetlands compared with those places that have not been designated (Bowman 2002; Castro et al. 2002).

The precept of conservation strategies in the twentieth century, particularly in the New World, was that there was a pre-development natural baseline, or reference condition of species distribution and ecological community composition that conservation management should aim to restore and maintain (Norris and Thoms 1999). The advent of climate change makes the concept of conservation to a reference condition impractical, requiring freshwater biodiversity proponents to articulate more sophisticated approaches to conservation (Catford et al. 2012; Finlayson et al. 2017; Kopf et al. 2015). With the importance of climate change impacts on freshwater ecosystems well established the next four sections discuss: (1) how the impacts of climate change on freshwater ecosystems may be modelled; (2) different approaches for managing freshwater ecosystems and resources; (3) options for climate change adaptation for freshwater ecosystems; and (4) how to manage the impacts of climate change responses in other sectors on freshwater ecosystems.

17.2 Modelling impacts of climate change on freshwater ecosystems

17.2.1 Introduction to modelling

The growth of modern industrial societies has extensively impacted upon freshwater ecosystems, such that climate is only the latest such driver of change (MEA 2005;

Strayer and Dudgeon 2010). Many non-climate drivers, such as the development of water-resources infrastructure, have considerably impacted on freshwater ecosystems (Poff et al. 2007). The societal responses to managing these non-climate drivers under climate change are further considered in a later section.

The Intergovernmental Panel on Climate Change has concluded that the impacts of climate change on freshwater ecosystems and resources include altered precipitation, evapotranspiration, temperature, extreme events such as droughts and floods, water quality, and geomorphology (Field et al. 2014). As freshwater species are dependent upon flows of the right quantity, quality, and timing, it is highly likely that climate change will cause major changes in the habitat used by these species (Arthington 2012). Increased temperatures may see ideal freshwater habitat for many species move closer to the poles, or to higher elevations, and change the water quality. However, watershed boundaries are likely to form a natural barrier to the movement of many freshwater species to more suitable habitats. These boundaries may be exacerbated by anthropogenic barriers, such as dams, or aided by canals and inter-basin transfers (Pringle 2003). The range and availability of habitats within such ecosystems is expected to change. As the climatic and hydrological thresholds of individual species differ, the species composition within habitats or within entire ecosystems is expected to change, creating new or novel ecosystems (Catford et al. 2012). New species are likely to move within and between these ecosystems, both regionally indigenous species, as well as exotic species, although our ability to ascertain the extent of such change will be affected by limited understanding of the range of natural variability due to climate extremes (Finlayson 2009).

A key question for managers is whether such changes can be modelled to aid management and adaptation. Here we discuss two contrasting and potentially complementary approaches; namely, 'top down' physical projections versus more 'bottom up' assessments of values to be conserved and options for doing so.

17.2.2 Models for climate, hydrology, species distribution, and prioritisation

For decades global climate change models have been used in processes like those of the Intergovernmental Panel on Climate Change to develop projections and scenarios for changes in temperature and precipitation. As computing power has increased, enabling more climatic variables to be included in models, and downscale modelling, more accurate projections are available and can increasingly be applied to smaller areas. A range of inexpensive models that can be used on desktop computers, like ANUCLIM (Xu and Hutchinson 2013) and online modelling services, are enabling wider use, including by organisations in developing countries. These climate models can be linked to hydrological models to project future water availability. A number of widely available down-scale climate models are shown in Table 17.1 and then discussed.

The simplest method to include climate data is to summarise existing ranges (available in Bioclim (Kriticos et al. 2014)) and run models based on extrapolation from these ranges. Only slightly more labour intensive is to calculate ranges from pre-summarised ensemble climate models—this is often used in freshwater conservation problems (see Bush et al. 2014) and seems like a parsimonious solution. If more detail about multiple sources of climate variation is needed, the free source data of the Coupled Model

Table 17.1. *Different levels of models for climate, hydrology, species distribution, and prioritisation.*

Model	Strengths	Limitations	References
Climate change modelling			
Static hindcast models; e.g., ANUCLIM/ BIOCLIM	Easily accessible, downloadable grids of all variables	Only hindcast—static summary variables	Xu and Hutchinson (2013), Kriticos et al. (2014)
Ensemble forecast models: MAGICC/ SCENGEN 5.3, CliMond	Easily accessible, downloadable grids	Needs deeper understanding of scenario choices	Fordham et al. (2012), Kriticos et al. (2012)
Coupled models: CMIP5	Flexible, can deal with multiple levels of uncertainty, freely available	Data and processing intensive, needs very good understanding of included scenarios	Collins et al. (2013)
Environmental flow methods	(adapted from Poff et al. 2017)		
Hydrological (Single flow indices)	Simple, can be calculated on very large scales (Norris et al. 2007), applicable for highly data-deficient systems	No flow regime dynamics included, spurious links to ecological function	Tennant (1976)
Hydraulic rating	Ecological method with relatively small data requirements, links habitat to species persistence	Not applicable for whole systems, mainly used on site level, low resolution/ confidence	
Habitat simulation (e.g., PHABSIM)	Increased ecological function included, modelled multiple locations and scenarios, communities instead of single species, higher resolution/confidence	Higher levels of expertise needed, higher data needs, higher cost	Stalnaker et al. (1995)
Holistic methods (e.g., BBM, ELOHA)	Models entire ecosystems, flexible data requirements, mix of data and expert judgement	Needs broad range of expertise, including ecology, hydrology, and social sciences	King et al. (2000), Poff et al. (2010)
Biodiversity mapping	For a review, see Linke et al. (2011)		
Species distribution modelling	Direct proxies, can incorporate functional responses	Data hungry, 'false presences' can lead to prioritisations that do not contain the desired features	Elith and Leathwick (2009)
Informed surrogates, community modelling (GDM)	Can deal with sparser data, still includes ecological function	Hard to include direct responses of species	Ferrier and Drielsma (2010), Drielsma et al. (2014)

(Continued)

Table 17.1 *Continued*

Physical surrogates	Easily available through global data sets, such as HydroATLAS	No information about species responses, often not representative in planning scenarios	Januchowski-Hartley et al. (2011)
Freshwater conservation planning tools			
'Binary'— reserved/ unreserved (Marxan, basic Zonation)	Easy to parameterise, outputs easily understandable	Limited realism, limited inclusion of ecological function	Moilanen et al. (2008), Linke et al. (2012)
Planning for multiple actions/ zones (Marxan with Zones, Zonation)	Includes multiple conservation actions; e.g., conservation/restoration/ zonal protection	Still limited realism, no interactions between threats, limited inclusion of species specific responses	Moilanen et al. (2011), Hermoso et al. (2015)
Planning for multiple zones/ flexible species responses	Includes ecological function, direct links to species benefits, realistic, flexible	Time intensive to parameterise	Cattarino et al. (2015), Cattarino et al. (2016)

Intercomparison Project (Taylor et al. 2012) can be custom-tailored (for a great review, see Collins et al. 2013)—however, deeper understanding about the different models is needed.

Hydrological methods to assess the effect of flow have emerged in the 1980s and have similar levels of complexity—from simple index-based models, which at least as regional scales are readily available (for an application see Norris et al. 2007) to highly complex three-dimensional hydraulic models. While simple models can be very effective—especially when coupled with advanced species distribution models, holistic flow methods have the advantage of maximising model parsimony. Holistic flow methods such as BBM or ELOHA (King et al. 2000; Poff et al. 2010) combine flexibility and function and are the choice of modern ecological applications. For a complete history and discussion of different environmental flow methods, see Poff et al. (2017).

Mapping of ecological features and conservation planning often go hand in hand—the former being a pre-requisite of the latter. In a review of freshwater conservation planning methods, Linke et al. (2011) highlighted that direct data for the conservation targets is desirable over surrogates—as long as the data are spatially comprehensive. Species distribution models (Elith and Leathwick 2009) are the preferred method to extrapolate for incomplete sampling, followed by stratified environmental surrogates. While still frequently employed in conservation planning, unstratified surrogates, such as climate or soil classes, lead to low (Grantham et al. 2010) or non-existent (Januchowski-Hartley et al. 2011) representation and are generally not desirable if they can be avoided.

Conservation planning tools have increased in realism lately; however, similarly to the previous three parameters, this comes as a trade-off with data requirements.

Criticisms of early conservation planning tools were largely based on the lack of including ecological function. This has even been remediated in simple conservation algorithms (Moilanen et al. 2008; Hermoso et al. 2011; Linke et al. 2012) that can now include connectivity, as well as accounting for disturbance and condition. Later developments include different management actions in still reasonably simple parameterisation frameworks (Watts et al. 2009; Moilanen et al. 2011), although note that including connectivity in zoning algorithms is not trivial (Hermoso et al. 2015). Recent developments inject process-based realism into tried and tested conservation concepts by considering single-species ecological responses in a multi-stressor environment (Cattarino et al. 2015; Cattarino et al. 2016). They are only feasible in data-rich areas in which ecological response have been studied in great detail.

Each step in the chain of modelling temperature, precipitation, evapotranspiration, and run-off increases uncertainty such that the final projections may at best indicate a range of possible outcomes and trends. However, these generally lack the precision that managers and policymakers seek to support management decisions, including those specifically dealing with adaptation measures (Matthews and Wickel 2009). As an example, the Australian government commissioned research into the Murray-Darling Basin that reported in 2008 that water availability in 2030 could be anything from 7 per cent greater to 37 per cent less on average (CSIRO 2008). Consequently, no allowance for climate change was included in the Basin-wide water plan that was adopted in 2012 for implementation through to 2026 (Pittock 2013).

In modelling climate change impacts on freshwater systems there is a risk of focusing on purported average annual future conditions, as the Australian government has done with regards the Murray-Darling Basin, rather than considering how to manage more catastrophic climate change and variability (Pittock 2013). Research suggests that extreme events such as droughts and floods are likely to be more frequent in future, but impact differently by region. For instance, Hirabayashi et al. (2013) project a large increase in flood frequency at the end of the century in South East Asia, Peninsular India, eastern Africa, and the northern half of the Andes, with a decrease in other areas of the world.

In practice, researchers and practitioners will have to make choices about model complexity for four input components—climate models, environmental flow assessments, species distribution models, and methods for setting priorities (see Table 17.1). All of the four components can range from simple, often non-dynamic, models to highly complex systems analysis.

17.2.3 A values-based approach to selecting conservation options

A different approach to modelling climate change impacts and responses for freshwater ecosystems starts with a societal discussion and agreement on the kinds of values and services that should be conserved, which might range from a healthy floodplain wetland to water for irrigated agriculture. For conservation of ecological values, there are a broad range of environmental flow models that can be used to identify the thresholds for water timing, volume, and quality required to sustain those values (Arthington 2012).

On regulated river systems, reoperation of water infrastructure may enable flows that meet these thresholds under a changing climate, or inform policy decisions on the points at which a changing climate may require a fall back strategy to be adopted for biodiversity conservation (Pittock and Finlayson 2011). Lukasiewicz et al. (2016) set out a Catchment Adaptation Framework as one means of working with stakeholders through a range of different climate change impact scenarios to select robust adaptation options which may provide benefits under a range of possible climate outcomes. Similarly for groundwater-dependent ecosystems, modelling of conservation values and recharge with climate change could be used to regulate water extraction to maintain selected biodiversity (Pittock et al. 2016).

The kinds of modelling described above need to feed into governance and management frameworks, such as those discussed next. In particular they will inform decisions about the types of ecosystems that wider society may want to replace those that predominated in the past (Kopf et al. 2015), with consequent implications for wetland managers and authorities.

17.3 Different approaches for managing risks to freshwater ecosystems and resources

The deepening crises in managing freshwater ecosystems and resources around the world has led to the development and application of different approaches for water management that are briefly reviewed here.

Integrated Water Resource Management (IWRM) has been defined by the Global Water Partnership as 'a process which promotes the co-ordinated development and management of water, land and related resources, in order to maximize the resultant economic and social welfare in an equitable manner without compromising the sustainability of vital ecosystems' (GWP 2000). Offshoots of this approach, Integrated River Basin Management and Integrated Groundwater Management, have been promoted as a means of managing competing uses of freshwater ecosystems and resources sustainably within natural, physical catchments (Pittock et al. 2016; WWF 2003). However, the concept has been extensively criticised, with Biswas (2004) describing it as 'amorphous' with 'no agreement on fundamental issues like what aspects should be integrated, how, by whom, or even if such integration in a wider sense is possible'. This criticism has led to different efforts to find more focused and tractable frameworks for managing the conservation of freshwater ecosystems and the trade-offs among different water users.

Water security has been promoted as a way of increasing the attention paid to water policy and engaging more powerful financial, security, and leadership elements of governments in debates on sustainable water management. Grey and Sadhoff (2007) define water security as 'the availability of an acceptable quantity and quality of water for health, livelihoods, ecosystems and production, coupled with an acceptable level of water-related risks to people, environments and economies'. The water security proposition that climate change impacts on water availability may lead to violent conflict is contested (Kallis and Zografos 2013). The climate, energy, and water nexus that emerged over the past decade came out of frustration that broader approaches like sustainable development and IWRM were not delivering better outcomes quickly. By singling out two, three, or more priority- linked sectors and focusing on tractable policies it is hoped that more tangible benefits may be generated (Hussey et al. 2015).

Marketisation has been promoted as a way of creating an incentive for more efficient use by putting a price on water (Olmstead 2010). However, pricing water has proven difficult in practice for reasons ranging from difficulties in measurement and enforcement, challenges defining private versus public benefits, and different perspectives on rights to water (Savenije and van der Zaag 2002). In some instances, such as Australia's Murray-Darling Basin, markets that capped water use have decoupled consumption from growth in socio-economic benefits, but the environmental outcomes remain uncertain at best (Grafton and Horne 2014).

Ecosystem services (see Chapter 15) is a concept intended to aid the identification and conservation of the functions of ecosystems that produce services valued by people, including supporting, regulating, provisioning, and cultural services (MEA 2005). This has been promoted as a means of linking biodiversity conservation and livelihoods of people, and is especially tractable with respect to freshwater ecosystems and resources. Around the world many examples of payments for (watershed) ecosystem services schemes have been established where a portion of fees paid by water users is dedicated to the conservation and restoration of freshwater catchments. This enticing concept has been criticised, however, as trading off the degradation of threatened, biodiversity-rich lowland river ecosystems for water supply and energy generation against conservation of more intact upland ecosystems (Pittock et al. 2015).

Non-government stewardship programmes are being promoted as new ways of doing business with environmental and social sectors of society collaborating to define and promote higher standards of water management than those mandated and enforced through government regulations. The Alliance for Water Stewardship (AWS 2014) defines this for its programme for certifying water users as: 'The use of water that is socially equitable, environmentally sustainable and economically beneficial, achieved through a stakeholder-inclusive process that involves site and catchment-based actions.' The concept behind this non-government approach is to: establish new societal norms and demonstrate how water can be better managed, recognise and reward those water users who do so more sustainably, and to provide support for governments to play a stronger role in water management.

Finally, ecosystem- or nature-based adaptation is being promoted through the Convention on Biological Diversity and the United Nations Framework Convention on Climate Change as a way of utilising 'biodiversity and ecosystem services to support climate change adaptation' (Chong 2014). Restoration of catchments and floodplains are common examples of ecosystem-based adaptation that benefit freshwater biodiversity conservation, reduce risks of floods and improve water quality (Pittock 2009).

This diversity of approaches speaks to the complexity and contested nature of management of freshwater ecosystems and resources. A preferred concept is not recommended here, but rather, managers need to be aware of the strengths and weaknesses of these different approaches, and be prepared to engage in places where one or other concept is dominant. The next section outlines why climate change impacts and adaptation for freshwaters will make this management task even more challenging. In the subsequent section, the likely societal responses to climate change in other sectors are discussed in terms of their impacts on water management. These conflicting sectoral requirements highlight the need to apply the kinds of management approaches listed above in order to develop cross-sectoral policies that sustain freshwater ecosystems and resources to an acceptable extent.

17.4 Climate change adaptation for freshwater ecosystems

Capon et al. (2013) discuss riparian ecosystems but their conclusions could apply to all freshwaters when they say that: (1) riparian ecosystems are likely to be highly vulnerable to climate change impacts; (2) given the strong links with human well-being that considerable means for adaptation to climate change also exist; and (3) that the importance of many ecosystem functions, goods, and services will grow under a changing climate. They designate these ecosystems as adaptation 'hotspots'.

The impacts of climate change are bringing recognition of the need to set freshwater ecosystem conservation targets based on socially defined values linked to states that are physically possible to maintain; in other words, designer ecosystems (Acreman et al. 2014; Kopf et al. 2015). An objective under these circumstances could be to manage for resilience; that is, ecosystems that have the capacity to respond to impacts and maintain similar functions, such as generation of ecosystem services (Fischer et al. 2009). Planning adaptation to climate change is constrained by the uncertainty as to the direction and scale of many of the impacts, posing the question of whether there are 'no regrets' and robust measures that may be applied for freshwater biodiversity conservation (Pittock and Finlayson 2011). In this context, how should the institutions for conservation of freshwater biodiversity be revised to account for the synergistic impacts of climate change with other pressures?

New approaches are needed for institutions to set resilient targets for freshwater biodiversity conservation under a changing climate. Climate change will involve the loss of some freshwater habitats, for example, to sea level rise, and expansion of others in areas with greater water availability. The barriers in human-dominated landscapes will prevent the natural migration of some species and ecological communities to favourable hydro-climatic niches under a changing climate, leading to their diminution. The range of species and resulting composition of ecosystems will change. While some suggest that society should substantially accept these changes (Catford et al. 2012; Poff and Matthews 2013), choosing strategic, purposeful adaptation interventions may enable society to retain a broader area and range of freshwater ecosystems, more species and also higher production of valued ecosystem services. We argue that given the changes underway in species and habitats that robust and achievable adaptation requires an emphasis on the benefits of protecting freshwater ecological processes and ecosystem services (Finlayson et al. 2017).

The location of the habitats of many species will change. In many instances, conserving freshwater species will require maintaining or restoring pathways for movement;

for instance, to enable migration along river corridors to high elevation and cooler habitats (Lukasiewicz et al. 2016). There may be a number of refugia where species can survive; for instance, where groundwater inflows or riparian forest restoration sustain cooler, higher-quality water in stream reaches (Davies 2010; Olden and Naiman 2009). In other instances, bio-physical barriers to movement will prompt the difficult question of whether some species should be translocated (Hannah 2010). There are wetland species for which all natural habitat is projected to disappear and *ex-situ* conservation could be the only option remaining; for instance, some aquatic fauna in south-western Australia (Davies 2010).

The species composition and distribution of freshwater ecosystems is changing with the climate, yet the diversity of wetland types as defined by structures and functions is likely to remain similar to those of today. Targets could be set for the conservation of structural wetland types, in terms of the area of habitat and the ecological services that they provide. For example, as some fens are lost to a drier climate other areas in Britain may become higher conservation priorities for these freshwater ecosystems (Acreman et al. 2009). The climate-induced changes to species ranges and composition of fresh-water ecosystems, and the contraction or expansion of freshwater ecosystems, makes it essential to set species and wetland structure targets for conservation at a regional rather than a site scale (Poff et al. 2010). The Freshwater and Marine Ecoregions of the World assessments (Abell et al. 2008; Spalding et al. 2007) provide globally consistent bio-logical regions for wetland ecosystems. As species ranges change with the climate, path-ways for movement need to be maintained or restored across the landscape to enable species to occupy newly suitable habitat. Along riverine systems, setting targets for maintaining and restoring connectivity may require: reserving remaining free-flowing rivers, providing environmental flows, controlling thermal pollution from dams, con-structing fish passages over dams, and restoring floodplains (Lukasiewicz et al. 2016; Pittock et al. 2015). Strategic restoration of riparian vegetation can play a key role shad-ing streams to reduce water temperatures in a warming world and improve other aspects of habitat quality (Davies 2010). Many wetland types exist as discrete sites across the landscape and cannot readily migrate to other sites, as shown in the example of British fens (Acreman et al. 2009). Elevational, latitudinal, and other climatic gradients can be assessed that may enable the identification of sites where particular wetlands and species may thrive under future climates. This may facilitate decisions to redirect conservation investments from sites that are becoming less viable under a changing climate to prior-ity sites that have the prospect of remaining as good habitat in the longer term.

An example of this approach is the 'living landscapes' programme of The Wildlife Trusts (2017) in the UK. This involves around 150 large-scale and decades-long pro-jects across the nation that aim to conserve whole river catchments and entire tracts of upland. These involve restoring and expanding core areas of high-quality wildlife habitat and connecting them, often along rivers and streams as nature's natural corridors. One of these living landscape projects is the restoration of 3,700 hectares of The Great Fen near Peterborough in eastern England (Chapter 4), one of the largest such restoration projects in Europe (Great Fen Team 2017). This is intended to re-establish a more resilient habitat, large enough to support populations of threatened fen wildlife, such as bitterns and otters.

More resilient landscapes are required. Some parts of the freshwater landscape have physical features that make them more diverse or resilient to climate change. In pre-historic times these would have been refugia for biodiversity in the face of climate change. In addition to the environmental gradients described above, these may include bio-physical attributes such as topographic shading, groundwater inflows, self-scouring pools; as well as management-based attributes—for example, remaining free-flowing rivers and largely intact catchments (Lukasiewicz et al. 2016). These places should be identified and prioritised for conservation. In many instances, the places that may be more physically resilient in the future are currently degraded and will require restoration to provide the desired habitat for biodiversity.

Finlayson et al. (2017) propose six key principles for freshwater ecosystem conserva-tion and management under climate change; namely, that:

1. Objectives and targets for management should look to accommodate and com-pensate for climate change, rather than accepting or avoiding impacts, especially in early phases of adaptation.
2. Objectives for management under climate change should include ecological, social, and economic targets across multiple scales and consider issues such as representativeness, connectivity, and refugial values.
3. Flexible governance and adaptive co-management frameworks across multiple scales and sectors are essential to managing freshwater ecosystems under climate change.
4. Easily reversed, no-regret, or low-regret adaptation options with multiple, cross-sectoral benefits should be implemented in the initial phases of adapting.
5. Long-term management strategies should identify triggers for new actions including novel or high-risk adaptation options (e.g., species translocations) and should plan for such eventualities.
6. Scientific monitoring and evaluation of management strategies are needed.

These principles for adaptation for freshwater biodiversity need to be implemented in a broader context that considers the other societal responses to climate change, and measures that may aid or hinder ecosystem conservation.

17.5 Managing the impacts of climate change responses in other sectors on freshwater ecosystems

Not only will climate change directly impact upon freshwater biodiversity, but so too will the measures that society implements to respond to climate change across a range of sectors. In this section we will consider climate mitigation and adaptation policy measures. Many of these measures are already enshrined in the policies of nations around the world with little consideration of the perverse impacts on freshwater eco-systems and resources (Pittock 2011).

Many climate change mitigation policies will have severe impacts on the availability of water and nature of river flows in different places. In terms of the provision of energy produced with lower greenhouse gas emissions, these technologies may increase water consumption or by replacing thermal power stations, make more available. For example, in Australia, the closure of coal-fired power stations in the water-stressed Hunter and Latrobe valleys may make more water available for the environment and people (Pittock et al. 2013). Turning to negative examples, first-generation biofuel crops have been supported through policies, such as those of the EU, but they tran-spire a lot of water (Dalla Marta et al. 2015). In another example, shale gas production in the USA increases water use for hydraulic fracking (Mauter et al. 2014), whereas coal seam gas production in Australia involves dewatering aquifers (Pittock et al. 2013). There are opportunities, however, such as solar photovoltaic energy replacing water-cooled coal-fired power stations, and when upgrading hydropower dams to add infra-structure to reduce impacts on freshwater biodiversity (Pittock 2015). Considering programmes for sequestering carbon through reforestation, these can dramatically reduce inflows into streams (Jackson et al. 2005). Similarly, a great many adaptation measures could negatively impact upon freshwater biodiversity. For example, greater hydrological variability with climate change is leading to calls for greater investment in water storage and in expanding irrigated agriculture in the cause of water and food security. These kinds of measures will severely impact freshwater biodiversity through habitat fragmentation and loss of water (Pittock 2015).

These sectoral conflicts raise the question as to how our societies can better conserve freshwater biodiversity while also providing the energy, food, and water that people need, and avoiding the costs of going too far (Gordon et al. 2010). While there will always be some level of trade-offs, solutions lie in four domains (Pittock 2015); namely:

1) Provision of more publicly accessible information in compatible formats across these sectors so that decision makers are informed of the potential conflicts and positive synergies; for example, in development planning.

2) Adoption of new technologies that may deliver services to people with lower impacts on freshwater ecosystems, such as off-river pumped storage hydropower.

3) Harmonisation of markets for natural resources management so that there is a financial cost for perverse impacts on other sectors; for instance, ensuring that markets for carbon credits require reforestation projects to purchase entitlements for the additional water consumed.

4) Implementation of governance measures to ensure that proposed climate change mitigation and adaptation measures assess the implications across sectors so that transparent decisions may be taken on any trade-offs; for example, through stra-tegic environmental assessments of proposed programmes.

This discussion highlights how freshwater biodiversity proponents need to actively engage decision-makers in sectors such as agriculture, energy, and water resources to ensure that the conservation needs of ecosystems are adequately incorporated in other policy realms.

Two examples illustrate the positive opportunities that exist to better harmonise sectoral policies. When South Africa adopted a new Water Act in 1998, exotic forest plantations were recognised as a 'stream flow reduction activity' and steps were taken to limit that impact by requiring plantation owners to acquire water entitlements within a catchment-wide cap and pay an annual fee for the water used (Mackay 2003). This approach enables forest plantations for carbon sequestration or biofuels to expand within sustainability limits, linking the forestry and water markets. The impact of this new law led to collaboration among researchers, businesses, and non-government organisations to define wetland soils where the forestry companies agreed to remove exotic plantations (Dickens et al. 2003). Not only did this limit transpiration but it also enabled restoration of extensive areas of wetlands along streams. Another example is the periodic relicensing of non-Federal Government-owned hydropower dams in the USA every 30 to 50 years by the Federal Energy Regulatory Commission (Pittock and Hartmann 2011). This institutional policy reform window enables dam performance to be publicly re-evaluated to enable changes to meet current social, economic, and environmental standards. It enables technologies such as fish passages, invented since the dams were constructed to be retrofitted. This institution may enable biodiversity conservation measures to be enacted to adapt to climate change.

17.6 Conclusion

Our changing climate brings great challenges for conservation of freshwater ecosystems. Fortunately, guidance is emerging for managers in the form of literature from the Ramsar Convention, the ecosystem services community of practice, and the academic community (Acreman et al. 2014; Pittock et al. 2015). The key conclusions and information presented in this chapter are synthesised to suggest a climate change adaptation process that freshwater ecosystem managers can apply to enhance biodiversity conservation (Figure 17.1). The steps are:

1. Identify key freshwater ecosystem values that managers and society wish to retain.
2. Project the range of potential impacts of climate change on the hydrology and maintenance of key freshwater ecosystem values.
3. Apply most locally tractable governance approach/es (as described above: IWRM, water security, marketisation, ecosystem services, water stewardship, and ecosystem-based adaptation).
4. Implement the six freshwater adaptation principles.
5. Manage the impacts of responses in other sectors (e.g., agriculture, energy) to climate change.

These steps should then be repeated to address further climate change, evolution of societal values, and to incorporate knowledge from implementation in an adaptive management cycle.

A number of tools exist that can be applied to model and select climate change adaptation measures. While the water sector is replete with competing management approaches (such as security and stewardship), the following principles can be implemented through any of those chosen in a particular jurisdiction.

Conservation strategies for freshwater ecosystems should look to accommodate and compensate for climate change and the objectives for management should focus on

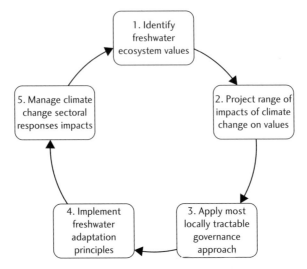

Figure 17.1 A climate change adaptation process that freshwater ecosystem managers can apply to enhance biodiversity conservation.

sustaining ecological processes and services across multiple scales. Conservation should target freshwater ecosystems for structural representativeness, connectivity, and refugial values. No-regret or low-regret adaptation options with multiple benefits can be identified and should be implemented as soon as practical. Longer-term strategies should identify triggers and plan for novel or high-risk adaptation options, such as species translocations. Scientific monitoring and evaluation is required to enhance interventions and communicate with stakeholders.

Conflicts and positive synergies with climate change responses in a range of sectors need to be managed to maximise freshwater biodiversity conservation. Cross-sectoral management tools include better information, technology, natural resources markets, and governance. In particular, good governance and adaptive co-management frameworks across multiple scales and sectors are required. Freshwater ecosystems will be impacted by climate change, but with smarter strategies and engagement of associated sectors, much biodiversity may be conserved. These outcomes will be facilitated by further consideration of the policy setting for wetland conservation and adaptation, including through the Ramsar Convention on Wetlands that has recognised the importance of climate change for wetlands, but not taken a central role in the global dialogue around wetlands and climate change mitigation and adaptation.

References

Abell, R., Allan, J.D., and Lehner, B. (2007). Unlocking the potential for protected areas in conserving freshwaters. *Biological Conservation*, 134, 48–63.

Abell, R., Thieme, M., Revenga, C., Bryer, M., Kottelat, M., Bogutskaya, N., Coad, B., Mandrak, N., Contreras-Balderas, S., Bussing, W., Stiassny, M.L.J., Skelton, P., Allen, G.R., Unmack, P., Naseka, A., Ng, R., Sindorf, N., Robertson, J., Armijo, E., Higgins, J., Heibel, T.J., Wikramanayake, E., Olson, D., Lopez, H.L., de Reis, R.E., Lundberg, J.G., Sabaj Perez,

M.H., and Petry, P. (2008). Freshwater ecoregions of the world: a new map of biogeographic units for freshwater biodiversity conservation. *BioScience*, 58, 403–14.

Acreman, M.C., Blake, J.R., Booker, D.J., Harding, R.J., Reynard, N., Mountford, J.O., and Stratford, C.J. (2009). A simple framework for evaluating regional wetland ecohydrological response to climate change with case studies from Great Britain. *Ecohydrology*, 2, 1–17.

Acreman, M.C., Overton, I.C., King, J., Wood, P.J., Cowx, I.G., Dunbar, M.J., Kendy, E., and Young, W.J. (2014). The changing role of ecohydrological science in guiding environmental flows. *Hydrological Sciences Journal*, 59, 433–50.

Appleyard, S. and Cook, T. (2009). Reassessing the management of groundwater use from sandy aquifers: acidification and base cation depletion exacerbated by drought and ground-water withdrawal on the Gnangara Mound, Western Australia. *Hydrogeology Journal*, 17, 579–88.

Arthington, A.H. (2012). Environmental flows. Saving rivers in the Third Millennium. University of California Press, Berkeley.

AWS. (2014). The AWS International Water Stewardship Standard. Alliance for Water Stewardship, Edinburgh.

Biswas, A.K. (2004). Integrated Water Resources Management: A Reassessment. *Water International*, 29, 248–56.

Bowman, M. (2002). The Ramsar Convention on Wetlands: Has it made a difference? In O.S. Stokke and O.B. Thommessen (eds), Yearbook of International Co-operation on Environment and Development 2002/2003, 61–8. Earthscan, London.

Bush, A., Hermoso, V., Linke, S., Nipperess, D., Turak, E., and Hughes, L. (2014). Freshwater conservation planning under climate change: demonstrating proactive approaches for Australian Odonata. *Journal of Applied Ecology*, 51,1273–81.

Capon, S., Chambers, L., Mac Nally, R., Naiman, R.J., Davies, P., Marshall, N., Pittock, J., Reid, M., Capon, T., Douglas, M., Catford, J., Baldwin, D.S., Stewardson, M., Roberts, J., Parsons, M., and Williams, S.E. (2013). Riparian ecosystems in the 21st century: Hotspots for climate change adaptation? *Ecosystems*, 3, 359–81.

Castro, G., Chomitz, K., and Thomas, T. (2002). The Ramsar Convention: Measuring its effectiveness for conserving Wetlands of International Importance. The World Bank and World Wildlife Fund, Washington DC.

Catford, J., Naiman, R., Chambers, L., Roberts, J., Douglas, M., and Davies, P. (2012). Predicting novel riparian ecosystems in a changing climate. *Ecosystems*,16, 382–400.

Cattarino, L., Hermoso, V., Bradford, L.W., Carwardine, J., Wilson, K.A., Kennard, M.J., and Linke, S. (2016). Accounting for continuous species' responses to management effort enhances cost-effectiveness of conservation decisions. *Biological Conservation*, 197,116–23.

Cattarino, L., Hermoso, V., Carwardine, J., Kennard, M.J., and Linke, S. (2015). Multi-action planning for threat management: a novel approach for the spatial prioritization of conservation actions. *Plos One*, 10, e0128027.

Chong, J. (2014). Ecosystem-based approaches to climate change adaptation: progress and challenges. *International Environmental Agreements: Politics, Law and Economics*, 14, 391–405.

Chopra, K., Leemans, R., Kumar, P., and Simons, H. (2005). Ecosystems and human well-being: Policy responses. Island Press, Washington DC.

Collins, M., Knutti, R. Arblaster, J., Dufresne, J.-L., Fichefet, T., Friedlingstein, P., Gao, X., Gutowski, W., Johns, T., and Krinner, G. (2013). Long-term climate change: Projections, commitments and irreversibility. In IPCC (ed.) Climate Change 2013: The Physical Science Basis. IPCC Working Group. I Contribution to AR5. Cambridge University Press, Cambridge.

CSIRO (2008). Water availability in the Murray-Darling Basin. A report from CSIRO to the Australian Government. CSIRO, Canberra.

Dalla Marta, A., Orlando, F., Mancici, M., and Orlandini, S. (2015). Water and biofuels. In J. Pittock, K. Hussey, and S. Dovers (eds) Climate, Energy and Water, 108–22. Cambridge University Press, Cambridge.

Daufresne, M. and Boet, P. (2007). Climate change impacts on structure and diversity of fish communities in rivers. *Global Change Biology*, 13, 2467–78.

Davidson, N.C. (2014). How much wetland has the world lost? Long-term and recent trends in global wetland area. *Marine and Freshwater Research*, 65, 934–41.

Davies, P.M. (2010). Climate Change Implications for River Restoration in Global Biodiversity Hotspots. *Restoration Ecology*, 3, 261–8.

Dickens, C.W.S., Kotze, D., Mashigo, S., MacKay, H., and Graham, M. (2003). Guidelines for integrating the protection, conservation and management of wetlands into catchment management planning. Water Research Commission, Pretoria.

Drielsma, M., Ferrier, S., Howling, G., Manion, G., Taylor, S., and Love, J. (2014). The Biodiversity Forecasting Toolkit: Answering the 'how much', 'what', and 'where' of planning for biodiversity persistence. *Ecological Modelling*, 274, 80–91.

Elith, J. and Leathwick, J.R. (2009). Species Distribution Models: Ecological Explanation and Prediction Across Space and Time. *Annual Review of Ecology Evolution and Systematics*, 40, 677–97.

Falkenmark, M., Finlayson, C.M., and Gordon, L. (2007). Agriculture, water, and ecosystems: avoiding the costs of going too far. In D. Molden (ed.), Water for Food, Water for Life: A Comprehensive Assessment of Water Management in Agriculture, 234–77. Earthscan, London.

Ferrier, S. and Drielsma, M. (2010). Synthesis of pattern and process in biodiversity conservation assessment: a flexible whole-landscape modelling framework. *Diversity and Distributions*, 16, 386–402.

Field, C.B., Barros, V.R., Mach, K., and Mastrandrea, M. (2014). Climate Change 2014: Impacts, Adaptation, and Vulnerability. Contribution of Working Group II to the Fifth Assessment Report of the Intergovernmental Panel on Climate Change. IPCC, Geneva.

Finlayson, C.M. (2009). Biotic pressures and their effect on wetland functioning. In E. Maltby and T. Barker (eds) The Wetlands Handbook, 667–88. Wiley-Blackwell, Oxford.

Finlayson, C.M., Capon, S.J., Rissik, D., Pittock, J., Fisk, G., Davidson, N.C., Bodmin, K.A., Papas, P., Robertson, H.A., Schallenberg, M., Saintilan, N., Edyvane, K., and Bino, G. (2017). Policy considerations for managing wetlands under a changing climate. *Marine and Freshwater Research*, 68, 1803–15.

Fischer, J., Peterson, G.D., Gardner, T.A., Gordon, L.J., Fazey, I., Elmqvist, T., Felton, A., Folke, C., and Dovers, S. (2009). Integrating resilience thinking and optimisation for conservation. *Trends in Ecology and Evolution*, 24, 549–54.

Fordham, D.A., Wigley, T.M.L., Watts, M.L., and Brook, B.W. (2012). Strengthening forecasts of climate change impacts with multi-model ensemble averaged projections using MAGICC/SCENGEN 5.3. *Ecography*, 35, 4–8.

Gordon, L.J., Finlayson, C.M., and Falkenmark, M. (2010). Managing water in agriculture for food production and other ecosystem services. *Agricultural Water Management*, 97, 512–19.

Grafton, R.Q. and Horne, J. (2014). Water markets in the Murray-Darling Basin. *Agricultural Water Management*, 145, 61–71.

Grantham, H.S., Pressey, R.L., Wells, J.A., and Beattie, A.J. (2010). Effectiveness of Biodiversity Surrogates for Conservation Planning: Different Measures of Effectiveness Generate a Kaleidoscope of Variation. *Plos One*, 5, e11430.

Great Fen Team (2017). Great Fen. Plans for the future. Great Fen Team, Ramsey Heights. Great Fen project website.

Grey, D. and Sadhoff, C. (2007). Sink or swim? Water security for growth and development. *Water Policy*, 9, 545–71.

Groves, C.R., Jensen, D.B., Valutis, L.L., Redford, K.H., Shaffer, M.L., Scott, J.M., Baumgartner, J.V., Higgins, J.V., Beck, M.W., and Anderson, M.G. (2002). Planning for biodiversity conservation: Putting conservation science into practice. *BioScience*, 52, 499–512.

GWP (2000). Integrated Water Resources Management. Global Water Partnership, Stockholm.

Hannah, L.E.E. (2010). A global conservation system for climate-change adaptation. *Conservation Biology*, 24, 70–7.

Hannah, L., Midgley, G., Andelman, S., Araújo, M., Hughes, G., Martinez-Meyer, E., Pearson, R., and Williams, P. (2007). Protected area needs in a changing climate. *Frontiers in Ecology and the Environment*, 5, 131–8.

Headley, R.M. (2017). Climate change: River redirected. *Nature Geoscience*, 10, 327–8.

Hermoso, V., Cattarino, L., Kennard, M.J., Watts, M., and Linke, S. (2015). Catchment zoning for freshwater conservation: refining plans to enhance action on the ground. *Journal of Applied Ecology*, 52, 940–9.

Hermoso, V., Linke, S. Prenda, J., and Possingham, H.P. (2011). Addressing longitudinal connectivity in the systematic conservation planning of fresh waters. *Freshwater Biology*, 56, 57–70.

Hirabayashi, Y., Mahendran, R., Koirala, S., Konoshima, L., Yamazaki, D., Watanabe, S., Kim, H., and Kanae, S. (2013). Global flood risk under climate change. *Nature Climate Change*, 3, 816–21.

Hole, D.G., Willis, S.G., Pain, D.J., Fishpool, L.D., Butchart, S.H.M., Collingham, Y.C., Rahbek, C., and Huntley, B. (2009). Projected impacts of climate change on a continent-wide protected area network. *Ecology Letters*, 12, 420–31.

Hussey, K., Pittock, J., and Dovers, S. (2015). Justifying, extending and applying 'nexus' thinking in the quest for sustainable development. In J. Pittock, K. Hussey, and S. Dovers (eds) *Climate, Energy and Water*, 1–5. Cambridge University Press, Cambridge.

Jackson, R.B., Jobbágy, E.G., Avissar, R., Roy, S.B., Barrett, D.J., Cook, C.W., Farley, K.A., le Maitre, D.C., McCarl, B.A., and Murray, B. (2005). Trading water for carbon with biological carbon sequestration. *Science*, 310, 1944–7.

Januchowski-Hartley, S.R., Hermoso, V., Pressey, R.L., Linke, S., Kool, J., Pearson, R.G., Pusey, B.J., and VanDerWal, J. (2011). Coarse-filter surrogates do not represent freshwater fish diversity at a regional scale in Queensland, Australia. *Biological Conservation*, 144, 2499–511.

Kallis, G. and Zografos, C. (2013). Hydro-climatic change, conflict and security. *Climatic Change*, 123, 69–82.

Kernan, M., Battarbee, R.W., and Moss, B.R. (2011). Climate change impacts on freshwater ecosystems. John Wiley & Sons, Hoboken.

King, J.M., Tharme, R.E., and De Villiers, M. (2000). Environmental flow assessments for rivers: manual for the Building Block Methodology. Water Research Commission, Pretoria.

Kopf, R.K., Finlayson, C.M., Humphries, P., Sims, N.C., and Hladyz, S. (2015). Anthropocene Baselines: Assessing Change and Managing Biodiversity in Human-Dominated Aquatic Ecosystems. *BioScience*, 65, 798–811.

Kriticos, D.J., Jarošik, V., and Ota, N. (2014). Extending the suite of bioclim variables: a proposed registry system and case study using principal components analysis. *Methods in Ecology and Evolution*, 5, 956–60.

Kriticos, D.J., Webber, B.L., Leriche, A., Ota, N., Macadam, I., Bathols, J., and Scott, J.K. (2012). CliMond: global high-resolution historical and future scenario climate surfaces for bioclimatic modelling. *Methods in Ecology and Evolution*, 3, 53–64.

Linke, S., Kennard, M.J., Hermoso, V., Olden, J.D., Stein, J., and Pusey, B.J. (2012). Merging connectivity rules and large-scale condition assessment improves conservation adequacy in river systems. *Journal of Applied Ecology*, 49, 1036–45.

Linke, S., Turak, E., and Nel, J. (2011). Freshwater conservation planning: the case for systematic approaches. *Freshwater Biology*, 56, 6–20.

Lukasiewicz, A., Pittock, J., and Finlayson, C.M. (2016). Are we adapting to climate change? A catchment-based adaptation assessment tool for freshwater ecosystems. *Climatic Change*, 138, 641–54.

Mackay, H. (2003). Water policies and practices. In D. Reed and M. de Wit (eds) Towards a Just South Africa: the Political Economy of Natural Resource Wealth, 49–83. World Wildlife Fund, Macroeconomics Program Office, Washington.

Matthews, J.H. and Wickel, B. (2009). Embracing uncertainty in freshwater climate change adaptation: a natural history approach. *Climate and Development*, 1, 269–79.

Mauter, M.S., Alvarez, P.J.J., Burton, A., Cafaro, D.C., Chen, W., Gregory, K.B., Jiang, G., Li, Q., Pittock, J., Reible, D., and Schnoor, J.L. (2014). Regional variation in water-related impacts of shale gas development and implications for emerging international plays. *Environmental Science and Technology*, 48, 8298–306.

McLeod, E., Salm, R., Green, A., and Almany, J. (2008). Designing marine protected area networks to address the impacts of climate change. *Frontiers in Ecology and the Environment*, 7, 362–70.

MEA. (2005). Millennium Ecosystem Assessment, Ecosystems and Human Well-Being: Wetlands and Water Synthesis. World Resources Institute, Washington DC.

Moilanen, A., Leathwick, J., and Elith, J. (2008). A method for spatial freshwater conservation prioritization. *Freshwater Biology*, 53, 577–92.

Moilanen, A., Leathwick, J.R., and Quinn, J.M. (2011). Spatial prioritization of conservation management. *Conservation Letters*, 4, 383–93.

Norris, R.H., Linke, S., Prosser, I., Young, W.J., Liston, P., Bauer, N., Sloane, N., Dyer, F., and Thoms, M. (2007). Very-broad-scale assessment of human impacts on river condition. *Freshwater Biology*, 52, 959–76.

Norris, R.H. and Thoms, M.C. (1999). What is river health? *Freshwater Biology*, 41, 197–209.

Olden, J.D. and Naiman, R.J. (2009). Incorporating thermal regimes into environmental flows assessments: modifying dam operations to restore freshwater ecosystem integrity. *Freshwater Biology*, 55, 86–107.

Olmstead, S.M. (2010). The economics of managing scarce water resources. *Review of Environmental Economics and Policy*, 4, 179–98.

Parmesan, C. and Yohe, G. (2003). A globally coherent fingerprint of climate change impacts across natural systems. *Nature*, 421, 37–42.

Pittock, J. (2009). Lessons for climate change adaptation from better management of rivers. *Climate and Development*, 1, 194–211.

Pittock, J. (2011). National climate change policies and sustainable water management: Conflicts and synergies. *Ecology and Society*, 16, 25.

Pittock, J. (2013). Lessons from adaptation to sustain freshwater environments in the Murray–Darling Basin, Australia. *Wiley Interdisciplinary Reviews: Climate Change*, 4, 429–38.

Pittock, J. (2015). The nexus, biodiversity and ecosystems. In: J. Pittock, K. Hussey, and S. Dovers (eds) Climate, Energy and Water, 283–302. Cambridge University Press, Cambridge.

Pittock, J. and Finlayson, C.M. (2011). Australia's Murray-Darling Basin: freshwater ecosystem conservation options in an era of climate change. *Marine and Freshwater Research*, 62, 232–43.

Pittock, J., Finlayson, M., Arthington, A.H., Roux, D., Matthews, J.H., Biggs, H., Harrison, I., Blom, E., Flitcroft, R., Froend, R., Hermoso, V., Junk, W., Kumar, R., Linke, S., Nel, J., Nunes da Cunha, C., Pattnaik, A., Pollard, S., Rast, W., Thieme, M., Turak, E., Turpie, J., van Niekerk, L., Willems, D., and Viers, J. (2015). Managing freshwater, river, wetland and estuarine protected areas. In: G.L. Worboys, M. Lockwood, A. Kothari, S. Feary, and I. Pulsford (eds) Protected Area Governance and Management, 569–608. ANU Press, Canberra.

Pittock, J. and Hartmann, J. (2011). Taking a second look: climate change, periodic re-licensing and better management of old dams. *Marine and Freshwater Research*, 62, 312–20.

Pittock, J., Hussey, K., and McGlennon, S. (2013). Australian climate, energy and water policies: Conflicts and synergies. *Australian Geographer*, 44, 3–22.

Pittock, J., Hussey, K., and Stone, A. (2016). Groundwater management under global change: Sustaining biodiversity, energy and food supplies in a changing climate. In: A.J. Jakeman, O. Barreteau, R. Hunt, J.-D. Rinaudo, and A. Ross (eds) Integrated Groundwater Management Concepts, Approaches and Challenges, 75–96. Springer Open.

Poff, N.L. and Matthews, J.H. (2013). Environmental flows in the Anthropocence: past progress and future prospects. *Current Opinion in Environmental Sustainability*, 5, 667–75.

Poff, N.L., Olden, J.D., Merritt, D.M., and Pepin, D.M. (2007). Homogenization of regional river dynamics by dams and global biodiversity implications. *Proceedings of the National Academy of Sciences*, 104, 5732–7.

Poff, N.L., Richter, B.D., Arthington, A., Bunn, S.E., Naiman, R.J., Kendy, E., Acreman, M., Apse, C., Bledsoe, B.P., Freeman, M.C., Henriksen, J., Jacobsen, R.B., Kennen, J.G., Merritt, D.M., O'Keeffe, J.H., Olden, J.D., Rogers, K., Tharme, R.E., and Warner, A. (2010). The ecological limits of hydrologic alteration (ELOHA): a new framework for developing regional environmental flow standards. *Freshwater Biology*, 55, 147–70.

Poff, N.L., Tharme, R.E., and Arthington, A.H. (2017). Evolution of Environmental Flows Assessment Science, Principles, and Methodologies. In: Horne, A.C., Webb, A., Stewardson, M.J., Richter, B., and Acreman, M. Water for the Environment from Science and Policy to Implementation and Management, 203–36. Academic Press, Waltham.

Pringle, C. (2003). What is hydrologic connectivity and why is it ecologically important? *Hydrological Processes*, 17, 2685–9.

Savenije, H. and van der Zaag, P. (2002). Water as an economic good and demand management. Paradigms with pitfalls. *Water International*, 27, 98–104.

Searle, J.A., McHugh, S., Paton, A.C., and Bathols, G. (2010). Perth Shallow Groundwater Systems Investigation: Lake Mariginiup, Hydrogeological Record Series Report No. HG 36. Western Australia Department of Water, Perth.

Spalding, M.D., Fox, H.E., Allen, G.R., Davidson, N., Ferdaña, Z.A., Finlayson, M., Halpern, B.S., Jorge, M.A., Lombana, A., and Lourie, S.A. (2007). Marine ecoregions of the world: a bioregionalization of coastal and shelf areas. *BioScience*, 57, 573–83.

Stalnaker, C., Lamb, B.L., Henriksen, J., Bovee, K., and Bartholow, J. (1995). The instream flow incremental methodology: a primer for IFIM. U.S. Department of the Interior, National Biological Service, Washington, DC.

Strayer, D.L. and Dudgeon, D. (2010). Freshwater biodiversity conservation: recent progress and future challenges. *Journal of the North American Benthological Society*, 29, 344–58.

Taylor, K.E., Stouffer, R.J., and Meehl, G.A. (2012). An Overview of CMIP5 and the Experiment Design. *Bulletin of the American Meteorological Society*, 93, 485–98.

Tennant, D.L. 1976. Instream flow regimens for fish, wildlife, recreation and related environmental resources. *Fisheries*, 1, 6–10.

The Wildlife Trusts (2017). What are living landscapes? The Wildlife Trusts, Newark, Wildlife Trusts website.

Watts, M.E., Ball, I.R., Stewart, R.S., Klein, C.J., Wilson, K., Steinback, C., Lourival, R., Kircher, L., and Possingham, H.P. (2009). Marxan with Zones: Software for optimal conservation based land- and sea-use zoning. *Environmental Modelling and Software*, 24, 1513–21.

WWF (2003). Managing Rivers Wisely: Lessons from WWF's Work for Integrated River Basin Management. WWF International, Gland.

WWF (2014). The Living Planet Report 2014. Species and Spaces, People and Places. WWF International, Gland.

Xu, T. and Hutchinson, M.F. (2013). New developments and applications in the ANUCLIM spatial climatic and bioclimatic modelling package. *Environmental Modelling and Software*, 40, 267–79.

18

Restoration of Freshwaters: Principles and Practice

Carl Sayer, Helen Bennion, Angela Gurnell, Emma Goodyer,
Donovan Kotze, and Richard Lindsay

Corresponding author: c.sayer@ucl.ac.uk

18.1 Introduction

On a global scale, and partly due to our huge dependence on them, freshwater ecosystems are both highly threatened and in major decline (Dudgeon et al. 2006). As a consequence and for our own sake in terms of human health and well-being, as well as for the fate of species which reside in freshwater habitats, we need to engage with ecosystem protection and restoration. Conservationists and scientists involved with the restoration of freshwaters face many challenges, including the competing demands of industry, agriculture and water supply, issues of cost, support by the public and government, and often the large geographical scale of environmental problems and their causes. Further, it is probably fair to say that restoration practitioners lack a sure-footed, well-communicated, and widely applicable science basis from which to design and implement successful projects. Nevertheless, although freshwater restoration science is a relatively new discipline, a substantial literature now exists such that some key principles that might underlie successful restoration can be broadly established. This chapter will outline some of these principles and is largely inspired by the classic restoration ecology paper of Bradshaw (1996) who introduces some '*Underlying Principles of Restoration*'. Bradshaw's restoration principles include the use of natural processes where possible, recognition that return to an original state may not be achievable but that ecosystem development should be on 'an unrestricted upward path', and the need to aim for whole ecosystem restoration which considers both structure and function. Some 20 years on, these principles remain fundamental to freshwater restoration ecology.

The habitats encompassed by this chapter are lakes, ponds, rivers, and various wetland systems including fens and bogs. It is not possible to be exhaustive in terms of habitats, environmental stressors, and biogeography in a short chapter and hence the subject matter covered herein will be largely drawn from the authors' own research fields in Northern Europe. Issues considered include freshwaters negatively affected by

Sayer, C., Bennion, H., Gurnell, A., Goodyer, E., Kotze, D., and Lindsay, R., *Restoration of freshwaters: Principles and practice*. In: *Freshwater Ecology and Conservation: Approaches and Techniques*. Edited by Jocelyne M. R. Hughes: Oxford University Press (2019). © Oxford University Press 2019. DOI: 10.1093/oso/9780198766384.003.0018 eutrophication and acidification (especially for lakes), habitat degradation and fragmentation (especially for rivers, ponds, and other wetlands), habitat conversion to other forms of land use (especially for wetlands), reductions in water quantity (all habitats), and invasive species (all habitats). Some tropical and dryland case studies from other parts of the world are included, however, and it should also be recognised that the principles and approaches covered in this chapter should be fairly widely applicable and will certainly have relevance to other biomes and aquatic habitats. In this chapter good restoration practice is seen to follow six major principles, which in turn form the basis of key sections and examples. These principles are given as follows:

1. Restoration targets should recognise the value of historical and pre-disturbance data, but should take into account projected changes in climate, water quantity, and other constraining factors.
2. Projects need to diagnose the problem and remove those factors that are restraining natural ecosystem re-development and recovery.
3. Projects should take account of landscape-scale influences and processes that inevitably impact on the success and sustainability of restoration outcomes.
4. Where possible, robust Before-After Control-Impact (BACI) style monitoring should be included in restoration projects to help judge success and to inform future work.
5. Where possible, restoration should allow and support the capacity of natural processes to repair degraded freshwater habitats and ecosystems.
6. Restoration projects should ideally bring scientists, conservationists, and stakeholders together using best practice participatory approaches to setting restoration targets and developing restoration designs.

18.2 Restoration targets

A fundamental question that needs to be addressed before restoration targets can be considered is what constitutes 'restoration'. Bradshaw (1996) and Brookes and Shields (1996) define 'restoration' as the act of returning an ecosystem to its, pre-disturbance, 'natural' state, with 'rehabilitation' conceptualised as partial progress towards the original state, which is nonetheless not achieved. In addition, other restoration-type activities are highlighted by this work including 'replacement' and 'creation' which encompass the development of a resource, property, or alternative ecosystem that did not previously exist at a location. It follows, given the constraints (see Section 18.3) associated with populated, modified, and heavily modified freshwater landscapes, that rehabilitation and replacement-creation are often the most widely used and practicable forms of restoration.

The types of targets used in freshwater restoration have been many and varied, with water chemistry, water quantity, individual species, biological assemblages, species diversity, habitat structure, ecosystem functioning, and ecosystem services all used individually or in combination with one another. In very heavily modified situations, restoration goals may relate to aesthetics and minimisation of risks to humans, whereas in other settings, goals related to the overall health of the ecosystem and the species it supports have been more common. In particular, for rivers, the term 'river restoration' has embraced several perspectives, often reflecting the development of restoration techniques ahead of targets. Initially, river restoration was strongly focused on the physical structure or morphology of the river channel and its mosaic of physical habitats, but then expanded to consider the morphology of the river's margins and floodplain together with their morphodynamics. More recently, aspects of biogeochemical and ecological functioning have been included (Palmer et al. 2014). By contrast, for turbid, plantless, shallow lakes affected by nutrient-enrichment, there has been a more stable aim, with most projects focused on returning clear-water, macrophyte-dominated conditions (Jeppesen et al. 2012).

But what needs to be considered when setting a restoration target? Is a target always needed? The advantages of having a target are many, allowing restoration practitioners something to aim towards, thus informing management techniques and demonstrating success or otherwise. Certainly the importance of gaining knowledge about earlier stages in the development of freshwater systems has come to the fore over recent decades with the introduction of water legislation such as the US Clean Water Act (CWA; Barbour et al. 2000) and the European Council Water Framework Directive (WFD; European Union 2000). Both of these legal instruments require that assessments of water quality and biological assembly be based on the degree to which present-day conditions deviate from those expected in the absence of significant anthropogenic influence, so-called 'reference conditions'. Similarly, ecological condition assessment of Ramsar Convention wetlands of international significance, and sites designated under the EC Habitats and Species Directive (European Union 1992), are based on comparisons with a baseline state (Gell et al. 2013). Nonetheless, there is no universal definition of what 'reference conditions' actually constitute, with definitions ranging from natural conditions in the absence of humans to those best achieved under the influence of humans (Johnson et al. 2010).

To assess pre-disturbance conditions in freshwater habitats, various approaches are available. In lakes, ponds, and wetlands that accumulate sediments, reference conditions can be derived using palaeoecology, whereby fossil remains of a range of biological groups (e.g., algae, bryophytes, macrophytes, zooplankton, invertebrates, and fish) found in dated core samples are used to indicate former environmental conditions, biological assembly, and in turn past mechanisms of ecological change (e.g., Hughes et al. 2000; Sayer et al. 2010). Further, historical records, sketches, photographs, and even paintings can be usefully employed to build up a picture of habitat structure in freshwater ecosystems that can subsequently be utilised as a 'guiding image' for what might be achieved by restoration actions (Willby 2011). For example, Madgwick et al. (2011) assembled a whole range of historical and palaeolimnological data for Barton Broad, a shallow lake in eastern England, to illustrate the spatial arrangement of aquatic vegetation and past aspects of plant sociology (Figure 18.1). In turn, this image was incorporated into a major review of lake restoration practices to help target and inspire future restoration action (Phillips et al. 2015). Similar work has been undertaken for wetland systems, including floodplain fens and peat bogs,

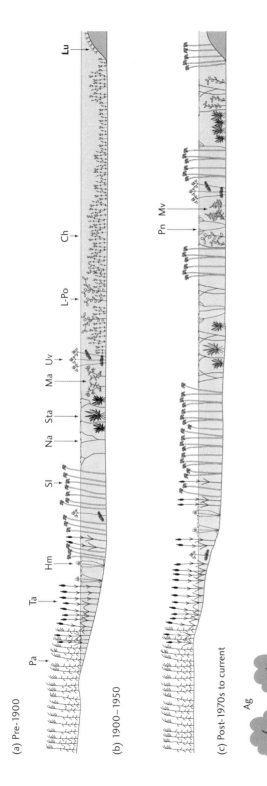

Figure 18.1. Reconstruction of macrophyte spatial relationships in Barton Broad over three time periods: (a) pre-1900; (b) 1900–1950; and (c) 1970s–current. Codes to plant names: Pa—*Phragmites australis*, Ta—*Typha angustifolia*, Hm—*Hydrocharis morsus-ranae*, Sl—*Schoenoplectus lacustris*, Na—*Nymphaea alba*, Sta—*Stratiotes aloides*, Ma—*Myriophyllum alterniflorum*, Uv—*Utricularia vulgaris*, L-Po—Large, broad-leaved *Potamogeton* taxa (e.g., *P. lucens*, *P. praelongus*, *P. alpinus*), Ch—*Chara* spp., Lu—*Littorella uniflora*, Pn—*Potamogeton natans*, Mv—*Myriophyllum verticillatum*, Ag—*Alnus glutinosa*, Fpz—Fine-leaved *Potamogeton* taxa (e.g., *P. pectinatus*, *P. pusillus*) and *Zannichellia palustris*, Cd—*Ceratophyllum demersum* (from Madgwick et al. 2011).

with the accumulation of water-logged semi-decomposed plant material as peat, at least where it is intact and undisturbed, affording an excellent record of site ecohydrological development (Whitehouse et al. 2008).

While palaeolimnology undoubtedly has much unrealised and future potential in river restoration (e.g., Howard et al. 2009), space-for-time substitution is a more frequently utilised approach, with reference sites from the best available river examples often used to guide restoration. If these examples are carefully chosen to ensure that they are drawn from a similar river 'type', such an approach can be applied in relation to river biology (Wright et al. 2000), as well as hydrology and geomorphology (Rinaldi et al. 2016). However, it is increasingly apparent that human impacts on river flow and sediment delivery-transport regimes in many parts of the world are so significant, that true river restoration is rarely feasible. Consequently, it is often more appropriate to help a river adjust its form in a semi-natural way in response to the most natural flow and sediment regime that is achievable given inevitable human pressures (see Section 18.3).

Although it is often not possible to restore a site to its natural, background state, knowledge of this limitation and of past conditions is highly desirable, helping to define some of the options for management whilst dictating achievable limits. Indeed, a historically derived reference state can act as a baseline against which future restoration targets might be assessed and framed. Nonetheless, it is important that a target is actually achievable. In this respect Bradshaw (1996) contends: *'What is crucial is that the development of the ecosystem should be on an upward path in terms of structure and function, and that no barriers to its long term further development can be envisaged.'*

There are, however, numerous restraining factors that limit the restoration potential of freshwaters and hence the achievability of a restoration target. These include factors such as high costs of remediation, insufficient water availability due to abstraction (especially for rivers and wetlands), reduced flood disturbance due to flow regulation, an increase in fine sediment delivery due to agricultural intensification (especially for rivers), extinct or declining populations of former species, dispersal limitation, and the influence of other confounding pressures, such as legacy pollutants, nitrogen deposition, and, importantly, climate change. It is now widely recognised that climate change may limit the use of historically derived restoration targets, as the future status of freshwater ecosystems will differ from the present even under 'do nothing' scenarios (Battarbee et al. 2012; Gell et al. 2013; Verdonschot et al. 2013); for example, as species become eliminated or migrate towards cooler habitats. While this does not invalidate historical targets, there is a need to re-define the reference state as boundary conditions change and, if necessary, adjust them to increase achievability (Battarbee et al. 2005). On this basis, in recent years the concept of 'shifting baselines' has gained traction (Duarte et al. 2009; Bennion et al. 2011; Battarbee et al. 2014), conceptualising the fact that reference conditions are not only dynamic, but are subject to directional change. In the case of shallow lakes, ponds, and wetlands, for example, it has been recognised that valued examples are often transitional states along a hydroseral pathway. Thus, restoration targeting needs to consider the point at which restoration actively switches to maintenance management in order to prevent further successional development (Tansley 1939; Sayer et al. 2012).

Many studies have demonstrated that, as a pressure is reduced, recovery does not follow a simple reverse pathway, such that ecosystems fail to return to a state that prevailed prior to impact (Duarte et al. 2009; Battarbee et al. 2014). This may be partly due to lag or hysteresis effects, but a range of confounding factors have also been implicated. In an analysis of nutrient and climate impacts on seven European lakes, Battarbee et al. (2012) attributed limited recovery to continuing eutrophication related to an increase in diffuse nutrient loading and/or internal P recycling, but there was also evidence for a climate change role in offsetting recovery. Based on the findings of this study, a conceptual diagram of past, present, and potential future trajectories of European lake systems experiencing nutrient-enrichment and climate change was constructed (Figure 18.2). Similarly, in studies of boreal lake recovery from acidification, declines in the richness of invertebrate assemblages have been observed unrelated to changes in acid deposition and more closely associated with climate-related influences on habitat quality, such as oxygen concentrations and temperature (Stendera and Johnson 2008). Thus, much evidence suggests that confounding factors, and climate change in particular, will increase the restoration challenge. In this respect a more dynamic and open-minded approach to restoration will be required—one that considers a range of approaches to deal with an increasingly uncertain future. Perhaps key considerations here are resilience and flexibility, in that restored ecosystems that are resilient and have sufficient room for movement (in the case of rivers in particular) and natural adjustment to changing conditions will likely perform better (see Section 18.3).

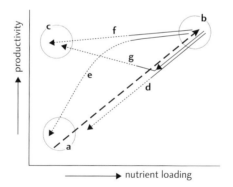

Figure 18.2 Conceptual diagram using a combination of palaeolimnological data (dashed line), contemporary long-term monitoring data (solid lines), and future conjecture (dotted lines) to show idealised changes in the past, present, and future relationship between nutrient loading and productivity for European lakes recovering from eutrophication. Point (a) indicates the reference state and the target endpoint following restoration; point (b) indicates the point of intervention to reduce nutrient loading; and (c) indicates a more probable potential endpoint in cases where recovery to the past reference (a) is prevented by the enriching effects of climate change. Arrow (d) represents a simple trajectory back towards the reference state, (e) represents delayed recovery towards the reference state (e.g., due to internal P loading), and (f) and (g) represent deflected trajectories away from the reference and towards a new endpoint (from Battarbee et al. 2012).

In this respect it is crucial to monitor the outcome of restoration projects (see Section 18.6) and, where appropriate, to use modelling approaches to predict restoration success under future conditions. For example, in the case of English chalk rivers, dynamic models have been used to simulate hydrology and water quality under a range of climate scenarios, revealing a strong association between cessation of drought periods and release of high nitrate loads into the river system (Whitehead et al. 2006). When restoration strategies were explored, models suggested that a combined management approach, involving land-use change or reduced fertiliser use, water meadow creation, and atmospheric pollution controls could reduce in-stream nitrate concentrations to those of the pre-1950s even under climate change. With such information to hand, restoration practitioners might be able to set more meaningful and achievable goals. Finally, it needs to be recognised that clearly defined restoration targets are not always useful, with this being especially true in the case of re-wilding style projects (see Section 18.4) where desired restoration endpoints are deliberately left much more fluid.

18.3 Diagnosing and tackling the problem

In order to develop appropriate and sustainable restoration designs it is essential to understand those processes, pressures, and interventions influencing restoration site(s), how these have changed, and how they are likely to change in the future. In turn, such a knowledge greatly assists the design of restoration schemes, enabling attention to be focused on removal of the key factors causing ecosystem degradation. Put simply, if key restraining factors are not tackled, a restoration project is unlikely to meet with success.

The importance of diagnosing and tackling major background problems in freshwater restoration is clearly emerging from the European shallow lakes literature, where, over the last two to three decades, many innovative in-lake restoration techniques, aimed at permanently shifting eutrophic lakes from turbid plant-free to clear macrophyte-dominated conditions, have been trialled and studied. These 'internal' measures include biomanipulation (e.g., removal of planktivorous fish, stocking of piscivorous fish, stocking of non-native mussels), which seeks to engineer clear-water conditions by enhancing rates of filter-feeding on phytoplankton, direct planting of macrophytes both within and without wild-fowl enclosures, and measures directed at affecting a reduction in internal P-loading, such as sediment removal by suction dredging and in-lake iron addition (see Jeppesen et al. 2012; Phillips et al. 2015; Phillips et al. 2016; Bakker et al. 2016). While there are complexities, exceptions and unanswered questions associated with all of these techniques, an emerging pattern is for a lack of long-term and sustained recovery. Biomanipulation is the most fully studied measure, especially in Denmark, where many parallel, multi-decadal studies suggest that positive lake recovery only occurs where nutrient concentrations have been appropriately reduced (perhaps below 50 µg/L for P) or where fish manipulations are regularly repeated, such that the planktivorous fish stock is permanently held in check (Jeppesen et al. 2012). Otherwise, although biomanipulation has frequently been shown to generate clear water conditions and macrophyte occupancy in lakes a few years after fish removal, with the recolonisation of fish, phytoplankton-dominated conditions have typically resumed resulting in plant decline after 5–10 years (Jeppesen et al. 2012). As a consequence it is emerging, perhaps unsurprisingly, that the key to sustainable restoration success in shallow, nutrient-enriched lakes is effective external nutrient reduction. Thus, nutrient budgets in combination with catchment walk-over surveys to locate nutrient sources, followed by the introduction of measures to reduce external nutrient influx, are probably essential.

Much evidence from other habitats also suggests that a restoration approach which diagnoses and then tackles key underlying causes of degradation as a priority is more likely to be successful. For fen peatlands and many other wetland habitats, it is clearly critical that restoration addresses water-quantity issues, with this especially true in arid and semi-arid regions where inflows generally constitute a major component of hydrological inputs to a wetland relative to direct precipitation. In the Murray–Darling Basin (south-eastern Australia), for example, the average time period between environmentally beneficial floods on the Murray River has approximately doubled as a result of surface water abstraction, thus severely disrupting wetland functioning (CSIRO 2008; Pittock and Finlayson 2011). A considerable challenge for these floodplain wetlands is to restore free-flowing tributaries and attain effective management of water releases in regulated portions of the basin to complement flows from free-flowing sections. If achieved, such measures will also enhance resilience to the additional impacts of climate change (Pittock and Finlayson 2011). For many floodplain wetlands, the alteration of geomorphological processes is a further pressure requiring diagnosis in the development of appropriate restoration designs. This issue is illustrated by the Seekoeivlei wetland, South Africa, where introduction of non-native willow (*Salix* spp.) trees to an historically treeless environment drastically altered the geomorphological dynamics of the wetland system leading to the abandonment of a former channel and rapid headward growth of a new channel (McCarthy et al. 2010). In this case, knowledge of the geomorphological processes underlying these changes, including rates of change, contributed greatly to the formulation, evaluation, and hence the sustainability of long-term intervention options.

Another important source of problems in wetlands of major relevance to the development of appropriate restoration designs is human-induced alterations to disturbance regimes. Domestic livestock have been linked to impacts such as increased soil erosion and a decline of sensitive plant species in many wetlands, and in this respect restoration typically involves excluding the influence of livestock, or substantially reducing their numbers (Ramstead et al. 2012). However, many wetlands have evolved under the influence of grazing by large indigenous ungulates and if these species are no longer present then the diversity of native plant species and habitats can decline at the expense of one or a few dominants. In such situations grazing by domestic livestock needs to closely simulate the effect of indigenous grazers, and thus part of restoring an 'over-protected' wetland might, in fact, be to introduce domestic grazers where it is not practical to re-introduce large indigenous grazers (Middleton 2013). Similarly, an important problem that often needs to be addressed in the restoration of fire-dependent wetlands may be anthropogenic exclusion of fire. This is illustrated by the KwaMbonambi, northern KwaZulu-Natal, South Africa where, in 1936, herbaceous vegetation comprised 25 per cent of the landscape, but by 2009 there had been a decline to just 2 per cent.

A key factor contributing to this change was suppression of fire by plantation forestry management, leading to colonisation by forest species. The KwaMbonambi wetlands naturally support a rich diversity of fire-dependent herbs and grasses, including the only known wild population of the critically endangered herb *Kniphofia leucocephala*. A priority for continued restoration of these wetlands and their associated species is therefore re-instating a regime of periodic burning (Luvuno et al. 2016).

A need for effective problem diagnosis is especially the case for river restoration where determining degradation causes and in turn appropriate restoration goals and methods depend upon understanding changing river processes, forms, and human activities that extend beyond restoration site(s) to encompass the upstream, and sometimes downstream, parts of a catchment. A prevalent form of river restoration in the late twentieth century, particularly within Europe, has been habitat-based, focusing on modifying river channel morphology (widening, narrowing, re-meandering), introducing stabilising structures (deflectors, boulders), artificially creating habitats (riffles, pools), and implementing planting schemes to provide habitat for specific species or species groups. An underlying assumption of this approach was that the 'renaturalised', more heterogeneous river habitat would lead to biological improvements (Palmer et al. 2010). Nevertheless, despite evidence for positive restoration within floodplain and riparian zones (e.g., floodplain plants and beetles—Kail et al. 2015; Friberg et al. 2016; Figure 18.3), overall the results of small-scale, habitat restoration projects for

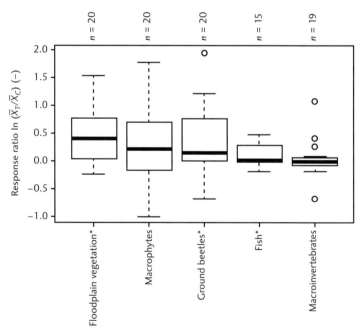

Figure 18.3 Effects of restoration on species richness (five organism groups) for European rivers as reflected by the response ratio of Osenberg et al. (2017), which relates the value of a restored section (X_T) to a degraded control section (X_C). Mean values that are significantly different to zero (*t*-test, $p<0.05$) are marked with an asterisk and positive values are indicative of positive restoration responses (from Friberg et al. 2016).

in-channel biological communities, have been less than convincing. Indeed, recent meta-studies of European river restoration (e.g., Palmer et al. 2010; Jahnig et al. 2010) have shown that, despite achieving measurable improvements in physical habitat diversity, evidence for significant, positive effects on in-river biology have been patchy, with this especially true for fishes and invertebrates.

The reasons for a lack of in-channel restoration success, and indeed a differential response of river and floodplain systems, may relate to a number of factors, not least methodological variation across studies (e.g., specific techniques, time since restoration), continued poor river water quality, dispersal barriers for river invertebrates and fishes, restricted species pools available for recolonisation and the small spatial scale of the restoration work in relation to catchment size. However, a major cause of failure, even among more recent, less engineered restoration projects, is a lack of understanding of the processes, forms and functions that the river is able to sustain following restoration actions, resulting in a non-optimal restoration design coupled with unrealistic restoration targets.

To fully appreciate the relevant processes that may constrain river restoration success, it is essential to go beyond particular river reaches to embrace larger-scale water, sediment, and organism transfer processes and their connectivity between river reaches, floodplains, and within catchments (e.g., Lake 2012). These are the key fluvial (river flow, sediment transport, water quality) and biological processes that affect and will continue to affect any reach that is to be restored. It is also crucial to recognise that the character of river systems is continually altering in response to changes in these 'natural' processes in combination with human pressures and interventions affecting both restored reaches and the upstream river that influences them. Consequently an understanding of the historical evolution of a river to its present state is essential to developing an appropriate and sustainable restoration design (Grabowski et al. 2014), as is an appreciation of how key controlling factors have changed in the past and may change in the future (e.g., Davies 2010; Perry et al. 2015). Through new process-based frameworks for supporting restoration design (Gurnell et al. 2016; Box 18.1), river restoration is moving into an era where objectives are being more clearly defined, diagnosis of underlying problems and processes is more robust, and a combination of designing with natural processes and incorporating adaptive restoration management is providing a pathway towards genuine and sustainable improvements in river health.

18.4 Good restoration encourages the natural repair of degraded systems

In his classic paper Bradshaw (1996) emphasises the huge advantage of utilising natural recovery processes in ecological restoration, pointing out that, in the North American Great Lakes region, pre-existing soils and vegetation were repeatedly destroyed by ice ages, but were subsequently able to build up ecosystems of high complexity and diversity. In other words, nature is a powerful force that should be allowed to heal ecosystems

Box 18.1 The REFORM project

The REFORM project (REstoring rivers FOR effective catchment Management—http://www.reformrivers.eu/home) was funded by the EU's 7th Framework Programme (2011–2015). The central aim of REFORM was to provide a series of tools to help improve the success of river restoration.

The REFORM framework (Gurnell et al. 2016) exemplifies recent open-ended, approaches to diagnosing the key hydrogeomorphological factors that influence river form and function. The framework (Figure 18.4; Gurnell et al. 2016) has been developed for application in a European context (e.g., England and Gurnell 2016) and helps to diagnose the key factors influencing form and function in river reaches. It considers individual river reaches in the context of the valley and river segment, landscape unit, and catchment within which they are located. Further, it incorporates three main stages of analysis: 'delineation' of spatial units; 'characterisation' of contemporary and historical key processes within the spatial units through the generation of indicators; and 'assessment-diagnosis'. Four types of assessment and diagnosis are conducted based on information, particularly the values of the indicators, assembled during the characterisation phase. First, the current form and function of individual reaches is assessed, including their channel sediments, morphology, vegetation and degree of human alteration; the character, function, and artificiality of their riparian corridor; and any evidence of current morphological adjustment. Second, past and present indicators of water and sediment production, transfer, and delivery from the catchment through the river network are assessed at all spatial scales. Third, reach-scale indicators of historical morphological adjustment are assembled. Finally, the results of the historical and contemporary analyses conducted in the three previous assessments are combined, to summarise space-time changes; to unravel causes and responses in order to understand trajectories of change that have occurred; and to consider likely responses to specific future scenarios such as the effects of climate and management change. Such a multi-scale approach to diagnosing why a river reach has a particular form and dynamics is absolutely essential to designing restoration interventions that 'work with the river' to achieve sustainable improvements.

wherever possible. It follows, therefore, that good restoration helps to stimulate and enhance natural physical and ecological recovery processes wherever possible. It is probably true that many freshwater habitats, especially rivers and wetlands, are capable of natural repair, with the removal of factors that hamper recovery, but frequently the longer time frames required for natural recovery have meant that such approaches have often been ignored.

There is mounting scientific evidence for the importance and potential of natural processes in freshwater restoration. For example, in restoring wetland vegetation, two common and opposing restoration practices are self-design vs. intensive revegetation (O'Connell et al. 2013). Self-design restores hydrogeomorphology without the artificial

Box 18.1 *Continued*

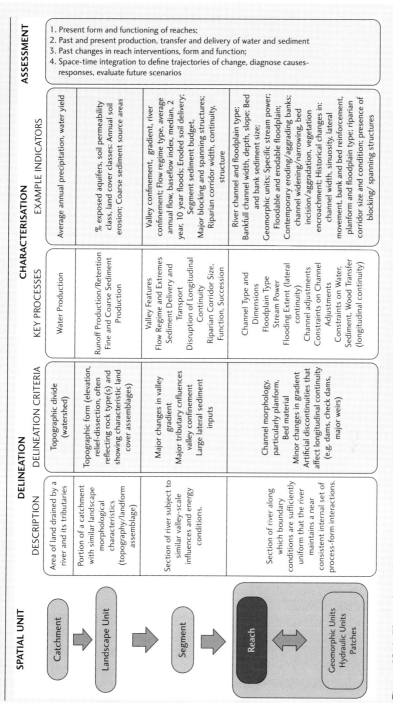

SPATIAL UNIT	DELINEATION		CHARACTERISATION		ASSESSMENT
	DESCRIPTION	DELINEATION CRITERIA	KEY PROCESSES	EXAMPLE INDICATORS	
Catchment	Area of land drained by a river and its tributaries	Topographic divide (watershed)	Water Production	Average annual precipitation, water yield	1. Present form and functioning of reaches; 2. Past and present production, transfer and delivery of water and sediment 3. Past changes in reach interventions, form and function; 4. Space-time integration to define trajectories of change, diagnose causes-responses, evaluate future scenarios
Landscape Unit	Portion of a catchment with similar landscape morphological characteristics (topography/landform assemblage)	Topographic form (elevation, relief-dissection, often reflecting rock type(s) and showing characteristic land cover assemblages)	Runoff Production/Retention Fine and Coarse Sediment Production	% exposed aquifers, soil permeability class, land cover classes: Annual soil erosion; Coarse sediment source areas	
Segment	Section of river subject to similar valley-scale influences and energy conditions.	Major changes in valley gradient Major tributary confluences valley confinement Large lateral sediment inputs	Valley Features Flow Regime and Extremes Sediment Delivery and Transport Disruption of Longitudinal Continuity Riparian Corridor Size, Function, Succession	Valley confinement, gradient, river confinement; Flow regime type, average annual flow, baseflow index, median, 2 year, 10 year floods; Eroded soil delivery; Segment sediment budget, Major blocking and spanning structures; Riparian corridor width, continuity, structure	
Reach	Section of river along which boundary conditions are sufficiently uniform that the river maintains a near consistent internal set of process-form interactions.	Channel morphology, particularly planform, Bed material Minor changes in gradient Artificial discontinuities that affect longitudinal continuity (e.g. dams, check dams, major weirs)	Channel Type and Dimensions Floodplain Type Stream Power Flooding Extent (lateral continuity) Channel adjustments Constraints on Channel Adjustments Constraints on Water, Sediment, Wood Transfer (longitudinal continuity)	River channel and floodplain type; Bankfull channel width, depth, slope; Bed and bank sediment size; Geomorphic units; Specific stream power; Floodable and erodable floodplain; Contemporary eroding/aggrading banks; channel widening/narrowing, bed incision/aggradation, vegetation encroachment; Historical changes in: channel width, sinuosity, lateral movement, bank and bed reinforcement, planform and floodplain type; riparian corridor size and condition; presence of blocking/ spanning structures	
Geomorphic Units Hydraulic Units Patches					

Figure 18.4 The multi-scale REFORM framework, that follows delineation, characterisation, and assessment-diagnosis phases in order to develop understanding of how the character of individual river reaches adjust to processes and pressures operating at catchment to patch scales and change from the past through the present to the future (based on concepts explained in Gurnell et al. 2016).

introduction of plants (e.g., via seeds and plug-plants) into restoration sites. By contrast, intensive revegetation does the same but involves the often costly and time-consuming inoculation of sites with plants. Self-design has many advantages, being cheaper and potentially resulting in a more natural, local vegetation, but there are risks and potential disadvantages (e.g., invasive species arrival, soil erosion) if vegetation fails to colonise rapidly due to a highly depleted seed bank and/or limited existing plants on site (Weinhold and van der Valk 1989). Further factors affecting unassisted recolonisation include the extent to which the native flora and fauna is dispersal-limited and the proximity of intact areas from which colonisation can occur.

Inspiring examples are emerging of cases where restoration based on natural plant recovery has led to the return of surprisingly high biodiversity to freshwater systems. For UK agricultural ponds lost to land consolidation (so-called 'Ghost Ponds'), Alderton et al. (2017) showed that many aquatic plants can colonise resurrected pond basins from long-lived (150+ years) seed banks. Equally rapid, pond-wide re-colonisation of aquatic vegetation for overgrown farmland ponds restored through scrub and mud removal strongly suggests the same effect (Sayer et al. 2012), as illustrated in Figure 18.5. A particularly spectacular example of rapid plant re-colonisation following restoration comes from Lake Fil in Denmark, a large shallow (mean depth 1.5 m) lake drained to permit agricultural land expansion from 1852 to 1952 (Baastrup-Spohr et al. 2016). Within just two years of re-establishing the lake it developed a remarkably high diversity of aquatic macrophytes (33 species), many of which were locally and indeed nationally rare. This occurred despite intense farming of the land for 50 years prior to restoration. Similarly, in the case of fen peatlands destroyed by agricultural land development, diverse fen assemblages have been restored within a very short amount of time by stripping off agriculturally enriched uppermost layers to expose seed banks and by allowing plants to self-establish (McBride et al. 2011).

In river systems, natural vegetation recolonisation of restored sections can also be successful if propagule delivery is efficient from upstream species pools. For example, restoration of a section of the River Cole, UK, involved cutting a new sinuous river channel in its historical location to bypass a realigned section (Gurnell et al. 2006). No soil, seed or plants were applied to the newly cut river banks and yet 145 plant taxa were identified within the seed bank and standing vegetation after just two years of the river being diverted into the new channel. Furthermore, the colonising vegetation did not include any widely occurring non-native invasive species commonly found along UK rivers (e.g., Himalayan balsam *Impatiens glandulifera*, Japanese knotweed *Fallopia japonica*) despite a predominantly urban upstream catchment (Gurnell et al. 2006). Thus, there are strong arguments for always considering the potential for natural re-vegetation of freshwater habitats in restoration projects, be they rivers, lakes, ponds, or wetlands. More research to address this theme is urgently needed in order to identify where and when natural re-revegetation is effective, particularly in relation to factors such as former land-use, hydrological setting, soil type and chemistry, and importantly propagule longevity for different plant species (e.g., Stroh et al. 2012; Bakker et al. 2013).

Figure 18.5 Shooting Close Pond, a small farmland pond in eastern England, UK.
(a) before, (b) during, and (c) two years after restoration by scrub and sediment removal
in September 2014.

A particularly promising approach to stimulating natural recovery in freshwater environments is to utilise 're-wilding', whereby natural, often missing processes are re-introduced and/or allowed to operate with minimal human interference. In freshwater environments this idea encompasses a range of examples, including land abandonment and the removal of management (Navarro and Pereira 2012), re-introduction of natural tree-fall into rivers (Thompson et al. 2018) and the re-introduction of apex predators (Beschta and Ripple 2016). The most extensively studied and perhaps the most dramatic demonstration of re-wilding benefits to fresh-water systems probably comes from the re-introduction of beavers. Both American beaver (*Castor canadensis*) and European beaver (*Castor fiber*) have started to recover in the wild during recent decades assisted by several re-introduction projects. Where they have returned to wetlands many positive changes in both hydrogeomorphology and ecology have been observed. In particular, beaver introduction (and the subsequent natural dispersal of individuals), together with consequent habitat modification via tree-felling, grazing, and dam building, has led to an extensification and diversification of wetland habitat with significant biodiversity increases demonstrated across a range of aquatic and semi-aquatic organisms including plants, invertebrates, amphibians, birds, and bats (Nummi and Holopainen 2014; Cunningham et al. 2007; Law et al. 2017; Figure 18.6) as well as increased flood attenuation and reduced diffuse pollution

Figure 18.6 Overview of results from a long-term European beaver (*Castor fiber*) release study in Blairgowrie, eastern Scotland showing the study site 1 year and 12 years post release (left) and accompanying changes to wetland vegetation (right). From Law et al. (2017).

(Puttock et al. 2017). Although more studies are needed, especially in terms of long-term responses and dynamics, as suggested by Law et al. (2017), in many cases beaver introduction has achieved successes that far out-weigh the benefits of more engineered approaches to freshwater restoration such as pond creation and river re-meandering. The mechanisms that drive beaver-benefits are still to be fully understood, however, as they often involve the re-instatement of hitherto little understood processes and indirect species interactions, including linkages across the aquatic–terrestrial interface (see McCaffrey and Eby 2016). For example, introduction of beavers may help to restore wetland vegetation to former agricultural land not only through lifting the water table but also through the disturbance and exposure of old seed banks (Law et al. 2017).

A re-wilding approach to freshwater and wetland restoration clearly has considerable potential, but constraints must also be recognised. For example, in the case of peat bogs, natural recovery is probably wholly possible, given the removal of adverse pressures and sufficient time. However, the timescales involved in the natural recovery of full ecosystem diversity and function following extensive peat extraction or severe erosion, for example, may span many centuries or even millennia and are thus often unacceptable in terms of lost ecosystem services over such a long period. Consequently there is generally a strong argument, or at least a strong incentive, for intervention in order to hasten the natural recovery process. In particular the setting of targets and timescales for delivery are common features of funding conditions which tend to conflict with re-wilding ideals, typically resulting in a more hands-on and short-term focus to system restoration actions. In the case of beavers and large predators (e.g., lynx and wolf), human-wildlife conflicts are also a key issue that delay or sometimes prohibit re-wilding attempts (Lorimer et al. 2015). Finally, in urban and intensely managed and populated areas, re-wilding approaches to restoration are often precluded due to severe constraints in terms of space and general flexibility. Nonetheless, even where full and natural hydrogeomorphic and ecological processes cannot be set in motion in an ideal way, elements of this ethos can be applied to most settings and one positive aspect of a re-wilding approach is its ability to inspire, encourage, and sometimes completely re-energise restoration practitioners.

18.5 Recognise the importance of landscape-scale influences and processes

Given the non-local nature of many freshwater environmental problems (e.g., diffuse pollution, catchment-scale barriers to species migration, over-abstraction), there have been growing calls for a restoration approach that considers larger spatial scales and looks outwards from the site-scale to the wider landscape and beyond (as summarised in the context of rivers in Box 18.1). Indeed, it is now fairly widely recognised that conservation and restoration need to extend beyond the boundaries of existing protected areas (Adams et al. 2016). To date, however, although there will be many exceptions, most freshwater restoration work has been undertaken on relatively small patches, for example individual river reaches (typically <1 km), floodplains and lakes and relatively small collections of ponds. In particular, where small units of river have been worked on, but where key negative catchment-scale influences have been not been sufficiently tackled, restoration projects have often failed to deliver significant measurably improved outcomes (Palmer et al. 2014), probably partly for this reason. For example, where a lowland river, lake, or wetland system is severely impacted by excessive nutrient and sediment inputs from an upstream agricultural catchment, local habitat restoration in rivers and within-lake restoration measures are likely to have relatively dampened benefits (see Section 18.3). Similarly, local river restoration activities directed at fish (e.g., gravel introduction, re-meandering), where barriers that prevent natural upstream-downstream fish migration have not been removed, are less likely to have positive effects on fish assemblages (Champkin et al. 2018). The spatial scale of a restoration project in relation to degree of influence from broader catchment-scale pressures is thus critical to restoration success.

There are strong arguments for freshwater restoration projects and strategies that are as ambitious as possible in the spatial scales they encompass and which recognise the importance of larger scale system connections both longitudinally (upstream—downstream) and laterally (links to riparian zones) for fluxes of water, pollutants, key substances (nutrients, sediment, carbon) and movements of propagules and species (Friberg et al. 2016; Fergus et al. 2017). From a biodiversity conservation perspective much research has shown that the connectivity between different aquatic habitat patches has an important, often positive, influence on biological structure and biodiversity. It follows, therefore, that a more open-minded approach to freshwater restoration, one that considers the full spectrum of aquatic habitat patches in a landscape—an 'aquatic landscapes' approach—is likely to be more successful (Sayer 2014). For example, river restoration projects which have facilitated enhancements to floodplain and associated pond and backwater habitats have been shown to have more substantial benefits than restoration of the river channel on its own (Sayer 2014; Friberg et al. 2016; Figure 18.3). Equally, by undertaking catchment-wide and strategic studies of barriers to fish passage, it is possible to ensure that restoration by barrier removal work has maximum effect (Perkin et al. 2015). In the case of wetlands it is becoming increasingly evident that successful restoration of native wetland vegetation depends, not only on re-creating favourable on-site conditions, but also the location of a wetland in relation to other neighbouring sites. Specifically, a wetland in close proximity to intact areas of native wetland vegetation is generally much better placed in terms of natural re-colonisation by native plants than are distant, isolated wetlands (Findlay and Houlahan 1997; O'Connell et al. 2013).

A catchment-scale approach to restoration which looks firstly to address issues of headwater pollution and which aims to facilitate freer movement of water, propagules, and species throughout an aquatic network may be the ideal (Sayer 2014), but of course there are a number of caveats. For example, it is not always possible to resolve upstream pollution problems in their entirety due to the predominantly anthropogenic nature of many catchments. In some cases, local-scale buffering may be sufficiently effective (Weisstenier et al. 2013) but it also needs to be recognised that larger-scale problems such as transboundary pollution and climate change cannot be addressed at the individual-project scale. Furthermore, removing hydrological barriers can increase the spread of invasive species, so decisions need to be made on the pros and cons of enhancing connectivity, especially where rare native taxa could come under threat. The important point here is that practitioners should at least seek to obtain sufficient knowledge of wider catchment and landscape influences, so that informed decisions about the most appropriate restoration approach can be made.

18.6 Importance of monitoring restoration

Monitoring is essential for providing assessments of the success or otherwise of freshwater restoration, as well as to assist the design of more effective restorations in the future and, importantly, to help make the case for future investment in this field. In order to judge the effectiveness of freshwater restoration activities it is crucially important that data are collected pre- and post-restoration and that they are quantitative; qualitative sampling is incapable of assessing changes to species population size and also makes biodiversity changes harder to assess. It is also important that restoration monitoring of biology incorporates measurement of key hydrogeomorphic and chemical variables, such that the underlying mechanisms which affect restoration success/ failure might also be inferred. Finally, it is essential that unrestored 'control' sites are included to account for natural and other background drivers of change. Such an experimental set-up represents the Before-After Control-Impact (BACI) approach. An even more ideal set up providing enhanced statistical power would be to undertake multiple concurrent restorations focused on the same type of intervention (hence MBACI; Thompson et al. 2018). However, while the rationale for BACI is strong, it is often the case that restoration projects have neglected, or do not have sufficient resources to incorporate appropriate control sites, making it impossible or at least difficult to arrive at clear conclusions about restoration success (Feld et al. 2011).

As well as adopting a BACI approach, a crucial aspect of restoration monitoring is that it is established and funded to cover appropriate time-periods. In a review of river restorations in Europe and North America, Feld et al. (2011) revealed that the majority of studies spanned a period of just 1–7 years, a period shorter than might often be required to detect biological recovery. Indeed, invertebrate response to river restoration that is limited by poor water quality (Kail et al. 2012), can take 5–10 years before any change is detectable, while fish populations often take decades to change in response to perturbations (Trexler 1995). For example, wetland trees such as willow and European alder (*Alnus glutinosa*) may require decades to mature, such that the full function of riparian buffer strips and trees in rivers (temperature modification, large wood recruitment) may take 30–40 years to be achieved. Similarly, although the restoration of peat-forming vegetation may occur relatively quickly as a result of restoration actions such as ditch and gully blocking, the processes involved in establishing full ecosystem diversity and function require more than just a few decades of monitoring.

Due to funding constraints, long-term monitoring is necessarily restricted to a few sites, but the evidence gathered from such projects is incredibly valuable. For example, 20–30 years of monitoring Danish shallow lakes is starkly revealing the shortcomings of in-lake restoration techniques such as biomanipulation (Jeppesen et al. 2012).

Further, in the UK, 30 years of the Acid Waters Monitoring Network (AWMN; now the Uplands WMN), based on 22 lakes and streams, has been instrumental in assessing the recovery of freshwaters from acidification, revealing that, while marked changes in deposition chemistry and water chemistry are consistent with recovery, the extent of biological recovery remains somewhat limited (Battarbee et al. 2014). Given the slow operation of some hydrogeomorphic processes, as well as lags and delays in responses of some long-lived freshwater components and, importantly, confounding and other background influences on freshwater systems, it is evident that short-term 'typical' restoration monitoring studies can only provide limited insight into system responses to restoration. It is therefore crucially important that scientists and indeed the public fight against an ever-growing trend of cutting funds for freshwater monitoring stations and networks. In essence, freshwater restoration science will only be as good as the quality and indeed quantity of system monitoring that takes place and without monitoring the evidence on which restoration designs are based will inevitably be weak.

Finally, and linked to the arguments for stakeholder working made in Section 18.7, it is important that knowledge gained from restoration studies be clearly and sensitively communicated to practitioners and stakeholders involved in restoration work so that the adoption of evidence-based approaches is taken up early and mistakes of the past are not continually re-made.

18.7 Working with stakeholders

Human well-being, both physically and psychologically, is intimately connected with freshwater. Consequently, freshwater restoration activities can have long-term benefits for humans as well as ecosystems, although this is not always fully appreciated (Chapter 2). This means that there is a crucial need to highlight to the public and indeed to governments the vital importance of clean freshwater and of healthy, species-rich, more naturally functioning freshwater systems. Restoration is also likely to be better funded (and for longer perhaps), supported and valued if such societal relevance is made clear. Equally, to achieve restoration success it is crucial that stakeholders are on board with a project's aims and see its importance right from the start while also being fully engaged with the restoration process. There is much evidence to suggest that partnership approaches to restoration which involve the local community and landowners as well as more formal stakeholder groups (e.g., conservation organisations) are most likely to achieve the most sustainable outcomes (e.g., Eden and Tunstall 2006; Åberg and Tapsell 2013).

Full engagement of all stakeholders requires early and effective communication of what is proposed to enable the multiple objectives and potential win-win aspects of restoration to be appreciated by a wide audience. Communication should not be one-way. It needs to proceed in multiple directions in order to integrate stakeholders fully into discussions and decision-making. To achieve this requires the adoption of diverse participatory techniques (e.g., Moran et al. 2016) in order to build mutual trust. An increasing emphasis on the many ecosystem services delivered by freshwater systems when they are functioning well (e.g., Acuna et al. 2013) can be an effective vehicle for quantifying and thus communicating restoration benefits (e.g., Vermaat et al. 2016). Such an approach can extend beyond individual small projects to support an integrated understanding of the potential benefits of multiple combined projects or very large individual restoration schemes. One model which has worked well is the establishment of projects under a branded partnership umbrella with funds secured for core staff to communicate both the value of the work with stakeholders and to manage the restoration works. Having these partnerships provides some longevity to a suite of short-term projects, allows for coordination of funding and provides a point of contact for stakeholders and the public. The partnership model of delivery is extensively used in the UK to deliver effective and long-term peatland restoration projects (Cris et al. 2011) and for river systems through the establishment of catchment partnerships.

Acknowledging that the general public may have perceptions which differ widely from scientists, and the possibility that memories of the past may not always be accurate due to some level of personal amnesia and shifting baseline syndrome (Papworth et al. 2009), local people nonetheless often have valuable knowledge of a site, sometimes accumulated over many generations. This knowledge should be harnessed and can be extremely valuable in guiding the restoration process. Furthermore, in recognising that stakeholders, including local people, may have widely divergent opinions on what constitutes a problem worthy of restoration and what the endpoint should be in addressing restoration problems, there is a need for open discussion of the perceived values of systems in different states in order to facilitate effective and hopefully more consensual decision-making (Hobbs 2009). Such desired open discussions can be difficult to achieve when divergent interests, norms, values, and perceptions are brought together, and approaches such as social learning offer useful means of providing people from different backgrounds a 'safe space', to share their experiences and to develop new knowledge, ways of thinking, and possibilities (Wals 2007).

Given a frequent need for public participation in freshwater restoration, a degree of compromise is often needed, with this being especially true in highly populated urban settings. For example, public perceptions on what a successful wetland restoration outcome is (typically neat, picturesque, and with open water) generally differ greatly from what would be the ecologically functional or reference wetland determined from a scientific basis (Nassauer 2004). Such 'ecologically directed' wetlands often lack open water and appear to the general public as 'untidy'. It is therefore suggested that, even if wetland restoration is being designed primarily to achieve a reference ecological state (whether pristine or functionally defined), the restoration plan may need to include recognisable and valued landscape characteristics to improve the likelihood of it being sustained through societal support over the long term (Nassauer 2004). It may also be important to take into account that public perceptions of what constitutes a threat requiring restoration/rehabilitation intervention may also differ greatly from that determined through scientific assessment (Schumm 1994).

18.8 Conclusions

Freshwater systems afford some of the most biodiverse and culturally important habitats on the planet and it is essential that we rise to the restoration challenge. While the science of freshwater restoration is developing at a pace, uncertainties about how we repair the hydrogeomorphology and ecology of freshwater habitats remain. More studies and, importantly, high-quality monitoring work are therefore required. Many advances have been made with regards to problem diagnosis, restoration targeting, and importantly the selection of appropriate restoration approaches. But one thing is clear restoration is easier when the extent of system damage is reduced. Thus, a key lesson is to identify and then protect and conserve existing high-quality sites and to enact restoration work before human-induced degradation is too severe. For example, it is much easier to restore a shallow lake that has not already lost its macrophytes. Furthermore, in the case of peat bogs, given the long timescales required for natural peat formation and accumulation, the key human action is not to damage such sites in the first place. We cannot rely on restoration activities to sort out all our problems and often rare and declining species cannot afford to wait until a large-scale ecosystem restoration programme is established. Early intervention is key, and the central messages of Bradshaw's classic paper undoubtedly hold true: good restoration should look to tackle the root cause of the issues that are degrading a freshwater system, but wherever possible we should give freshwater systems the space, time, and flexibility needed to repair themselves via natural processes.

References

Åberg, E.U. and Tapsell, S.M. (2013). Revisiting the River Skerne: The long-term social benefits of river rehabilitation. *Landscape & Urban Planning*, 113, 94–103.

Acuna, V., Diez, J.R., Flores, L., Meleason, M., and Elosegi, A. (2013). Does it make economic sense to restore rivers for their ecosystem services? *Journal of Applied Ecology*, 50, 988–99.

Adams, W.A., Hodge, I.D., Macgregor, N.A., and Sandbrook, L. (2016) Creating restoration landscapes: partnerships in large-scale conservation in the UK. *Ecology & Society*, 21(3), 1.

Alderton, E., Sayer, C.D., Davies, R., Lambert, S.J., and Axmacher, J.C. (2017). Buried alive: Aquatic plants survive in 'ghost ponds' under agricultural fields. *Biological Conservation*, 212, 105–10.

Baastrup-Spohr, L., Kragh, T., Petersen, K., Moeslund, B., Schou, J.C., and Sand-Jensen, K. (2016). Remarkable richness of aquatic macrophytes in 3-year old re-established Lake Fil, Denmark. *Ecological Engineering*, 95, 375–83.

Bakker, E.S., Sarneel, J.M., Gulati, R.D., Liu, Z., and van Donk, E. (2013). Restoring macrophyte diversity in shallow temperate lakes vs. biotic versus abiotic constraints. *Hydrobiologia*, 710, 23–37.

Bakker, E.S., van Donk, E., and Immers, A.K. (2016). Lake restoration by in-lake iron addition: a synopsis of iron impact on aquatic organisms and shallow lake ecosystems. *Aquatic Ecology*, 50, 121–35.

Barbour, M.T., Swietlik, W.F., Jackson, S.K., Courtemanch, D.L., Davies, S.P., and Yoder, C.O. (2000). Measuring the attainment of biological integrity in the USA: a critical element of ecological integrity. *Hydrobiologia*, 422–3, 453–64.

Battarbee, R.W., Anderson, N.J., Bennion, H., and Simpson, G.L. (2012). Combining limnological and palaeolimnological data to disentangle the effects of nutrient pollution and climate change on lake ecosystems: problems and potential. *Freshwater Biology*, 27, 2091–106.

Battarbee, R.W., Anderson, N.J., Jeppesen, E., and Leavitt, P.R. (2005). Combining palaeolimnological and limnological approaches in assessing lake ecosystem response to nutrient reduction. *Freshwater Biology*, 50, 1772–80.

Battarbee, R.W., Simpson, G.L., Shilland, E.M., Flower, R.J., Kreiser, A., Yang, H., and Clarke, G. (2014). Recovery of UK lakes from acidification: An assessment using combined palaeoecological and contemporary diatom assemblage data. *Ecological Indicators*, 37, 365–80.

Beschta, R.L. and Ripple, W.J. (2016). Riparian vegetation recovery in Yellowstone: The first two decades after wolf reintroduction. *Biological Conservation*, 198, 93–103.

Bradshaw, A.D. (1996). Underlying principles of restoration. *Canadian Journal of Fisheries & Aquatic Sciences*, 53, 3–9.

Brookes, A. and Shields, F.D. (1996). *River channel restoration: Guiding principles for sustainable projects*. Wiley, Chichester, UK.

Champkin, J., Copp, G.H., Sayer C.D., Clilverd, H.M., George, L., Vilizzi, L., Godard, M.J., and Clarke, J. (in press 2018). Responses of fishes and lampreys to the re-creation of meanders in a small English chalk stream. *River Research & Applications*, 34, 34–43.

Cris, R., Buckmaster, S., Bain, C., and Bonn, A. (eds) (2011). *UK Peatland Restoration—Demonstrating Success*. IUCN UK National Committee Peatland Programme, Edinburgh.

CSIRO (2008). Water availability in the Murray–Darling Basin. Report from CSIRO to the Australian Government. CSIRO, Canberra. Available at http://www.csiro.au/files/files/po0n.pdf

Cunningham, J.M., Calhoun, A.J.K., and Glanz, W.E. (2007). Pond-breeding amphibian species richness and habitat selection in a beaver-modified landscape. *Journal of Wildlife Management*, 71, 2517–26.

Davies, P.M. (2010). Climate change implications for river restoration in global biodiversity hotspots. *Restoration Ecology*, 18, 261–8.

Duarte, C., Conley, D., Carstensen, J., and Sánchez-Camacho, M. (2009). Return to *Neverland*: Shifting baselines affect eutrophication restoration targets. *Estuaries & Coasts*, 32, 29–36.

Dudgeon, D., Arthington, A.H., Gessner, M.O., Kawabata, Z.-I., Knowler, D.J., Lévêque, C., Naiman, R.J., Prieur-Richard, A.-H., Soto, D., Stiassny, M.L.J., and Sullivan, C.A. (2006). Freshwater biodiversity: importance, threats, status and conservation challenges. *Biological Reviews*, 81, 163–82.

Eden, S. and Tunstall, S. (2006). Ecological versus social restoration? How urban river restoration challenges but also fails to challenge the science – Policy Nexus in the United Kingdom. *Environment and Planning C: Government and Policy*, 24, 661–80.

England, J. and Gurnell, A.M. (2016). Incorporating catchment to reach scale processes into hydromorphology assessment in the UK. *Water & Environment Journal*, 30, 22–30.

European Union (1992). Council Directive 92/43/EEC of 21 May 1992 on the conservation of natural habitats and of wild fauna and flora. *Official Journal of the European Communities*, L206, 7–50.

European Union (2000). Directive 2000/60/EC of the European Parliament and the Council of 23 October 2000 establishing a framework for Community action in the field of water policy. *Official Journal of the European Communities*, L327, 1–72.

Feld, C.K., Birk, S., Bradley, D.C., Hering, D., Kail, J., Marzin, A., Melcher, A., Nemitz, D., Pedersen, M.L., Pletterbauer, F., Pond, D., Verdonschot, F.M., and Friberg, N. (2011). From natural to degraded rivers and back again: a test of restoration ecology theory and practice. *Advances in Ecological Research*, 44, 119–209.

Fergus, C.E., Lapierre, J.F., Oliver, S.K., Skaff, N.K., Cheruvelil, K.S., Webster, K., Scott, C., and Soranno, P. (2017). The freshwater landscape: lake, wetland and stream abundance and connectivity at macroscales. *Ecosphere*, 8 (8).

Findlay, C.S. and Houlahan, J. (1997). Anthropogenic correlates of species richness in southeastern Ontario wetlands. *Conservation Biology*, 11, 1000–9.

Friberg, N., Angelopoulos, N.V., Buijse, A.D., Cowx, I.G., Kail, J., Moe, T.F., Moir, H., O'Hare, M.T., Verdonschot, P.F.M., and Wolter, C. (2016). Effective river restoration in the 21st

Century: From trial and error to novel evidence-based approaches. *Advances in Ecological Research*, 55, 535–611.

Gell, P., Mills, K., and Grundell, R. (2013). A legacy of climate and catchment change: the real challenge for wetland management. *Hydrobiologia*, 708, 133–44.

Grabowski, R.C., Surian, N., and Gurnell, A.M. (2014). Characterizing geomorphological change to support sustainable river restoration and management. *WIREs Water*, 1, 483–512.

Gurnell, A.M., Boitsidis, A.J., Thompson, K., and Clifford, N.J. (2006). Seed bank, seed dispersal and vegetation cover: Colonization along a newly-created river channel. *Journal of Vegetation Science*, 17, 665–74.

Gurnell, A.M., Rinaldi, M., Belletti, B., Bizzi, S., Blamauer, B., Braca, G., Buijse, T., Bussettini, M., Camenen, B., Comiti, F., Demarchi, L., García de Jalón, D., González del Tánago, M., Grabowski, R.C., Gunn, I.D.M., Habersack, H., Hendriks, D., Henshaw, A.J., Klösch, M., Lastoria, B., Latapie, A., Marcinkowski, P., Martínez-Fernández, V., Mosselman, E., Mountford, J.O., Nardi, L., Okruszko, T., O'Hare, M.T., Palma, M., Percopo, C., Surian, N., van de Bund, W., Weissteiner, C., and Ziliani, L. (2016). A multi-scale hierarchical framework for developing understanding of river behaviour to support river management. *Aquatic Sciences*, 78, 1–16.

Hobbs, R.J., Higgs, E.., and Harris, J.A. (2009). Novel ecosystems: implications for conservation and restoration. *Trends in Ecology & Evolution*, 24, 599–605.

Howard, L.C., Wood, P.J., Greenwood, M.T., and Rendell, H.M. (2009). Reconstructing riverine paleo-flow regimes using subfossil insects (Coleoptera and Trichoptera): the application of the LIFE methodology to paleochannel sediments, *Journal of Paleolimnology*, 42, 453–66.

Hughes, P.D.M., Mauquoy, D., Barber, K.E., and Langdon, P.G. (2000). Mire-development pathways and palaeoclimatic records from a full Holocene peat archive at Walton Moss, Cumbria, England. *The Holocene*, 10, 465–79.

Jähnig, S.C., Brabec, K., Buffagni, A., Erba, S., Lorenz, A.W., Ofenböck, T., Verdonschot, P.F.M., and Hering, D. (2010). A comparative analysis of restoration measures and their effects on hydromorphology and benthic invertebrates in 26 central and southern European rivers. *Journal of Applied Ecology*, 47, 671–80.

Jeppesen E., Søndergaard, M., Lauridsen, T.L. Davidson, T.A., Liu, Z., Mazzeo, N., Trochine, C., Özkan, K., Jensen, H.S., Trolle, D., Starling, F., Lazzaro, X., Johansson, L.S., Bjerring, R., Liboriussen, L., Larsen, S.E., Landkildehus, F., and Meerhoff, M. (2012). Biomanipulation as a restoration tool to combat eutrophication: recent advances and future challenges. *Advances in Ecological Research*, 47, 411–87.

Johnson, R.K., Battarbee, R.W., Bennion, H., Hering, D., Soons, M.B., and Verhoeven, J.T.A. (2010). Climate change: defining reference conditions and restoring freshwater ecosystems. In Kernan, M., Battarbee, R.W., and Moss, M. (eds) Climate Change Impacts on Freshwater Ecosystems. Wiley-Blackwell, Chichester, UK.

Johnson, R.K., Battarbee, R.W., Bennion, H., Hering, D., Soons, M.B., and Verhoeven, J.T.A. (2010). Climate change: defining reference conditions and restoring freshwater ecosystems. In Kernan, M., Battarbee, R.W., and Moss, M. (eds) Climate Change Impacts on Freshwater Ecosystems. Wiley-Blackwell, Chichester, UK.

Kail J., Arle, J., and Jähnig, S.C. (2012). Limiting factors and thresholds for macroinvertebrate assemblages in European rivers: Empirical evidence from three datasets on water quality, catchment urbanization, and river restoration. *Ecological Indicators*, 18, 63–72.

Kail, J., Brabec, K., Poppe, M., and Januschke, K. (2015). The effect of river restoration on fish, macroinvertebrates and aquatic macrophytes: a meta-analysis. *Ecological Indicators*, 58, 311–21.

Lake, P.S. (2012). Flows, floods, floodplains and river restoration. *Ecological Management & Restoration*, 13, 210–11.

Law, A., Gaywood, M.J., Jones, K.C., Ramsay, P., and Willby, N.J. (2017). Using ecosystem engineers as tools in habitat restoration and rewilding: beaver and wetlands. *Science of the Total Environment*, 605–6, 1021–30.

Lorimer, J., Sandom, C., Jepson, P., Doughty, C., Barua, M., and Kirby, K.J. (2015). Rewilding: science, practice, and politics. *Annual Review of Environment & Resources*, 40, 39–62.

Luvuno L.B., Kotze, D.C., and Kirkman, K.P. (2016). Long-term landscape changes in vegetation structure: fire management in the wetlands of KwaMbonambi, South Africa, *African Journal of Aquatic Science*, 41, 279–88.

Madgwick, G., Emson, D., Sayer, C.D., Willby, N.J., Rose, N., Jackson, M.J., and Kelly, A. (2011). Centennial-scale changes to the aquatic vegetation structure of a shallow eutrophic lake and implications for restoration. *Freshwater Biology*, 56, 2620–36.

McBride, A., Diack, I., Droy, N., Hamill, B., Jones, P., Schutten, J., Skinner, A., and Street, M. (2011). The Fen Management Handbook 1st edn, Scottish Natural Heritage, Perth.

McCaffrey, M. and Eby, L. (2016). Beaver activity increases aquatic subsidies to terrestrial consumers. *Freshwater Biology*, 61, 518–32.

McCarthy, T.S., Tooth, S., Kotze, D.C., Collins, N.B., Wandrag, G., and Pike, T. (2010). The role of geomorphology in evaluating remediation options for floodplain wetlands: the case of Ramsar-listed Seekoeivlei, eastern South Africa. *Wetlands Ecology & Management*, 18, 119–34.

Middleton, B.A. (2013). Rediscovering traditional vegetation management in preserves: Trading experiences between cultures and continents *Biological Conservation*, 158, 271–9.

Moran, S., Perreault, M., and Smardon, R. (2016). Finding our way: A case study of urban waterway restoration and participatory process. *Landscape & Urban Planning*, October.

Nassauer, J.I. (2004). Monitoring the success of metropolitan wetland restorations: cultural sustainability and ecological function. *Wetlands*, 24, 756–65.

Navarro, L.M. and Pereira, H.M. (2012). Rewilding abandoned landscapes in Europe. *Ecosystems*, 15, 900–12.

Nummi, P. and Holopainen, S. (2014). Whole-community facilitation by beaver: ecosystem engineer increases waterbird diversity. *Aquatic Conservation: Marine and Freshwater Ecosystems*, 24, 623–33.

O'Connell, J.L., Johnson, L.A., Beas, B.J., Smith, L.M., McMurry, S.T., and Haukos, D.A. (2013). Predicting dispersal-limitation in plants: Optimizing planting decisions for isolated wetland restoration in agricultural landscapes. *Biological Conservation*, 159, 343–54.

Osenberg, C.W., Sarnelle, O., and Cooper, S.D. (1997). Effect size in ecological experiments: the application of biological models in meta-analysis. *American Naturalist*, 150, 798–812.

Palmer, M.A., Hondula, K.L., and Koch, B.J. (2014). Ecological restoration of streams and rivers: Shifting strategies and shifting goals. *Annual Review of Ecology, Evolution & Systematics*, 45, 247–69.

Palmer, M.A., Menninger, H.L., and Bernhardt, E. (2010). River restoration, habitat heterogeneity and biodiversity: a failure of theory or practice? *Freshwater Biology*, 55, 205–22.

Papworth, S.K., Rist, J., Coad, L., and Milner-Gulland, E.J. (2009). Evidence for shifting baseline syndrome in conservation. *Conservation Letters*, 2, 93–100.

Perkin, J.S., Gido, B.K., Cooper, A.R., Turner, T.F., Osborne, M.J., Johnson, E.R., and Mayes, K.B. (2015). Fragmentation and de-watering transform Great Plains stream fish communities. *Ecological Monographs*, 85, 73–92.

Perry, L.G., Reynolds, L.V., Beechie, T.J., Collins, M.J., and Shafroth, P.B. (2015). Incorporating climate change projections into riparian restoration planning and design. *Ecohydrology*, 8, 863–79.

Phillips, G., Bennion, H., Perrow, M., Sayer, C.D., Spears, B., and Willby, N. (2015). A review of lake restoration practices and their performance in the Broads National Park, 1980–2013. Report for Broads Authority & Natural England, Norwich.

Phillips, G., Willby, N., and Moss, B. (2016). Submerged macrophyte decline in shallow lakes: What have we learnt in the last forty years? *Aquatic Botany*, 135, 37–45.

Pittock, J. and Finlayson, C.M. (2011). Australia's Murray Darling Basin: freshwater ecosystem conservation options in an era of climate change. *Marine & Freshwater Research*, 62, 232–43.

Puttock, A., Graham, H.A., Cunliffe, A.M., Elliott, M., and Brazier, R.E. (2017). Eurasian beaver activity increases water storage, attenuates flow and mitigates diffuse pollution from intensively-managed grasslands. *Science of the Total Environment*, 576, 430–43.

Ramstead, K.M., Allen, J.A., and Springer, A.E. (2012). Have wet meadow restoration projects in the Southwestern U.S. been effective in restoring geomorphology, hydrology, soils, and plant species composition? *Environmental Evidence*, 1, 11.

Rinaldi, M., Gurnell, A.M., González del Tánago, M., Bussettini, M., and Hendriks, D. (2016). Classification of river morphology and hydrology to support management and restoration. *Aquatic Sciences*, 78, 17–33.

Sayer, C.D. (2014). Conservation of aquatic landscapes: ponds, lakes, and rivers as integrated systems. *WIRE's: Water*, 1, 573–85.

Sayer, C.D., Andrews, K., Shilland, E., Edmonds, N., Edmonds-Brown, R., Patmore, I., Emson, D., and Axmacher, J.A. (2012). The role of pond management for biodiversity conservation in an agricultural landscape. *Aquatic Conservation: Marine & Freshwater Ecosystems*, 22, 626–38.

Sayer C.D., Davidson, T.A., Jones, I.J., and Langdon, P.G. (2010). Combining contemporary ecology and palaeolimnology to understand shallow lake ecosystem change. *Freshwater Biology*, 55, 487–99.

Schumm, S.A. (1994). Erroneous perceptions of fluvial hazards. *Geomorphology*, 10, 129–38.

Stendera, S. and Johnson, R.K. (2008). Tracking recovery trends of boreal lakes: use of multiple indicators and habitats. *Journal of the North American Benthological Society*, 27, 529–40.

Stroh, P.A., Hughes, F.M.R., Sparks, T.H., Mountford, O., and Mountford, J.O. (2012). The influence of time on the soil seed bank and vegetation across a landscape-scale wetland restoration project. *Restoration Ecology*, 20, 103–12.

Tansley, A.G. (1939). The British Islands and Their Vegetation. Cambridge University Press, Cambridge.

Thompson, M.S.A., Brooks, S.J., Sayer, C.D., Woodward, G., Axmacher, J.C., Perkins, D., and Gray, C. (in press 2018). Large woody debris 'rewilding' restores biodiversity in riverine food webs. *Journal of Applied Ecology,*, 55, 895–904.

Trexler, J.C. (1995). Restoration of the Kissimmee River: A conceptual model of past and present fish communities and its consequences for evaluating restoration success. *Restoration Ecology*, 3, 195–210.

Verdonschot, P.F.M., Spears, B.M., Feld, C.K., Brucet, S., Keizer-Vlek, H., Borja, A., Elliott, M., Kernan, M., and Johnson, R.K. (2013). A comparative review of recovery processes in rivers, lakes, estuarine and coastal waters. *Hydrobiologia*, 704, 453–74.

Vermaat, J., Wagtendonk, A.J., Brouwer, R., Sheremet, O., Ansink, E., Brockhoff, T., Plug, M., Hellsten, S., Aroviita, J., Tylec, L., Gielczewski, M., Kohut, L., Brabec, K., Haverkamp, J., Poppe, M., Boeck, K., Coerssen, M., Segersten, J., and Hering, D. (2016). Assessing the societal benefits of river restoration using the ecosystem services approach. *Hydrobiologia*, 769, 121–35.

Wals, A.E.J. (2007). Learning in a Changing World and Changing in a Learning World: Reflexively Fumbling Towards Sustainability. *South African Journal of Environmental Education*, 24, 35–45.

Weinhold, C.E. and van der Valk, A.G. (1989). The impact of drainage on the seed banks of northern prairie wetlands. *Canadian Journal of Botany*, 67, 1878–84.

Weisstenier, C.J., Bouraoui, F., and Aloe, A. (2013). Reduction of nitrogen and phosphorus loads to European rivers by riparian buffer zones. *Knowledge & Management of Aquatic Ecosystems*, 408, 08.

Whitehead, P.G., Wilby, R.L., Butterfield, D., and Wade, A.J. (2006). Impacts of climate change on in-stream nitrogen in a lowland chalk stream: An appraisal of adaptation strategies. *Science of the Total Environment*, 365, 260–73.

Whitehouse, N., Langdon, P., Bustin, R., and Galsworthy, S. (2008). Fossil insects and ecosystem dynamics in wetlands: implications for conservation and management. *Biodiversity & Conservation*, 17, 2055–78.

Willby, N.J. (2011). From metrics to Monet: The need for an ecologically meaningful guiding image (Editorial). *Aquatic Conservation: Marine & Freshwater Ecosystems*, 21, 601–3.

Wright J.F., Sutcliffe, D.W., and Furse, M.T. (eds) (2000). Assessing the biological quality of fresh waters. RIVPACS and other techniques. Proceedings of an International Workshop, Oxford, September 1998. Freshwater Biological Association, Ambleside, UK.

Wetland Landscapes and Catchment Management

Caroline A. Sullivan, C. Max Finlayson, Elizabeth Heagney,
Marie-Chantale Pelletier, Mike C. Acreman, and Jocelyne M.R. Hughes

Corresponding author: caroline.sullivan@scu.edu.au

19.1 Evolving world views

As available and accessible freshwater remains a tiny fraction of global water resources, with wetlands, lakes, and rivers just a small part of this, it becomes increasingly important that these resources are managed effectively and equitably, not only for human use, but also for all other ecological components of the earth system (Naiman and Dudgeon 2011; Schmutz and Sendzimir 2018). For several decades, it has been recognised that human appropriation of global biological resources, including rivers (Gregory 2006), is out of proportion with our species niche (Meadows et al. 1972). More recently the concept of the *Anthropocene* (Crutzen 2006; Vörösmarty et al. 2013) has gained acceptance, underlying the fact that human impacts on the Earth's natural biological, geomorphological, and chemical systems is now measurable, marking the beginning of a new epoch characterised by deforestation, mass extinctions, unregulated waste and plastic pollution, carbon dioxide emissions, and sea level rise.

There are two main causes for the imbalance humans have with their natural environment: exponential human population growth and its consequent consumption, and rapid technological change. These factors can clearly be attributed to human intelligence and ingenuity, but also to what was described by the early economist Adam Smith as our '*unlimited wants*', (Smith 1776), and by philosophers such as La Rochefoucauld as '*amour-propre*', or *self-interest* (La Rochefoucauld 1665; Box 19.1). Although hundreds of years have passed since the concepts were first recognised, they are still highly pertinent, and at the heart of the modern environmental crisis. Human nature and its insatiable pursuit of personal gain, poor environmental stewardship, and economic development remain the major threat to the future functionality of freshwater systems and the vital services they provide to human populations (Green et al. 2015).

More recently, these long-established perspectives have been reinforced by a growing body of science, neatly encapsulated in the concept of 'Planetary Boundaries' (Rockström et al. 2009; Steffen et al. 2015). In response, the global community has constructed a set

Sullivan, C. A., Finlayson, C. M., Heagney, E., Pelletier, M. -C., Acreman, M. C., and Hughes, J. M. R., *Wetland landscapes and catchment management*. In: *Freshwater Ecology and Conservation: Approaches and Techniques*. Edited by Jocelyne M. R. Hughes: Oxford University Press (2019). © Oxford University Press 2019.
DOI: 10.1093/oso/9780198766384.003.0019

Box 19.1 Old philosophies still apply to drivers of human nature

---

Unlimited wants: 'the pursuit of social recognition, self-worth, success and power'
  Adam Smith (1776)

Self-interest: 'Man's natural tendency for self-love and self-interest can distort his perception of himself, and the world around him, leading him to become blind and insensitive to others, and prone to misguided judgements.'
  La Rochefoucauld (1665)

Box 19.2 Sustainable Development Goals and freshwater systems

---

SDG goal 6
  Ensure availability and sustainable management of water and sanitation for all.
  Target 6.5. By 2030, implement integrated water resources management at all levels, including through transboundary cooperation as appropriate.
  Target 6.6. By 2020, protect and restore water-related ecosystems, including mountains, forests, wetlands, rivers, aquifers, and lakes.
SDG goal 15
  Protect, restore and promote sustainable use of terrestrial ecosystems, sustainably manage forests, combat desertification, and halt and reverse land degradation and halt biodiversity loss.
  Target 15.1. By 2020, ensure the conservation, restoration and sustainable use of terrestrial and inland freshwater ecosystems and their services, in particular forests, wetlands, mountains, and drylands, in line with obligations under international agreements.
  Target 15.5. Take urgent and significant action to reduce the degradation of natural habitats, halt the loss of biodiversity and, by 2020, protect and prevent the extinction of threatened species.

---

of Sustainable Development Goals (UN 2015) through which human behaviour, and our impacts on the Earth System, can be moderated. Two of these goals (Box 19.2) are highly relevant to the management of freshwater systems (Vörösmarty et al. 2013).

19.2 Freshwater governance

Hydrologically, a catchment is an integrated, interlinked waterbody, but in practice, this is not the case in social or political terms. Few river systems in the world come under the jurisdiction of a single geopolitical, institutional or governance body. Within a single catchment, most rivers can flow through several different administrative areas, including local, state, or national authorities, irrigation boards, electricity companies, landowners, and agricultural organisations, as well as different national jurisdictions in neighbouring countries.

Water governance has been defined as the '*range of political, social, economic and administrative systems that are in place to develop and manage water resources, and the delivery of water services*' (GWP 2009). When considering, however, the challenges of managing water at the catchment scale, these multiple systems, and the potential discord between them, can severely compromise efforts in catchment management or prove extremely expensive to sustain.

The term 'catchment management' is used widely in the literature. Specifically, it refers to the fact that *hydrological systems within any single watershed are fully connected, both above and below ground, from source to estuary.* Countries across the world have adopted the catchment system as the foundation for managing their freshwater and wetland resources. For example, in the countries of the EU, this provides the foundation of the EU *Water Framework Directive* (EC 2000), and as a result, catchment management plans have become a mandatory component of EU water policy.

In other parts of the world, governments have adopted this approach: catchment management agencies have been created and catchment plans put in place. In countries as diverse as Canada, Mozambique, Israel, Ukraine, and Chile, catchment management strategies have been implemented with varying levels of success under the dominant water management paradigm; namely, *Integrated Water Resource Management* (IWRM) (GWP 2009). Many of the world's largest rivers have *River Basin Commissions*, with a view to implementing IWRM and supporting cooperation between the multiple countries that rely on water resources in transboundary basins.

19.2.1 Challenges of management at the catchment scale

For those involved in whole catchment management efforts, data availability and scale are often major problems. Data definitions and methods of measurement can vary in both space and time, resulting in incommensurability of information. This has resulted from data being generated from a series of short-term independent projects rather than as part of long-term research programmes. Furthermore, in some places, where long-term water monitoring programmes have been undertaken, institutional change, and changes in information technology over the last 20 years have resulted in some long-term data sets being abandoned, with valuable data being lost.

This data loss has often been the unintended consequence of political change. For example, when the EU Water Framework Directive was put in place in 2000, a number of countries abandoned some of their long-term legacy data sets, due to resource limitations. In some less-developed countries, hydrological data sets that had been built up over decades in colonial times, have since been abandoned. In the case of hydro-meteorological data, problems have arisen due to equipment failure, theft of solar panels and copper cabling, and inadequate support to maintain crucial globally important monitoring networks.

A second challenge for those involved in catchment management arises from the existence of imprecise legislative frameworks. For example, although globally over 50 per cent of potable water comes from groundwater, water law often relates only to surface water. Furthermore, wetlands are often poorly defined in national water law, and thus may be overlooked in management regimes and decision-making (Sullivan and Fisher 2011). The legal principles of '*equitable and reasonable use*' and '*do no harm*' have become accepted as important underpinnings for water policy and, indeed, current efforts to strengthen water law by building in protection and joint management of ecosystems, are encouraging (McIntyre 2014).

At the operational level, the need for environmental water requirements (Horne et al. 2017; Acreman 2016) has become a major issue worldwide. Several countries have expanded their legislation to support environmental flows, particularly in South Africa, Tanzania, Costa Rica, and Australia. Elsewhere, however, although environmental flow assessments have been undertaken, few have been implemented due to methodological, operational, and institutional barriers—see assessment and implementation framework proposed by Opperman et al. (2018).

Finally, recent work has shown that land–atmosphere teleconnections exist which may result in changes in water quantity and quality in any river basin influenced by land-use changes in other areas outside the specific basin (Wang-Erlandsson et al. 2017), suggesting that the concept of managing water resources at the catchment scale may be ambiguous. Atmospheric teleconnections are, however, still poorly understood, and until more progress is made in this aspect of atmospheric science, the use of the catchment scale remains valid for the purposes of water management.

19.2.2 Beyond the catchment—challenges in transboundary basins

Globally, some 60 per cent of human populations depend on water from 261 shared river basins, with an urgent need for more streamlined transboundary water governance at the catchment scale (UN Water 2015). Although globally there have been over 3,600 water-sharing treaties agreed on since 805 AD, there is little widespread legislation to support water management across international boundaries. The 1997 UN *Convention on the Law of the Non-Navigational Uses of International Watercourses* has been signed by only 36 countries, and although this means it has now been ratified into law, there is no guarantee that the majority of the world's transboundary riparian states will adhere to it. The UN *Water Convention* (UNECE 2013) goes some way to address the type of water conflicts that often arise in transboundary catchments (McIntyre 2014), but international water law needs to be further strengthened through a greater uptake of these two conventions, and stronger linkages with the *Ramsar Convention on Wetlands*.

The importance of wetlands in transboundary cooperation can be well demonstrated in the case of the Okovango basin. In this huge wetland area (530,000 km²), which provides a crucial hydrological resource for Botswana and its neighbours, the formation of the basin commission OKACOM has the objective to secure sustainable management of the basin by building cooperation between the riparian states and their local government agencies (Philip et al. 2008).

Another important example of a transboundary basin is the Mekong river, which flows through six countries in Asia, providing livelihoods for over 70 million people (Mekong River Commission website http://www.mrcmekong.org/). Along most of its length (5,000 km), hundreds of wetlands provide water storage, flow regulation and flood prevention functions, and support valuable fish habitat. Today, however, these wetland ecosystems are being impacted by reductions in flows, increases in sedimentation, and changes in water temperatures. Large-scale dam construction in all the riparian countries have contributed to this degradation, with potentially irreversible impacts. The Mekong River Commission has not been able to reverse the trend of river degradation, which is a serious threat to food security across the region (see '10 rivers at risk' and 'Securing water for people and environment' at https://www.worldwildlife.org/).

Initial optimism about these river basin commissions has been relatively short-lived. In the Mekong, for example, failure to engage upstream riparian countries has resulted in increased threats to the vast wetlands of the Mekong Delta in Vietnam, and around Lake Tonle Sap in Cambodia, due to the number of large dams being constructed upstream. In the Nile basin, little real progress has been achieved in spite of the millions of dollars that have been spent to develop benefit sharing under the Nile Basin Initiative and the debates surrounding the construction and filling of the Grand Ethiopian Renaissance Dam (Wheeler et al. 2016). In Australia, where the Murray–Darling system has been managed for over two decades, conflict between stakeholders has increased over the way water resources should be used (Sullivan 2013), and in recent years, a number of incidences of illegal water acquisition, involving thousands of gigalitres of environmental water, have been reported.

19.3 Ecosystem services

19.3.1 Benefits from ecosystem services

Wetlands generate many ecosystem services (MEA 2005; Ramsar 2015; Sullivan et al. 2009). The ecosystem services concept is generally understood to describe the *benefits provided by nature to households, communities, and economies* (Gómez-Baggethun et al. 2010). The concept has been further developed for wetlands than for any other ecosystem (Maltby and Acreman 2011; See chapter 15 of this volume), with provisioning and regulating services being the most dominant. Provisioning services provided by wetlands include the supply of clean drinking water both for humans and animals, irrigation water, and food and fibre from other wetland resources (Russi et al. 2013). Cultural ecosystem services such as recreation and tourism, generation of scientific knowledge, educational opportunity, and spirituality are all important in wetlands (Chan et al. 2011), along with the provision of habitat for biodiversity (Harper and Quigley 2005).

Wetlands also provide regulating services such as flood mitigation, storm protection, water purification, sediment transport, erosion control, and carbon storage. Global reviews have found that wetlands consistently improve a range of water-quality indicators, including total phosphorus, nitrogen, nitrate and nitrites, ammonia, reactive phosphorus, chemical and biological oxygen demand, and suspended sediments (Fisher and Acreman 2004; Newman et al. 2015).

Many examples of these processes can be found across the world. In the Nakivubo Swamp, on the edge of Lake Victoria in Uganda, waste water and untreated sewage are received from the capital Kampala (Kansiime and Nalubega 1999), and acting as a buffer, the wetland removes nutrients and other pollutants so effectively that the city's water supply intake can be sited just 3 km away along the lake. Similarly, in the Orange River basin in South Africa, as the river flows slowly for 20 km through the Seekleivoi wetlands, in the heavily populated area of Soweto, important water purification functions are performed. Such benefits can be valued financially through comparison with costs incurred through the construction and operation of a wastewater treatment plant (Bonjean and Sullivan 2009). In spite of all these benefits, the rate of global wetland loss has become acute (Davidson 2014).

19.3.2 Disbenefits and trade-offs of wetland services

Scientific evidence for water-based ecosystem services may, however, be confusing or contradictory. Primarily, this has arisen because the term 'wetland' covers many land-cover and land-use types. Different wetland types give rise to a variety of hydrological functions across the landscape, making it difficult to generalise across different studies and scales (Bullock and Acreman 2003). Floodplain wetlands, for example, have been found to reduce or delay floods, but wetlands located in the headwaters of river systems, can increase flow rates and flood peaks.

In many places, floodplain wetlands have also been found to be particularly effective for flood mitigation (Blackwell and Maltby 2006; Acreman et al. 2003). In the UK (Acreman et al. 2003), Africa (McCartney et al. 2013), and India (Nielsen et al. 1991) wetlands were found to act as flood-water storage areas (Acreman and Holden 2013). In contrast, many headwater wetlands positioned at the foot of slopes or flat land adjacent to rivers are permanently or frequently saturated, so they tend to contribute to flood peaks (Holden and Burt 2003; Hewlett and Hibbert 1967). Temporal variability is also important in the functioning of wetlands, as shown in headwater wetlands (dambos) in Zimbabwe (McCartney 2000) where at the start of the wet season when water tables are low, wetlands can absorb rainfall, but soon become saturated and then contribute to flood run-off (Nilsson et al. 2005).

While natural systems such as wetlands often provide benefits to people, disbenefits can also occur. Temporal variability can be both negative as well as positive, as illustrated by the Somerset Levels, a large UK wetland system. Here, trade-offs between positive and negative (dis-)services were found, depending on how the wetlands are managed (Acreman et al. 2011). For example, restoring wetland hydrology by raising water levels reduced carbon dioxide release, at the same time increasing biodiversity, and the cultural, aesthetic, and recreational values associated with that. However, this enhanced wetness led to the negative services of reduced agriculture production (less nutritious grazing), increased methane production, and increased likelihood of insect-borne pathogens for people and animals (Table 19.1).

Table 19.1 *Trade-off in ecosystem services under different water management actions on the Somerset Levels and Moors, UK (after Acreman et al. 2011).*

Management action	Positive service effects	Negative service effects
Increasing wetness by raising water levels	· Increased biodiversity and aesthetic, cultural, and recreation values · Reduced release of carbon dioxide	· Increased release of methane · Increased insect-borne diseases · Diminished agricultural production
Decreasing wetness by lowering water levels	· Improved agricultural production · Reduced release of methane	· Increased release of carbon dioxide · Loss of biodiversity

19.4 Valuing wetlands

The social and economic values of wetlands are many and diverse. The concept of Total Economic Value (TEV) (Turner et al. 1994) serves as a good framework for scoping values and ensuring that, wherever possible, the broadest range of values are considered in decisions relating to the management of wetlands and other natural assets (Brander et al. 2006). Many of the ecosystem services and non-use values described above are 'non-market' goods, and are difficult to quantify in monetary terms. As such, they can often be overlooked when decisions about ecosystem management are made in an economic decision-making forum. This means that some environmental attributes remain as free goods, as they are not perceived to be limited in supply, and much land and water degradation has occurred due to the open-access nature of such attributes.

Key limitations remain around incorporating the monetary values of ecosystem attributes and services into the evaluation of management options for wetlands (Hanley and Barbier 2009; Rolfe et al. 2000). There is a critical need for a deeper understanding of the link between ecosystem form and function—or in economic terms—the marginal change in value arising from different levels of ecosystem condition. Some attempts have been done to achieve this using a Choice Experiment approach (Birol et al. 2006, Hearne and Salinas 2002; Morrison et al. 1999). Without such an understanding, it is difficult, if not impossible, to predict how ecosystem service values might change under different management regimes, especially when considering the range of services that exist (Figure 19.1).

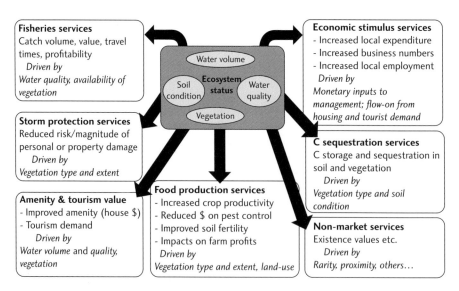

Figure 19.1 Ecosystem services from wetlands—potential monetary values (in bold font) and functional drivers (in italics, bottom lines in each box).

19.5 Tools and techniques for conserving wetland landscapes

19.5.1 Market-based instruments for water and wetlands

The need to engage private firms and individuals in biodiversity conservation has given rise to market-based instruments (MBIs), which, to achieve conservation outcomes, rely on market forces, rather than direct management or regulation (Stavins 2003; Kroeger and Casey 2007). MBIs seek to establish a market signal for environmental goods and services, and to create an incentive structure that favours conservation behaviours. In essence, they are designed to endow private firms and individuals with a greater personal financial interest in maintaining healthy, functioning ecosystems. In some cases, MBIs have been applied directly to wetland management, through their direct effects on ecosystem processes and health. Even where this is not the case, the implementation of MBIs at the broader catchment or landscape scale can have important implications for freshwater health. The following sections provide additional detail about three more common types of MBIs applied to freshwaters and wetlands across catchments.

19.5.2 Payment for ecosystem services

To date, Payments for Ecosystem Services (PES) schemes are the most commonly and widely implemented MBI for conservation management. In PES schemes, private landholders are paid to conserve, maintain or manage the environmental quality of their land. They can be considered a 'beneficiary pays' system, whereby the public benefit from healthy ecosystems are paid for by the general public (through government payments) or by a smaller number of interested individuals (represented by NGOs or philanthropic organizations). To date, most PES schemes have operated at a national or sub-national scale, but there is increasing attention being paid to international payment schemes (Wünscher and Engel 2012; Chichilnisky and Proctor 2010), particularly those that involve payments from developed- to developing- countries (Wunder 2007; Porras et al. 2008).

One of the primary benefits of PES, from an ecological perspective, is that they can be used to target diffuse sources of pollution (e.g., agriculture, which has been a key shortcoming in traditional command and control approaches). This is an important contribution: traditional policies, regulations, and tools have typically been directed towards point source pollution problems (Greiber 2009), but, as Fisher-Vanden and Olmstead (2013) report, 'even if all point sources were to achieve zero discharge, only 10 per cent of USA river and stream miles would rise one step or more on the USA's EPA water quality ladder'.

19.5.3 Biodiversity (and ecosystem service) offsetting

In biodiversity offsetting schemes, also variously termed bio-banking, biodiversity banking, or mitigation banking (Burgin 2008; Office of Environment and Heritage, Australia 2014; Environmental Protection Agency, USA 2017), developers who stand to benefit economically from filling, draining, or dredging wetlands are required to offset the loss of wetland function by improving the functionality of other similar wetlands. In biodiversity offsetting, paid environmental management work at an 'offset site' is undertaken to compensate for negative environmental impacts from development elsewhere (Sullivan and Hannis 2015). Price signals are designed to direct development away from ecologically sensitive areas (which will require more extensive, and more expensive mitigation works) towards less-sensitive areas. Where development does proceed, offsetting policies seek to balance biodiversity losses at the development site and biodiversity gains at the offset site, typically by following 'no net loss' or similar principles (e.g., Ten Kate et al. 2004; BBOP 2009). Most offsetting schemes also follow 'like-for-like' trading principles, where the development and offset sites must be matched in terms of ecosystem type and other selected biodiversity metrics in order to ensure that losses are not disproportionately focused on a small number of ecosystem types within any catchment.

Many have questioned the potential for offsetting to deliver no-net-loss outcomes, citing issues around timing (especially where losses precede gains), equivalence, and additionality (Burgin 200; Levrel et al. 2012; Gardner et al. 2013; Maron et al. 2016). Perhaps the key concern for the design of offsetting schemes is the need for good 'additionality'—that is, the offsetting process must provide better environmental outcomes than would otherwise be achieved. This requires that offsetting schemes focus on sites that are at high risk of imminent loss or degradation (Wunder 2007), and that management intensity and associated management outcomes are improved beyond existing levels.

The largest and most successful example of a wetland offsetting scheme (Madsen et al. 2010; Robertson 2009) comes from the USA. Wetland mitigation was first introduced into the USA in the 1970s, and by 2013 the wetland credit market was flourishing, with mitigation banks having registered more than 1,800 sites being made available for mitigation banking (Environment Protection Agency, USA 2017). Early schemes in Illinois, Florida, and Chicago addressed some of the key concerns around offsetting (lack of additionality and species lag) by focusing on the restoration of former wetlands as the offset site, where its restoration had to be completed ahead of any works at the development site (Robertson 2004).

19.5.4 Water-quality trading schemes

Pollution trading schemes based on 'cap and trade' were first developed in the context of managing point source air pollution. Such polluter-pays approaches are increasingly applied to the management of water pollutants. A report by Breetz et al. (2004), identified water-quality trading schemes in 14 US states, targeting a variety of pollutants, ranging from phosphorus and nitrogen to dissolved solids and heavy metals. In Australia, trading schemes are also currently operating for the management of salinity (Quinn 2011).

Water-quality trading seeks to internalise the cost of environmental damage (i.e., the externalities) arising from specific target pollutants, giving businesses or other polluters a financial incentive to reduce their pollution loads. Water-quality trading schemes proceed via the establishment of a 'cap'—that is, a maximum acceptable pollutant load for a specific river or wetland area, based on scientifically, socially (or otherwise determined) acceptable standards for water quality. The acceptable pollution load is then shared amongst the relevant population of polluters via the allocation of quotas. Quotas may be allocated in a variety of ways, but evidence from a range of examples suggest that when tradeable permits are initially given away free, schemes are less cost effective, and less likely to bring about a change in behaviour. When the permits are sold to buyers (via auction or other means), making their value explicit, and invested in voluntarily by the buyer, behavioural change is more likely to occur. Having acquired a quota, firms may then choose to either use their pollution quota, or to sell it (or some portion of it) to another firm or landowner. This is considered an economically efficient means of achieving pollution abatement: those firms that can cheaply and easily reduce their pollution loads will do so, while those with higher abatement costs will elect to purchase pollution quotas instead (Whitten et al. 2003; Goulder 2013).

19.5.5 How wetland values can be used to support regional development

Through the implementation of a selection of MBIs and other incentive measures, it is possible to design socially sensitive regional rejuvenation schemes in degraded wetland areas (Box 19.3).

Box 19.3 Linking ecological and social capital to recapture wetland benefits

A valuation study has been carried out for the 5,000-hectare Tuckean wetland located on the coastal floodplain of the Richmond River in northern New South Wales, Australia. Between the 1820s and 1970s, the Tuckean was gradually cleared and drained for agricultural production, resulting in wetland degradation and an increased risk of acid flushes into the river, leading to fish kills in the lower Richmond catchment.

A 'Choice Experiment' survey was used to elicit the value placed on the Tuckean wetland by the people of New South Wales. A proposal is being made to buy back the flood-prone land from farmers in the Tuckean, enabling them to retain their homes at least as smallholdings, while allowing the water to return to promote wetland recovery in high flood-risk areas. Potentially, this purchased area would be managed by an existing agency charged with reinstating wetland functions, improving biodiversity, and achieving water-quality outcomes and values for the broader community.

Existing landholders would remain in their neighbourhoods, retaining cohesive social capital and ensuring future viability of the community. Given the fact that many of the farmers in the region are considering retirement, this land buy-back has the potential to release private financial capital, providing a stimulus to economic growth, while protecting the natural and social capital of the region.

From the results of this survey, it was found that the total value of the wetland to the NSW community is almost twice the estimated costs of implementing the highest level of restoration and management needed to achieve this buy-back option. Notably, these survey results demonstrated how non-use values for water quality and native vegetation were seen as more important than the use values associated with walking tracks and other recreational facilities.

Source: Pelletier et al. (2018)

19.6 Wetland law and policy

The law relating to wetlands includes a wide range of legislation, such as statutes, acts, decrees, and ordinances. In addition, other rules and regulations implemented by agencies may also have the force of law. Depending on the jurisdiction, policies may provide principles or rules that guide the decision-making process, but also may be legally binding (Gardner et al. 2012). Judicial decisions that either apply to, or interpret the legislation, regulations, and policies that have been agreed to within a jurisdiction, can be used to guide the way wetlands are managed. Laws and policies may address activities that benefit both wetlands and people, or cause them harm, but legal frameworks are always needed to support complex resource-management policies, such as the implementation of IWRM (Schoeman et al. 2014; McIntyre 2014).

At the international level, the main policy and law setting is provided by decisions made through the *Ramsar Convention on Wetlands*, particularly in relation to the wise use or sustainable development of wetlands (Finlayson et al. 2011). The Convention has prepared extensive policy guidance for countries on the wise use of wetlands (CEC 1995), including how to prepare national policies and management plans (Finlayson and Gardner 2016). Recently, this has been extended to account for climate uncertainty, leading to the development of key principles to provide wetland managers with guidance on how to respond to the declining condition of wetlands now being exacerbated by climate change (Finlayson et al. 2017). These include the following:

(1) Wetland management should accommodate and compensate for climate change, rather than accepting or avoiding impacts.
(2) Ecological, social, and economic targets across multiple scales must be assessed in the context of ecological issues such as representativeness, connectivity, and refugial values.
(3) Flexible governance and adaptive co-management frameworks across multiple scales and sectors are essential.
(4) In the initial phases of adapting wetland management, easily reversed, low-regret adaptation options with multiple, cross-sectoral benefits should be implemented.
(5) Long-term management strategies should identify triggers for new actions, including novel adaptation options (e.g., species translocations), and plans for them.
(6) Integrated scientific monitoring and evaluation of management strategies are needed.

If such principles can be put in place, it is anticipated that wider landscape and catchment management processes will be better supported and conserved.

19.7 Challenges and solutions in urban wetlands

With more than 54 per cent of the global population living in cities or urban habitats, it is not surprising that the management of urban catchments has now to consider freshwater and wetlands ecosystem management more explicitly (McDonald and Shemie 2014). As a result, green infrastructure for ecosystem services and human well-being (e.g. Sustainable Urban Drainage Systems; Low Impact Developments; Natural Flood Management) is a key aspect of urban planning with examples of good practice in China's 'Sponge Cities'.

As in rural catchments, city catchments must be managed holistically, with the involvement of stakeholders, consumers, planning authorities, and other appropriate governance bodies. City catchments can be quantified in the same way as rural catchments, by measuring and monitoring the flow pathways, storages, and fluxes that make up the urban hydrological cycle (see Chapter 4) and by sampling and monitoring water quality and nutrient loads (see Chapter 5). Generally, cities are less permeable than rural areas, resulting in larger volumes of surface flow, lower rates of infiltration, and poor water quality with excessive nutrient loads (e.g., road run-off; untreated effluent).

An example of the value of managing and conserving wetlands to maintain key services within a city catchment, is in Kolkata, India. The East Kolkata Wetland (EKW) complex is a Ramsar site covering some 3000 ha. This wetland area not only provides a natural sewage treatment plant for removing phosphorus, but it is also habitat for fish cultivation and irrigation (Das Gupta et al. 2016). From this highly urban catchment, the wetlands process 550,000 m^3 of raw sewage and storm water per day, while generating some 16 per cent of Kolkata's fish sales. For decades, these wetlands have been maintained in dynamic equilibrium via integrated policy, social and environmental mechanisms, and government subsidies for labour, canal construction, and land acquisition (Carlisle 2013). In order to maintain the effective functionality of the wetland system, it is important that action be taken now to protect it from the urban encroachment that is beginning to make its mark.

In contrast with this traditional natural system, the city of Melbourne in Australia provides an example of how new governance structures can be developed to apply Integrated Urban Water Management (IUWM) and Water Sensitive Urban Design (WSUD), to the management of all components of the urban water cycle (Furlong et al. 2016; City of Melbourne 2014). Part of the approach for managing the chemical and nutrient loading of stormwater entering Melbourne's waterways and Port Philip Bay has required the construction of over 600 wetlands across the city. In addition to treating Melbourne's stormwater, this network of urban wetlands provides a rich environment for wildlife, but careful account of urban land use must be taken to ensure the functional efficiency of such constructed wetlands, and the sustainability of their use (Sharley et al. 2017).

19.8 The urgency of inaction

Human impact on the earth system is now exceeding its carrying capacity (Rockström et al. 2009), and during the twentieth century, wetland extent declined by as much as 70 per cent, continuing today at a similar pace (Gardner et al. 2015). Robust and consistent measures must be taken to reduce wetland degradation, and powerful coercive and persuasive techniques are needed to get the message through to those who have the power to influence landholder behaviour and its ecological outcomes across catchments.

19.8.1 Reversing the decline

While the uptake of different management tools can make a useful contribution towards behavioural change, there is still an important need to rebuild human respect for the earth system and all it provides. Lessons learned from such policy outcomes as the Montreal Protocol can guide this public education process, and deliver recognisable and manageable actions that can be widely implemented. Use of space technology and social media have great potential to support this objective through persuasive and coercive messages delivered at unprecedented scales across the globe.

At all levels of governance, the most important factor in achieving a reduction in wetland and floodplain degradation is through mobilisation of political will. Through the development of more integrated data and analytical frameworks, human impacts on freshwater systems can be more cost-effectively monitored, so interventions can be implemented and enforced, before thresholds of decline are reached. More sustainable long-term financing of under-resourced wetland management can be achieved through the use of market-based instruments, but these must be supported by clear property rights and appropriate legal and regulatory frameworks. Extending the use of market instruments more widely in the public and private sector has great potential to promote sustainable development in catchment systems.

19.9 Concluding remarks

Wetlands and other freshwater systems are key providers of vital ecosystem services that underpin our very life support system. The integrated, holistic management actions which are needed now are difficult to achieve when external pressure is brought to bear from vested interests such as large industrial organisations, mining companies, or large-scale commercial forestry or agricultural operations. In some countries, these organisations weald such political influence that they are able to bring about a veto on any action that may not be in their own self-interest, undermining water management efforts at the catchment scale. If effective action is to be taken to reverse the current rate of decline of wetland and freshwater systems across the world, tackling commercial self-interest and fragmented management must be seen as an urgent priority for all involved.

Acknowledgements

Support for this work has been provided by the Australian Research Council, under Grant Number LP130100498.

References

Acreman, M.C., Booker, D.J., and Riddington, R. (2003). Hydrological impacts of floodplain restoration: a case study of the river Cherwell, UK. *Hydrology and Earth System Sciences*, 7, 75–86.

Acreman, M.C., Harding, R.J., Lloyd, C., McNamara, N.P., Mountford, J.O., Mould, D.J., Purse, B.V., Heard, M.S., Stratford, C.J., and Dury, S. (2011). Trade-off in ecosystem services of the Somerset Levels and Moors wetlands. *Hydrological Sciences Journal*, 56, 1543–65.

Acreman, M.C. and Holden, J. (2013). Do wetlands reduce floods? *Wetlands* 33: 773–86.

Acreman, M. (2016). Environmental flows—basics for novices WIREs Water 3: 622–8, DOI: 10.1002/WAT2.1160

Business and Biodiversity Offsets Programme (2009). *Biodiversity Offset Design Handbook*. Washington, DC.

Birol, E., Karousakis, K., and Koundouri, P. (2006). Using a choice experiment to account for preference heterogeneity in wetland attributes: The case of Cheimaditida wetland in Greece. *Ecological Economics*, 60, 145–56.

Blackwell, M.S.A. and Maltby, E. (2006). Ecoflood Guidelines: How to use floodplains for flood risk reduction. European Commission D.G. Research, Brussels. Available online.

Bonjean, M. and Sullivan, C.A. (2009). Valuation of wetland functioning for Water Quality Improvement: an Example from the Klip River, Gauteng. Newater Report, Oxford University, UK.

Brander, L.M., Florax, R.J., and Vermaat, J.E. (2006). The empirics of wetland valuation: a comprehensive summary and a meta-analysis of the literature. *Environmental and Resource Economics*, 33, 223–50.

Breetz, H.L., Fisher-Vanden, K., Garzon, L., Jacobs, H., Kroetz, K., and Terry R. (2004). Water quality trading and offset initiatives in the US: A comprehensive survey. Dartmouth College and the Rockefeller Center for the US Environmental Protection Agency.

Bullock A. and Acreman, M.C. (2003). The role of wetlands in the hydrological cycle. *Hydrology and Earth System Sciences*, 7, 3, 75–86.

Burgin, S. (2008). BioBanking: an environmental scientist's view of the role of biodiversity banking offsets in conservation. *Biodiversity and Conservation* 17: 807–16.

Carlisle, S. (2013). Productive filtration: living system infrastructure in Calcutta. Scenario 03: Rethinking Infrastructure. Available online.

Chan, K.M., Goldstein, J., Satterfield, T., Hannahs, N., Kikiloi, K., Naidoo, R., and Woodside, U. (2011). Cultural services and non-use values in Natural Capital: Theory and Practice of Mapping Ecosystem Services, edited by Peter Kareiva, Oxford University Press, Oxford, UK.

Chichilnisky, G. and Proctor, W. (2010). International Payments for Ecosystem Services (IPES). United Nations Environment Programme (UNEP) and International Union for Conservation of Nature (IUCN) with the Secretariat of the Convention on Biological Diversity (CBD).

City of Melbourne Council (2014). Total Watermark: City as a Catchment Strategy. Melbourne, Australia. Available online.

Commission of the European Communities (1995). Wise use and conservation of wetlands. Communication from the Commission to the Council and the European Parliament, COM (95) 189 (final).

Crutzen, P.J. (2006). The 'Anthropocene'. In: Ehlers, E. and Krafft, T. (eds) Earth System Science in the Anthropocene. Springer, Berlin, Heidelberg.

Das Gupta, A., Sarkar, S., Singh, J., Saha, T., and Kumar, A. (2016). Nitrogen dynamics of the aquatic system is an important driving force for efficient sewage purification in single pond natural treatment wetlands at East Kolkata Wetland. *Chemosphere*, Volume 164, December 2016, 576–84.

Davidson, N.C. (2014). How much wetland has the world lost? Long-term and recent trends in global wetland area. *Marine and Freshwater Research*, 65, 934–41.

Environmental Protection Agency, USA (2017). Mitigation Banking Factsheet. Available online: https://www.epa.gov/cwa-404/compensatory-mitigation-factsheet

European Community (2000). Directive 2000/60/EC of the European parliament and of the Council of 23 October 2000, establishing a framework for community action in the field of water policy. Official Journal of the European Communities, L327: 1–72.

Finlayson, C.M. and Gardner, R.C. (2016). Wetland law and policy: overview. In: Finlayson, C.M., Everard, M., Irvine, K, McInnes, R.J., Middleton, B.A., van Dam, A.A., and Davidson, N.C. (eds) The Wetland Book I: Structure and Function, Management and Methods. Springer, Dordrecht, The Netherlands.

Finlayson, C.M., Capon, S.J., Rissik, D., Pittock, J., Fisk, G., Davidson, N.C., Bodmin, K.A., Papas, P., Robertson, H.A., Schallenberg, M., Saintilan, N., Edyvane, K., and Bino, G. (2017). Adapting policy and management for the conservation of important wetlands under a changing climate. Marine and Freshwater Research 68, 1803–15.

Finlayson, C.M., Davidson, N., Pritchard, D., Milton, G.R., and MacKay, H. (2011). The Ramsar Convention and ecosystem-based approaches to the wise use and sustainable development of wetlands. *Journal of International Wildlife Law and Policy* 14, 176–98.

Fisher, J. and Acreman, M.C. (2004). Water quality functions of wetlands. Hydrology and Earth System Sciences 8, 4, 673–85.

Fisher-Vanden, K., and Olmstead, S. (2013). Moving pollution trading from air to water: potential, problems, and prognosis. *The Journal of Economic Perspectives* 27: 147–71.

Furlong, C., Gan, K., and De Silva, S. (2016). Governance of Integrated Urban Water Management in Melbourne, Australia. *Utilities Policy*, 43, 48–58.

Gardner, R.C., Bonells, M., Okuno, E., and Zarama, J.M. (2012). Avoiding, mitigating, and compensating for loss and degradation of wetlands in national laws and policies. Ramsar Scientific and Technical Briefing Note 3. Ramsar Convention on Wetlands Secretariat. Gland, Switzerland.

Gardner, R.C., Barchiesi, S., Beltrame, C., Finlayson, C.M., Galewski, T., Harrison, I., Paganini, M., Perennou, C., Pritchard, D.E., Rosenqvist, A., and Walpole, M. (2015). State of the World's Wetlands and their Services to People: A compilation of recent analyses. Ramsar Scientific and Technical Briefing Note 7. Ramsar Convention on Wetlands Secretariat. Gland, Switzerland.

Gardner, T.A., VON Hase, A., Brownlie, S., Ekstrom, J.M., Pilgrim, J.D., Savy, C.E., Stephens, R.T., Treweek, J., Ussher, G.T., and Ward, G. (2013). Biodiversity offsets and the challenge of achieving no net loss. *Conservation Biology* 27: 1254–64.

Global Water Partnership (2009). Implementing the GWP Strategy 2009–2013. Report on the GWP Consulting Partners Meeting 15–16 August 2009, Stockholm, Sweden.

Global Water Partnership (2009). Integrated Water Resources Management in Practice: Better Water Management for Development. Earthscan (2009).

Gómez-Baggethun, E., De Groot, R., Lomas, P.L., and Montes, C. (2010). The history of ecosystem services in economic theory and practice: from early notions to markets and payment schemes. *Ecological Economics* 69: 1209–18.

Goulder, L.H. (2013). Markets for pollution allowances: what are the (new) lessons? *The Journal of Economic Perspectives* 27: 87–102.

Green, P.A., Vorosmarty, C.J., Harrison, I. Farrell, T., Saenz, L., and Fekete, B.M. (2015). Freshwater ecosystem services supporting humans: Pivoting from water crisis to water solutions. *Global Environmental Change*, 34, 108–18.

Gregory, K.J. (2006). The human role in changing river channels. *Geomorphology* 79:172–91.

Greiber, T. (2009). Payments for ecosystem services: Legal and institutional frameworks. International Union for Conservation of Nature and Natural Resources. Gland, Switzerland.

Hanley, N. and Barbier, E. (2009). Pricing nature: cost-benefit analysis and environmental policy. Edward Elgar Publishing, Cheltenham, UK.

Harper, D. and Quigley, J. (2005). No net loss of fish habitat: a review and analysis of habitat compensation in Canada. *Environmental Management* 36: 343–55.

Hearne, R.R. and Salinas, Z.M. (2002). The use of choice experiments in the analysis of tourist preferences for ecotourism development in Costa Rica. *Journal of Environmental Management*, 65, 153–63.

Hewlett, J.D. and Hibbert, A.R. (1967). Factors affecting the response of small watersheds to precipitation in humid regions. In:, W.E. Sopper and H.W. Lull (eds). Forest Hydrology. Pergamon Press, Oxford, UK.

Holden, J. and Burt, T.P. (2003). Runoff production in blanket peat covered catchments. *Water Resources Research*, 39, 1191–8.

Horne, A., Webb, J.A., Stewardson, M., Richter, B.M., Acreman, M.C. (eds) (2017). Water for the environment: from policy and science to implementation and management. Elsevier.

Kansiime, F. and Nalubega, M. (1999). Waste water treatment by a natural wetland: the Nakivubo Swamp, Uganda. Processes and implications. PhD Thesis, UNESCO-IHE Institute for Water Education, Delft, The Netherlands.

Kroeger, T. and Casey, F. (2007). An assessment of market-based approaches to providing ecosystem services on agricultural lands. *Ecological Economics*, 64, 321–32.

La Rochefoucauld, L.R. (1665). Réflexions ou Sentences et Maximes Morales. Claude Barbin, Paris.

Levrel, H., Pioch, S., and Spieler, R. (2012). Compensatory mitigation in marine ecosystems: Which indicators for assessing the 'no net loss' goal of ecosystem services and ecological functions? *Marine Policy* 36: 1202–10.

McIntyre, O. (2014). The protection of freshwater ecosystems revisited: Towards a common understanding of the 'ecosystems approach' to the protection of transboundary water resources. *Review of European, Comparative and International Environmental Law, Special Issue on International Water Law*, 23, 88–95.

Madsen, B., Carrol, N., and Moore Brands, K. (2010). State of biodiversity markets report: offset and compensation programs worldwide. Ecosystem Marketplace, Washington, DC.

Maltby, E. and Acreman, M.C. (2011). Ecosystem Services of Wetlands: pathfinder for a new paradigm. *Hydrological Sciences Journal* 56, 8, 1–19.

Maron, M., Ives, C.D., Kujala, H., Bull, J.W., Maseyk, F.J., Bekessy, S., Gordon, A., Watson, J.E., Lentini, P.E., and Gibbons, P. (2016). Taming a wicked problem: resolving controversies in biodiversity offsetting. *BioScience* 66: 489–98.

McCartney, M.P. (2000). The water budget of a headwater catchment containing a dambo. *Physics and Chemistry of the Earth* 25, 611–16.

McCartney, M.P., Cai, X., and Smakhtin, V. (2013). Evaluating the flow regulating functions of natural ecosystems in the Zambezi Basin. International Water Management Institute Research Report 148. Colombo, Sri Lanka.

McDonald, R.I. and Shemie, D. (2014). Urban Water Blueprint: Mapping Conservation Solutions to the Global Water Challenge. Nature Conservancy, Washington, DC. Available online.

Meadows, D.H., Meadows, D.L., Randers, J. and Behrens, W. (1972). Limits to Growth, Universe Books, New York.

Millennium Ecosystem Assessment (2005). Ecosystems and Human Well-Being: Wetlands and Water Synthesis. World Resources Institute, Washington, DC.

Morrison, M., Bennett, J., and Blamey, R. (1999). Valuing improved wetland quality using choice modeling. *Water Resources Research*, 35, 2805–14.

Naiman J. and Dudgeon, D. (2011). Global alteration of freshwaters: Influences on human and environmental well-being. *Ecological Research* 26: 865–73.

Newman, J.R., Duenas-Lopez, M., Acreman, M.C., Palmer-Felgate, E.J., Verhoeven, J.T.A., Scholz, M., and Maltby, E. (2015). Do on-farm natural, restored, managed and constructed wetlands mitigate agricultural pollution in Great Britain and Ireland? A Systematic Review. Report WT0989. Department for Environment, Food and Rural Affairs, London.

Nielsen, S.A., Refsgaard, J.C., Mathur, V.K. (1991). Conceptual modelling of water loss on floodplains and its application to River Yamuna upstream of Delhi. *Nordic Hydrology* 22: 265–74.

Nilsson, C., Rediy, C.A., Dynesius, M., and Revenga, C. (2005). Fragmentation and flow regulation of the world's largest rivers systems. *Science*, 308, 5720, 405–8.

Office of Environment and Heritage (2014). Biobanking Assessment Methodology. New South Wales, Australia.

Opperman, J.J., Kendy, E., Tharme, R.E., Warner, A.T., Barrios, E., and Richter, B.D. (2018). A three-level framework for assessing and implementing environmental flows. *Frontiers in Environmental Science*. DOI: 10.3389/fenvs.2018.00076

Pelletier, M.C., Hatton Macdonald, D., Rose, J., and Sullivan, C.A. (2018). Does information matter in forming social values for wetlands? AARES, Adelaide, Australia.

Philip, R., Anton, B., Bonjean, M., Bromley, J., Cox, D., Dickens, C., Nyagwambo, L., Smits, S., Sullivan, C.A., Van Nierkerk, K., Chonguiça, E., Monggae, F., Pule, R., and Berraondo, M. (2008). Engaging in IWRM—Practical Steps and Tools for Local Governments: training handbook. Southern Cross University publications, Local Governments for Sustainability (ICLEI), Cape Town, South Africa.

Porras, I.T., Grieg-Gran, M., and Neves, N. (2008). All that glitters: A review of payments for watershed services in developing countries. International Institute of Environment and Development, London.

Quinn, N.W. (2011). Adaptive implementation of information technology for real-time, basin-scale salinity management in the San Joaquin Basin, USA, and Hunter River Basin, Australia. *Agricultural Water Management*, 98: 930–40.

Ramsar (2015). State of the World's Wetlands and their Services to People: A compilation of recent analyses. Ramsar Briefing Note 7. Ramsar Convention on Wetlands Secretariat, Gland, Switzerland.

Robertson, M. (2009). The work of wetland credit markets: two cases in entrepreneurial wetland banking. *Wetlands Ecology and Management*, 17, 35–51.

Robertson, M.M. (2004). The neoliberalization of ecosystem services: wetland mitigation banking and problems in environmental governance. *Geoforum*, 35: 361–73.

Rockström, J., Steffen, W.L., Noone, K. Persson, Å., Chapin III, F.S., Lambin, E.F., Lenton, T.M., Scheffer, M., et al. (2009). Planetary Boundaries: Exploring the Safe Operating Space for Humanity. *Ecology and Society*, 14: 32.

Rolfe, J., Bennett, J., and Louviere, J. (2000). Choice modelling and its potential application to tropical rainforest preservation. *Ecological Economics*, 35, 289–302.

Russi, D., ten Brink, P., Farmer, A., Badura, T., Coates, D., Förster, J., Kumar, R., and Davidson, N. (2013). The Economics of Ecosystems and Biodiversity for Water and Wetlands. Institute for European Environmental Policy, Brussels; Ramsar Secretariat, Gland, Switzerland.

Schmutz, S. and Sendzimir, J. (2018). Riverine Ecosystem Management – Science for Governing Towards a Sustainable Future. Springer, Dordrecht.

Schoeman, J., Allan, C., and Finlayson, C.M. (2014). A new paradigm for water? A comparative review of integrated, adaptive and ecosystem-based water management in the Anthropocene. *International Journal of Water Resources Development*, 30, 377–90.

Sharley, D.J., Sharp, S.M., Marshall, S., Jeppe, K., and Pettigrove, V.J. (2017). Linking urban land use to pollutants in constructed wetlands: Implications for stormwater and urban planning. *Landscape and Urban Planning*, 162, 80–91.

Smith, A. (1776). *An Inquiry into the Nature and Causes of the Wealth of Nations*, Strathan, London.

Stavins, R.N. (2003). Experience with market-based environmental policy instruments. *Handbook of Environmental Economics*, 1: 355–435.

Steffen, W., Richardson, K., Rockstrom, J., Cornell, S.E., Fetzer, I. Bennett, E.M., Biggs, R., Carpenter, S.R., de Vries, W., de Wit, C.A., Folke, C., Gerten, D., Heinke, J., Mace, G.M., Persson, L.M., Ramanathan, V., Reyers, B. and Sorlin, S. (2015). Planetary boundaries: Guiding human development on a changing planet. Science, 347, 1259855.

Sullivan, C.A. and Fisher, D.E. (2011). Managing wetlands: integrating natural and human processes according to law. *Hydrological Sciences Journal*, 56.

Sullivan C.A., Macfarlane, D., Dickens, C., Mander, M., Teixeira-Leita, A., Pringle, K., and Bonjean, M. (2008). Keeping the Benefits Flowing and Growing: Quantifying the Benefits of Wetlands in the upper Orange Senqu Basin. Report for the European Union NeWater Project, Institute of Natural Resources, South Africa.

Sullivan, C.A. (2013). Planning for the Murray-Darling Basin: Lessons from Transboundary Basins around the World. Stochastic Environmental Research and Risk Assessment. DOI: 10.1007/s00477-013-0789-8

Sullivan, S. and Hannis, M. (2015). Nets and frames, losses and gains: Value struggles in engagements with biodiversity offsetting policy in England. *Ecosystem Services*, 15: 162–73.

Ten Kate, K., Bishop, J., and Bayon, R. (2004). Biodiversity offsets: Views, experience, and the business case. International Union for Conservation of Nature, Gland, Switzerland, and Cambridge, UK.

United Nations Economic Commission for Europe (2013). The UN water convention. UNECE, Geneva.

United Nations (2015). Transforming our world: the 2030 Agenda for Sustainable Development. Resolution adopted by the United Nations General Assembly on 25 September, 2015, New York.

UN-Water (2015). Water for a Sustainable World, The United Nations World Water Development Report 2015. United Nations Educational, Scientific and Cultural Organization (UNESCO), Paris, France.

Vörösmarty, C.J., Pahl-Wostl, C., Bunn, S.E., and Lawford, R. (2013). Global water, the anthropocene and the transformation of a science. *Current Opinion in Environmental Sustainability*, 5, 539–50.

Wang-Erlandsson, L., Fetzer, I., Keys, P.W., van der Ent, R.J., Savenije, H.G., and Gordon, L.J. (2017). Remote land use impacts on river flows through atmospheric teleconnections. *Hydrology and Earth System Sciences*, Discussion 494. Preprint under review.

Wheeler, K.G., Basheer, M., Zelalem T., Mekonnen, T., Eltoum, S.O., Mersha, A., Abdo, G.M., Zagona, E.A., Hall, J.W., and Dadson, S.J. (2016). Cooperative filling approaches for the Grand Ethiopian Renaissance Dam. *Water International*, 41, 611–34.

Whitten, S., Van Bueren M., and Collins, D. (2003). An overview of market-based instruments and environmental policy in Australia. Proceedings of the 6th annual Australian Agricultural and Resource Economics Society national symposium. AARES, Canberra.

Wunder, S. (2007). The efficiency of payments for environmental services in tropical conservation. *Conservation Biology*, 21: 48–58.

Wünscher, T. and Engel, S. (2012). International payments for biodiversity services: Review and evaluation of conservation targeting approaches. *Biological Conservation*, 152: 222–30.

Index

Note: references to tables, figures, and boxes are indicated by *t*, *f*, and *b* after the page number. For example, 145*f* refers to a figure on page 145.